FIRST COURSE
in
FUNCTIONAL ANALYSIS

David Cardan

FIRST COURSE
in
FUNCTIONAL ANALYSIS

CASPER GOFFMAN

Professor of Mathematics
Purdue University

GEORGE PEDRICK

Professor of Mathematics
California State University, Hayward

CHELSEA PUBLISHING COMPANY
NEW YORK, N.Y. 10016

SECOND EDITION

Library of Congress Catalog Card Number 82-74164
International Standard Book Number 0-8284-0319-8
Printed on 'long-life' acid-free paper

Printed in the United States of America

To Gertrude and Dorothy

PREFACE TO THE SECOND EDITION

During the several years that our book, *A First Course in Functional Analysis,* has been out of print, a rather large number of mathematics teachers have expressed a wish to see it republished. It is gratifying that this is now being done by the Chelsea Publishing Company. This new edition is basically the same as the previous one but incorporates the correction of a number of typographical errors. The bibliography has been revised.

CASPER GOFFMAN GEORGE PEDRICK

PREFACE TO THE FIRST EDITION

Portions of this book have been used by the authors in beginning graduate courses at Purdue University during the past several years.

Our view is that analysis itself is basic and that the abstract theories to which it leads are primarily of interest as tools which may be used in treating problems in analysis. This includes the theories of metric spaces, compactness and lower semicontinuity, normed vector spaces, Hilbert spaces, ordered vector spaces, Banach algebras, etc. Thus, we consider the abstract theories themselves only in part and then indicate their roles in analysis by giving various applications. In this regard, many important applications have not been considered because they are too complicated to be treated in an introductory study.

We assume that the reader is familiar with the symbols of set theory. For the most part, then, the book is self-contained.

There are many exercises at the end of each chapter. Each exercise is numbered so as to indicate the section to which it refers. Thus the exercise numbered 8.12 is concerned with the material of section 8 of the chapter. Some of the exercises are simple applications of the material in the text, whereas others should challenge the reader's ingenuity.

We take this occasion to express our thanks to Judy Wolf for her excellent typing of the manuscript.

We are grateful to R. C. Buck and J. J. Price for reading the text. Several improvements in form and content were made at their suggestion.

CASPER GOFFMAN GEORGE PEDRICK

CONTENTS

FIRST COURSE
in
FUNCTIONAL ANALYSIS

CHAPTER 1

METRIC SPACE

1.1 Definitions and Examples

A **metric space** consists of a set X, together with a notion of distance between pairs of elements of X, satisfying certain conditions. Thus, if X is a set, a **metric** in X is defined by a mapping

$$\rho: X \times X \to R,$$

of the space $X \times X$ into the real numbers, such that

(a) $\rho(x, x) = 0$ for every $x \in X$, and
$\rho(x, y) > 0$ for every $x, y \in X$, $x \neq y$,
(b) $\rho(x, y) = \rho(y, x)$ for every $x, y \in X$, and, **the triangle inequality**,
(c) $\rho(x, z) \leq \rho(x, y) + \rho(y, z)$ for every $x, y, z \in X$.

We give several examples of metric spaces, some of which are already familiar to the reader.

(i) Let X be any set of real numbers, and let ρ be defined by

$$\rho(x, y) = |x - y|.$$

(ii) Let X be any subset of the plane $R \times R$, and let ρ be defined by

$$\rho(x, y) = [(x_1 - y_1)^2 + (x_2 - y_2)^2]^{1/2},$$

where $x = (x_1, x_2)$ and $y = (y_1, y_2)$.

(iii) Let X be any set, and let ρ be defined by

$$\rho(x, y) = \begin{cases} 0 \text{ if } x = y \\ 1 \text{ if } x \neq y. \end{cases}$$

This is called the **discrete** metric in the set X.

We shall sometimes designate a space X with metric ρ as the **metric space** (X, ρ).

(iv) If (X, ρ) is a metric space and $Y \subset X$, then the restriction of ρ to

$Y \times Y$ determines a metric in Y which we shall also designate by ρ. The metric space (Y, ρ) is said to be a **subspace** of (X, ρ).

(v) Let X be real n-space (complex n-space) for a positive integer n. Then X consists of all n-tuples of real numbers (respectively, complex numbers),

$$x = (x_1, x_2, \ldots, x_n).$$

We shall consider a variety of metrics in this set. The most important metric is the euclidean metric given by

$$\rho_2(x, y) = \left[\sum_{i=1}^{n} |x_i - y_i|^2\right]^{1/2}$$

where $x = (x_1, x_2, \ldots, x_n)$ and $y = (y_1, y_2, \ldots, y_n)$.

In analogous fashion, for every $p \geq 1$, a metric is given by

$$\rho_p(x, y) = \left[\sum_{i=1}^{n} |x_i - y_i|^p\right]^{1/p}.$$

That these functions ρ_p satisfy the triangle inequality is the content of the Minkowski inequality, whose proof follows from another inequality—the Hölder inequality. In the next section, we present proofs of these inequalities.

1.2 Inequalities of Hölder and Minkowski

THEOREM 1 (Hölder's inequality). If $p > 1$ and q is defined by $1/p + 1/q = 1$,

$$\sum_{i=1}^{n} |x_i y_i| \leq \left[\sum_{i=1}^{n} |x_i|^p\right]^{1/p} \left[\sum_{i=1}^{n} |y_i|^q\right]^{1/q}$$

for any complex numbers $x_1, x_2, \ldots, x_n; y_1, y_2, \ldots, y_n$.

Proof. We first prove the inequality

$$a^{1/p} b^{1/q} \leq \frac{a}{p} + \frac{b}{q}, \quad a \geq 0, b \geq 0.$$

In order to prove this, we consider the function

$$f(t) = t^\alpha - \alpha t + \alpha - 1$$

defined for $0 < \alpha < 1$, $t \geq 0$.

Then

$$f'(t) = \alpha(t^{\alpha-1} - 1),$$

so that

$$f(1) = f'(1) = 0,$$

$$f'(t) > 0 \quad \text{for } 0 < t < 1,$$

and

$$f'(t) < 0 \quad \text{for } t > 1.$$

It follows that

$$f(t) \le 0 \quad \text{for } t \ge 0.$$

The inequality is obvious for $b = 0$. Suppose $b > 0$, and let $t = a/b$ and $\alpha = 1/p$. Then

$$f\left(\frac{a}{b}\right) = \left(\frac{a}{b}\right)^{1/p} - \frac{1}{p}\frac{a}{b} + \frac{1}{p} - 1 \le 0.$$

Multiplying by b, we obtain

$$a^{1/p}b^{1-1/p} \le \frac{a}{p} + b\left(1 - \frac{1}{p}\right),$$

which is our inequality.

Applying this to the numbers

$$a_j = \frac{|x_j|^p}{\sum_{i=1}^{n}|x_i|^p},$$

$$b_j = \frac{|y_j|^q}{\sum_{i=1}^{n}|y_i|^q},$$

for each $j = 1, 2, \ldots, n$, we get

$$\frac{|x_j y_j|}{\left[\sum_{i=1}^{n}|x_i|^p\right]^{1/p}\left[\sum_{i=1}^{n}|y_i|^q\right]^{1/q}} \le \frac{a_j}{p} + \frac{b_j}{q}, \quad j = 1, 2, \ldots, n.$$

By adding these n inequalities, we obtain the Hölder inequality, since

$$\sum_{j=1}^{n}\left(\frac{a_j}{p} + \frac{b_j}{q}\right) = \frac{\sum_{j=1}^{n} a_j}{p} + \frac{\sum_{j=1}^{n} b_j}{q} = \frac{1}{p} + \frac{1}{q} = 1.$$

We prove now the

THEOREM 2 (Minkowski's inequality). If $p \ge 1$, then

$$\left[\sum_{i=1}^{n}|x_i + y_i|^p\right]^{1/p} \le \left[\sum_{i=1}^{n}|x_i|^p\right]^{1/p} + \left[\sum_{i=1}^{n}|y_i|^p\right]^{1/p}$$

for any complex numbers $x_1, x_2, \ldots, x_n; y_1, y_2, \ldots, y_n$.

Proof. The case $p = 1$ is obvious. Suppose $p > 1$. It is obvious that

$$\left[\sum_{i=1}^{n}|x_i + y_i|^p\right]^{1/p} \le \left[\sum_{i=1}^{n}(|x_i| + |y_i|)^p\right]^{1/p}.$$

Now

$$(|x_i| + |y_i|)^p = (|x_i| + |y_i|)^{p-1}\,|x_i| + (|x_i| + |y_i|)^{p-1}\,|y_i|.$$

Summing these identities for $i = 1, 2, \ldots, n$,

$$\sum_{i=1}^{n}(|x_i| + |y_i|)^p = \sum_{i=1}^{n}(|x_i| + |y_i|)^{p-1}|x_i| + \sum_{i=1}^{n}(|x_i| + |y_i|)^{p-1}|y_i|.$$

The first sum on the right may be estimated by Hölder's inequality as follows (similarly for the second sum).

$$\sum_{i=1}^{n}(|x_i| + |y_i|)^{p-1}|x_i| \leq \left[\sum_{i=1}^{n}|x_i|^p\right]^{1/p}\left[\sum_{i=1}^{n}((|x_i| + |y_i|)^{p-1})^q\right]^{1/q}$$
$$= \left[\sum_{i=1}^{n}|x_i|^p\right]^{1/p}\left[\sum_{i=1}^{n}(|x_i| + |y_i|)^p\right]^{1/q}.$$

Thus,

$$\sum_{i=1}^{n}(|x_i| + |y_i|)^p \leq \left\{\left[\sum_{i=1}^{n}|x_i|^p\right]^{1/p} + \left[\sum_{i=1}^{n}|y_i|^p\right]^{1/p}\right\}\left[\sum_{i=1}^{n}(|x_i| + |y_i|)^p\right]^{1/q}.$$

Dividing $\left(\text{we may suppose } \sum_{i=1}^{n}(|x_i| + |y_i|)^p \neq 0\right)$, we obtain

$$\left[\sum_{i=1}^{n}(|x_i| + |y_i|)^p\right]^{1/p} \leq \left[\sum_{i=1}^{n}|x_i|^p\right]^{1/p} + \left[\sum_{i=1}^{n}|y_i|^p\right]^{1/p},$$

from which Minkowski's inequality follows, since

$$\left[\sum_{i=1}^{n}|x_i + y_i|^p\right]^{1/p} \leq \left[\sum_{i=1}^{n}(|x_i| + |y_i|)^p\right]^{1/p}.$$

The triangle inequality for the metric ρ_p is obtained by putting $x_i - z_i$ in place of x_i and $z_i - y_i$ in place of y_i in the statement of Minkowski's inequality.

The special case of Hölder's inequality with $p = q = 2$ is known as the **Cauchy-Schwarz** inequality:

$$\left[\sum_{i=1}^{n}|x_i y_i|\right]^2 \leq \left[\sum_{i=1}^{n}|x_i|^2\right]\left[\sum_{i=1}^{n}|y_i|^2\right].$$

1.3 Examples Continued; l_p Spaces

Continuing the examples of Sec. 1.1, we mention an additional metric for n-space, namely,

$$\rho_\infty(x, y) = \max_{1\leq i\leq n}|x_i - y_i|.$$

The notation is motivated by the fact that for any x and y,

$$\lim_{p\to\infty}\rho_p(x, y) = \max_{1\leq i\leq n}|x_i - y_i|.$$

Spaces of sequences of real (complex) numbers provide natural extensions of the foregoing examples. These are obtained by considering the expression

$$\left[\sum_{i=1}^{\infty} |x_i - y_i|^p\right]^{1/p}, \quad p \geq 1,$$

for two sequences $x = (x_1, x_2, \ldots)$ and $y = (y_1, y_2, \ldots)$.

(vi) The space l_p, for $1 \leq p < \infty$, is the set of all sequences $x = (x_1, x_2, \ldots)$ satisfying

$$\sum_{i=1}^{\infty} |x_i|^p < \infty,$$

with the metric

$$d_p(x, y) = \left[\sum_{i=1}^{\infty} |x_i - y_i|^p\right]^{1/p}, \quad y = (y_1, y_2, \ldots).$$

To see that the series for d_p converges under the assumptions on x and y, and that the triangle inequality holds, we extend Minkowski's inequality as follows.

THEOREM 1 (Minkowski's inequality). If $1 \leq p < \infty$ and if $x, y \in l_p$, then

$$\left[\sum_{i=1}^{\infty} |x_i + y_i|^p\right]^{1/p} \leq \left[\sum_{i=1}^{\infty} |x_i|^p\right]^{1/p} + \left[\sum_{i=1}^{\infty} |y_i|^p\right]^{1/p}.$$

Proof. For any n we have

$$\begin{aligned}\left[\sum_{i=1}^{n} |x_i + y_i|^p\right]^{1/p} &\leq \left[\sum_{i=1}^{n} |x_i|^p\right]^{1/p} + \left[\sum_{i=1}^{n} |y_i|^p\right]^{1/p} \\ &\leq \left[\sum_{i=1}^{\infty} |x_i|^p\right]^{1/p} + \left[\sum_{i=1}^{\infty} |y_i|^p\right]^{1/p}.\end{aligned}$$

The convergence of the series $\sum_{i=1}^{\infty} |x_i + y_i|^p$ and the inequality follow at once.

It is now an easy matter to verify that l_p, $1 \leq p < \infty$, is in fact a metric space.

(vii) The space l_∞ is defined to be the space of all bounded sequences, i.e., all $x = (x_1, x_2, \ldots)$ for which $\sup_{1 \leq i < \infty} |x_i| < \infty$, with the metric

$$d_\infty(x, y) = \sup_{1 \leq i < \infty} |x_i - y_i|.$$

The Hölder inequality has an extension to sequence spaces, which we state in terms of the following notations.

If $1 \leq p \leq \infty$, then we call q the **conjugate index** of p if

$$\begin{aligned} \frac{1}{p} + \frac{1}{q} &= 1 \quad \text{for } 1 < p < \infty \\ q &= \infty \quad \text{for } p = 1 \\ q &= 1 \quad \text{for } p = \infty. \end{aligned}$$

The distance $d_p(x, 0)$ of x from the sequence $0 = (0, 0, 0, \ldots)$ is called the **norm** of the sequence x and denoted $\|x\|_p$.

THEOREM 2 (Hölder's inequality). If $1 \leq p \leq \infty$, q is conjugate to p, $x \in l_p$ and $y \in l_q$, then

$$(x_1 \cdot y_1, x_2 \cdot y_2, \ldots) \in l_1,$$

and

$$\sum_{i=1}^{\infty} |x_i y_i| \leq \|x\|_p \, \|y\|_q.$$

The proof is easy and is left to the reader.

The space l_2 is of special importance; it is called **sequential Hilbert space.** The above statement with $p = q = 2$ gives the Cauchy-Schwarz inequality:

$$\left[\sum_{i=1}^{\infty} |x_i y_i|\right]^2 \leq \left(\sum_{i=1}^{\infty} |x_i|^2\right) \left(\sum_{i=1}^{\infty} |y_i|^2\right).$$

1.4 Examples Continued; Function Spaces

Many important examples of metric spaces whose elements are functions (or equivalence classes of functions) occur naturally in the various branches of analysis. We shall mention only three examples of such function spaces here, restricting ourselves to real-valued functions defined on a closed interval $[a, b]$ of the real line. Further examples may be supplied by the reader.

(viii) $C[a, b]$ is the space of all continuous functions on $[a, b]$ with the metric

$$\rho(x, y) = \sup_{a \leq t \leq b} |x(t) - y(t)|.$$

(ix) $M[a, b]$ is the space of all bounded functions on $[a, b]$ with the above metric.

Of course, $C[a, b]$ is a subspace of $M[a, b]$.

(x) Let $BV[a, b]$ denote the class of all functions of bounded variation on $[a, b]$, i.e., all f for which the total variation

$$V(f) = \sup \sum_{i=1}^{n} |f(x_i) - f(x_{i-1})|$$

is finite, where the supremum is taken over all partitions $a = x_0 < x_1 < \ldots < x_n = b$. It is easily seen that

$$\rho(f, g) = V(f - g)$$

has all the properties of a metric except that $V(f - g)$ may be zero for distinct functions f and g. Indeed, $V(f - g) = 0$ if and only if f and g differ by a constant.

If BV is decomposed into equivalence classes according to the equivalence relation defined by

$$f \cong g \quad \text{if} \quad f(t) - g(t) \quad \text{is constant on} \quad [a, b],$$

then this $\rho(f, g)$ determines a metric $\tilde{\rho}$ on the space $\widetilde{BV}$ of such equivalence classes in an obvious way.

Alternatively, we may modify the definition of ρ so as to obtain a metric on the original class BV. For example,

$$\rho(f, g) = |f(a) - g(a)| + V(f - g)$$

is a metric on BV. The subspace of this metric space, consisting of all $f \in BV$ for which $f(a) = 0$, can be identified in a natural way with the space $\widetilde{BV}$.

(xi) An example of a metric space which is of a somewhat different nature than those discussed so far is the space of non-empty, closed, bounded, convex subsets of the plane.

A set is **convex** if for each pair of its points it contains the line segment joining them. This segment, for points P and Q, has the parametric representation

$$tP + (1 - t)Q, \quad 0 \leq t \leq 1,$$

where tP denotes the point with coordinates (tx, ty) if P has the coordinates (x, y) relative to a cartesian coordinate system in the plane.

Let Γ denote the collection of all non-empty, closed, bounded, convex sets. For $A \in \Gamma$ and $\delta > 0$, denote by A_δ the union of all closed disks of radius δ centered at points of A.

Let $d(A, B)$ be defined by

$$d(A, B) = \inf \{\delta > 0 \colon A_\delta \supset B \text{ and } B_\delta \supset A\}.$$

We show that d is a metric on Γ. Clearly $d(A, A) = 0$, $d(A, B) \geq 0$ and $d(A, B) = d(B, A)$. If $d(A, B) = 0$, then for every $\delta > 0$, $A \subset B_\delta$. It follows that A is a subset of the closure of B. Since B is closed, $A \subset B$. By symmetry, $B \subset A$, so that $B = A$.

For the triangle inequality we first observe that

$$A \in \Gamma \text{ implies } A_\delta \in \Gamma \text{ for any } \delta > 0 \text{ and } (A_\delta)_\epsilon = A_{\delta+\epsilon}, \text{ for any } \epsilon > 0.$$

Now let

$$\begin{aligned} R &= \{\rho > 0 : A_\rho \supset B \text{ and } B_\rho \supset A\} \\ S &= \{\sigma > 0 : A_\sigma \supset C \text{ and } C_\sigma \supset A\} \\ T &= \{\tau > 0 : C_\tau \supset B \text{ and } B_\tau \supset C\}. \end{aligned}$$

We must show

$$\inf R \leq \inf S + \inf T.$$

The right-hand side equals

$$\inf \{\sigma + \tau : \sigma \in S \text{ and } \tau \in T\}.$$

The result will follow if we show that $\sigma + \tau \in R$ for all $\sigma \in S$ and $\tau \in T$. But this is the case since

$$A_{\sigma+\tau} = (A_\sigma)_\tau \supset (C)_\tau \supset B$$

and

$$B_{\sigma+\tau} = (B_\tau)_\sigma \supset (C)_\sigma \supset A.$$

1.5 Convergence and Related Notions

The metric space provides an abstract setting for a discussion of the convergence of sequences and related notions. By a sequence $\{x_n\}$ of elements of X is meant a mapping $n \to x_n$ of the positive integers into X.

A sequence $\{x_n\}$ **converges** to an element $x \in X$ if $\lim_n \rho(x_n, x) = 0$. We shall write $x = \lim_n x_n$ when $\{x_n\}$ converges to x.

It is immediately seen that a sequence can have at most one limit:

$$\lim_n x_n = x \text{ and } \lim_n x_n = y \text{ imply } x = y.$$

Application of this definition in a particular case may yield a familiar notion of convergence, as in the space $C[a, b]$, for example, where we have the

PROPOSITION 1. In $C[a, b]$, $\lim_n x_n = x$ if and only if $\{x_n(t)\}$ converges to $x(t)$ uniformly on $[a, b]$.

In view of this fact, the metric in $C[a, b]$

$$\rho(x, y) = \sup_{a \le t \le b} |x(t) - y(t)|$$

is called the **uniform** metric.

In other particular metric spaces, the above definition yields a new notion of convergence. Thus, in $l_p, p \ge 1$, there appears a notion of the convergence "in mean of order p" of a sequence of sequences of numbers.

Finally, it should be noted that some fundamental notions of convergence are not derivable from a metric space convergence. Indeed, the ordinary notion of point-wise convergence of a sequence of functions on $[a, b]$,

$$\lim_n f_n = f \quad \text{if} \quad \lim_n f_n(t) = f(t) \quad \text{for each } t \in [a, b]$$

cannot be described by any metric on a non-trivial class of functions (see Sec. 5.1).

In a metric space (X, ρ) the set

$$S(x_0, r) = \{x \in X : \rho(x_0, x) < r\}, \quad r > 0, x_0 \in X,$$

is called the **open sphere** of radius r and center x_0. We speak of the **closed sphere** if the inequality is replaced by $\rho(x_0, x) \le r$ and denote the set by $\bar{S}(x_0, r)$.

A subset $G \subset X$ is called **open** if it contains a sphere about each of its points. A set F is **closed** if its complement $X \sim F$ is open. An open sphere is an open set and a closed sphere is a closed set.

The collection $\mathcal{G}$ of all open sets has the following properties.

(1) $0 \in \mathcal{G}$ and $X \in \mathcal{G}$, where 0 is the empty set.
(2) Any union of members of $\mathcal{G}$ is a member of $\mathcal{G}$.
(3) The intersection of finitely many members of $\mathcal{G}$ is a member of $\mathcal{G}$.

These facts are easily proved, as are the dual facts about closed sets, namely,

(1′) 0 and X are closed.
(2′) Any intersection of closed sets is closed.
(3′) A finite union of closed sets is closed.

A **neighborhood** of a point $x_0 \in X$ is any set which contains an open set to which x_0 belongs. We say that x_0 is an **interior** point of a set A if A is a neighborhood of x_0. The **interior** of a set A consists of all interior points of A and can easily be seen to coincide with the largest open set contained in A.

If $A \subset X$ and $x_0 \in X$, then x_0 is called a **limit point** of A if every neighborhood of x_0 contains points of A distinct from x_0 (x_0 may or may not be in A). One easily sees that x_0 is a limit point of A if, and only if, A contains an infinite sequence of distinct points which converges to x_0.

For each $A \subset X$, the set $\bar{A}$ consisting of all points which are either points of A or limit points of A is called the **closure** of the set A. The closure of A is a closed set and is in fact the smallest closed set containing A.

1.6 Separable Space, Examples

If A and B are subsets of X, then we say that ***A* is dense in *B*** if $\bar{A} \supset B$; every element of B is a point or a limit point of A. In particular, A is said to be **everywhere dense** if $\bar{A} = X$. This means that for every $x \in X$ and $\epsilon > 0$ there is a $y \in A$ such that $\rho(x, y) < \epsilon$.

If the metric space X has a countable subset which is everywhere dense, it is called a **separable** metric space.

Since the rational numbers form a countable set which is everywhere dense in the real line, the latter is a separable space.

In n-space with any of the metrics we have introduced, the countable set consisting of all $x = (x_1, x_2, \ldots, x_n)$ for which each coordinate x_i is rational, is easily seen to be everywhere dense (a complex number is called rational in case its real and imaginary parts are both rational).

The spaces l_p for $1 \leq p < \infty$ are separable, but l_∞ is not separable. If e_k denotes the sequence consisting of 1 in the kth place and zeros elsewhere, then the set of all finite sums of rational multiples of the e_k's (i.e., all sequences with only finitely many non-zero coordinates, each of which is rational) is dense in l_p, $1 \leq p < \infty$. For, if $x = (x_1, x_2, \ldots) \in l_p$, and $r_1, r_2, \ldots, r_k$ are rational, then

$$d_p^p\left(x, \sum_{i=1}^{k} r_i e_i\right) = \sum_{i=1}^{k} |x_i - r_i|^p + \sum_{i=k+1}^{\infty} |x_i|^p$$

and, for any $\epsilon > 0$, k may be chosen so that

$$\sum_{i=k+1}^{\infty} |x_i|^p < \frac{\epsilon^p}{2}.$$

Then $r_1, r_2, \ldots, r_k$ may be chosen to give

$$|x_i - r_i|^p < \frac{\epsilon^p}{2k}, \qquad i = 1, 2, \ldots, k,$$

with the result,

$$d_p\left(x, \sum_{i=1}^{k} r_i e_i\right) < \epsilon$$

from which the separability follows in case $1 \leq p < \infty$.

In the case $p = \infty$, we observe that the set of all sequences of zeros and ones is an uncountable set (may be put in one-one correspondence with the interval [0, 1] by means of binary representation) and that any two members of this set have distance 1 apart in l_∞. The impossibility of a countable dense set is then apparent.

That the space $C[a, b]$ is separable can be seen by means of the Weierstrass approximation theorem (see Sec. 1.13), which says that any element of $C[a, b]$ can be approximated uniformly on $[a, b]$ by a polynomial. But any polynomial can be uniformly approximated by a polynomial with rational coefficients. Since there are countably many polynomials with rational coefficients, the result follows.

The property of separability has the following reformulation.

Proposition 1. *X is separable if, and only if, there is a countable collection B of open sets such that an arbitrary open set can be expressed as a union of members of B.*

The proof is left as an exercise.

If $\mathcal{C}$ is a collection of open sets in a metric space X with the property that every $x \in X$ is a member of at least one set $G \in \mathcal{C}$, then $\mathcal{C}$ is called an **open covering** of X. A **subcovering** of the open covering $\mathcal{C}$ is any subcollection $\mathcal{C}' \subset \mathcal{C}$ which is also an open covering of X. It is often of use, for technical reasons, to know when a space X has the property that every open covering contains a countable subcovering. We have the

PROPOSITION 2 (Lindelöf). If X is separable and $\mathcal{C}$ is an open covering of X, then there is a countable subcovering $\mathcal{C}' \subset \mathcal{C}$.

The proof, based on the preceding proposition, is left as an exercise.

1.7 Complete Space, Examples

In a metric space, certain sequences, called Cauchy sequences, or fundamental sequences, are of primary importance. A sequence $\{x_n\}$ in a metric space $X = (X, \rho)$ is said to be a **Cauchy sequence** if, for every $\epsilon > 0$, there is an N such that

$$m > N \text{ and } n > N \text{ implies } \rho(x_m, x_n) < \epsilon.$$

More briefly, $\{x_n\}$ is a Cauchy sequence if

$$\lim_{m,n\to\infty} \rho(x_m, x_n) = 0.$$

An equivalent statement is that $\rho(x_{n+k}, x_n)$ converges to 0 uniformly in k as n tends to infinity.

Every convergent sequence is a Cauchy sequence. For, suppose $\{x_n\}$ converges to x. Let $\epsilon > 0$. Then there is an N such that $n > N$ implies $\rho(x, x_n) < \epsilon/2$. But this implies

$$\rho(x_n, x_m) \le \rho(x_n, x) + \rho(x, x_m) < \epsilon,$$

for every $m > N$, $n > N$, so that $\{x_n\}$ is a Cauchy sequence.

The converse is not always true. In some metric spaces there are Cauchy sequences which do not converge. A classical example is the space of rational numbers with

$$\rho(x, y) = |x - y|.$$

The sequence .1, .101, .101001, .1010010001, . . . is easily seen to be a Cauchy sequence which does not converge.

A metric space (X, ρ) is said to be **complete** if every Cauchy sequence in the space converges. The space of rational numbers is, accordingly, not complete. It turns out that every metric space may be imbedded in a complete metric space in which it is dense. For the space of rational numbers, the corresponding complete metric space is the space of real numbers. However, the reals play a special role and their completeness must already be known before the general imbedding theorem can be established.

In many basic developments in analysis, the completeness of the reals is used in the form of the so-called nested intervals theorem. It has the following counterpart in general metric spaces.

PROPOSITION 1. A metric space (X, ρ) is complete if and only if for every sequence $\{\bar{S}_n\}$ of closed spheres, with

$$\bar{S}_{n+1} \subset \bar{S}_n, \quad n = 1, 2, \ldots$$

and

$$\lim_{n\to\infty} r_n = 0,$$

where r_n is the radius of $\bar{S}_n$, the intersection

$$\bigcap_{n=1}^{\infty} \bar{S}_n$$

consists of exactly one point.

The proof is left to the reader.

We consider some examples.

(i) The space $C[a, b]$ is complete.

Let $\{x_n\}$ be a Cauchy sequence. Then for every $\epsilon > 0$ there is an N such that $m, n > N$ implies $\rho(x_n, x_m) < \epsilon$. But this means that

$$\max\{|x_n(t) - x_m(t)| : t \in [a, b]\} < \epsilon.$$

In particular, for every $t \in [a, b]$, $\{x_n(t)\}$ is a Cauchy sequence of real numbers. By the completeness of the real numbers, $\{x_n(t)\}$ converges to a number $x(t)$, so that we have determined a function x on $[a, b]$. Now, for every $n > N$

$$\max\{|x(t) - x_n(t)| : x \in [a, b]\} \leq \epsilon.$$

This implies that $\{x_n\}$ converges uniformly to x, that x is continuous, and that $\lim\limits_{n\to\infty} \rho(x, x_n) = 0$. Thus $C[a, b]$ is complete.

(ii) Again, let S be the set of continuous functions on a closed interval $[a, b]$. But now let

$$\rho(x, y) = \int_a^b |x(t) - y(t)|\, dt.$$

Then (S, ρ) is a metric space. We observe that it is not complete.

Let $a < c < b$, and for every n so large that $a < c - 1/n$, define x_n as:

$$x_n(t) = \begin{cases} 0 & \text{if } a \leq t \leq c - \dfrac{1}{n} \\ nt - nc + 1 & \text{if } c - \dfrac{1}{n} \leq t \leq c \\ 1 & \text{if } c \leq t \leq b. \end{cases}$$

Then $\rho(x_n, x_m) \leq \dfrac{1}{n} + \dfrac{1}{m}$, from which it is obvious that $\{x_n\}$ is a Cauchy sequence.

Now, let $x \in S$. Then

$$\rho(x_n, x) = \int_a^{c-(1/n)} |x(t)|\, dt + \int_{c-(1/n)}^{c} |x_n(t) - x(t)|\, dt + \int_c^b |1 - x(t)|\, dt$$

and $\lim_{n\to\infty} \rho(x_n, x) = 0$ implies

$$x(t) = 0,\ t \in [a, c) \quad \text{and} \quad x(t) = 1,\ t \in (c, b].$$

Since it is impossible for a continuous function to have this property, $\{x_n\}$ does not have a limit.

This example suggests the consideration of $\int_a^b |x(t) - y(t)|\, dt$ as a metric for a wider class of functions; in particular, on the class of all absolutely integrable functions. The sense in which the term integrable is to be understood depends on the definition of the integral being used. We wish to accept the integral in the Riemann sense and make some observations about the associated metric space. For simplicity, we take the interval $[a, b]$ to be $[0, 1]$, which is typical of the general case.

It is possible for $\int_0^1 |x(t) - y(t)|\, dt = 0$ with $x \neq y$. For example, let

$$x(t) = 0 \quad \text{if } x \in [0, 1]$$

$$y(t) = \begin{cases} 0 & \text{if } x \in [0, 1) \\ 1 & \text{if } x = 1. \end{cases}$$

We shall call x and y equivalent if $\rho(x, y) = 0$ and consider ρ on the equivalence classes of functions so determined, i.e., the distance of two equivalence classes is $\rho(x, y)$ where x and y are, respectively, any representatives of these classes. In this way, a metric is determined on the set of equivalence classes of absolutely integrable functions.

(iii) The resulting metric space is not complete. A simple example shows this. Let

$$x_n(t) = \begin{cases} \sqrt{n} & \text{if } 0 \leq t \leq \dfrac{1}{n} \\ \dfrac{1}{\sqrt{t}} & \text{if } \dfrac{1}{n} \leq t \leq 1\,. \end{cases}$$

Then $\{x_n\}$ is easily seen to be a Cauchy sequence which does not converge in our space. However, if we were to extend the interpretation of integrability to include unbounded functions for which an improper integral exists and extend the definition of $\rho(x, y)$ as an improper integral, then this sequence would have the limit

$$x(t) = \begin{cases} t^{-1/2} & \text{if } 0 < t \leq 1 \\ 0 & \text{if } x = 0. \end{cases}$$

Our next example shows that the completeness property breaks down in a more serious way.

(iv) Define a sequence $\{S_n\}$ of sets as follows:

$$S_0 = [0, 1]$$

$S_1 = S_0$ with the middle open interval of length $1/4$ removed,

$S_2 = S_1$ with the middle open intervals of the component intervals of S_1 removed, each of length $1/4^2$.

Then, by induction, having already defined S_n so as to consist of 2^n disjoint closed intervals, of equal length, let

$S_{n+1} = S_n$ with the middle open intervals of the component intervals of S_n removed, each of length $1/4^{n+1}$.

For each $n = 1, 2, \ldots$, the sum of the lengths of the component open intervals of S_n is given by

$$m(S_n) = 1 - \sum_{i=0}^{n-1} \frac{1}{4^{i+1}} \cdot 2^i = \frac{1}{2} + \frac{1}{2^{n+1}}.$$

For every $n = 1, 2, \ldots$, let x_n be defined by

$$x_n(t) = \begin{cases} 1 & \text{if } t \in S_n \\ 0 & \text{if } t \notin S_n. \end{cases}$$

We note that $\{x_n\}$ is a Cauchy sequence. For, let $m > n$. Then

$$\begin{aligned} \rho(x_n, x_m) &= \int_0^1 |x_n(t) - x_m(t)|\, dt \\ &= m(S_n) - m(S_m) \\ &= \frac{1}{2^{n+1}} - \frac{1}{2^{m+1}}. \end{aligned}$$

We shall show that $\{x_n\}$ does not converge to any Riemann integrable function. Let x be a Riemann integrable function and suppose

$$(*) \quad \lim_{n \to \infty} \int_0^1 |x(t) - x_n(t)|\, dt = 0.$$

Let I_1 be the open interval removed in forming S_1; I_1, I_2, I_3 the open intervals removed in forming S_2, etc.

For each $k = 1, 2, \ldots$ there is an N so that $n > N$ implies $x_n(t) = 0$, $t \in I_k$. It follows that x is equivalent to a function which is identically zero on

$$T = \bigcup_{k=1}^{\infty} I_k.$$

But the lower Riemann integral of such a function is zero. Since x is integrable,

$$\int_0^1 x(t)\,dt = 0.$$

But

$$\int_0^1 x_n(t)\,dt > \tfrac{1}{2}, \quad n = 1, 2, \ldots.$$

This contradicts (*).

The failure of the space of absolutely integrable functions to be complete is to be regarded as a defect in the definition of the integral. The definition of Lebesgue, which will be discussed later, corrects this and other defects of the Riemann integral.

(v) The spaces l_p, $1 \le p < \infty$, are complete. Let $\{x^{(n)}\}$ be a Cauchy sequence in l_p, where

$$x^{(n)} = \{x_1^{(n)}, x_2^{(n)}, \ldots\}, \quad n = 1, 2, \ldots.$$

For each $\epsilon > 0$, there is an N such that $m, n > N$ implies

$$\sum_{i=1}^{\infty} |x_i^{(n)} - x_i^{(m)}|^p < \epsilon^p.$$

It follows that, for each i, the sequence of numbers $\{x_i^{(n)}\}$ is a Cauchy sequence, since $m, n > N$ implies

$$|x_i^{(n)} - x_i^{(m)}| < \epsilon.$$

Let

$$x_i = \lim_{n\to\infty} x_i^{(n)}, \quad i = 1, 2, 3, \ldots,$$

and let

$$x = \{x_1, x_2, \ldots\}.$$

We show that $x \in l_p$ and that $\{x^{(n)}\}$ converges to x in l_p. Let $n > N$ and fix k. Then

$$\sum_{i=1}^{k} |x_i|^p \le \left\{\left[\sum_{i=1}^{k} |x_i^{(n)} - x_i|^p\right]^{1/p} + \left[\sum_{i=1}^{k} |x_i^{(n)}|^p\right]^{1/p}\right\}^p$$
$$\le \{\epsilon + \|x^{(n)}\|\}^p.$$

It follows that $\sum_{i=1}^{\infty} |x_i|^p$ converges, so that $x \in l_p$.

Moreover, the observation that for every $n > N$ and every k,

$$\left[\sum_{i=1}^{k} |x_i^{(n)} - x_i|^p\right]^{1/p} \le \epsilon$$

implies that

$$\|x^{(n)} - x\| \le \epsilon, \quad n > N,$$

so that $\{x^{(n)}\}$ converges to x in the l_p norm.

(vi) The space l_∞, of bounded sequences with sup norm, is complete.

Let $\{x^{(n)}\}$ be a Cauchy sequence in l_∞. Then for each $\epsilon > 0$, there is an N such that $m, n > N$ implies

$$\sup_i |x_i^{(n)} - x_i^{(m)}| < \epsilon.$$

It follows that, for each i, $\{x_i^{(n)}\}$ is a Cauchy sequence. Let

$$x_i = \lim_{i \to \infty} x_i^{(n)}$$

and let

$$x = \{x_1, x_2, \ldots\}.$$

Now, for each i and $n > N$, it follows that

$$|x_i - x_i^{(n)}| \leq \epsilon,$$

from which both facts, that $x \in l_\infty$ and that $\{x^{(n)}\}$ converges to x in the l_∞ norm, follow.

The n-space is complete under the various metrics we have introduced.

PROPOSITION 2. A subspace of a complete metric space is complete if and only if it is closed.

The proof is left for the reader.

(vii) The Baire-classes of bounded functions form a family of closed subspaces of the space of bounded functions with the uniform metric.

For real-valued functions on [0, 1], let M be the metric space of bounded functions with

$$d(f, g) = \sup [|f(x) - g(x)| : x \in [0, 1]].$$

The Baire subspaces of M are defined as follows.

$$B_0 = C[0, 1].$$

B_1 is the set of bounded functions which are point-wise limits of sequences of functions in B_0. B_2 is the set of all bounded functions which are point-wise limits of sequences of functions in B_1.

In general, if α is any ordinal number, of countable cardinal, then B_α is the set of all bounded functions which are point-wise limits of sequences of functions in $\bigcup [B_\beta : \beta < \alpha]$.

Each B_α is a closed subspace of M and so is complete (for the proof see Goffman [1]).

For every ordinal α, of countable cardinal, $\bigcup [B_\beta : \beta < \alpha]$ is a proper subspace of B_α (for the proof see Hausdorff [1]).

1.8 Contractions, Applications to Differential and Integral Equations

In this section, we show how various existence theorems in analysis follow from a very simple fact about mappings in a metric space.

Let $X = (X, \rho)$ be a metric space. A mapping $T : X \to X$ is called a **contraction in** X if there is a number K, with $0 < K < 1$, such that x, $y \in X$, $x \neq y$, implies

$$\rho(Tx, Ty) \leq K\rho(x, y).$$

THEOREM 1. If X is a complete metric space, and T is a contraction in X, then T has a fixed point and it is unique. That is to say, there is a unique $x_0 \in X$ such that $Tx_0 = x_0$.

Proof. We first prove uniqueness. Suppose $Tx = x$ and $Ty = y$. Then, if $x \neq y$, $\rho(x, y) = \rho(Tx, Ty) < \rho(x, y)$, which is impossible. Hence $x = y$.

For the existence of a fixed point, we may start with any $x \in X$ and consider the sequence of images of x under repeated application of T,

$$x,\ Tx,\ T^2x = T(Tx), \ldots,\ T^nx = T(T^{n-1}x), \ldots .$$

We show that $\{T^nx\}$ is a Cauchy sequence. For every $n > 1$ and $p > 0$,

$$\begin{aligned}
\rho(T^{n+p}x, T^nx) &\leq K\rho(T^{n+p-1}x, T^{n-1}x) \leq \ldots \leq K^n\rho(T^px, x)\\
&\leq K^n[\rho(T^px, T^{p-1}x) + \ldots + \rho(Tx, x)]\\
&\leq K^n\rho(Tx, x)(K^{p-1} + K^{p-2} + \ldots + K + 1)\\
&< \frac{K^n}{1 - K}\,\rho(Tx, x).
\end{aligned}$$

Since $K < 1$ and $\rho(Tx, x)$ are fixed, it follows that $\{T^nx\}$ is Cauchy.

Let $x_0 = \lim_{n\to\infty} T^nx$. Since T is a contraction, the interchange of operations in

$$Tx_0 = T \lim T^nx = \lim T^{n+1}x = x_0$$

is easily justified so that x_0 is a fixed point.

As a first example of the application of this theorem, we prove a standard existence and uniqueness theorem for ordinary differential equations.

Let D be a connected open set in the plane, and let f be a real function defined on D. We shall say that f satisfies a **Lipschitz condition** in y on D, with Lipschitz constant M, if for every (x, y_1) and (x, y_2) in D, we have

$$|f(x, y_2) - f(x, y_1)| \leq M\,|y_2 - y_1|.$$

We shall show that if f satisfies a Lipschitz condition, then there exists a unique solution, through each point in D, to the differential equation

$$\frac{dy}{dx} = f(x, y),$$

and this solution extends across the domain.

We shall make the above statement precise and shall give the proof in steps. Let $(x_0, y_0) \in D$. By a local solution passing through (x_0, y_0) we mean a function φ defined on an open interval I such that $x_0 \in I$, $\varphi(x_0) = y_0$, $(x, \varphi(x)) \in D$ for every $x \in I$, and $\varphi'(x) = f(x, \varphi(x))$ for every $x \in I$.

THEOREM 2. If f is continuous on an open connected set D and satisfies a Lipschitz condition in y on D, then for every $(x_0, y_0) \in D$, the differential equation $dy/dx = f(x, y)$ has a unique local solution passing through (x_0, y_0).

Proof. We leave it for the reader to verify that a function φ defined on an open interval I is such that $\varphi(x_0) = y_0$ and $\varphi'(x) = f(x, \varphi(x))$ for every $x \in I$ if and only if

$$\varphi(x) = y_0 + \int_{x_0}^{x} f(t, \varphi(t))\, dt$$

for every $x \in I$.

Now, let $D' \subset D$ be open, connected, containing (x_0, y_0), and such that f is bounded on D', and let $|f(x, y)| < A$ for all $(x, y) \in D'$.

Let $d > 0$ be such that

(a) the rectangle $R \subset D'$, where

$$R = [x_0 - d, x_0 + d] \times [y_0 - dA, y_0 + dA],$$

(b) $Md < 1$, where M is the Lipschitz constant for f in D.

Let $J = [x_0 - d,\ x_0 + d]$. The set B of continuous functions ψ on J such that $\psi(x_0) = y_0$ and $|\psi(x) - y_0| \le dA$ for every $x \in J$ is a complete metric space under the uniform metric ρ. Consider the mapping T defined by

$$(T\psi)(x) = y_0 + \int_{x_0}^{x} f(t, \psi(t))\, dt \quad \text{for } \psi \in B \text{ and } x \in J.$$

Now, $(T\psi)(x_0) = y_0$, $T\psi$ is continuous, and for every $x \in J$,

$$|(T\psi)(x) - y_0| = \left| \int_{x_0}^{x} f(t, \psi(t))\, dt \right| \le \int_{x_0}^{x} |f(t, \psi(t))|\, dt \le dA.$$

Hence $T\psi \in B$. Thus T maps B into B.

We show that T is a contraction. Let $\psi_1, \psi_2 \in B$. Then, for every $x \in J$,

$$\begin{aligned} |(T\psi_1)(x) - (T\psi_2)(x)| &= \left| \int_{x_0}^{x} \{f(t, \psi_1(t)) - f(t, \psi_2(t))\}\, dt \right| \\ &\le \int_{x_0}^{x} |f(t, \psi_1(t)) - f(t, \psi_2(t))|\, dt \\ &\le Md \max\, [|\psi_1(t) - \psi_2(t)| : t \in J], \end{aligned}$$

so that $\rho(T\psi_1, T\psi_2) \le Md\rho(\psi_1, \psi_2)$. Hence T is a contraction. By theorem 1, we obtain theorem 2.

We next show that the local solution can be extended across D'. Let $J = J_1$, $d = d_1$ and $x_0 + d = x_1$, $\varphi(x_1) = y_1$. By theorem 2, applied to (x_1, y_1), we obtain J_2, d_2 and (x_2, y_2), and the solution functions $\varphi_1 = \varphi$ on J_1 and φ_2 on J_2 agree on an interval and so yield a solution on $J_1 \cup J_2$. In this way, we obtain a sequence $\{(x_n, y_n)\}$ with $x_{n+1} > x_n$, $n = 1, 2, \ldots$. We assume D' bounded and show that the distance of (x_n, y_n) from the boundary of D' converges to zero. If this distance is δ_n, the above $d_1, d_2, \ldots, d_n, \ldots$ may be chosen so that

$$d_n = \min\left[\frac{\delta_n}{A^2 + 1}, \frac{1}{2M}\right].$$

Since $\sum_{n=1}^{\infty} d_n < \infty$ it follows that $\sum_{n=1}^{\infty} \delta_n < \infty$ and $\lim \delta_n = 0$. We thus have (using the fact that D is the union of an increasing sequence of sets each having the above properties of D'):

THEOREM 3. If f is continuous on an open connected set D and satisfies a Lipschitz condition in y on D, then for every $(x_0, y_0) \in D$ the differential equation $dy/dx = f(x, y)$ has a unique solution $y = \varphi(x)$ such that $y_0 = \varphi(x_0)$ and such that the curve given by the solution passes across D from boundary to boundary.

Thus the solutions to the equation form a simple covering of D.

The condition that f satisfies a Lipschitz condition may be localized, but we shall not discuss these details.

We next consider an integral operator. Let K be defined and continuous on the closed square $[a, b] \times [a, b]$, let g be continuous on $[a, b]$, and let λ be a real number. We seek a function f satisfying

$$f(x) = \lambda \int_a^b K(x, y)f(y)\, dy + g(x), \quad x \in [a, b].$$

In other words, we seek a fixed point, in the space $C[a, b]$, of the mapping T, where

$$(T\psi)(x) = \lambda \int_a^b K(x, y)\psi(y)\, dy + g(x), \quad x \in [a, b].$$

Clearly, T maps $C[a, b]$ into itself and a sufficient condition for the existence and uniqueness of a solution is that T be a contraction in $C[a, b]$. If M is an upper bound for $|K(x, y)|$ on the square, then

$$\begin{aligned}\rho(T\psi_1, T\psi_2) &= \sup\left\{\left|\lambda \int_a^b K(x, y)\{\psi_1(y) - \psi_2(y)\}\, dy\right| : x \in [a, b]\right\} \\ &\le |\lambda|\, M(b - a) \sup\{|\psi_1(y) - \psi_2(y)| : y \in [a, b]\} \\ &= |\lambda|\, M(b - a)\rho(\psi_1, \psi_2).\end{aligned}$$

The condition thus takes the form

$$|\lambda| < \frac{1}{M(b - a)}.$$

1.9 Completion

By following a procedure similar to that which yields the real number system from the rationals in terms of equivalence classes of Cauchy sequences, it can be shown that any metric space (X, ρ) can be imbedded in a complete metric space, as a dense subspace, in essentially a unique way. The enlarged space is called the **completion** of the space (X, ρ).

This section is devoted to a precise formulation of this statement and an outline of the proof.

A transformation of a metric space (X, ρ) into a metric space (Y, σ) is called an **isometry** if, for every $x, y \in X$,

$$\sigma(Tx, Ty) = \rho(x, y).$$

Such an isometry is said to **imbed** (X, ρ) in (Y, σ). (X, ρ) and (Y, σ) are **isomorphic** as metric spaces if there is an isometry of (X, ρ) onto (Y, σ).

THEOREM 1. If (X, ρ) is any metric space, it can be imbedded, as a dense subspace, in a complete metric space $(\tilde{X}, \tilde{\rho})$. Two complete spaces in which (X, ρ) can be so imbedded are isomorphic.

Outline of Proof. We assume that the real numbers are a complete metric space. Let $\mathfrak{X}$ be the set of all Cauchy sequences in X. Define an equivalence relation in $\mathfrak{X}$ by letting $\{x_n\} \cong \{y_n\}$ if $\{\rho(x_n, y_n)\}$ converges to 0. The set of equivalence classes so formed is denoted by $\tilde{X}$ and a metric $\tilde{\rho}$ is defined in $\tilde{X}$ as follows:

If $\{x_n\}, \{y_n\} \in \mathfrak{X}$, then $\{\rho(x_n, y_n)\}$ is a Cauchy sequence of real numbers. Let $\tilde{\rho}$ be defined as $\lim_{n \to \infty} \rho(x_n, y_n)$. This limit is unchanged if $\{x_n\}$ and $\{y_n\}$ are replaced by equivalent sequences. Then

$$\tilde{\rho}(\tilde{x}, \tilde{y}) = \lim_{n \to \infty} \rho(x_n, y_n)$$

when $\{x_n\} \in \tilde{x} \in \tilde{X}$ and $\{y_n\} \in \tilde{y} \in \tilde{X}$.

We omit the proof that $(\tilde{X}, \tilde{\rho})$ is a complete metric space.

The imbedding is accomplished by associating with each $x \in X$ the equivalence class Tx in $\tilde{X}$ to which the Cauchy sequence $\{x, x, x, \ldots\}$ belongs. Then T is easily shown to be an isometry.

To see that $T(X)$ is dense in $\tilde{X}$, let $\{x_n\} \in \tilde{x} \in \tilde{X}$, and observe that the sequence $\{Tx_n\}$ converges to $\tilde{x}$ in $\tilde{X}$. This shows that a completion of (X, ρ) exists.

For the uniqueness, we must realize that a completion of (X, ρ) consists of both a metric space $(\tilde{X}, \tilde{\rho})$ and an imbedding map T. Thus, we must show that if $(\tilde{X}, \tilde{\rho}, T)$ and $(\tilde{Y}, \tilde{\sigma}, S)$ are completions of (X, ρ), then an isomorphism exists between $(\tilde{X}, \tilde{\rho})$ and $(\tilde{Y}, \tilde{\sigma})$.

Now T^{-1} is an isomorphism of $T(X)$ onto X, and S is an isomorphism of

X onto $S(X)$. The composite mapping ST^{-1} is then an isomorphism of $T(X)$ onto $S(X)$. We obtain a mapping φ by extending this mapping to all of $\tilde{X}$ from the dense subset $T(X)$. Choose $\tilde{x} \in \tilde{X}$ and $\{Tx_n\}$, a sequence in $T(X)$ converging to $\tilde{x}$. Then $\{ST^{-1}Tx_n\} = \{Sx_n\}$ is a Cauchy sequence in Y whose limit is defined to be $\varphi\tilde{x}$. The mapping φ defined in this way can be shown to be an isomorphism of $\tilde{X}$ onto $\tilde{Y}$.

1.10 Category, Nowhere Differentiable Continuous Functions

An important fact about complete metric spaces, which has many interesting applications to analysis, is that they are of the second category.

If (X, ρ) is a metric space, a set $A \subset X$ is said to be **nowhere dense** if its closure $\bar{A}$ contains no sphere, or equivalently if $\bar{A}$ has no interior points.

The union of any finite set of nowhere dense sets is nowhere dense. However, the union of countably many nowhere dense sets need not be nowhere dense.

A set $A \subset X$ is said to be of the **first category** in X if it is the union of countably many nowhere dense sets in X. Otherwise, A is said to be of the **second category** in X.

PROPOSITION 1. (X, ρ) is of the second category in itself, if and only if, in any representation of X as the union

$$X = \bigcup_{i=1}^{\infty} F_i$$

of countably many closed sets F_i, at least one of the F_i contains a sphere.

The proof is easy and will be omitted.

THEOREM 1. Every complete metric space (X, ρ) is of the second category in itself.

Proof. Suppose, on the contrary, that

$$X = \bigcup_{i=1}^{\infty} F_i,$$

where each F_i is closed and contains no sphere. Then $X \sim F_1$ is a non-empty open set. Choose x_1 and $\epsilon_1 < 1$ so that $x_1 \in X \sim F_1$ and

$$S(x_1, \epsilon_1) \cap F_1 = 0.$$

The sphere $S(x_1, \epsilon_1/2)$ is not contained in F_2. Hence there is an $x_2 \in S(x_1, \epsilon_1/2)$ and an $\epsilon_2 < \frac{1}{2}$ such that

$$S(x_2, \epsilon_2) \cap F_2 = 0 \quad \text{and} \quad S(x_2, \epsilon_2) \subset S\left(x_1, \frac{\epsilon_1}{2}\right).$$

Proceeding inductively, we can choose a sequence $\{x_n\}$ of points and a sequence $\{\epsilon_n\}$ of radii for which

(a) $S(x_{n+1}, \epsilon_{n+1}) \subset S\left(x_n, \dfrac{\epsilon_n}{2}\right)$,

(b) $\epsilon_{n+1} < \dfrac{1}{2^n}$,

and

(c) $S(x_{n+1}, \epsilon_{n+1}) \cap F_{n+1} = 0$.

The sequence $\{x_n\}$ is Cauchy, for

$$\begin{aligned}\rho(x_{n+k}, x_n) &\leq \rho(x_n, x_{n+1}) + \rho(x_{n+1}, x_{n+2}) + \ldots + \rho(x_{n+k-1}, x_{n+k})\\ &< \frac{\epsilon_n}{2} + \frac{\epsilon_{n+1}}{2} + \ldots + \frac{\epsilon_{n+k-1}}{2}\\ &< \frac{1}{2^n} + \frac{1}{2^{n+1}} + \ldots + \frac{1}{2^{n+k-1}}\\ &< \frac{1}{2^{n-1}}.\end{aligned}$$

Since X is complete, there is an x such that

$$x = \lim_{n\to\infty} x_n.$$

Now, fix n. Then $x \in S(x_n, \epsilon_n)$, since $x_{n+k} \in S(x_n, \epsilon_n/2)$ for every $k = 1, 2, \ldots$. Hence, by (c), $x \notin F_n$. Thus $x \notin \bigcup_{n=1}^{\infty} F_n = X$, an obvious contradiction.

As an application of theorem 1, we prove the existence of continuous functions on [0, 1] which are nowhere differentiable. For this purpose, we consider the metric space $C_0[0, 1]$ of continuous functions f for which $f(0) = f(1)$, with

$$d(f, g) = \max [|f(x) - g(x)| : x \in [0, 1]].$$

Then $C_0[0, 1]$ is a complete metric space. We show that those functions in $C_0[0, 1]$ which are somewhere differentiable form a subset of the first category. Since the space $C_0[0, 1]$ is of the second category, this proves the existence of nowhere differentiable functions.

It will be convenient to extend the functions of $C_0[0, 1]$ to the entire axis by periodicity and to treat the space Γ of such extensions with the metric d defined above.

Let $K \subset \Gamma$ be the set of functions f such that, for some ξ, the set of numbers

$$\left[\frac{f(\xi + h) - f(\xi)}{h} : h > 0\right]$$

is bounded. K contains the set of functions which are somewhere differentiable. We shall show that K is of the first category in Γ. Let

$$K_n = \left[f \in \Gamma : \text{for some } \xi, \left| \frac{f(\xi + h) - f(\xi)}{h} \right| \le n, \text{ for all } h > 0 \right].$$

Then

$$K = \bigcup_{n=1}^{\infty} K_n.$$

We shall show that for every $n = 1, 2, \ldots,$

(a) K_n is closed,
(b) $\Gamma \sim K_n$ is dense.

Then (a) and (b) imply that each K_n is nowhere dense, so that K is of the first category.

For (a), let f be a limit point of K_n and let $\{f_k\}$ be a sequence in K_n converging to f.

For each $k = 1, 2, \ldots$ let ξ_k be in $[0, 1]$ and such that

$$\left| \frac{f_k(\xi_k + h) - f_k(\xi_k)}{h} \right| \le n \text{ for all } h > 0.$$

Let ξ be a limit point of $\{\xi_k\}$ and let $\{\xi_{k_i}\}$ converge to ξ.

We may suppose, by considering the proper sequence $\{f_k\}$, that $\{\xi_k\}$ converges to ξ.

For $h > 0$ and $\epsilon > 0$,

$$\begin{aligned} \left| \frac{f(\xi + h) - f(\xi)}{h} \right| & \\ \le & \left| \frac{f_k(\xi_k + h) - f_k(\xi_k)}{h} \right| \\ & + \frac{1}{h} \{ |f(\xi + h) - f(\xi_k + h)| + |f(\xi_k + h) - f_k(\xi_k + h)| \\ & + |f_k(\xi_k) - f(\xi_k)| + |f(\xi_k) - f(\xi)| \}. \end{aligned}$$

There is an $N = N(\epsilon, h)$ such that $k > N$ implies

$$\sup |f_k(t) - f(t)| < \frac{\epsilon h}{4}.$$

By continuity of f, there is an $M > N$ such that $k > M$ implies

$$|f(\xi + h) - f(\xi_k + h)| < \frac{\epsilon h}{4} \quad \text{and} \quad |f(\xi_k) - f(\xi)| < \frac{\epsilon h}{4},$$

since $\lim_{k \to \infty} \xi_k = \xi$. Hence, if $k > M$, we have

$$\left| \frac{f(\xi + h) - f(\xi)}{h} \right| < \left| \frac{f_k(\xi_k + h) - f_k(\xi_k)}{h} \right| + \epsilon \le n + \epsilon.$$

It follows that

$$\left|\frac{f(\xi + h) - f(\xi)}{h}\right| \leq n$$

for all $h > 0$. Thus, $f \in K_n$ and K_n is closed.

For (b), suppose $g \in \Gamma$. Let $\epsilon > 0$. We may partition [0, 1] into k equal intervals so that if x, x' are in the same interval of the partitioning then

$$|g(x) - g(x')| < \frac{\epsilon}{2}.$$

Thus, in the ith subinterval $\frac{i-1}{k} \leq x \leq \frac{i}{k}$, if we consider the rectangle formed with

$$g\left(\frac{i-1}{k}\right) - \frac{\epsilon}{2} \leq y \leq g\left(\frac{i-1}{k}\right) + \frac{\epsilon}{2},$$

then the point $\left(\frac{i}{k}, g\left(\frac{i}{k}\right)\right)$ lies on the right-hand side and the point

$$\left(\frac{i-1}{k}, g\left(\frac{i-1}{k}\right)\right)$$

lies on the left-hand side of the rectangle. By joining the two points by a

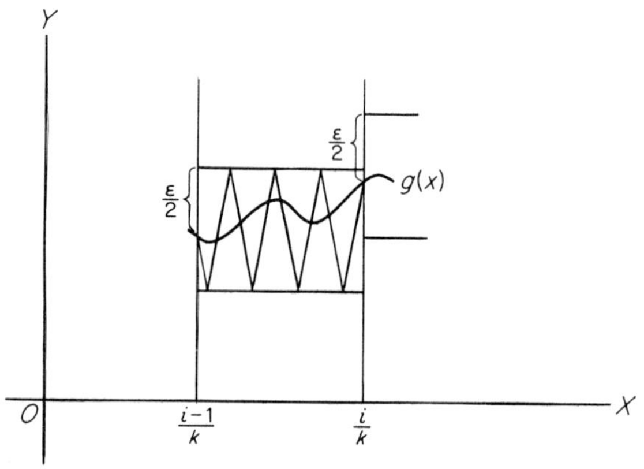

Figure 1

polygonal graph which remains within the rectangle and whose line segments have slopes exceeding n in absolute value, we obtain a continuous function which is within ϵ of g and is in $\Gamma \sim K_n$. This proves (b).

1.11 Compactness, Continuity

A metric space (X, ρ) is said to be **compact** if every infinite subset of X has at least one limit point.

A set $K \subset X$ is then compact if the space (K, ρ) is compact.

PROPOSITION 1. K is compact if and only if every sequence with values in K has a subsequence which converges to a point in K.

The proof is easy after a distinction is made between sequences which assume infinitely many values and those assuming only finitely many values.

Another fact, easy to prove and whose proof we omit, is given by

PROPOSITION 2. Every compact set in a metric space X is a closed, bounded subset of X. If X itself is compact then every closed subset of X is compact.

If X is a metric space and K is a subset, such that its closure $\bar{K}$ is compact, then K is said to be **relatively compact.**

The property of being closed and bounded is characteristic of the compact sets in euclidean n-space. This is the familiar Bolzano-Weierstrass theorem. In a general metric space, however, there may exist closed and bounded sets which are not compact. An example is the subset of l_2 consisting of the elements $e_1 = (1, 0, 0, \ldots)$, $e_2 = (0, 1, 0, 0, \ldots)$, $e_3 = (0, 0, 1, 0, 0, \ldots)$, ... This set is obviously bounded and it is closed since it has no limit points. For the same reason, it is not compact.

PROPOSITION 3. Every compact metric space is complete.

The proof is obvious.

Among the complete metric spaces, the compact ones are distinguished by a property called **total boundedness.**

A metric space X is said to be totally bounded if, for every $\epsilon > 0$, X contains a finite set, called an **ϵ-net**, such that the finite set of open spheres of radius ϵ and centers in the ϵ-net covers X.

THEOREM 1. A metric space X is compact if and only if it is complete and totally bounded.

We shall first show that the theorem follows from a lemma.

LEMMA 1. X is totally bounded if and only if every sequence in X has a Cauchy subsequence.

Proof of theorem 1. Suppose X is complete and totally bounded. By lemma 1, every sequence in X has a Cauchy subsequence, which converges since X is complete. Hence X is compact.

Suppose X is compact. Then X is complete. Moreover, every sequence in X has a convergent, hence a Cauchy, subsequence.

Proof of lemma 1. (Sufficiency) Let $\epsilon > 0$, and let $x_1 \in X$. If $X \sim S(x_1, \epsilon)$ is empty, we have found an ϵ-net. Otherwise, choose $x_2 \in X \sim S(x_1, \epsilon)$. If $X \sim [S(x_1, \epsilon) \cup S(x_2, \epsilon)]$ is empty, we have an ϵ-net. Otherwise, choose $x_3 \in X \sim [S(x_1, \epsilon) \cup S(x_2, \epsilon)]$. Clearly, it is only necessary to show that this process terminates. If it did not terminate, then $x_1, x_2, x_3, \ldots$ would be an infinite sequence each pair of which would have distance at least ϵ apart, and so no subsequence could be Cauchy, contrary to hypothesis.

(Necessity) Let $\{x_n\}$ be a sequence in X whose value set may be assumed infinite. Choose a $\frac{1}{2}$-net in X. Then one of the spheres of radius $\frac{1}{2}$, with center in the net, contains the values of an infinitely many-valued subsequence of the sequence. Proceeding in this way, we obtain a sequence of sequences, each a subsequence of the preceding, call them

$$\{x_n\}, \{x_n^{(1)}\}, \ldots, \{x_n^{(k)}\}, \ldots,$$

so that for every $k = 1, 2, \ldots, \{x_n^{(k)}\}$ is infinite-valued and all the values lie in a sphere of radius $\dfrac{1}{2^k}$.

Now, $\{x_n^{(n)}\}$ is a subsequence of $\{x_n\}$. Let $\epsilon > 0$. Choose N so that

$$\frac{1}{2^{N-1}} < \epsilon.$$

Then

$$n > N,\ p > 0 \text{ implies } \rho(x_{n+p}^{(n+p)}, x_n^{(n)}) < \frac{1}{2^{n-1}} < \epsilon,$$

so that $\{x_n^{(n)}\}$ is a Cauchy sequence.

PROPOSITION 4. Every totally bounded metric space is separable.

Proof. The union of $1/n$-nets for $n = 1, 2, \ldots$ is a countable dense set.

COROLLARY 1. Every compact metric space is separable.

Compactness is characterized in metric spaces by the property of the familiar Borel covering theorem; namely, we have the

PROPOSITION 5. A metric space X is compact if and only if for every collection $\mathcal{G}$ of open sets which covers X there is a finite subset $G_1, \ldots, G_n$ of $\mathcal{G}$ which covers X.

Proof. Suppose X is compact, and $\mathcal{G}$ is a collection of open sets which covers X. By corollary 1 and the Lindelöf theorem, there is a countable subset $G_1, G_2, \ldots$ of $\mathcal{G}$ which covers X. Suppose

$$X \sim \bigcup_{i=1}^{N} G_i$$

is non-empty for every N and let x_N be a point of this set. By compactness, a subsequence $\{x_{N_k}\}$ converges to some $x \in X$. Now, $x \in G_j$ for some j, and since G_j is open it contains all x_{N_k} with sufficiently large index. But this contradicts the way these points were chosen.

Conversely, suppose X is not compact. Let A be an infinite set in X which has no limit point. Then the distance d_x of x from $A \sim \{x\}$ is positive for every $x \in X$. The spheres $S(x, d_x)$ cover X. If a finite number of them were to cover X, then at least one of them would contain an infinite number of points of A. But these spheres were chosen to contain at most one point of A. Hence, there is an open covering of X which has no finite subcovering.

For mappings between metric spaces, the notions of continuity and uniform continuity have meaning. Let T map (X, ρ) into (Y, σ). T is said to be **continuous at $x \in X$** if for every sequence $\{x_n\}$ converging to x, $\{Tx_n\}$ converges to Tx in Y. T is said to be **continuous** if it is continuous at every $x \in X$.

A useful reformulation of the latter property is given by the

PROPOSITION 6. T is continuous if and only if for every open $G \subset Y$ the inverse image

$$T^{-1}(G) = [x : x \in X, Tx \in G]$$

is open in X.

We leave the proof for the reader.

T is said to be **uniformly continuous** on X if for every $\epsilon > 0$ there is a $\delta > 0$ such that

$$\rho(x, x') < \delta \text{ implies } \sigma(Tx, Tx') < \epsilon$$

for all $x, x' \in X$.

The proofs of the next two propositions are not difficult and are left as exercises for the reader.

PROPOSITION 7. If X is compact, then every T which is continuous on X is uniformly continuous.

PROPOSITION 8. If T is continuous, then the image of a compact set is compact.

In particular, if (Y, σ) is the real line, then the image of a compact set $A \subset X$ is closed and bounded, hence contains its supremum and infimum. This yields

COROLLARY 1. If f is a real-valued continuous function defined on a metric space (X, ρ), then for every compact $A \subset X$ the values

$$\sup [f(x) : x \in A], \quad \inf [f(x) : x \in A]$$

are finite and are attained by f at some points of A.

Another important property of continuous functions of a real variable, the intermediate-value property, also has a generalization to arbitrary metric spaces.

A set A in a metric space (X, ρ) is said to be **connected** if it cannot be represented as the union of two sets, each of which is disjoint from the closure of the other.

PROPOSITION 9. If T is continuous, then the image of a connected set is connected.

1.12 Equicontinuity, Application to Differential Equations

There are many applications of compactness in analysis, some of which we shall give below. For this reason, it is important to characterize the compact sets in the spaces which appear most often in analysis. We do this for C here and for other spaces later.

Let X be a compact metric space. The metric space of continuous real-valued functions on X with the metric

$$\rho(f, g) = \max\,[|f(x) - g(x)| : x \in X]$$

is a complete metric space, which we denote by $C(X)$. We have already discussed the special case $C[a, b]$.

A collection $\mathcal{F}$ of functions on a set X is said to be **uniformly bounded** if there is an $M > 0$ such that

$$|f(x)| \leq M \text{ for all } x \in X \text{ and all } f \in \mathcal{F}.$$

For subsets of $C(X)$, uniform boundedness agrees with boundedness in the metric space, i.e., a set $\mathcal{F}$ is uniformly bounded if and only if it is contained in a sphere.

A collection $\mathcal{F}$ of functions defined on a metric space X is called **equicontinuous** if for each $\epsilon > 0$ there is a $\delta > 0$ such that $\rho(x, x') < \delta$ implies $|f(x) - f(x')| < \epsilon$ for all $x, x' \in X$ and all $f \in \mathcal{F}$.

According to this definition, the functions belonging to the equicontinuous collection are uniformly continuous.

The following theorem is of great historical importance in the development of general topology and of functional analysis, as well as being of great use.

THEOREM 1 (Arzelà-Ascoli). If X is a compact metric space, a subset $K \subset C(X)$ is relatively compact if and only if it is uniformly bounded and equicontinuous.

Proof. Suppose K is relatively compact. It then follows that K is totally bounded. Let $\epsilon > 0$ and let $f_1, f_2, \ldots, f_n$ be an $\epsilon/3$-net in K. Now, let $f \in K$ and $x, x' \in X$. For each $i = 1, 2, \ldots, n$,

$$|f(x) - f(x')| \leq |f(x) - f_i(x)| + |f_i(x) - f_i(x')| + |f_i(x') - f(x')|.$$

Choose j so that

$$\sup [|f(x) - f_j(x)| : x \in X] < \frac{\epsilon}{3}.$$

Then

$$|f(x) - f(x')| < |f_j(x) - f_j(x')| + \frac{2\epsilon}{3}.$$

Since X is compact, the functions f_i, $i = 1, \ldots, n$, are uniformly continuous. There is then a $\delta > 0$ so that $\rho(x, x') < \delta$ implies

$$|f_i(x) - f_i(x')| < \frac{\epsilon}{3}, \quad i = 1, \ldots, n.$$

It follows that $\rho(x, x') < \delta$ implies

$$|f(x) - f(x')| < \epsilon \text{ for every } f \in K.$$

Hence, K is equicontinuous.

Suppose, conversely, that K is uniformly bounded and equicontinuous. Suppose M is an integer such that

$$|f(x)| \leq M \text{ for all } x \in X \text{ and } f \in K.$$

Let $\epsilon > 0$. Choose δ so that $\rho(x, x') < \delta$ implies $|f(x) - f(x')| < \epsilon/4$ for all $f \in K$. Since X is compact, it has a δ-net $x_1, \ldots, x_n$. Choose a positive integer m such that $1/m < \epsilon/4$ and divide $[-M, M]$ into $2Mm$ equal parts by the points

$$y_0 = -M < y_1 < y_2 < \ldots < y_k = M,$$

where $k = 2Mm$.

Consider those n-tuples $(y_{i_1}, y_{i_2}, \ldots, y_{i_n})$ of the numbers y_i, $i = 0, 1, \ldots, k$, such that some $f \in K$ has the property

$$|f(x_j) - y_{i_j}| < \frac{\epsilon}{4}, \quad j = 1, 2, \ldots, n$$

and choose one such $f \in K$ for each such n-tuple.

We shall show that the resulting finite subset E of K is an ϵ-net for K.

If $f \in K$ then we may choose $y_{i_1}, y_{i_2}, \ldots y_{i_n}$ so that

$$|f(x_j) - y_{i_j}| < \frac{\epsilon}{4}, \quad j = 1, 2, \ldots, n,$$

and so there is a corresponding $e \in E$.

Let $x \in X$ and choose j so that $\rho(x, x_j) < \delta$. Then

$$|f(x) - e(x)| \leq |f(x) - f(x_j)| + |f(x_j) - y_{i_j}| + |y_{i_j} - e(x_j)| + |e(x_j) - e(x)| < \epsilon.$$

Hence

$$\sup_{x \in X} |f(x) - e(x)| < \epsilon.$$

An application of the Arzelà-Ascoli theorem yields the existence of solutions to the differential equation $y' = f(x, y)$ in case f is only assumed to be continuous. To be precise we have the

THEOREM 2. If f is continuous on an open set D, then for every $(x_0, y_0) \in D$ the differential equation $dy/dx = f(x, y)$ has a local solution passing through (x_0, y_0).

Proof. Let R be a closed, non-degenerate rectangle contained in D having center (x_0, y_0) and sides parallel to the axes. Let $M > 0$ be such that

$$|f(x, y)| < M \text{ for all } (x, y) \in R.$$

The two lines through (x_0, y_0) with slopes $\pm M$ may intersect the horizontal sides of R. If so, take I to be the projection on the x-axis of the interval between the points of intersection. Otherwise, I is the projection of the entire horizontal side. In either case, I has the form $[x_0 - a, x_0 + a]$ for some

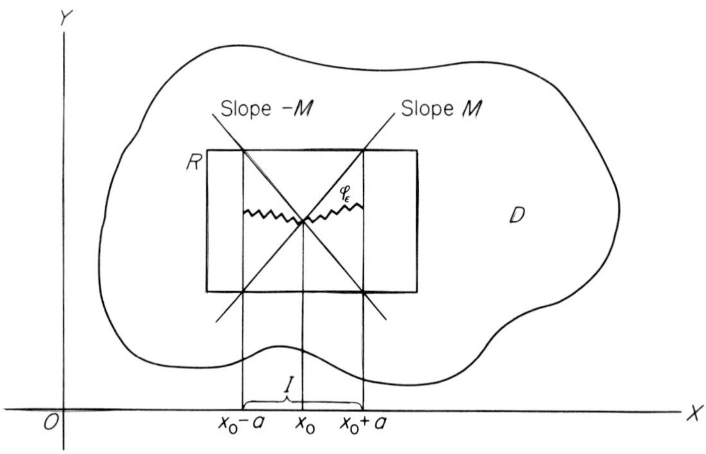

Figure 2

$a > 0$. Our solution φ will be defined on I. It will be obtained by applying the Arzelà-Ascoli theorem in $C(I)$ to a set of "approximate" solutions.

For $\epsilon > 0$ we shall call a function φ_ϵ defined on I an ϵ-approximate solution if its graph is a polygonal arc through (x_0, y_0), lying between the two lines of slope $\pm M$ and the lines $x = x_0 \pm a$ and satisfying

$$|\varphi_\epsilon'(x) - f(x, \varphi_\epsilon(x))| < \epsilon$$

at each $x \in I$ where φ_ϵ' exists (i.e., except at the finite number of points where the vertices occur).

Since f is uniformly continuous on R, there is a $\delta > 0$ such that, for (x, y) and $(\bar{x}, \bar{y})$ in R,

$$|x - \bar{x}| < \delta \text{ and } |y - \bar{y}| < \delta \text{ imply } |f(x, y) - f(\bar{x}, \bar{y})| < \epsilon.$$

It is enough to describe φ_ϵ on the interval $[x_0, x_0 + a]$. Divide this interval by the points

$$x_0 < x_1 < x_2 < \ldots < x_n = x_0 + a$$

so that

$$|x_i - x_{i-1}| < \min\left(\delta, \frac{\delta}{M}\right), \quad i = 1, \ldots, n.$$

Let $\varphi_\epsilon(x_0) = y_0$ and, in $[x_0, x_1]$ let φ_ϵ be linear with slope $f(x_0, y_0)$. Then $\varphi_\epsilon(x_1)$ is defined. In $[x_1, x_2]$, take φ_ϵ to be linear with slope $f(x_1, \varphi_\epsilon(x_1))$. Continuing in a way which should now be clear, φ_ϵ is defined on $[x_0, x_0 + a]$.

Since the slope of φ_ϵ nowhere exceeds M in absolute value, we obtain

$$|\varphi_\epsilon(x) - \varphi_\epsilon(\bar{x})| \leq M\,|x - \bar{x}| \text{ for all } x, \bar{x} \in I.$$

In other words, the functions φ_ϵ, where we consider one such function for each $0 < \epsilon < a$, are an equicontinuous collection.

Fix ϵ. If $x \in I$ and $x \neq x_i$, $i = 0, 1, \ldots, n$, then $x_{j-1} < x < x_j$ for some j and

$$|\varphi_\epsilon(x) - \varphi_\epsilon(x_{j-1})| \leq M\,|x - x_{j-1}| < M \cdot \frac{\delta}{M} = \delta.$$

But this implies

$$|f(x_{j-1}, \varphi_\epsilon(x_{j-1})) - f(x, \varphi_\epsilon(x))| < \epsilon.$$

But $\varphi_\epsilon'(x)$ exists and equals $f(x_{j-1}, \varphi_\epsilon(x_{j-1}))$. It follows that φ_ϵ is an ϵ-approximate solution,

$$|\varphi_\epsilon'(x) - f(x, \varphi_\epsilon(x))| < \epsilon, \quad x \in I \text{ and } x \neq x_i, \quad i = 1, \ldots, n.$$

We now write

$$(*) \qquad \varphi_\epsilon(x) = y_0 + \int_{x_0}^{x} \{f(u, \varphi_\epsilon(u)) + [\varphi_\epsilon'(u) - f(u, \varphi_\epsilon(u))]\}\, du,$$

where φ_ϵ' can have any values at the vertices. Choose them so that the term in brackets vanishes there.

The collection φ_ϵ, $a > \epsilon > 0$, is uniformly bounded on I (since the graphs all lie in R) and is equicontinuous. Hence if $\{\epsilon_n\}$ is a null sequence, it has a subsequence, which we again call $\{\epsilon_n\}$, such that $\{\varphi_{\epsilon_n}\}$ converges uniformly to a function φ on I.

Since f is uniformly continuous, $\{f(u, \varphi_{\epsilon_n}(u))\}$ converges uniformly to $f(u, \varphi(u))$ on I. Moreover, $\{\varphi'_{\epsilon_n}(u) - f(u, \varphi(u))\}$ converges uniformly to zero on I. It then follows by $(*)$ that

$$\varphi(x) = y_0 + \int_{x_0}^{x} f(u, \varphi(u))\, du.$$

But this implies that φ is differentiable on I and is the required solution.

1.13 Stone-Weierstrass Theorems

Our next topic is the Weierstrass approximation theorem and its generalization by Stone. The Weierstrass theorem asserts that if $[a, b]$ is a closed interval on the real line, then the polynomials are dense in $C[a, b]$. In other words, every continuous function on $[a, b]$ is the limit of a uniformly convergent sequence of polynomials. In order to prove this we need only show that for every continuous f and $\epsilon > 0$ there is a polynomial p such that

$$\max \{|f(x) - p(x)| : x \in [a, b]\} < \epsilon.$$

It follows from the fact that f is uniformly continuous on $[a, b]$ that there is a g, whose graph is a polygonal arc, such that

$$\max \{|f(x) - g(x)| : x \in [a, b]\} < \frac{\epsilon}{3}.$$

It is easy to prove that the Fourier series of g converges uniformly to g on $[a, b]$. We assume that $f(b) = f(a)$ and $g(b) = g(a)$, leaving it to the reader to note that this entails no loss in generality. There accordingly exists a trigonometric polynomial

$$t(x) = a_0 + \sum_{k=1}^{n} (a_k \cos k\alpha x + b_k \sin k\alpha x),$$

where α is the proper positive number, such that

$$\max \{|g(x) - t(x)| : x \in [a, b]\} < \frac{\epsilon}{3}.$$

But the Taylor series of the functions $\cos k\alpha x$, $\sin k\alpha x$, $k = 1, \ldots, n$ converge uniformly to them, so there is a polynomial p such that

$$\max \{|t(x) - p(x)| : x \in [a, b]\} < \frac{\epsilon}{3}.$$

It follows that

$$\max \{|f(x) - p(x)| : x \in [a, b]\} < \epsilon$$

and the theorem is proved.

This remarkable theorem has a large number of proofs. There are two famous proofs where the polynomials are given directly in terms of the function f. These proofs are relegated to the exercises. Another proof shows that the theorem follows almost immediately after it is proved for the function $|x|$. This fact is used also in the generalization given by Stone, which we now discuss.

Let (X, ρ) be a compact metric space, and let $C(X)$ be the space of continuous real functions on X, with the usual metric

$$\rho(f, g) = \max [|f(x) - g(x)| : x \in X].$$

We define algebraic operations in $C(X)$ as follows: If $f, g \in C(X)$ and a is real, then, for every $x \in X$,

$$\begin{aligned}(f + g)(x) &= f(x) + g(x),\\ (fg)(x) &= f(x)g(x),\\ (af)(x) &= a f(x).\end{aligned}$$

A set $A \subset C(X)$ is called an **algebra** if $f, g \in A$ and a real imply $f + g \in A$, $fg \in A$ and $af \in A$.

For example, the polynomials and the trigonometric polynomials are algebras in $C[a, b]$.

If $E \subset C(X)$, the intersection of all algebras in $C(X)$ containing E, which is itself an algebra containing E, is denoted by $A(E)$ and is called the **algebra generated by *E*.**

For example, the polynomials on $[a, b]$ are generated by the pair $1, x$. We shall need the fact, easy to prove, that if $A \subset C(X)$ is an algebra then its closure $\bar{A}$ is also an algebra.

We also make use of the Weierstrass theorem for $[a, b] = [-1, 1]$. We shall not need the full force of the theorem, only the fact that the function $|x|$ is the limit of a uniformly convergent sequence of polynomials. In order to prove this directly, we note that for every $\eta > 0$ the Taylor séries of

$$(t + \eta^2)^{1/2}$$

converges uniformly to the function on $[0, 1]$. Hence for $\epsilon > 0$ there is a polynomial $p(x)$ such that

$$|(t + \eta^2)^{1/2} - p(t)| < \frac{\epsilon}{2} \text{ for all } t \in [0, 1].$$

For $x \in [-1, 1]$, let $|x|^2 = t$. Then

$$\left|(|x|^2 + \eta^2)^{1/2} - p(|x|^2)\right| < \frac{\epsilon}{2} \text{ for all } x \in [-1, 1].$$

Thus if $\eta = \epsilon/2$,

$$\begin{aligned}\left||x| - p(x^2)\right| &\le \left||x| - (|x|^2 + \eta^2)^{1/2}\right|\\ &\quad + \left|(|x|^2 + \eta^2)^{1/2} - p(x^2)\right| < \epsilon.\end{aligned}$$

The idea of basing the proof of the Weierstrass theorem on this special case is due to Lebesgue.

We now state Stone's theorem.

THEOREM 1. Let A be a closed algebra in $C(X)$, X a compact metric space, such that $1 \in A$ and if $x, y \in X$, $x \neq y$, there is an $f \in A$ for which $f(x) \neq f(y)$. Then $A = C(X)$.

Proof. We first show that $f \in A$ implies $|f| \in A$. Suppose at first that $\max [|f(x)| : x \in X] \leq 1$. Let $\epsilon > 0$ and let $p(t) = a_0 + a_1 t + \ldots + a_n t^n$ be a polynomial such that

$$\big||t| - p(t)\big| < \epsilon \text{ for every } t \in [-1, 1].$$

Then if $p(f) = a_0 + a_1 f + \ldots + a_n f^n$, $p \in A$ and

$$\big||f(x)| - p(f(x))\big| < \epsilon \text{ for every } x \in X.$$

Thus $|f| \in A$, since A is closed in $C(X)$. For any $f \in A$ there is an $a \neq 0$ with $|af(x)| \leq 1$ for every $x \in X$ and since $|af|$ is in A, so is $|f|$.

We next note that if $f, g \in A$ then $\min(f, g)$ and $\max(f, g)$ are in A, where, for example,

$$[\min(f, g)](x) = \min(f(x), g(x)).$$

This follows since

$$\min(f, g) = \tfrac{1}{2}(f + g) - \tfrac{1}{2}|f - g|,$$

and $|f - g| \in A$ if $f, g \in A$.

Let $x, y \in X$ with $x \neq y$. Let $f \in C(X)$. The constant function g with the value $f(x)$ is in A. Let $h \in A$ be such that $h(x) \neq h(y)$. There is no loss in generality in assuming $h(x) = 0$. There is a constant a such that the function f_{xy} given by

$$f_{xy} = g + ah$$

satisfies $f_{xy}(x) = f(x)$ and $f_{xy}(y) = f(y)$ and belongs to A. Let $\epsilon > 0$. For every $y \in X$ there is an open sphere S_y such that $y \in S_y$ and $f_{xy}(z) < f(z) + \epsilon$ for every $z \in S_y$. Since the space X is compact, a finite number of these spheres covers X; call them $S_{y_1}, \ldots, S_{y_n}$. Let

$$f_x = \min(f_{xy_1}, \ldots, f_{xy_n}).$$

Then $f_x \in A$, $f_x(x) = f(x)$ and, for every $z \in X$, $f_x(z) < f(z) + \epsilon$. Now for every $x \in X$, x is in an open sphere T_x such that, for every $z \in T_x$,

$$f_x(z) > f(z) - \epsilon.$$

Since X is compact, a finite number of these spheres $T_{x_1}, \ldots, T_{x_m}$ covers X. Let

$$F = \max(f_{x_1}, \ldots, f_{x_m}).$$

Then $F \in A$ and, for every $z \in X$,

$$|f(z) - F(z)| < \epsilon.$$

This proves the theorem.

1.14 Normal Families

Let G be a connected open set in the plane, considered as the complex $z = x + iy$ plane. For functions defined in G, a natural and fruitful notion of convergence is that of convergence in G, uniformly on each compact subset of G. For example, according to a theorem of Weierstrass, if a sequence of functions holomorphic in G converges in this sense, then the limit function is holomorphic in G.

We assume the functions involved are bounded on compact sets. The above convergence may then be obtained from a metric $d(f, g)$ as follows. Let K_n, $n = 1, 2, \ldots$, be a sequence of compact subsets of G such that

$$K_n \subset K_{n+1}, \quad n = 1, 2, \ldots,$$

and

$$G = \bigcup_{n=1}^{\infty} K_n.$$

Let d_n be the uniform metric on K_n, $d_n(f, g) = \sup [|f(z) - g(z)| : z \in K_n]$, and put

$$d(f, g) = \sum_{n=1}^{\infty} \frac{1}{2^n} \frac{d_n(f, g)}{1 + d_n(f, g)}.$$

It can be shown that d is a metric. It is not hard to prove the

Proposition 1. $\{f_n\}$ converges to f in the metric d if and only if $\{f_n\}$ converges to f uniformly on each compact subset of G.

The theorem of Weierstrass mentioned above gives at once the completeness of the space of all functions holomorphic in G, under the metric d. The space of all functions continuous on G is also complete under d. These facts are easily seen and are left to the reader.

We consider the question of compactness in these spaces. The term **normal family** is used for a collection of functions which is relatively compact for the above notion of convergence. Such families have important and surprising applications in complex variable theory.

The Arzelà-Ascoli theorem yields the

Proposition 2. *F is a normal family of continuous functions on G if and only if F is uniformly bounded and equicontinuous on each compact $K \subset G$.*

Proof. We only discuss the sufficiency. Let K_n, $n = 1, 2, \ldots$, be compact subsets of G as above. Let $\{f_n\}$ be any sequence in F. A subsequence $\{f_n^{(1)}\}$ of $\{f_n\}$ converges uniformly on K_1. A subsequence $\{f_n^{(2)}\}$ of $\{f_n^{(1)}\}$ converges uniformly on K_2. Continuing in this way and then choosing the diagonal sequence $\{f_k^{(k)}\}$, we obtain a subsequence of $\{f_n\}$ which converges uniformly on each compact $K \subset G$.

It is a noteworthy fact that if F consists of holomorphic functions on G, then the equicontinuity assumption need not be made explicitly.

THEOREM 1 (Montel). A family F of functions holomorphic in a domain G is normal if and only if F is uniformly bounded on each compact subset of G.

Proof. Let $K \subset G$ be compact. We shall show that F is equicontinuous on K.

Let $r > 0$ be so small that the set $K^{(r)}$ of points whose distance from K does not exceed r is a subset of G. $K^{(r)}$ is compact, so that F is uniformly bounded there. Let M be a uniform bound, i.e.,

$$|f(z)| \leq M, \quad f \in F, \quad z \in K^{(r)}.$$

Let $0 < \epsilon < M$, and set

$$\delta = \frac{\epsilon r}{2M}.$$

Then $\delta < r/2$.

Now, if $z, z' \in K$ and $|z - z'| < \delta$, then the circle C, with center z and radius r, is in $K^{(r)}$. Hence, for every $f \in F$,

$$|f(z) - f(z')| = \left| \frac{1}{2\pi i} \int_C f(t) \left[\frac{1}{t - z} - \frac{1}{t - z'} \right] dt \right|$$

$$\leq \frac{1}{2\pi} \int_C \frac{|f(t)|\,|z - z'|}{|t - z|\,|t - z'|} |dt|$$

$$< \frac{M\delta \cdot 2\pi r}{2\pi \cdot r \cdot r/2} = \epsilon.$$

We consider next a variant of proposition 2 in which equicontinuity is replaced by a condition on the convergence of the integral means of the functions. An analogous treatment will be given other spaces in Chapter 3.

Let K be a compact set in G and let $d > 0$ be less than half the distance from K to the boundary of G. Let K' be the set of all points whose distance from K does not exceed d. K' is compact.

If f is defined in G and $z \in K'$, then for $0 < r \leq d$, let

$$M_r(f; z) = \frac{1}{\pi r^2} \int_0^{2\pi} \int_0^r f(z + \rho e^{i\theta}) \rho \, d\rho \, d\theta.$$

Of course, the function f is assumed to be integrable. The function M_r is called the **integral mean** of f.

The function M_r is continuous; if f is continuous, then M_r has partial derivatives, which are continuous, on the interior of K', and are given by the formulas:

$$\frac{\partial M_r(f; z)}{\partial x} = \frac{1}{\pi r} \int_0^{2\pi} f(z + re^{i\theta}) \cos\theta \, d\theta$$

$$\frac{\partial M_r(f; z)}{\partial y} = \frac{1}{\pi r} \int_0^{2\pi} f(z + re^{i\theta}) \sin\theta \, d\theta.$$

We leave the details of the proofs to the reader.

Let f be continuous on G. The identity

$$M_r(f; z) - f(z) = \frac{1}{\pi r^2} \int_0^{2\pi}\int_0^r [f(z + \rho e^{i\theta}) - f(z)]\rho \, d\rho \, d\theta$$

shows that $M_r(f)$ converges to f, as r goes to zero, uniformly on compact subsets of G. Moreover, if F is a normal family, then it is equicontinuous and it follows, by the same identity, that the convergence is uniform for $f \in F$, i.e.,

$$\sup\,[|M_r(f; z) - f(z)| : z \in K, \quad f \in F]$$

converges to zero as r converges to zero. This proves the necessity in the

THEOREM 2. The family F of continuous functions in G is normal if and only if, for each compact $K \subset G$, F is uniformly bounded on K and the convergence of $M_r(f)$ to f, uniform on K as r goes to zero, is uniform with respect to $f \in F$.

Proof. We need only show the equicontinuity of F on K under the above conditions.

The functions f will be supposed real-valued. The complex case then follows.

Let M denote a uniform bound for $|f(z)|$ for $f \in F$ and $z \in K'$, where K' is defined as above. Let $\epsilon > 0$.

For any $f \in F$ and $z, z' \in K$,

$$|f(z) - f(z')| \leq |f(z) - M_r(f; z)| + |M_r(f; z) - M_r(f; z')| + |M_r(f; z') - f(z')| < |M_r(f; z) - M_r(f; z')| + \frac{\epsilon}{2},$$

for r sufficiently small, independent of f, z and z'.

Choose $\delta > 0$ so that $\dfrac{2M}{r}\delta < \dfrac{\epsilon}{2}$ and if $z, z' \in K$ with $|z - z'| < \delta$, then the segment joining z to z' and the circles of radius r with centers on this segment are contained in K'. By the mean value theorem, there is a $\tilde{z}$ on the segment for which

$$M_r(f; z) - M_r(f; z') = (z - z')\left\{\frac{\partial M_r(f; \tilde{z})}{\partial x} + i\,\frac{\partial M_r(f; \tilde{z})}{\partial y}\right\}$$

$$= \frac{z - z'}{\pi r} \int_0^{2\pi} f(\tilde{z} + re^{i\theta})(\cos\theta + i\sin\theta)\, d\theta.$$

Thus, for $|z - z'| < \delta$, $z, z' \in K$ and $f \in F$,

$$|f(z) - f(z')| < \frac{2\delta M}{r} + \frac{\epsilon}{2} < \epsilon.$$

It should be remarked that the uniform boundedness is essential to the above proof of equicontinuity. Simple examples can be given of families for which the convergence of the integral means has the above uniformity but which are not equicontinuous.

1.15 Semi-continuity, Application to Arc Length

It is known that if f is a continuous real function on a closed interval $[a, b]$, then f has an absolute minimum on $[a, b]$, i.e., there is an $x \in [a, b]$ such that $f(y) \geq f(x)$ for every $y \in [a, b]$.

This theorem remains true for continuous functions on compact metric spaces. However, in most applications the function which arises is not continuous. Fortunately, the more general result, given in proposition 1 below, which is of great importance, is valid. We first must define semi-continuity.

If X is a metric space and f is a real function on X then f is said to be **lower semi-continuous** at $x \in X$ if, for every $\epsilon > 0$, there is a $\delta > 0$ such that $\rho(x, y) < \delta$ implies

$$f(y) > f(x) - \epsilon.$$

Upper semi-continuity is defined in a similar fashion, and it is then easy to show that f is continuous at x if and only if it is both upper semi-continuous at x and lower semi-continuous at x.

f is said to be **lower semi-continuous on** X if it is lower semi-continuous at every $x \in X$. We now state the

PROPOSITION 1. If X is a compact metric space and f is a lower semi-continuous real function on X, then f has an absolute minimum on X.

Proof. We first note that the set of values of f on X has a lower bound. Otherwise, there would be an $x_n \in X$, for every $n = 1, 2, \ldots$, such that $f(x_n) < -n$. Since X is compact, $\{x_n\}$ has a convergent subsequence. Let $\{y_n\}$ be this subsequence and let $y = \lim_n y_n$. Then $\lim_n f(y_n) = -\infty$ and, since $f(y)$ is real, this is easily seen to contradict the lower semi-continuity of f.

Now, let

$$L = \inf\,[f(x) : x \in X].$$

There is a $\xi_n \in X$, for every $n = 1, 2, \ldots$ such that $f(\xi_n) < L + \dfrac{1}{n}$. Since X is compact, $\{\xi_n\}$ has a convergent subsequence. Let $\{\eta_n\}$ be this subsequence, and let $x = \lim \eta_n$. Then $\lim_n f(\eta_n) = L$. But it follows from the definition of lower semi-continuity that $f(x) \le \lim_n f(\eta_n) = L$. However, by the definition of L, $f(x) \ge L$. This proves the proposition.

We shall give one simple application. For this purpose it is necessary to define the length of a curve. We shall consider continuous mappings $T = (x, y, z)$ on the unit interval $I = [0, 1]$ into euclidean 3-space E_3, i.e., the functions x, y, z are continuous real functions. Let ρ be the euclidean metric. For each $t \in I$, $T(t) = (x(t), y(t), z(t))$ is a point in E_3. For mappings T_1 and T_2 let

$$d(T_1, T_2) = \max [\rho(T_1(t), T_2(t)) : t \in I].$$

It is easy to show that if $\mathcal{T}$ is the set of all T then $(\mathcal{T}, d)$ is a complete metric space. Curves will be defined as certain equivalence classes in $\mathcal{T}$. In the first place T_1 will be equivalent to T_2 if there is a homeomorphism h of I onto itself such that T_2 is the composite

$$T_2 = T_1 \circ h.$$

It is not satisfactory to take these equivalence classes as curves since they will not form a metric space (with the natural metric defined from d). However, the closures of these equivalence classes also form equivalence classes, and these will be called curves.

Precisely, if T_1 and T_2 are in $\mathcal{T}$ we shall say that T_1 is **equivalent to** T_2 if, for every $\epsilon > 0$, there is a homeomorphism h of I onto itself such that

$$d(T_2, T_1 \circ h) < \epsilon.$$

It is fairly routine to show that this is an equivalence relation in T. We shall designate the equivalence classes as $C, C_1, C_2, \ldots$, and let $\mathcal{C}$ be the set of all equivalence classes, or **curves**. A member T of an equivalence class C will be called a **representation** of the curve C.

We define a metric δ in $\mathcal{C}$ as

$$\delta(C_1, C_2) = \inf [d(T_1, T_2) : T_1 \in C_1, T_2 \in C_2].$$

We leave it to the reader to verify that $(\mathcal{C}, \delta)$ is a metric space.

Having defined curves, we now define the length of a curve. Let C be a curve and let $T \in C$. For every partition

$$\pi = [0 = t_0 < t_1 < \ldots < t_n = 1]$$

of $[0, 1]$, let

$$l(T, \pi) = \sum_{i=1}^{n} \rho(T(t_i), T(t_{i-1}))$$

and let

$$L(T) = \sup l(T, \pi),$$

where the supremum is taken for all partitions π, and may assume the value $+\infty$.

It is not hard to show that if T_1 is equivalent to T_2 then $L(T_1) = L(T_2)$. We may thus define the **length $L(C)$ of the curve C** to be the common value of $L(T)$ for all $T \in C$.

We prove

PROPOSITION 2. $L(C)$ is a lower semi-continuous non-negative (extended) real-valued function on the metric space $\mathcal{C}$.

Proof. Let $C \in \mathcal{C}$ and suppose $L(C) < \infty$. Let $T \in C$. Fix $\epsilon > 0$. There is a partition

$$\pi = [0 = t_0 < t_1 < \ldots < t_n = 1]$$

of [0, 1] such that

$$l(T, \pi) > L(C) - \frac{\epsilon}{2}.$$

But, there is a $\delta > 0$ such that $d(T', T) < \delta$ implies

$$l(T', \pi) > l(T, \pi) - \frac{\epsilon}{2}$$

so that

$$L(C') = L(T') \geq l(T', \pi) > L(C) - \epsilon.$$

Now, let C' be such that $\delta(C', C) < \delta$. There is then a $T' \in C'$ such that $d(T', T) < \delta$. We have thus shown that $\delta(C', C) < \delta$ implies

$$L(C') > L(C) - \epsilon.$$

For the case $L(C) = +\infty$, lower semi-continuity at C means that for every $M > 0$ there is a $\delta > 0$ such that $\delta(C, C') < \delta$ implies $L(C') > M$. The proof for this case is similar to the finite case and is left to the reader.

Now let A be the closed cube of points (x, y, z) where $|x| \leq a$, $|y| \leq a$, $|z| \leq a$, $a > 0$. We let $\mathcal{C}_A$ be the set of curves in A, i.e., for every $C \in \mathcal{C}_A$ and $T \in C$, $T(t) \in A$ for every $t \in I$.

Next, let $M > 0$ and let $\mathcal{C}_{AM}$ be the space of curves in $\mathcal{C}_A$ of length not greater than M. We now prove

PROPOSITION 3. The metric space $\mathcal{C}_{AM}$ is compact.

Proof. In order to prove this result, it is convenient to consider a particular representation $R \in C$ for every $C \in \mathcal{C}_{AM}$. Let $T \in C$. For every $t \in I$ we consider the mapping T restricted to the interval $[0, t]$. Call this mapping T_t. Consider $L(T_t)$. This is a monotonically non-decreasing function on [0, 1] and is easily seen to be continuous. For every $t \in [0, 1]$, there is a unique $u \in [0, 1]$ such that $L(T_t) = uL(T)$. Define R so that

$$R(u) = T(t).$$

Then R is defined everywhere on $[0, 1]$, and $L(R_u) = uL(T)$. The case where T is constant must be treated separately; here, $R = T$.

It is not hard to show that $R \in C$ and we leave the proof to the reader. We shall call R the **arc length representation** of C. Now for every $C \in \mathcal{C}_{AM}$, with $R \in C$

$$\rho(R(t_1), R(t_2)) \leq M\,|t_1 - t_2|.$$

Thus the collection of arc length representations of curves in $\mathcal{C}_{AM}$ is equicontinuous. Since it is uniformly bounded, every sequence has a Cauchy subsequence. Since

$$\delta(C, C') \leq d(R, R'),$$

the same holds for $\mathcal{C}_{AM}$.

It remains only to show that $\mathcal{C}_{AM}$ is complete. But this is an obvious consequence of the lower semi-continuity of L.

Since L is a lower semi-continuous functional on the compact space C_{AM}, it has a minimum. Since L is 0 for the curves consisting of one point, this result is trivial. However, we may consider closed subspaces of $\mathcal{C}_{AM}$. For example, consider two points (x_1, y_1, z_1) and (x_2, y_2, z_2) and consider those curves for which these are the end points. This is a closed subspace of $\mathcal{C}_{AM}$ and so it is compact. Moreover, L is still lower semi-continuous. This yields the result that among the curves in A joining two points, there is one of shortest length. This result is still not very exciting. However, let S be an arbitrary closed subset of A, and let (x_1, y_1, z_1) and (x_2, y_2, z_2) be any points in S. Suppose there is a curve of length not greater than M, with these points as end points, which lies in S. It is easy to see that the set of all curves lying in S, with these points as end points, and of length not greater than M, is a non-empty closed subspace of $\mathcal{C}_{AM}$. It then follows that we have

THEOREM 1. Let S be a closed bounded set in euclidean 3-space, let (x_1, y_1, z_1) and (x_2, y_2, z_2) be in S, and suppose there is a curve in S, with these end points, which has finite length. Then there is a curve in S, among those with these end points, for which the length is a minimum.

For example, if S is a torus, every pair of points in S is joined by an arc in S of minimum length.

1.16 Space of Compact, Convex Sets

To conclude this chapter, we return to some considerations in the space (Γ, d) of all non-empty, compact, convex subsets of the plane introduced in example (xi), Sec. 1.3.

We shall need the following lemma, whose proof is left to the reader.

LEMMA 1. If $\{A_n\}$ is a decreasing sequence of sets in Γ, then $A = \bigcap_{n=1}^{\infty} A_n$ is in Γ and $\{A_n\}$ converges to A in the metric of (Γ, d).

We now prove the

PROPOSITION 1. (Γ, d) is complete.

Proof. Let $\{A_n\}$ be a Cauchy sequence in Γ. For $\epsilon > 0$, there is an N such that

$$d(A_n, A_{n+p}) < \epsilon,\ n \geq N, \quad p = 1, 2, 3, \ldots.$$

For each $n = 1, 2, 3, \ldots$, let H_n be the smallest closed convex set containing the set

$$A_n \cup A_{n+1} \cup A_{n+2} \cup \ldots,$$

i.e., the intersection of all closed convex sets containing this set. Then, for every n, $H_n \supset H_{n+1}$, and it follows that

$$H = \bigcap_{n=1}^{\infty} H_n$$

is a non-empty, closed, convex set.

Now, for $\epsilon > 0$, choose N as above. We obtain, for $n \geq N$,

$$(A_n)_\epsilon \supset A_{n+p}, \quad p = 1, 2, 3, \ldots$$

so that

$$(A_n)_\epsilon \supset A_{n+1} \cup A_{n+2} \cup \ldots,$$

and

$$(A_n)_\epsilon \supset H_{n+1} \supset H.$$

In particular, this shows that $H \in \Gamma$, since it is bounded.

By lemma 1, there is an M such that $n \geq M$ implies $H_n \subset H_\epsilon$, so that, since $A_n \subset H_n$, we have $A_n \subset H_\epsilon$. It follows that, for $n \geq \max(N, M)$,

$$H_\epsilon \subset A_n \quad \text{and} \quad (A_n)_\epsilon \subset H,$$

so that

$$d(A_n, H) < \epsilon$$

and the proof is complete.

The space (Γ, d) is introduced as a tool in the general theory of convex sets (all of our considerations are valid for sets in n-space). The main fact is the following compactness criterion.

THEOREM 1. Let Q be a closed square. The subset Γ_Q of Γ, of those sets belonging to Γ which are contained in Q, is a compact metric space.

Proof. In view of proposition 1, we need only show that Γ_Q is totally bounded, i.e., if $A_n \in \Gamma_Q$, $n = 1, 2, \ldots$, then $\{A_n\}$ has a Cauchy subsequence.

Let s denote the length of the sides of Q, and for each $i = 1, 2, 3, \ldots$, consider the subdivision of Q into 4^i equal squares of side length $s/2^i$. For each n and i, we associate with A_n the set of those squares of the ith subdivision which meet A_n. There will be a subsequence $\{A_{1n}\}$ of $\{A_n\}$ for each of whose terms the associated set of squares in the first subdivision is always the same set σ_1. There will then be a subsequence $\{A_{2n}\}$ of $\{A_{1n}\}$ for each of whose terms the associated set of squares in the second subdivision is always the same set σ_2. We then obtain

$$\{A_{1n}\}, \{A_{2n}\}, \ldots, \{A_{in}\}, \ldots,$$

each sequence being a subsequence of the preceding and such that for each term of $\{A_{in}\}$, $i = 1, 2, \ldots$, the associated set of squares in the ith subdivision is always the same set σ_i.

The diagonal sequence $\{A_{nn}\}$ is a subsequence of $\{A_n\}$. We show that it is a Cauchy sequence. We first note that if S_i is the set consisting of the union of squares in σ_i, then

$$d(S_i, A_{in}) \leq \frac{s}{2^i}\sqrt{2}, \quad n = 1, 2, 3, \ldots.$$

But the sets of the diagonal sequence with index not less than i are all among the A_{in}. Hence,

$$d(S_i, A_{kk}) \leq \frac{s}{2^i}\sqrt{2}, \quad k \geq i.$$

It follows that $m, n \geq i$ implies

$$d(A_{nn}, A_{mm}) \leq d(S_i, A_{nn}) + d(S_i, A_{mm}) \leq \frac{s}{2^{i-1}}\sqrt{2},$$

and $\{A_{nn}\}$ is a Cauchy sequence.

We consider the area $a(A)$ and the perimeter $c(A)$ of $A \in \Gamma$. It is easily seen that the first of these functionals is continuous on Γ, and the second is continuous at any $A \in \Gamma$ for which $a(A) > 0$.

Let $\Gamma_{QL\alpha}$ be the subset of sets A in Γ_Q for which $c(A) = L$ and $a(A) \geq \alpha > 0$. (We suppose L, α chosen so that this set is non-empty.) Then $\Gamma_{QL\alpha}$ is closed, and so it is compact.

It follows that among all convex sets of a given perimeter, there is one which encloses the largest area. By means of another consideration, which we do not discuss here, it can be shown that this convex set must be enclosed by a circle.

EXERCISES

1.1 Show that examples (i) and (ii) satisfy the conditions of a metric space.

1.2 Let X be the set of continuous functions on $(0, 1)$ and for every $x, y \in X$, let

$$E(x, y) = [t \in (0, 1) : x(t) \neq y(t)].$$

$E(x, y)$ is the union of disjoint open intervals. Define $\rho(x, y)$ to be the sum of the lengths of these intervals. Show that (X, ρ) is a metric space.

1.3 The real line with

$$\rho(x, y) = \frac{|x - y|}{1 + |x - y|}$$

is a metric space.

1.4 For what values of p is $|x - y|^p$ a metric on the real line?

2.1 Give necessary and sufficient conditions for equality to hold in Hölder's inequality.

2.2 Is n-space a metric space if ρ is taken to be

$$\rho(x, y) = \left[\sum_{i=1}^{n} |x_i - y_i|^p\right]^{1/p}$$

with $0 < p < 1$?

3.1 Prove that

$$\lim_{p\to\infty} \rho_p(x, y) = \max\,[|x_i - y_i| : 1 \leq i \leq n].$$

3.2 Prove that if $x \in l_p$ and $y \in l_p$, then the series for $d_p(x, y)$ converges, without using Minkowki's inequality.

3.3 Let X consist of all sequences of complex numbers. Show that

$$\rho(x, y) = \sum_{i=1}^{\infty} \frac{1}{2^i} \frac{|x_i - y_i|}{1 + |x_i - y_i|}$$

is a metric in X.

3.4 Show that if $p > p' \geq 1$, then $l_p \supset l_{p'}$ and the inclusion is proper.

3.5 Show that for $p > 1$, the union of all the $l_{p'}$ spaces, $p > p' > 1$, is a proper subspace of l_p.

3.6 Give an example of a sequence which converges to 0 but is not in any l_p space, $1 \leq p < \infty$.

3.7 Prove theorem 2.

4.1 Show that $C[a, b]$ and $M[a, b]$ are metric spaces.

4.2 Define the metric $\tilde{\rho}$ on $\widetilde{BV}$ and show that it is indeed a metric.

4.3 Let X be the set of all continuous functions on $(-\infty, \infty)$ which vanish outside of some interval (the interval depending on the particular function). Show that X is a metric space under

$$\rho(x, y) = \max\,[|x(t) - y(t)| : -\infty < t < \infty].$$

4.4 Extending the definition of S_δ to arbitrary sets S in the plane, show that $\bigcap\,[S_\delta : \delta > 0] = \bar{S}$, where $\bar{S}$ is the closure of S. Defining $d(S, T)$ for arbitrary sets S and T as in the text, show that $d(S, T) = 0$ if and only if $\bar{S} = \bar{T}$.

4.5 For the class of all bounded sets in the plane, an equivalence relation is given by S equivalent to T if $\bar{S} = \bar{T}$. Show that $d(S, T)$ is a metric on these equivalence classes of sets.

5.1 Prove proposition 1.

5.2 Prove that the collection $\mathscr{G}$ of open sets satisfies the properties 1, 2 and 3.

5.3 Prove that the closure of a set A is the intersection of all closed sets containing A.

5.4 Is the space of exercise 1.2 separable?

5.5 In the space l_2, let

$$A = \left[x = (x_1, x_2, \ldots) : |x_n| \leq \frac{1}{n}, \quad n = 1, 2, \ldots\right].$$

Prove that A is closed.

6.1 Show that $C[0, 1]$ is separable by exhibiting a countable set of functions, which is dense in $C[0, 1]$, each of whose graphs consists of a finite number of line segments.

6.2 Is the space of all bounded continuous functions on $(-\infty, \infty)$ with $\rho(x, y) = \sup\,[|x(t) - y(t)| : -\infty < t < \infty]$ separable?

6.3 Prove proposition 1.

6.4 Prove proposition 2.

7.1 Show that $\{x_n\}$ is a Cauchy sequence if and only if $\rho(x_{n+k}, x_n)$ converges to zero uniformly in k.

7.2 Prove that the sequence

$$.1,\ .101,\ .101001,\ .1010010001, \ldots$$

is a Cauchy sequence of rationals which does not converge in the space of rationals.

7.3 Exhibit a non-convergent Cauchy sequence in the space of polynomials on $[0, 1]$ with the uniform metric.

7.4 Exhibit a non-convergent Cauchy sequence in the space of exercise 1.2.

7.5 Prove proposition 1.

7.6 In example (ii), give the details of the proof that the limit function x must be $x(t) = 0,\ t \in [a, c)$ and $x(t) = 1,\ t \in (c, b]$.

7.7 Show that a necessary and sufficient condition that

$$\int_0^1 |x(t)|\, dt = 0$$

is that x be bounded and that, for every $\epsilon > 0$, the set

$$[t \in [0, 1] : |x(t)| < \epsilon]$$

contains a finite set of intervals the sum of whose lengths exceeds $1 - \epsilon$.

7.8 Show that the sequence $\{x_n\}$ in example (iii) is a Cauchy sequence but is not convergent.

7.9 For the set $S = \bigcap_{n=1}^{\infty} S_n$ of example (iv), show that for every partition of

[0, 1], the sum of the lengths of the intervals of the partition, which contain points of S, exceeds $\frac{1}{2}$.

7.10 Show that the bounded functions on an arbitrary set, with the uniform metric, form a complete metric space.

7.11 Show that the set of continuous functions on $(-\infty, \infty)$ with the metric

$$\rho(x, y) = \sum_{n=1}^{\infty} \frac{1}{2^n} \frac{\max\,[|x(t) - y(t)| : |t| \leq n]}{1 + \max\,[|x(t) - y(t)| : |t| \leq n]}$$

is a complete metric space.

8.1 Show that theorem 1 fails to hold if T has only the property

$$\rho(Tx, Ty) < \rho(x, y).$$

8.2 Prove that if T is a contraction in a complete metric space X and $x \in X$, then

$$T \lim_n T^n x = \lim_n T^{n+1} x.$$

8.3 Show that a function φ, defined on an interval I containing x_0, is such that $\varphi(x_0) = y_0$ and $\varphi'(x) = f(x, \varphi(x))$, for all $x \in I$, where φ is differentiable on I and f is continuous, if and only if

$$\varphi(x) = y_0 + \int_{x_0}^{x} f(t, \varphi(t))\, dt$$

for every $x \in I$.

8.4 Show that if D is an open connected set, and a differential equation

$$\frac{dy}{dx} = f(x,y)$$

is such that its solutions form a simple covering of D, then f is the limit of a sequence of continuous functions.

8.5 Show that the system

$$\begin{aligned} x &= \tfrac{1}{3}x - \tfrac{1}{4}y + \tfrac{1}{4}z - 1 \\ y &= -\tfrac{1}{2}x + \tfrac{1}{5}y + \tfrac{1}{4}z + 2 \\ z &= \tfrac{1}{5}x - \tfrac{1}{3}y + \tfrac{1}{4}z - 2 \end{aligned}$$

has a unique solution by using the contraction mapping theorem.

9.1 Give the details of the proof that the space $(\tilde{X}, \tilde{\rho})$, associated with (X, ρ), is a complete metric space.

9.2 Show that the extension φ of ST^{-1} is indeed an isomorphism of the two completions $(\tilde{X}, \tilde{\rho})$ and $(\tilde{Y}, \tilde{\sigma})$.

10.1 Prove proposition 1.

10.2 Let X be a separable metric space and let $S \subset X$. Call $x \in X$ a point of the first category relative to S if some nieghborhood of x meets S in a set of the first category in X. Show that the set composed of those points of S, which are points of the first category relative to S, is a set of the first category.

10.3 A point is of the strong second category relative to S if it has a neighborhood none of whose points are of the first category relative to S. Show that the subset of S, of points that are not of the strong second category relative to S, is a set of the first category.

10.4 Prove the assertions in exercises 10.2 and 10.3 without the separability assumption.

10.5 Let X be the space of functions on [0, 1] which are differentiable and have a continuous first derivative (one-sided at the end points) with the metric

$$\rho(x, y) = \max [|x(t) - y(t)| : t \in [0, 1]] + \max [|x'(t) - y'(t)| : t \in [0, 1]].$$

Show that the set of all polynomials is of the first category in X.

10.6 In the space $BV[0, 1]$ with

$$\rho(x, y) = |x(0) - y(0)| + V(x - y)$$

show that the continuous functions of bounded variation form a set of the first category.

11.1 Let us say that metric spaces (X, ρ) and (Y, σ) are equivalent if there is a one-one mapping between them which preserves open sets. Show that a space equivalent to a complete space need not be complete. Show that a space equivalent to a compact space is necessarily compact.

11.2 Prove proposition 1.

11.3 Prove proposition 2.

11.4 Show that the subset of l_2 of points $\{x_n\}$ such that

$$|x_n| \leq \frac{1}{n}, \quad n = 1, 2, \ldots$$

is compact.

11.5 Show that the unit sphere in $C[0, 1]$ of points

$$[x : \max |x(t)| \leq 1, t \in [0, 1]]$$

is not compact.

11.6 Show that X is compact if and only if for any collection $\mathcal{F}$ of closed sets with the property that $\bigcap_{i=1}^{n} F_i \neq 0$ for any finite collection in $\mathcal{F}$, it follows that $\bigcap [F : F \in \mathcal{F}] \neq 0$.

11.7 Prove proposition 6.

11.8 Prove proposition 7.

11.9 Prove proposition 9.

11.10 Give necessary and sufficient conditions for a subset of l_p, $p > 1$, to be compact.

11.11 If X is a compact metric space, show that $C(X)$ is separable.

12.1 Show that a uniformly bounded sequence of functions has a subsequence which converges point-wise on a prescribed countable set.

12.2 By using the result of exercise 12.1, give an alternate proof of the sufficiency part of theorem 1.

12.3 Give an example of a continuous function f on an open set D and an $(x_0, y_0) \in D$ such that the differential equation

$$\frac{dy}{dx} = f(x, y)$$

has several distinct local solutions through (x_0, y_0).

12.4 Prove theorem 2 by considering auxiliary differential equations

$$\frac{dy}{dx} = p_n(x, y), \quad n = 1, 2, \ldots,$$

where the p_n are polynomials which converge uniformly to f on an appropriate rectangle.

13.1 If P_n, $n = 1, 2, \ldots$, is the polynomial $P_n(x) = c_n(1 - x^2)^n$, where c_n is chosen so that

$$\int_{-1}^{1} P_n(x)\,dx = 1,$$

show that for any continuous f such that $f(x) = 0$ for $x \leq 0$ and for $x \geq 1$, the polynomials

$$L_n(x) = \int_{-1}^{1} P_n(x + t) f(t)\,dt$$

converge uniformly to f on $[0, 1]$.

13.2 Prove an analogous result in the two dimensional case.

13.3 The polynomials

$$B_n(f, x) = \sum_{k=0}^{n} \binom{n}{k} f\left(\frac{k}{n}\right) x^k (1 - x)^{n-k},$$

where $\binom{n}{k} = \dfrac{n!}{k!(n - k)!}$, are called the Bernstein polynomials of f. If $0 \leq x \leq 1$, these polynomials have a probabilistic interpretation. Discuss.

13.4 Show that if f is continuous on $[0, 1]$ then $B_n(f, x)$ converges uniformly to f on $[0, 1]$.

13.5 Prove that the closure $\bar{A}$ of an algebra $A \subset C(X)$ is an algebra.

13.6 Show that the Weierstrass theorem holds for $|x|$ on $[-1, 1]$ by showing that the Fourier series of $|x|$ converges uniformly to $|x|$ on $[-1, 1]$.

13.7 Show that the Stone theorem is false when $C(X)$ is the space of complex-valued continuous functions on X. (Let X be the set $[z : |z| \leq 1]$ in the complex plane and let A be the set of polynomials in z.)

13.8 Show that the additional hypothesis $f \in A$ implies $\bar{f} \in A$ corrects the above deficiency.

13.9 Show that the closed algebra generated by the pair $1, x^2$ is $C[0, 1]$ but is not $C[-1, 1]$.

13.10 For each natural number k, for which closed real intervals I is the closed algebra generated by 1, x^k the space $C(I)$?

13.11 Show that $\int_a^b f(x) \cdot x^n \, dx = 0$, $n = 1, 2, \ldots$, implies $f = 0$.

14.1 Prove proposition 1.

14.2 Prove that the space of all functions continuous on G is complete under the metric d.

15.1 Show that f is lower semi-continuous at x if and only if $\lim\limits_n x_n = x$ implies $\underline{\lim}\, f(x_n) \geq f(x)$.

15.2 Show that if we define T_1 to be equivalent to T_2 if, for every $\epsilon > 0$, there is a homeomorphism h of I onto itself such that

$$d(T_2, T_1 \circ h) < \epsilon,$$

then this is an equivalence relation.

15.3 Verify that $(\mathcal{C}, \delta)$ is a metric space.

15.4 Prove the lower semi-continuity of L at C when $L(C) = \infty$.

15.5 For any C, with $L(C) < \infty$, show that the arc length representation R belongs to C.

16.1 Prove lemma 1.

16.2 Extend theorem 1 to n-space.

16.3 Show that area is a continuous function on Γ.

16.4 Show that perimeter is continuous at those $A \in \Gamma$ for which $a(A) > 0$.

CHAPTER 2

BANACH SPACES

2.1 Vector Space

Many of the metric spaces which arise in analysis are endowed with a vector space structure, and the metrics are derived from norms related to this structure.

The first few sections of this chapter are concerned with vector spaces. A **vector space X over a field F** consists of the set X, a mapping $(x, y) \to x + y$ of X^2 into X, and a mapping $(a, x) \to ax$ of $F \times X$ into X, such that

(a) X is an abelian group with group operation $(x, y) \to x + y$,
(b) the associative law $a(bx) = (ab)x$ holds,
(c) the distributive laws $(a + b)x = ax + bx$, $a(x + y) = ax + ay$ hold,
(d) $1x = x$.

It is easy to show that $0 \cdot x = 0$, where 0 is the scalar zero on the left, and is the group identity on the right. It is also easy to show that $(-1)x = -x$, where $-x$ is the group inverse of x.

The most important cases, by far, are those for which the field F is either R, the real number field, or C, the complex number field. We shall consider these cases only. Most of our general discussion will involve R, although the complex case is often more important, and so will be considered at the appropriate times.

We list some examples of vector spaces over R.

(i) R itself is a vector space over R. In this case, $X = R$ is the real additive group and $F = R$ is the real field. The operation ax is multiplication for the real field.

(ii) The set R^n of n-tuples of real numbers $x = (x_1, x_2, \ldots, x_n)$, with the standard operations $(x_1, x_2, \ldots, x_n) + (y_1, y_2, \ldots, y_n) = (x_1 + y_1,\ x_2 + y_2, \ldots, x_n + y_n)$ and $a(x_1, x_2, \ldots, x_n) = (ax_1, ax_2, \ldots, ax_n)$, is a vector space.

(iii) More generally, for any set A, the set R^A of all $[x_\alpha]$, $\alpha \in A$, where $[x_\alpha] + [y_\alpha] = [x_\alpha + y_\alpha]$ and $a[x_\alpha] = [ax_\alpha]$, is a vector space. This is the vector space of real functions on A. The notation $x(\alpha)$ may be used instead of $[x_\alpha]$. This is also an example of the direct product of vector spaces, which we now define.

Let X_α, $\alpha \in A$, be a set of vector spaces. The **direct product**

$$X = \Pi[X_\alpha : \alpha \in A]$$

has as elements the functions $[x_\alpha]$, where $x_\alpha \in X_\alpha$ for each $\alpha \in A$, and addition and scalar multiplication are given by $[x_\alpha] + [y_\alpha] = [x_\alpha + y_\alpha]$ and $a[x_\alpha] = [ax_\alpha]$. It is easy to verify that X is a vector space. Example (ii) above is the special case where A is a finite set and $X_\alpha = R$ for every $\alpha \in A$, and example (iii) is the case where A is an arbitrary set and $X_\alpha = R$ for every $\alpha \in A$.

Now, let X_α, $\alpha \in A$, again be a set of vector spaces whose identity elements are all designated as 0. Let

$$\tilde{X} = \sum[X_\alpha : \alpha \in A]$$

consist of those $[x_\alpha] \in \Pi[X_\alpha : \alpha \in A]$ for which $x_\alpha = 0$ for all but a finite number of the $\alpha \in A$. It is easily verified that $\Sigma[X_\alpha : \alpha \in A]$ is a vector subspace of $\Pi[X_\alpha : \alpha \in A]$, and that the two spaces are identical if and only if A is finite. $\tilde{X}$ is called the **direct sum** of the vector spaces X_α, $\alpha \in A$.

In each of the above examples, the values of the coordinates may be complex. In that case the scalars may be taken to be real as well as complex. In general, for any complex vector space, we obtain the **associated real space** by restricting the scalars to the real subfield of the complex field.

It is not always immediately obvious that a certain system is a vector space. For example, let X consist of all real sequences $(x_1, x_2, \ldots, x_n, \ldots)$ for which the corresponding series $\sum_{n=1}^{\infty} x_n^2$ converges. In order to know that X is a vector space, with the usual operations, we must show that $\sum_{n=1}^{\infty} x_n^2 < \infty$ and $\sum_{n=1}^{\infty} y_n^2 < \infty$ implies $\sum_{n=1}^{\infty} (x_n + y_n)^2 < \infty$. We have seen in Chapter 1 that this is true but not immediately obvious.

Indeed, if we allow functions as integrable if an improper integral exists, then the functions on [0, 1] for which

$$\int_0^1 (e^{|f(x)|} - 1)\, dx < \infty$$

may be shown to not form a vector space.

Examples of important vector spaces of functions are

(α) the continuous functions on a metric space,
(β) the differentiable functions of a real variable,
(γ) the analytic functions,
(δ) the functions which are integrable in some sense.

For any two vector spaces X and Y, the set of linear transformations of

X into Y is also a vector space, which is designated by $L(X, Y)$. A **linear transformation** f is a mapping

$$f : X \to Y$$

on X into Y such that $f(ax_1 + bx_2) = af(x_1) + bf(x_2)$ for all $x_1, x_2 \in X$ and $a, b \in R$. It is evident that $L(X, Y)$ is a vector space if the operations are defined point-wise. Of special interest is the case where $Y = R$, i.e., Y is the vector space of real numbers. Then the elements of $L(X, R)$ are called **linear functionals** (also linear forms) on X. We call $L(X, R)$ the **algebraic dual** of X, and designate it by X^*. The qualifying adjective "algebraic" is introduced in order to distinguish this space from the various dual spaces of continuous linear functionals on X associated with the possible topologies in X.

2.2 Subspace

Let X be a vector space. A subset $Y \subset X$ is called a vector subspace, or simply a **subspace** of X if it is itself a vector space with the same operations as for X. A necessary and sufficient condition that a subset $Y \subset X$ should be a vector subspace of X is that for every $x_1, x_2 \in X$ and $a, b \in R$ the element $ax_1 + bx_2 \in Y$. The condition is evidently necessary. For the sufficiency, we need only observe that if Y is closed under the above operations, then, since $0x = 0$ and $(-1)x = -x$, it follows that $0 \in Y$ and the group inverse of every element in Y is also in Y. Two special subspaces of a vector space X are X itself and the null subspace $\{0\}$ of X whose only member is 0. These are called **improper** subspaces of X. All other subspaces of X are called **proper.**

We now consider operations on subspaces of a vector space X which yield other subspaces of X.

If Y_α, $\alpha \in A$, is a set of subspaces of X, then the set $Y = \bigcap[Y_\alpha : \alpha \in A]$ is also a subspace of X. It is called the **intersection** of the subspaces Y_α, $\alpha \in A$, and is evidently the largest vector space contained in all of them.

The union of a set of subspaces of X need not be a subspace of X. Indeed, if Y and Z are subspaces of X, then the set $Y \cup Z$ is a subspace of X if and only if $Y \subset Z$ or $Z \subset Y$. For suppose $y \in Y$, $y \notin Z$ and $z \in Z$, $z \notin Y$. Then $y + z \notin Y \cup Z$. For, if $u = y + z \in Y$, then $z = u - y \in Y$, contradicting the assumption that $z \notin Y$. Similarly, $u \in Z$ implies $y \in Z$. It follows that $Y \cup Z$ is not a vector space.

There is, however, an operation on vector spaces which does yield a vector space and is, in a sense, the dual of the intersection operation. Let Y_α, $\alpha \in A$, be a set of subspaces of X. Let Y be the intersection of all the subspaces of X which contain all the Y_α, $\alpha \in A$. X itself is a subspace of X which contains all the Y_α, so that Y, as the intersection of a non-empty set of subspaces of X, is also a subspace of X. It is the smallest subspace of X which contains all the Y_α, $\alpha \in A$.

We shall use the notation $\vee[Y_\alpha : \alpha \in A]$ for this space. It is the least upper bound, or supremum, of the spaces Y_α, $\alpha \in A$, in the partial ordering of all subspaces of X, ordered by means of set inclusion. We sometimes use the notation $\wedge[Y_\alpha : \alpha \in A]$ instead of $\bigcap[Y_\alpha : \alpha \in A]$ for the intersection operation, in order to emphasize the fact that it is the greatest lower bound, or infimum, of the spaces Y_α, $\alpha \in A$. It follows that, with these operations, the subspaces of a vector space X form a complete lattice, i.e., every set of subspaces of X has a supremum and an infimum.

Let S be any subset of X. By the subspace $(S) \subset X$ **generated by** S we mean the smallest subspace of X which contains S. Then (S) is the intersection of all subspaces of X which contain S. We note that (S) is the set L of all linear combinations $\sum_{i=1}^{n} a_i x_i$, of finite sets in S. It is obvious that L is a vector space containing S. Moreover, every subspace containing S contains all elements of L.

Two subspaces Y and Z of X are called **disjoint** if $Y \cap Z = \{0\}$, the subspace of X which consists of the one element 0. Two disjoint subspaces Y and Z of X are called **complementary** if $Y \vee Z = X$. We shall show that every subspace of X has complementary subspaces which, however, are not unique. In this, we need Zorn's lemma, which we now state without proof.

A **partially ordered set** is a set S together with a relation $x \leq y$, between pairs of elements of S, satisfying: $x \leq x$, $x \leq y$ and $y \leq x$ implies $x = y$, and $x \leq z$ whenever $x \leq y$ and $y \leq z$. If every pair of elements of a subset C of S are comparable, i.e., $x \in C$ and $y \in C$ implies either $x \leq y$ or $y \leq x$, then C is called a **chain** or a **totally ordered** subset of S. An **upper bound** of a set $A \subset S$ is any $y \in S$ such that $x \leq y$ for all $x \in A$. A **maximal element** of S is any $y \in S$ such that $y \leq x$ implies $y = x$.

ZORN'S LEMMA. If S is a partially ordered set in which every totally ordered subset has an upper bound, then S has a maximal element.

Zorn's lemma was first proved by Hausdorff, but its importance as a tool was first recognized by Zorn. For further discussion see Kelley [1].

Now, let X be a vector space and Y a subspace of X. We consider the partially ordered set $\chi(Y)$ of all subspaces of X which are disjoint with Y, ordered by inclusion. Let Z_α, $\alpha \in A$, be a totally ordered subset of $\chi(Y)$. Then $Z = \bigcup [Z_\alpha : \alpha \in A]$ is also a subspace of X. Moreover $Y \wedge Z = Y \cap Z = \bigcup [Y \cap Z_\alpha : \alpha \in A] = \{0\}$, so that $Z \in \chi(Y)$. Hence, $\chi(Y)$ satisfies the condition of Zorn's lemma so that it has a maximal element M. We show that M is a complement of Y. For, let $W = Y \vee M$. Suppose there is an $x \in X$, $x \notin W$. Let M' be the subspace of X generated by M and x. Then M' is also disjoint with Y, contradicting the fact that M is maximal in $\chi(Y)$. Accordingly, $X = Y \vee M$, so that M is a complement of Y.

PROPOSITION 1. If X is a vector space, and Y and Z are complementary subspaces of X, then for every $x \in X$ there are $y \in Y$, $z \in Z$ such that $x = y + z$ and this representation is unique.

Proof. That every $x \in X$ has a representation $x = y + z$, $y \in Y$, $z \in Z$ follows from the fact that $X = Y \vee Z$. Suppose now that $x = u + v$, $u \in Y, v \in Z$. Then $y + z = u + v$, so that $w = y - u = z - v$ is in $Y \cap Z$. However, since $Y \cap Z = \{0\}$, we have $y = u$ and $z = v$.

2.3 Quotient Space

Let X be a vector space and Y a subspace of X. Then Y is a subgroup of the abelian group X, so that we may consider the quotient group X/Y. The elements of X/Y are the cosets modulo Y in X. These are the translations $x + Y$ of Y as x varies over X, where $x + Y = [z : z = x + y, y \in Y]$. It is well known, and easily shown, that the cosets of Y are disjoint sets whose union is X. It is also easy to show that if A and B are cosets modulo Y and $a \in R$ then $A + B$ and aA are also cosets modulo Y where $S + T = [x + y : x \in S, y \in T]$, and $aS = [ax : x \in S]$. This allows us to define vector space operations in X/Y, and it may be readily verified that X/Y is then a vector space. We leave all these easy details to the reader.

The notion of a homomorphism between vector spaces is of importance in connection with subspaces and quotient spaces. Let X and Z be vector spaces. By a **homomorphism** h of X **into** Z we mean a function defined on X with values in Z such that $h(x + y) = h(x) + h(y)$ and $h(ax) = ah(x)$. Thus a homomorphism is merely a linear mapping on X into Z. The homomorphism h is said to be **onto** Z if, for every $z \in Z$, there is an $x \in X$ such that $h(x) = z$. The set $K \subset X$ of all x for which $h(x) = 0$ is called the **kernel** of the homomorphism. The kernel K of a homomorphism is evidently a subspace of X. If a homomorphism is onto and its kernel contains only 0, then it is called an **isomorphism.**

Let X be a vector space and Y a subspace of X. Let φ be the mapping of X onto X/Y which takes every element of X into the element of X/Y which contains it when considered as a subset of X. The mapping φ is a homomorphism and is known as the **canonical homomorphism** of X onto X/Y.

We state one further fact. Let h be a homomorphism of X onto Z whose kernel is Y. Then the inverse images in X of the elements of Z are cosets modulo Y, so that h induces an isomorphism ψ of X/Y onto Z. It follows that the homomorphism h is the product of the canonical homomorphism φ of X onto X/Y followed by the induced isomorphism ψ of X/Y onto Z.

We now obtain a connection between the quotient space X/Y and any subspace Z of X complementary to Y.

PROPOSITION 1. If X is a vector space, Y and Z are subspaces of X, and Y is complementary to Z, then every element of X/Y contains exactly one element of Z.

Proof. By proposition 1, Sec. 2.2, every $x \in X$ has a unique representation $x = y + z$, where $y \in Y$ and $z \in Z$. Thus the cosets $z + Y$, $z \in Z$, are the totality of cosets of Y. That each of them contains exactly one element of Z follows since $z + y \in Z$, $z \in Z$, $y \in Y$ implies $y = 0$.

We also have the

COROLLARY. If X is a vector space, Y is a subspace of X, and Z a complement of Y in X, then X/Y is isomorphic to Z under the mapping which takes each element of X/Y into the element of Z which it contains.

2.4 Dimension, Hamel Basis

Let X be a vector space, and S a subset of X. S generates a subspace (S) of X. We say that the elements of S are **linearly independent** if the subspace (T) generated by any proper subset $T \subset S$ is a proper subspace of (S).

We prove the

PROPOSITION 1. Every vector space X contains a set of linearly independent elements which generates X.

Proof. Let $\mathcal{S}$ be the totality of linearly independent subsets of X, partially ordered by inclusion. We show that for every totally ordered subset S_α, $\alpha \in A$, of $\mathcal{S}$, the set $S = \bigcup [S_\alpha : \alpha \in A]$ is also in $\mathcal{S}$. For, if not, (S) would be generated by a proper subset $T \subset S$. Then for every $\alpha \in A$, (S_α) is generated by $T_\alpha = T \cap S_\alpha$. But the independence of S_α implies $T_\alpha = S_\alpha$. Thus

$$T = \bigcup [T \cap S_\alpha : \alpha \in A] = \bigcup [T_\alpha : \alpha \in A] = \bigcup [S_\alpha : \alpha \in A] = S,$$

contradicting the assumption that T is a proper subset S. The condition of Zorn's lemma being satisfied, there is a maximal $M \in \mathcal{S}$. Suppose (M) a proper subspace of X. Let $y \in X$, $y \notin (M)$. The subspace Y of X generated by M and y then contains (M) as a proper subspace. If for any proper subset $T \subset M$, T and y also generate Y, it follows that T also generates (M), thus contradicting the fact that M is linearly independent. There is thus no $y \in X$, $y \notin (M)$. Hence M generates X.

A set S of linearly independent elements which generates a vector space X is called a **Hamel basis** in X. Proposition 1 says that every vector space has a Hamel basis.

Although a given vector space has many Hamel bases, we note that they all have the same number (cardinal) of elements. Indeed, let S and T be Hamel bases of X.

Suppose S is finite and has n elements $x_1, x_2, \ldots, x_n$. If there were $n + 1$ elements $y_1, y_2, \ldots, y_n, y_{n+1}$ in T, then

$$y_i = \sum_{j=1}^{n} a_{ij} x_j, \quad i = 1, \ldots, n$$

and since $y_1, y_2, \ldots, y_n$ are linearly independent, the matrix (a_{ij}) has rank n. But

$$y_{n+1} = \sum_{j=1}^{n} a_{n+1,j} x_j$$

and the vector $(a_{n+1,1}, \ldots, a_{n+1,n})$ is a linear combination of the row vectors of the matrix (a_{ij}). But this implies $y_1, y_2, \ldots, y_n, y_{n+1}$ not linearly independent. Hence T has $m \leq n$ elements. By the same argument $n \leq m$.

Suppose S is infinite and has cardinal number a. Let b be the cardinal number of T. Every $y \in T$ is a linear combination, with non-zero coefficients, of a finite number $x_1, x_2, \ldots, x_n$ of elements of S, and only a finite number n of elements of T are associated in this way with the same set $x_1, x_2, \ldots, x_n$ or some subset of it. Since the cardinal number of the set of finite subsets of S is the same as that of S itself, it follows that $b \leq \aleph_0 \cdot a = a$. Similarily, $a \leq b$, so that $a = b$. This completes the proof of

PROPOSITION 2. If X is a vector space, all Hamel bases of X have the same cardinal number.

This cardinal number is called the **dimension** of X.

PROPOSITION 3. If S is a Hamel basis of X, then every $x \in X$ has a unique representation as a linear combination of a finite number of elements of S.

The proof is left to the reader.

PROPOSITION 4. If the dimension of X is at least as great as the cardinal number of the reals, then the dimension of X and the cardinal number of X are equal.

The proof is left to the reader.

We recall that vector spaces X and Y are called **isomorphic** if there is a mapping φ which is one-one and onto such that

$$\varphi(ax_1 + by_1) = a\varphi(x_1) + b\varphi(x_2) \text{ for all } x_1, x_2 \in X \text{ and } a, b \in R.$$

We prove

PROPOSITION 5. Two vector spaces X and Y are isomorphic if and only if they have the same dimension.

Proof. Suppose X and Y have the same dimension and S and T are Hamel bases of X and Y, respectively. Let φ be a one-one correspondence between S and T. We extend φ to X by letting

$$\varphi(x) = a_1\varphi(x_1) + \ldots + a_n\varphi(x_n)$$

for $x = a_1x_1 + \ldots + a_nx_n$, $x \in X$, $x_1, \ldots, x_n \in S$. It readily follows that this is an isomorphism between X and Y. The converse is obvious.

There are thus many different one-one correspondences between X and Y, called isomorphisms, which establish the fact that X and Y are isomorphic. We notice that a vector space can be isomorphic to a proper subspace.

By the **deficiency** of a subspace Y of a vector space X we mean the dimension of X/Y. The deficiency of Y is then the dimension of any subspace Z of X which is complementary to Y.

In particular, a subspace of X of deficiency 1 is called a **hyperplane** in X; its cosets are also called hyperplanes. We note that if S is a Hamel basis of X and $x \in S$, then $S \sim \{x\}$ is a Hamel basis of a hyperplane in X.

In the same way as we showed that every vector space X has a Hamel basis, we may show that if Y is a subspace of X then X has a Hamel basis S, which has a subset T which is a Hamel basis of Y. We omit the proof. It follows that every proper subspace of a vector space X is contained in a hyperplane.

2.5 Algebraic Dual, Second Dual

Let X be a vector space and let X^* be the algebraic dual of X. Then X^* is a vector space with operations $f + g$ and af defined by

$$(f + g)(x) = f(x) + g(x), \quad x \in X,$$

$$(af)(x) = af(x), \quad x \in X.$$

For any $f \in X^*$, the set

$$Z(f) = [x : x \in X, f(x) = 0]$$

is a vector subspace of X and is called the **null space** of f.

PROPOSITION 1. If $f \in X^*$, then $Z(f)$ has deficiency 0 or 1 in X. Conversely, if Z is a subspace of X of deficiency 0 or 1, there is an $f \in X^*$ such that $Z = Z(f)$.

Proof. Suppose $x \in X$ and $f(x) \neq 0$. Then, for every $y \in X$ there is a real number a such that $f(y) = af(x) = f(ax)$, or $f(y - ax) = 0$. Thus $y - ax \in Z(f)$. In other words, X is generated by $Z(f)$ and x, so that the deficiency of $Z(f)$ is 1. If $f(x) \equiv 0$ on X, then $Z(f) = X$ and its deficiency is 0.

For the converse, if $Z = X$, i.e., Z has deficiency 0, then $Z(f) = Z$ for $f(x) \equiv 0$. Suppose, Z has deficiency 1. Let $x_0 \notin Z$. Then every $y \in X$ has a unique representation $y = x + ax_0$, where $x \in Z$. Let $f(y) = a$. A simple calculation shows that f is linear and $Z(f) = Z$.

Now, let $S^* \subset X^*$ be a set of linear functionals on X. By the **null space** $Z(S^*)$ of S^* we shall mean the set of all $x \in X$ for which $f(x) = 0$ for all $f \in S^*$. That $Z(S^*)$ is a subspace of X follows since it is the intersection of the subspaces $Z(f), f \in S^*$.

A subspace $Y^* \subset X^*$ is called **total** if $Z(Y^*) = \{0\}$. Thus Y^* is total if and only if for every $x \neq 0$ in X there is an $f \in Y^*$ such that $f(x) \neq 0$. Most important subspaces are total.

PROPOSITION 2. For every vector space X, with algebraic dual X^*, $\dim X^* \geq \dim X$.

The proof is left to the reader.

For any vector space X, the dual X^* is also a vector space and so it has its own algebraic dual X^{**}. There is a natural isomorphism between a certain subspace of X^{**} and X itself, which we now discuss.

Fix $x_0 \in X$. For every $f \in X^*$, let $x_0(f) = f(x_0)$. Then $x_0(f)$ is a linear functional on X^*. For,

$$x_0(f + g) = (f + g)(x_0) = f(x_0) + g(x_0) = x_0(f) + x_0(g)$$

and

$$x_0(af) = (af)(x_0) = af(x_0) = ax_0(f).$$

Hence, for every $x_0 \in X$, $x_0(f) \in X^{**}$. Consider the mapping

$$\varphi : X \to X^{**}$$

defined by $\varphi(x_0) = x_0(f), f \in X^*$. We show that this is an isomorphism between X and a subspace of X^{**}. For,

$$(x_0 + y_0)(f) = f(x_0 + y_0) = x_0(f) + y_0(f)$$

and

$$(ax_0)(f) = f(ax_0) = af(x_0) = ax_0(f).$$

Moreover, the mate of 0 is the functional on X^* which is identically zero, and for any $x_0 \neq 0$, there is an $f \in X^*$ such that $f(x_0) \neq 0$, whence $x_0(f) \neq 0$.

Now, let $Y^* \subset X^*$ be any total subspace of X^*. Then for every $x_0 \in X$, $x_0(f)$ is a linear functional on Y^* so that it belongs to Y^{**}. The mapping defined by $\varphi(x_0) = x_0(f)$, $f \in Y^*$ is an isomorphism between X and Y^{**}, by the same discussion as above. Indeed, all of the interesting applications are concerned with this isomorphism for special subspaces Y^* of X^*.

If X is finite dimensional, the natural isomorphism $\varphi(x_0) = x_0(f), f \in X^*$ is between X and all of X^{**}. This is never true if X is infinite dimensional.

2.6 Convex Sets

Let X be a vector space. A set $S \subset X$ is called **convex** if for every $x, y \in S$ and every $a \in R$, with $0 \leq a \leq 1$, the element $ax + (1 - a)y \in S$. In other words, the line segment joining x and y is contained in S.

We first state some elementary properties of convex sets.

Every subspace Y of X is convex. For if $x, y \in Y$ then $ax + by \in Y$ for every $a, b \in R$.

The intersection, $S = \bigcap [S_\alpha : \alpha \in A]$ of any collection S_α, $\alpha \in A$, of convex sets is a convex set. For, let $x, y \in S$. Then $x, y \in S_\alpha$ for every $\alpha \in A$. Then for every $a \in R$, $0 \leq a \leq 1$, $ax + (1 - a)y \in S_\alpha$, for every $\alpha \in A$, so that $ax + (1 - a)y \in S$.

For every set $S \subset X$, there is a smallest convex set $C(S)$ such that $S \subset C(S)$. $C(S)$ is called the **convex hull** of S. In order to prove this, we observe that there are convex sets which contain S. For example, X itself is such a set. The intersection of all the convex sets containing S is convex. This is the convex hull of S.

The translations $x + S$ of convex sets and the products aS of convex sets by real numbers are convex sets. We leave the proofs to the reader.

If S is a convex set, and $x_0 \in S$, we shall say that x_0 is an **interior** point of S if every line through x_0 meets S in a line segment which has x_0 as interior point. Precisely, this means that for every $x \in X$, $x \neq 0$, there is a $k > 0$ such that $|a| < k$ implies $x_0 + ax \in S$.

Let H be a hyperplane in X such that $0 \in H$. Then H is a subspace of deficiency 1. There is a linear functional f whose null space is H. The sets for which $f(x) \geq 0$ and $f(x) \leq 0$ are called the two **sides** of H.

A **ray** with vertex x_0, determined by $x \neq 0$, is the set of all $x_0 + ax$, $a \geq 0$. A **cone** with vertex x_0 is the union of a set of rays with vertex x_0. For every $x_0 \in X$ and $S \subset X$, the cone $c(x_0, S)$ with vertex x_0 determined by S, is the union of the set of all rays with vertex x_0 determined by points of S.

PROPOSITION 1. If $x_0 \in X$ and S is convex, then $c(x_0, S)$ is convex.

Proof. Let $x, y \in c(x_0, S)$ and $0 \leq a \leq 1$. We note that $x = x_0 + a_1x_1$ $y = x_0 + a_2x_2$ where $a_1 \geq 0$, $a_2 \geq 0$ and $x_1, x_2 \in S$. Then,

$$\begin{aligned} ax + (1-a)y &= a(x_0 + a_1x_1) + (1-a)(x_0 + a_2x_2) \\ &= x_0 + aa_1x_1 + (1-a)a_2x_2 \\ &= x_0 + [aa_1 + (1-a)a_2] \\ &\quad \times \left[\frac{aa_1}{aa_1 + (1-a)a_2}x_1 + \frac{(1-a)a_2}{aa_1 + (1-a)a_2}x_2\right] \end{aligned}$$

so that $ax + (1 - a)y \in c(x_0, S)$.

2.7 Ordered Groups

We shall give a proof of the so-called Hahn-Banach theorem. Our proof depends on some elementary facts concerning ordered groups which we present in this section.

Let G be an abelian group with identity 0. A set $P \subset G$ is called a **positive cone** for G if $0 \in P$ and $x, y \in P$ implies $x + y \in P$. A positive cone P induces an "order relation" in G for which $x \geq y$ if $x - y \in P$. An abelian group G together with an order relation given by a positive cone in G is called a **preordered group.** It is called an **ordered group** if $P \cap (-P) = \{0\}$.

If $P \cap (-P) = \{0\}$ and $P \cup (-P) = G$, it is called a **totally ordered group.** G is evidently totally ordered if and only if, for every $x, y \in G$, either $x \leq y$ or $y \leq x$ and both $x \leq y$ and $y \leq x$ implies $x = y$. Instead of $x \geq y$ we may write $y \leq x$, as we have done above.

If G is a preordered group, the set $H = P \cap (-P)$ is a subgroup. Let G/H be the quotient group, and φ the canonical mapping $\varphi(x) = \xi$, where ξ is the coset modulo H to which x belongs. Let $\mathfrak{P}$ be the image in G/H of P under the mapping φ. It is easy to see that G/H is an ordered group with positive cone $\mathfrak{P}$ which is totally ordered if $P \cup (-P) = G$.

A totally ordered group is **archimedean** if for every $x, y \in P \sim \{0\}$ there are positive integers n, m such that $nx - y \in P$ and $my - x \in P$. Two totally ordered groups G and H, with positive cones P and Q, respectively, are said to be isomorphic if there is a group isomorphism between them which maps P onto Q. We then have

PROPOSITION 1. If G is an archimedean totally ordered group, then G is isomorphic to a subgroup of the totally ordered group of real numbers.

Proof. Let 0 be the identity in G, and P the positive cone of G. We let $\varphi(0) = 0$, and $\varphi(x_0) = 1$ where $x_0 \in P \sim \{0\}$. We now associate a positive real number with every $x \in P$ as follows: The rational number n/m belongs to the upper segment of the Dedekind cut defining $\varphi(x)$ if and only if $mx \leq nx_0$. That $x \in P$ actually defines a Dedekind cut in this way follows since G is archimedean, for this implies that the upper segment is non-empty. It is an exercise to show that $x \neq y$ implies $\varphi(x) \neq \varphi(y)$. We let $\varphi(-x) = -\varphi(x)$, so that we now have a one-one correspondence between G and a subset of the reals. It is a routine calculation, which we leave for the reader, to show that this mapping is an isomorphism.

2.8 Hahn-Banach Theorem, Separation Form

A variety of results, which are closely related to each other, and deal either with the separation of sets by hyperplanes, or the extension of linear functionals, are known as the Hahn-Banach theorem. The form which we first give is due to J. Dieudonné.

THEOREM 1. If S is a convex set which contains an interior point and $0 \notin S$, there is a hyperplane H, passing through 0, such that all of S is on the same side of H.

Proof. The cone $c(0, S)$ with vertex 0, determined by S, is convex. Let x_0 be an interior point of S. Then $-x_0 \notin c(0, S)$. Otherwise, there would be an $x \in S$ and $a > 0$ such that $-x_0 = ax$, but then $x_0 + ax = 0$, so that

$$\frac{1}{1+a}\,x_0 + \frac{a}{1+a}\,x = 0$$

and, since $\dfrac{1}{1+a} + \dfrac{a}{1+a} = 1$, we have $0 \in S$. The condition for Zorn's lemma is evidently satisfied by the collection of convex cones, ordered by inclusion, which contain $c(0, S)$ but do not contain $-x_0$. Let C be a maximal cone in this collection.

We first show that $X = C \cup (-C)$. Suppose then that $x \notin C$. Then the convex cone generated by C and x contains $-x_0$ since C is maximal among the convex cones not containing $-x_0$. Hence $-x_0 = ax + by$, where $a > 0$, $b \geq 0$ and $y \in C$. Thus $-x = \dfrac{1}{a} x_0 + \dfrac{b}{a} y \in C$, so that $x \in -C$.

Now $x, y \in C$ implies $x + y \in C$. For since C is convex, $\frac{1}{2}(x + y) \in C$, and since C is a cone, $x + y = 2(\frac{1}{2}(x + y)) \in C$. Thus the abelian group X is preordered with C as positive cone. Let $H = C \cap (-C)$. H is easily seen to be a subgroup, indeed a subspace, of X and the quotient group X/H is totally ordered.

We finally prove that X/H is archimedean, so that H has deficiency 1 and our theorem is proved. For this, we show that for every $x \in C \sim H$ there are $a, b > 0$ such that $x - ax_0 \in -C$ and $x - bx_0 \in C$. Since x_0 is an interior point of C, there is an $\alpha > 0$ such that $y = x_0 - \alpha(x - x_0) \in C$. Then $x - \dfrac{2+\alpha}{\alpha} x_0 \in -C$. For, if it were in C, then

$$x_0 - \alpha(x - x_0) + \alpha\left(x - \frac{2+\alpha}{\alpha} x_0\right) \in C,$$

but this last term is just $-x_0$. We need only let $a = \dfrac{2+\alpha}{\alpha} > 0$. Moreover, $-x \notin C$, so that the smallest convex cone containing C and $-x$ contains $-x_0$, since C is maximal among the convex cones not containing $-x_0$. Hence there are $\gamma > 0$, $\delta \geq 0$, with $-x_0 = \gamma(-x) + \delta y$, $y \in C$. Thus $x - \dfrac{1}{\gamma} x_0 \in C$. We need only let $b = 1/\gamma > 0$.

The condition that the convex set S should contain an interior point cannot be dropped entirely. Thus, let X have dimension $\aleph_0$ with basis

$$x_1, x_2, \ldots, x_n, \ldots.$$

Every $x \in X$, $x \neq 0$, has a unique representation $x = a_1x_1 + \ldots + a_mx_m$, where $m > 0$ and $a_m \neq 0$. Let S be the set of those $x \in X$ whose last coefficient in this representation is positive. Then $0 \notin S$. The set S is convex. For, if $x = a_1x_1 + \ldots + a_mx_m$, $a_m > 0$ and $y = b_1x_1 + \ldots + b_nx_n$, $b_n > 0$, if $0 \leq a \leq 1$, and $z = ax + (1 - a)y$, then clearly $z = c_1x_1 + \ldots + c_kx_k$ with $c_k > 0$. Let f be a linear functional on X which is non-negative on S. For every n and $a \in R$, we have $ax_n + x_{n+1} \in S$, so that $af(x_n) + f(x_{n+1}) \geq 0$.

This implies $f(x_n) = 0$ for every n, so that $f(x) \equiv 0$ on X. This means that there is no hyperplane H such that all of S is on one side of H. Of course, the above set S contains no interior points.

As a corollary to theorem 1 we have

THEOREM 2. If X is a vector space and S is a convex set in X which contains an interior point, and if Y is a subspace of X which has no points in common with S, then there is a hyperplane H such that $Y \subset H$ and all points of S are on the same side of H.

Proof. Let x_0 be an interior point of S. Then $-x_0$ is not in the cone of vertex 0 generated by $S \cup Y$. For, if it were, there would be a $y \in Y$, $x \in S$, $a > 0$, $b > 0$, with $-x_0 = ax + by$. Hence, $-y = \frac{1}{a} x_0 + \frac{b}{a} x$ and so $-\frac{a}{b+1} y \in S$. This means that $S \cap Y$ is non-empty, contradicting our original assumption. We may now complete the proof as for theorem 1.

The above theorem has the

COROLLARY. If X and S are as in the theorem, and Y is a translation of a subspace of X which has no points in common with S, then there is a hyperplane H (not necessarily a subspace) containing Y, such that all points in S are on the same side of H.

This means there is a linear functional f and $k \geq 0$ such that $f(x) \geq k$ for every $x \in S$.

Theorem 2 and its corollary are the separation form of the Hahn-Banach theorem. The following extension form follows from it.

THEOREM 3. Let X be a vector space, Y a subspace of X, and p a real function on X such that $p(x) \geq 0$, $p(x + y) \leq p(x) + p(y)$, and $p(ax) = |a| \, p(x)$. If f is a linear functional on Y such that $|f(x)| \leq p(x)$ for every $x \in Y$, there is a linear F on X such that $F(x) = f(x)$ on Y and $|F(x)| \leq p(x)$ on X.

The proof is left to the reader. In the next section we prove an important extension theorem of which theorem 3 is also an immediate consequence.

2.9 Hahn-Banach Theorem, Extension Form

We now prove

THEOREM 1. Let X be a (real) vector space, p a real-valued function on X such that $p(x + y) \leq p(x) + p(y)$ and $p(ax) = ap(x)$ for $a \geq 0$, and Y a subspace of X. If f is linear on Y and $f(x) \leq p(x)$ for every $x \in Y$, there is a linear F on X such that $F(x) = f(x)$ on Y and $F(x) \leq p(x)$ on X.

Proof. The main problem is to extend f to a φ defined on $(Y \cup \{x_0\})$ for any $x_0 \notin Y$ so that

$$\varphi(x) \leq p(x) \text{ for all } x \in (Y \cup \{x_0\}).$$

Assuming this can be done, the existence of F is obtained using Zorn's lemma as follows:

Let E be the class of all linear extensions of f which are dominated by p on their domains. Partially order E by

$$F_1 \leq F_2 \text{ if } F_2 \text{ is an extension of } F_1,$$

i.e., the domain of F_2 contains that of F_1 and $F_2(x) = F_1(x)$ for every x in the domain of F_1. Let $\mathcal{C}$ be a chain in E and define a linear functional U by

(a) the domain of U is the union of the domains of the F for all $F \in \mathcal{C}$,

(b) if x is in the domain of U, define $U(x) = F(x)$, where $F \in \mathcal{C}$ and x is in the domain of F. This is consistent, since $\mathcal{C}$ is a chain.

It is clear that U is a linear extension of f and $U(x) \leq p(x)$ on its domain, so that $U \in E$ and it is an upper bound of $\mathcal{C}$. By Zorn's lemma, E has a maximal element F. F must be an extension to X since if its domain were a proper subspace M and $x_0 \in X \sim M$, then an extension to $(M \cup \{x_0\})$ would be possible.

We now suppose f defined on Y and extend it to φ on $(Y \cup \{x_0\}) = Z$. Since $x \in Z$ has a unique representation $x = y + cx_0$, $y \in Y$, c real, it follows that $\varphi(x) = f(y) + ca_0$ is such that $\varphi(x) = f(x)$ for every $x \in Y$ no matter how we choose a_0. The problem is to show that a_0 may be chosen so that $\varphi(x) \leq p(x)$ for all $x \in Z$, i.e., so that $f(y) + ca_0 \leq p(y + cx_0)$. In other words, $ca_0 \leq p(y + cx_0) - f(y)$.

For $c = 0$, this always holds. For $c > 0$, the relation $ca_0 \leq p(y + cx_0) - f(y)$ is equivalent to

$$a_0 \leq \frac{1}{c} p(y + cx_0) - \frac{1}{c} f(y) = p\left(\frac{y}{c} + x_0\right) - f\left(\frac{y}{c}\right).$$

For $c < 0$, it is equivalent to

$$a_0 \geq -p\left(-\frac{y}{c} - x_0\right) - f\left(\frac{y}{c}\right).$$

Hence, we must show the existence of an a_0 such that, for every $y \in Y$,

$$-p(-y - x_0) - f(y) \leq a_0 \leq p(y + x_0) - f(y).$$

But this follows from the fact that, for every $y, z \in Y$,

$$-p(-y - x_0) - f(y) \leq p(z + x_0) - f(z),$$

since

$$f(z) - f(y) = f(z - y) \leq p[(z + x_0) + (-y - x_0)]$$
$$\leq p(z + x_0) + p(-y - x_0).$$

A corollary to this theorem is theorem 3, section 2.8.

2.10 Applications, Banach Limits, Invariant Measure

The Hahn-Banach theorem will be seen to have many applications in the development of the general theory of normed linear spaces and, more generally, locally convex topological vector spaces.

We give two direct applications of the theorem. The first is to the generalized notion of the limit of a sequence.

Let m denote the vector space of all bounded sequences of real numbers, c the subspace of all convergent sequences. If $x \in c$, then $l(x) = \lim_n x_n$ is defined, and l is a linear functional on c.

A **Banach limit** is any linear functional L defined on m such that

(a) $L(x) \geq 0$ if $x_n \geq 0$ for all n,
(b) $L(x) = L(\sigma x)$ where σ denotes the shift

$$\sigma x = \sigma(x_1, x_2, x_3, \ldots) = (x_2, x_3, \ldots),$$

(c) $L(x) = 1$ if $x = (1, 1, 1, \ldots)$.

These are properties which a generalized limit may be expected to have.

PROPOSITION 1. If L is a Banach limit, then

$$\underline{\lim}\, x_n \leq L(x) \leq \overline{\lim}\, x_n \text{ for all } x \in m.$$

Proof. Since $\underline{\lim}\, x_n = \lim_k \inf_{n \geq k} x_n$ and $\overline{\lim}\, x_n = \lim_k \sup_{n \geq k} x_n$, it follows by (b) that we need only show that

$$\inf x_n \leq L(x) \leq \sup x_n.$$

For $\epsilon > 0$, choose n_0 so that

$$\inf x_n \leq x_{n_0} < \inf x_n + \epsilon,$$

whence

$$x_n + \epsilon - x_{n_0} > 0 \text{ for all } n.$$

By (a) and (c), we get

$$L(x) + \epsilon \geq x_{n_0} \geq \inf x_n,$$

so that $L(x) \geq \inf x_n$. Similarly $L(x) \leq \sup x_n$.

In particular, for $x \in c$, the Banach limit equals the ordinary limit.
Define, on m,

$$q(x) = \overline{\lim_{n \to \infty}}\, \frac{x_1 + x_2 + \ldots + x_n}{n}.$$

Then

$$-q(-x) = \varliminf_{n\to\infty} \frac{x_1 + x_2 + \ldots + x_n}{n}.$$

If $\{x_n\}$ is convergent, it is easy to show that

$$\lim_{n\to\infty} x_n = \lim_{n\to\infty} \frac{x_1 + x_2 + \ldots + x_n}{n}.$$

We leave this for the reader to prove.

Thus, for $x \in c$,

$$q(x) = l(x).$$

Moreover, $q(x + y) \leq q(x) + q(y)$ and $q(ax) = aq(x)$, $a \geq 0$.

Let L be an extension of l from c to m given by the Hahn-Banach theorem (theorem 1, Sec. 2.9). Then

$$-q(-x) \leq L(x) \leq q(x)$$

for all $x \in m$.

In order to show that L is a Banach limit, it is only necessary to show that (b) holds. For this,

$$q(x - \sigma x) = \varlimsup_{n\to\infty} \left(\frac{x_{n+1} - x_1}{n}\right) = 0.$$

We thus have

THEOREM 1. Banach limits exist.

In order to characterize the sequences for which all Banach limits agree, it is convenient to give another proof of theorem 1.

Define, on m,

$$p(x) = \inf \varlimsup_{j\to\infty} \frac{1}{k} \sum_{i=1}^{k} x_{n_i+j}$$

where the infimum is taken over all finite sets of integers $n_1, n_2, \ldots, n_k$. Also define

$$p'(x) = -p(-x), \qquad x \in m.$$

Observe that

$$p'(x) = \sup \varliminf_{j\to\infty} \frac{1}{k} \sum_{i=1}^{k} x_{n_i+j}.$$

PROPOSITION 2. For any $x \in m$,

$$\varliminf x_n \leq p'(x) \leq p(x) \leq \varlimsup x_n.$$

In particular, $p(x) = l(x)$ for $x \in c$.

PROPOSITION 3. For any Banach limit L and $x \in m$,

$$p'(x) \leq L(x) \leq p(x).$$

Proof. Apply proposition 1 to the sequence $\left\{\frac{1}{k}\sum_{i=1}^{k} x_{n_i+j}\right\}$ for fixed k, $n_1, \ldots, n_k$ to get

$$\underline{\lim}_{j\to\infty} \frac{1}{k}\sum_{i=1}^{k} x_{n_i+j} \le L\left(\left\{\frac{1}{k}\sum_{i=1}^{k} x_{n_i+j}\right\}\right) = L(x) \le \overline{\lim}_{j\to\infty} \frac{1}{k}\sum_{i=1}^{k} x_{n_i+j}.$$

Proposition 4. $p(x)$ is such that $p(x + y) \le p(x) + p(y)$, and $p(ax) = ap(x)$ if $a \ge 0$.

Proof. We need only prove $p(x + y) \le p(x) + p(y)$. Let x, $y \in m$ and $\epsilon > 0$. There are $k, n_1, \ldots, n_k$ such that

$$\overline{\lim}_{j\to\infty} \frac{1}{k}\sum_{i=1}^{k} x_{n_i+j} < p(x) + \frac{\epsilon}{2}$$

and $l, m_1, \ldots, m_l$ such that

$$\overline{\lim}_{j\to\infty} \frac{1}{l}\sum_{i=1}^{l} x_{m_i+j} < p(y) + \frac{\epsilon}{2}.$$

Consider the kl integers $n_r + m_s$, $r = 1, \ldots, k$, $s = 1, \ldots, l$. Using the subadditive property of $\overline{\lim}$ we get

$$p(x + y) \le \overline{\lim}_{j\to\infty} \frac{1}{kl}\sum_{r=1}^{k}\sum_{s=1}^{l}(x_{n_r+m_s+j} + y_{n_r+m_s+j})$$

$$\le \overline{\lim}_{j\to\infty} \frac{1}{kl}\sum_{r=1}^{k}\sum_{s=1}^{l} x_{n_r+m_s+j} + \overline{\lim}_{j\to\infty} \frac{1}{kl}\sum_{r=1}^{k}\sum_{s=1}^{l} y_{n_r+m_s+j}.$$

Continuing with the first term only, it is no larger than

$$\frac{1}{l}\sum_{s=1}^{l} \overline{\lim}_{j\to\infty} \frac{1}{k}\sum_{r=1}^{k} x_{n_r+m_s+j}.$$

But each

$$\overline{\lim}_{j\to\infty} \frac{1}{k}\sum_{r=1}^{k} x_{n_r+m_s+j} < p(x) + \frac{\epsilon}{2},$$

so that the average also has this property. Similarly, the second term is less than $p(y) + \frac{\epsilon}{2}$ and

$$p(x + y) < p(x) + p(y) + \epsilon,$$

for every $\epsilon > 0$.

We now apply the Hahn-Banach theorem to $l(x)$ on c to obtain linear functionals $L(x)$ on m satisfying $L(x) \le p(x)$, $x \in m$. Putting $-x$ for x we obtain the other part of

$$p'(x) \le L(x) \le p(x), \qquad x \in m.$$

It remains to show that L satisfies (b). For any $x \in m$,

$$L(\sigma x) - L(x) = L(\sigma x - x) \le p(\sigma x - x)$$

and

$$p(\sigma x - x) \leq \overline{\lim_{j\to\infty}} \frac{1}{k} \sum_{i=1}^{k} (x_{n_i+1+j} - x_{n_i+j}).$$

So, if we take $n_i = i$, $i = 1, \ldots, k$, we have

$$\begin{aligned} p(\sigma x - x) &\leq \overline{\lim_{j\to\infty}} \frac{1}{k} \sum_{i=1}^{k} (x_{i+1+j} - x_{i+j}) \\ &= \frac{1}{k} \overline{\lim_{j\to\infty}} (x_{k+1+j} - x_{1+j}). \end{aligned}$$

If M is an upper bound for $|x_n|$, for all n, then for arbitrary k we have

$$p(\sigma x - x) \leq \frac{2M}{k},$$

so that $p(\sigma x - x) \leq 0$, and

$$L(\sigma x) \leq L(x).$$

Similarly,

$$L(x) - L(\sigma x) \leq p(x - \sigma x) \leq 0.$$

Hence

$$L(\sigma x) = L(x).$$

This completes the second proof of the existence of Banach limits.

All Banach limits are the same for the convergent sequences. However, there are other sequences which have this property; in particular, $\{1, 0, 1, 0, \ldots\}$. $x \in m$ is called **almost convergent** and the number s is called the ***F*-limit** of x if $L(x) = s$ for all Banach limits L. This notion was introduced and studied by G. Lorentz.

PROPOSITION 5. x is almost convergent if and only if $p'(x) = p(x)$.

Proof. The sufficiency follows by proposition 3. If $p'(x) < p(x)$ then $x \notin c$. There are distinct extensions of $l(x)$ from c to $(c \cup \{x\})$, since in the proof of the Hahn-Banach theorem the extensions can have any value at x on the interval

$$\left(\sup_{x\in c} [-p(-y - x) - l(y)], \qquad \inf_{x\in c} [p(y + x) - l(y)]\right)$$

and these end points reduce to $p'(x)$ and $p(x)$, respectively.

PROPOSITION 6. x is almost convergent and the F-limit of x is s if and only if

$$\lim_{p\to\infty} \frac{1}{p} (x_n + x_{n+1} + \ldots + x_{n+p-1}) = s$$

holds uniformly in n.

Proof. Under the condition of the theorem, for $\epsilon > 0$, we have

$$s - \epsilon < \frac{1}{p}(x_n + x_{n+1} + \ldots + x_{n+p-1}) < s + \epsilon$$

for all n and all sufficiently large p. It then readily follows that

$$s - \epsilon \leq p'(x) \leq p(x) \leq s + \epsilon,$$

so that F-lim $x_n = s$.

Conversely, suppose F-lim $x_n = s$. Let $\epsilon > 0$. Since $s = p(x)$, there are $k, n_1, \ldots, n_k$ such that

$$\overline{\lim_{j\to\infty}} \frac{1}{k} \sum_{i=1}^{k} x_{n_i+j} < s + \epsilon.$$

Thus, if j is sufficiently large, say $j \geq N$,

$$\frac{1}{k} \sum_{i=1}^{k} x_{n_i+j} < s + \epsilon.$$

Replacing j by $j - N$ and n_i by $m_i = n_i + N$, for $i = 1, 2, \ldots, k$, we obtain

$$\frac{1}{k} \sum_{i=1}^{k} x_{m_i+j} < s + \epsilon, \quad j = 0, 1, 2, \ldots.$$

Now, for any $n = 1, 2, 3, \ldots$ and $p = 1, 2, 3, \ldots,$

$$\frac{1}{p} \sum_{j=1}^{p} \frac{1}{k} \sum_{i=1}^{k} x_{m_i+j+n} < s + \epsilon.$$

This sum can be evaluated as

$$\frac{1}{pk} \sum_{i=1}^{k} \sum_{j=m_i+1}^{m_i+p} x_{j+n} = \frac{1}{pk} \sum_{i=1}^{k} \sum_{j=1}^{p-1} [x_{j+n} + (x_{m_i+1+j+n} - x_{j+n})]$$

$$= \frac{1}{p} \sum_{j=0}^{p-1} x_{j+n} + \frac{1}{p} \left\{ \frac{1}{k} \sum_{i=0}^{k} \sum_{j=0}^{p-1} (x_{m_i+1+j+n} - x_{j+n}) \right\}.$$

The quantity in brackets is bounded for large p, uniformly in n. In fact, if kp exceeds m_i for all $i = 1, 2, \ldots, k$, then cancellations occur in the inner sum, leaving only $2m_i$ terms. Thus, if M is a bound for $|x_n|$ for all $n = 1, 2, 3, \ldots,$ the quantity in brackets does not exceed in absolute value the quantity

$$\frac{2M}{k} \sum_{i=1}^{k} m_i.$$

It follows that p can be chosen so large that

$$\frac{1}{p} \sum_{j=0}^{p-1} x_{j+n} < s + 2\epsilon$$

uniformly in n.

The reverse inequality is obtained by a similar argument using $s = p'(x)$.

As a second application, we show the existence of a finitely additive invariant measure for the set of all subsets of a circle σ. By this we mean a non-negative function $m(S)$ defined for all subsets of σ such that

(α) $m(\sigma) = 1$,
(β) if $S \cap T$ is empty, then $m(S \cup T) = m(S) + m(T)$,
(γ) for every rotation f of σ and every S, $m(fS) = m(S)$.

It is equivalent to prove the existence of a generalized integral for the space X of bounded real functions $x(t)$ on σ; in other words, the existence of a $\varphi(x)$ on X such that

(α') $\varphi(1) = 1$, where $1(t) = 1$, $t \in \sigma$,
(β') $\varphi(ax + by) = a\varphi(x) + b\varphi(y)$ for all $x, y \in X$ and $a, b \in R$,
(γ') $\varphi(x) \geq 0$ if $x(t) \geq 0$,
(δ') $\varphi(x_\tau) = \varphi(x)$, where $x_\tau(t) = x(t + \tau)$.

We now define a special function p on X with the following properties:

(α'') $p(1) = 1$, $p(-1) = -1$,
(β'') $p(x + y) \leq p(x) + p(y)$ for all $x, y \in X$,
(γ'') $p(ax) = ap(x)$ for all $a \geq 0$ and $x \in X$,
(δ'') $p(x_\tau - x) \leq 0$ for all $x \in X$ and τ.

That the upper Riemann integral cannot serve as the p is clear if we consider the function $x(t) = 1$, t rational, $x(t) = 0$, t irrational, and let $\tau = \sqrt{2}$. Then $x_\tau - x$ is equal to 1 on a dense set so that its upper Riemann integral is equal to 1. However, Banach found a p which has the desired properties. For every $x \in X$ and $t_1, t_2, \ldots, t_n$ consider the set of mean values

$$\frac{1}{n}\sum_{j=1}^{n} x(t + t_j)$$

obtained as t varies over σ. Let

$$U(x; t_1, \ldots, t_n) = \sup_t \frac{1}{n}\sum_{j=1}^{n} x(t + t_j),$$

and let $p(x) = \inf U(x\,; t_1, \ldots, t_n)$ for all n and $t_1, \ldots, t_n$.

(α'') That $p(1) = 1$ and $p(-1) = -1$ are obvious since $1(t + t_j) = 1$ for all choices of t_j and t.

(β'') Let $x, y \in X$. For every $\epsilon > 0$, there are $t_1, \ldots, t_n$; $s_1, \ldots, s_m$ for which

$$p(x) \geq U(x; t_1, \ldots, t_n) - \frac{\epsilon}{2}$$

and

$$p(y) \geq U(y; s_1, \ldots, s_m) - \frac{\epsilon}{2}\,.$$

Now, for every t,

$$\frac{1}{nm}\sum_{i=1}^{n}\sum_{j=1}^{m}[x(t+t_i+s_j)+y(t+t_i+s_j)]$$

$$=\frac{1}{m}\left\{\frac{1}{n}\sum_{i=1}^{n}x((t+s_1)+t_i)+\dots+\frac{1}{n}\sum_{i=1}^{n}x((t+s_m)+t_i)\right\}$$

$$+\frac{1}{n}\left\{\frac{1}{m}\sum_{j=1}^{m}y((t+t_1)+s_j)+\dots+\frac{1}{m}\sum_{j=1}^{m}y((t+t_n)+s_j)\right\}$$

$$\leq U(x;t_1,\dots,t_n)+U(y;s_1,\dots,s_m),$$

so that

$$U(x+y;t_1+s_1,\dots,t_n+s_m)\leq U(x;t_1,\dots,t_n)$$
$$+U(y;s_1,\dots,s_m)\leq p(x)+p(y)+\epsilon.$$

Since $\epsilon>0$ is arbitrary, it follows that $p(x+y)\leq p(x)+p(y)$.

(γ'') Since, for every $a\geq 0$ and $x\in X$ we have, for every n, t and $t_1,\dots,t_n$,

$$\frac{1}{n}\sum_{i=1}^{n}ax(t+t_i)=a\left\{\frac{1}{n}\sum_{i=1}^{n}x(t+t_i)\right\},$$

it follows that $p(ax)=ap(x)$.

(δ'') Consider $x\in X$ and fix τ. For a given n, let $t_1=\tau$, $t_2=2\tau,\dots,$ $t_n=n\tau$. Then

$$\frac{1}{n}\sum_{j=1}^{n}[x(t+\tau+j\tau)-x(t+j\tau)]=\frac{1}{n}[x(t+(n+1)\tau)-x(t+\tau)]\leq\frac{2M}{n},$$

where $M=\sup_t|x(t)|$. It follows that $p(x_\tau-x)\leq\inf_n\frac{2M}{n}=0$.

We now apply theorem 1, Sec. 2.9, to the above space X and functional p, where Y is the null space $\{0\}$. There is a linear functional $\varphi(x)$ on X such that $\varphi(x)\leq p(x)$. We show that this is the required functional. By theorem 1, Sec. 2.9, φ satisfies (β'). Now, $\varphi(1)\leq p(1)=1$ and $\varphi(-1)\leq p(-1)=-1$, so that $1\geq\varphi(1)=-\varphi(-1)\geq 1$. Hence $\varphi(1)=1$, so that (α') holds. Suppose $x=x(t)\geq 0$. Then $p(-x)\leq 0$. Hence

$$\varphi(x)=-\varphi(-x)\geq -p(-x)\geq 0$$

so that (γ') holds. Consider any $x\in X$ and any τ. Now $\varphi(x_\tau-x)\leq p(x_\tau-x)\leq 0$. Moreover, $\varphi(x_\tau-x)=-\varphi(x-x_\tau)\geq -p(x-x_\tau)\geq 0$. Hence $\varphi(x_\tau)-\varphi(x)=\varphi(x_\tau-x)=0$, so that ($\delta'$) holds.

We observe that in proving that p satisfies (β''), the t_i and s_j were interchanged. In other words, we used the fact that the group of rotations of the circle is abelian. The above argument cannot be applied to the group of rotations of a 2-sphere since it is not an abelian group. Indeed, Hausdorff showed that there is no non-trivial finitely additive invariant measure for all the subsets of the 2-sphere.

2.11 Banach Space, Dual Space

We now turn to a discussion of normed vector spaces. A **norm** is a non-negative real function on a vector space X, written $\|x\|$, such that

(a) $x \neq 0$ implies $\|x\| > 0$,
(b) $\|ax\| = |a| \, \|x\|$,
(c) $\|x + y\| \leq \|x\| + \|y\|$.

Thus the function p of theorem 3, Sec. 2.8, would be a norm if $x \neq 0$ implied $p(x) > 0$. Such a function on X is called a **seminorm.**

We present some simple properties of normed spaces.

First of all a normed space may be given a natural metric by means of

$$d(x, y) = \|x - y\|.$$

It is easy to verify that d is a metric. Moreover, it is an invariant metric, i.e., $d(x, y) = d(x + z, y + z)$.

The norm is then a continuous function on this metric space.

We leave these proofs to the reader.

Since a normed vector space X is a metric space, it has a completion $\tilde{X}$ as a metric space. It is easy to define vector space operations and a norm in $\tilde{X}$ so that it is a normed vector space in which X is a subspace.

A complete (with the associated metric) normed vector space is called a **Banach space.**

We give some examples of normed vector spaces, some of which are Banach spaces.

(a) The vector space $C[0, 1]$ with the norm

$$\|x\| = \max\,[|x(t)| : t \in [0, 1]]$$

is a Banach space.

(b) The vector space $P[0, 1]$ of polynomials on $[0, 1]$ with the norm

$$\|x\| = \max\,[|x(t)| : t \in [0, 1]]$$

is a normed vector space which is not complete.

(c) The vector space m of bounded sequences with the norm

$$\|x\| = \sup\,[|x_n| : n = 1, 2, \ldots]$$

is a Banach space.

(d) The vector space c of convergent sequences with the norm

$$\|x\| = \sup\,[|x_n| : n = 1, 2, \ldots]$$

is a Banach space.

(e) The vector space l_p, $p \geq 1$, of sequences for which $\sum_{n=1}^{\infty} |x_n|^p < \infty$ with the norm

$$\|x\| = \left[\sum_{n=1}^{\infty} |x_n|^p\right]^{1/p}$$

is a Banach space.

Let X and Y be normed vector spaces. Linear transformations T of X into Y have been defined in Sec. 2.1. Since X and Y are metric spaces, we may discuss the continuity of such transformations. We prove

PROPOSITION 1. *If T is continuous at $x = 0$, then it is continuous everywhere and the continuity is uniform.*

Proof. Since T is continuous at $x = 0$, for every $\epsilon > 0$ there is a $\delta > 0$ such that $\|x\| < \delta$ implies $\|Tx\| < \epsilon$. But for every $x \in X$, $\|x - y\| < \delta$ then implies $\|Tx - Ty\| = \|T(x - y)\| < \epsilon$.

The point $x = 0$ could be replaced by any other point of X. Thus a linear transformation is either uniformly continuous or everywhere discontinuous. We shall accordingly refer to the property simply as continuity.

We shall refer to a linear transformation T as **bounded** if there is an M such that $\|Tx\| \leq M\,\|x\|$ for all $x \in X$.

PROPOSITION 2. *T is bounded if and only if T is continuous.*

Proof. Suppose T is bounded and M is a bound. Let $\epsilon > 0$ and $\delta = \epsilon/M$. Then $\|x\| < \delta$ implies $\|Tx\| < \epsilon$, so that T is continuous. Suppose T is continuous. Then there is a $\delta > 0$ such that $\|x\| = \delta$ implies $\|Tx\| < 1$. Let $x \in X$, then

$$\|Tx\| = \left\| T\left(\frac{\delta x}{\|x\|}\right)\right\| \cdot \frac{\|x\|}{\delta} < \frac{1}{\delta}\,\|x\|$$

so that T is bounded.

Now let T be a bounded linear transformation and let

$$\|T\| = \inf\,[M : \|Tx\| \leq M\,\|x\|, \quad x \in X].$$

Then $\|T\|$ is called the **norm** or **bound** of T. Evidently, $\|Tx\| \leq \|T\|\,\|x\|$ for every $x \in X$.

Let $B(X, Y)$ be the set of bounded linear transformations on X into Y. We shall show that $B(X, Y)$ is a normed vector space which is a Banach space if Y is a Banach space.

Let $T_1, T_2 \in B(X, Y)$. Define $T_1 + T_2$ by

$$(T_1 + T_2)(x) = T_1(x) + T_2(x)$$

and aT by

$$(aT)(x) = aT(x),$$

for every $x \in X$. Then

$$\|T_1 + T_2\| \leq \|T_1\| + \|T_2\|$$

and

$$\|aT\| = |a|\,\|T\|.$$

These imply that $B(X, Y)$ is a vector space. Moreover, since $\|T\| = 0$ implies $T = 0$, it is a normed vector space.

Suppose now that Y is complete. Let $\{T_n\}$ be a Cauchy sequence in $B(X, Y)$. Then, for any $x \in X$,

$$\|T_m x - T_n x\| \leq \|T_m - T_n\| \, \|x\|,$$

so that $\{T_n x\}$ is a Cauchy sequence in Y. Since Y is complete we may define

$$Tx = \lim_n T_n x \quad \text{for } x \in X.$$

It is obvious that T is linear. Since $\{T_n\}$ is Cauchy, $\{\|T_n\|\}$ is bounded, i.e., there is an M such that $\|T_n\| \leq M$, $n = 1, 2, \ldots$. For every x, $\|T_n x\| \leq M \|x\|$. So, $\|Tx\| \leq M \|x\|$. Hence $T \in B(X, Y)$.

Let $\epsilon > 0$. Choose n_0 so that $\|T_m - T_n\| < \epsilon$ if $m, n > n_0$. Then

$$\|T_m x - T_n x\| < \epsilon \|x\| \text{ for } m, n > n_0, \quad x \in X.$$

Letting n converge to infinity, we get

$$\|T_m x - Tx\| \leq \epsilon \|x\| \text{ for } m > n_0, \quad x \in X.$$

This implies $\|T_m - T\| \leq \epsilon$ for $m > n_0$, so that $\{T_n\}$ converges to T in $B(X, Y)$.

In particular, if Y is one-dimensional, we write X' instead of $B(X, Y)$. If X is a normed vector space, the space X' of continuous linear functionals on X, with norm

$$\begin{aligned} \|x'\| &= \sup\,[|x'(x)| : \|x\| = 1] \\ &= \inf\,[M : |x'(x)| \leq M \|x\|, \quad x \in X] \end{aligned}$$

is called the **dual** of X. We then have the

COROLLARY. The dual of a normed vector space is a Banach space.

2.12 Hahn-Banach Theorem in Normed Space

In a normed vector space, the Hahn-Banach theorem (theorem 3, Sec. 2.8) assumes the following form.

PROPOSITION 1. If X is a normed vector space, Y a subspace of X, and f a bounded linear functional on Y with bound $\|f\|$, relative to Y, then f has a continuous linear extension to an $x' \in X'$ with $\|x'\| = \|f\|$.

Proof. By theorem 3, Sec. 2.8 with $p(x) = \|f\| \, \|x\|$.

The following proposition assures that X' is a total subspace of X^*.

PROPOSITION 2. For any $x \neq 0$ in X there is an $x' \in X'$ such that $x'(x) = \|x\|$ and $\|x'\| = 1$.

Proof. Let Y be the one-dimensional space generated by x, let $x'(ax) = a\|x\|$ and apply proposition 1.

So far, we have only given the Hahn-Banach theorem for real vector spaces. The theorem also holds for complex vector spaces. In proving the analog of proposition 1 for complex vector spaces, we first recall that every complex vector space is also a real vector space. Let f be a complex linear functional on the complex vector space X. Then

$$f(x) = f_1(x) + if_2(x),$$

where f_1 and f_2 are real linear functionals on the real vector space X.

Now, let X be a complex normed vector space, Y a subspace of X, and $f = f_1 + if_2$ a linear functional on Y. Then $f(ix) = f_1(ix) + if_2(ix)$ and $f(ix) = if(x) = -f_2(x) + if_1(x)$, so that $f_2(x) = -f_1(ix)$. Hence, $f(x) = f_1(x) - if_1(ix)$. We suppose that f is bounded on Y and let $\| f \|$ be its norm on Y. If $p(x) = \| f \| \cdot \|x\|$ on X, then $|f(x)| \leq p(x)$ on Y so that $|f_1(x)| \leq p(x)$ on Y. By proposition 1, f_1 may be extended to an F_1 such that $|F_1(x)| \leq p(x)$ on X. We consider the extension $F(x) = F_1(x) - iF_1(ix)$ of f to X. For every real φ,

$$|\text{Re}\,(e^{i\varphi}F(x))| = |\text{Re}\,F(e^{i\varphi}x)| = |F_1(e^{i\varphi}x)| \leq p(e^{i\varphi}x) = p(x).$$

But there is a real φ for which $\text{Re}\,(e^{i\varphi}F(x)) = |F(x)|$. Hence $|F(x)| \leq p(x)$ on X, so that $\|F\| = \| f \|$. This proves the

PROPOSITION 3. If X is a complex normed vector space, and f is a continuous linear functional on Y, a subspace of X, then f may be extended to a continuous linear functional F on X such that $\|F\| = \| f \|$.

Proposition 2 readily yields a representation theorem for normed vector spaces.

PROPOSITION 4. For every normed vector space X there is a set A such that X is isomorphic with a subspace of the Banach space of bounded functions f on A with norm $\| f \| = \sup\,[|f(t)| : t \in A]$.

Proof. Let x_α, $\alpha \in A$, be a dense set in X. For every $\alpha \in A$, let $f(x, \alpha)$ be a linear functional on X such that

$$\|f(x, \alpha)\| = 1 \text{ and } f(x_\alpha, \alpha) = \|x_\alpha\|.$$

Now, for every $x \in X$, $x \neq 0$, and $\alpha \in A$, we have

$$1 = \|f(x, \alpha)\| \geq \frac{|f(x, \alpha)|}{\|x\|}$$

so that $\|x\| \geq |f(x, \alpha)|$. The function $[f(x, \alpha)]$, x fixed, $\alpha \in A$, is thus bounded. Let φ be defined on X by

$$\varphi(x) = [f(x, \alpha)].$$

Then evidently $\varphi(x + y) = \varphi(x) + \varphi(y)$ and $\varphi(ax) = a\varphi(x)$. We need only show that

$$\|x\| = \sup [|f(x, \alpha)| : \alpha \in A]$$

in order to complete the proof that φ is a norm preserving isomorphism. We already know that $\|x\| \geq \sup [|f(x, \alpha)| : \alpha \in A]$. Now,

$$\big|\, |f(x_\alpha, \alpha)| - \|x\| \,\big| = \big|\, \|x_\alpha\| - \|x\| \,\big| \leq \|x_\alpha - x\|$$

and

$$\big|\, |f(x, \alpha)| - |f(x_\alpha, \alpha)| \,\big| \leq |f(x, \alpha) - f(x_\alpha, \alpha)| \leq \|x_\alpha - x\|,$$

so that

$$\big|\, |f(x, \alpha)| - \|x\| \,\big| \leq 2\, \|x_\alpha - x\|.$$

Since, for every $\epsilon > 0$ there is an $\alpha \in A$ such that $\|x_\alpha - x\| < \epsilon$, it follows that $\|x\| \leq \sup [|f(x, \alpha)| : \alpha \in A]$. This completes the proof.

This theorem also follows from the imbedding of X in X'', with the unit sphere of X' as A (see Sec. 2.17), but this proof does not yield the interesting

COROLLARY. If X is a separable normed vector space, X is isomorphic to a subspace of the Banach space m of bounded sequences with the norm $\|x\| = \sup [|x_n| : n = 1, 2, \ldots]$.

As another application of the Hahn-Banach theorem we have

PROPOSITION 5. If X is an infinite dimensional Banach space, then $\dim X \geq c$.

Proof. Let x_1' be a non-zero continuous linear functional on X and let Z_1 be its null space. The existence of such a functional is assured by proposition 2. Let $x_1 \in X \sim Z_1$ with $\|x_1\| = 1$. The subspace Z_1 has deficiency 1 and so is also infinite dimensional. Let x_2' be a non-zero continuous linear functional on Z_1 and let Z_2 be its null space. Let $x_2 \in Z_1 \sim Z_2$ be such that $\|x_2\| = \frac{1}{2}$. By induction, we obtain a sequence $\{x_n\}$ such that, for every n, $\|x_n\| = 1/2^{n-1}$, and a decreasing sequence of closed subspaces

$$X \supset Z_1 \supset Z_2 \supset \ldots \supset Z_n \supset \ldots$$

such that, for every n, $x_1, \ldots, x_n \notin Z_n$.

We now establish a vector space isomorphism between the vector space m of bounded sequences and a subspace of X. Let $\{a_n\} \in m$. The series $\sum_{n=1}^{\infty} a_n x_n$ has as its sequence of partial sums a Cauchy sequence and so, since X is complete, it converges. The mapping

$$\varphi : \{a_n\} \to \sum_{n=1}^{\infty} a_n x_n$$

is clearly a vector space homomorphism of m into X. In order to show that it is an isomorphism, we must show that $\sum_{n=1}^{\infty} a_n x_n = 0$ implies all the $a_n = 0$. Now, $\sum_{n=1}^{\infty} a_n x_n = 0$ implies $a_1 x_1 = \sum_{n=2}^{\infty} -a_n x_n$, $\sum_{n=2}^{\infty} -a_n x_n \in Z_1$ but $x_1 \notin Z_1$. Hence $\sum_{n=2}^{\infty} -a_n x_n = 0$ and $a_1 = 0$. In similar fashion $a_2 = a_3 = \ldots = 0$. This proves that the dimension of X is at least as great as that of m.

It remains to be shown that m has dimension c. For every $\xi \in (0, 1)$ the sequence $(\xi, \xi^2, \ldots, \xi^n, \ldots)$ is in m. It is easy to show that these sequences are linearly independent. Hence $\dim m \geq c$. But the cardinality of m is c.

2.13 Uniform Boundedness Principle, Applications

We turn now to one of the most important properties of Banach spaces, the so-called principle of **uniform boundedness.** Although this principle was discovered by Lebesgue in 1908 in investigations on Fourier series, it was isolated as a general principle by Banach and Steinhaus. It states that if $\{f_n\}$ is a sequence of continuous linear functionals and, for every $x \in X$, the sequence $\{|f_n(x)|\}$ is bounded, then the sequence $\{\|f_n\|\}$ of norms is bounded.

We first observe that this is not true for every normed vector space X. For this, let X be the space of polynomials $x = x(t) = \sum_{n=0}^{\infty} a_n t^n$ where $a_n = 0$ for all $n > N(x)$, with $\|x\| = \max\,[|a_n|, n = 1, 2, \ldots]$. Let $f_n(x) = \sum_{k=0}^{n-1} a_k$. The functions f_n are continuous linear functionals on X. Moreover, for every $x = a_0 + a_1 t + \ldots + a_m t^m$, it is clear that for every n, $|f_n(x)| \leq (m + 1)\,\|x\|$, so that $\{|f_n(x)|\}$ is bounded. On the other hand, $\|f_n\| \geq n$, since for $x = 1 + t + \ldots + t^{n-1}$, $\|x\| = 1$ and $|f_n(x)| = n$. Hence $\{\|f_n\|\}$ is unbounded. We now state

THEOREM 1. If X is a Banach space, and $\{f_n(x)\}$ a sequence of continuous linear functionals on X such that $\{|f_n(x)|\}$ is bounded for every $x \in X$, then the sequence $\{\|f_n\|\}$ is bounded.

Proof. For every m, let $X_m \subset X$ be the set of all x such that $|f_n(x)| \leq m$ for all n. Then X_m, as the intersection of closed sets, is closed. Moreover, $X = \bigcup_{m=1}^{\infty} X_m$, so that, since X, as a complete metric space, is of the second category, some X_m contains a sphere. There is thus an $x_0 \in X$ and a $k > 0$ such that $\|x - x_0\| \leq k$ implies $|f_n(x)| \leq m$ for every n. Since $\{|f_n(x_0)|\}$ is bounded, there is a p such that $\|x\| \leq k$ implies $|f_n(x)| \leq m + p$ for every n, so that $\|f_n\| \leq \dfrac{m + p}{k}$ for every n.

Theorem 1 also holds for linear transformations from a Banach space X to a normed vector space Y. The proof is almost identical and will be omitted.

THEOREM 1′. If X is a Banach space, Y a normed vector space, and $\{T_n\}$ a sequence of continuous linear transformations on X into Y such that for every $x \in X$ the sequence $\{\|T_n(x)\|\}$ is bounded, then the sequence $\{\|T_n\|\}$ of norms is bounded.

An interesting application of the principle of uniform boundedness is to a proof that there are continuous functions of period 2π whose Fourier series diverge at a given point. For this purpose, let X be the Banach space of such functions with $\|x\| = \max |x(t)|$. The nth partial sum of the Fourier series of $x = x(t)$ at $t = 0$ may be written in the form

$$f_n(x) = \frac{1}{2\pi} \int_0^{2\pi} x(t) D_n(t)\, dt,$$

where

$$D_n(t) = \frac{\sin (n + \frac{1}{2})t}{\sin \frac{1}{2}t}.$$

Now, for each $x \in X$,

$$|f_n(x)| \le \frac{1}{2\pi} \int_0^{2\pi} |D_n(t)|\, dt \cdot \|x\|.$$

Evidently, f_n is linear and it is bounded with

$$\|f_n\| \le \frac{1}{2\pi} \int_0^{2\pi} |D_n(t)|\, dt.$$

Let y be the function which is $+1$ for those t for which $D_n(t) \ge 0$ and which is -1 for those t for which $D_n(t) < 0$. For every $\epsilon > 0$, y may be modified to a continuous x of norm 1 so that

$$\left| f_n(x) - \frac{1}{2\pi} \int_0^{2\pi} |D_n(t)|\, dt \right| = \frac{1}{2\pi} \left| \int_0^{2\pi} (x(t) - y(t)) D_n(t)\, dt \right| < \epsilon.$$

It follows that

$$\|f_n\| = \frac{1}{2\pi} \int_0^{2\pi} |D_n(t)|\, dt.$$

We show that the sequence

$$\left\{ \int_0^{2\pi} \left| \frac{\sin (n + \frac{1}{2})t}{\sin \frac{1}{2}t} \right| dt \right\}$$

is unbounded. Since $\sin u \le u$, $0 \le u \le \pi$, this follows if we show that

$$\left\{ \int_0^{\pi} \left| \frac{\sin (2n + 1)u}{u} \right| du \right\}$$

is unbounded. Now,

$$\int_0^\pi \left|\frac{\sin(2n+1)u}{u}\right| du = \sum_{k=0}^{2n} \int_{\frac{k\pi}{2n+1}}^{\frac{(k+1)\pi}{2n+1}} \left|\frac{\sin(2n+1)u}{u}\right| du$$

$$\geq \sum_{k=0}^{2n} \frac{2n+1}{(k+1)\pi} \int_{\frac{k\pi}{2n+1}}^{\frac{(k+1)\pi}{2n+1}} |\sin(2n+1)u|\, du$$

$$= \sum_{k=0}^{2n} \frac{1}{(k+1)\pi} \int_{k\pi}^{(k+1)\pi} |\sin x|\, dx = \frac{2}{\pi} \sum_{k=0}^{2n} \frac{1}{k+1}.$$

But the series $\sum_{n=1}^{\infty} \frac{1}{n}$ diverges, so that the assertion is proved.

Thus, the sequence $\{\|f_n\|\}$ is unbounded. By theorem 1, there is thus an $x \in X$ whose Fourier series diverges at $t = 0$.

The uniform boundedness theorem has a similar application to interpolation theory. For any function x defined on the interval [0, 1] and any partition

$$0 \leq t_1 < t_2 < \ldots < t_n \leq 1$$

of [0, 1], there is a polynomial of degree $n - 1$ which interpolates to x at the given points, i.e., takes the values $x(t_i)$ at t_i, $i = 1, \ldots, n$. This is called the Lagrange interpolation polynomial and is given by

$$L_{t_1 \ldots t_n}(x, t) = \sum_{i=1}^{n} x(t_i) \frac{(t - t_1) \ldots (t - t_{i-1})(t - t_{i+1}) \ldots (t - t_n)}{(t_i - t_1) \ldots (t_i - t_{i-1})(t_i - t_{i+1}) \ldots (t_i - t_n)}.$$

Consider any sequence of partitions

$$(*) \qquad \begin{array}{l} t_{1,1} \\ t_{2,1}, t_{2,2} \\ \cdots \\ t_{n,1}, t_{n,2}, \ldots, t_{n,n} \\ \cdots \end{array}$$

and the corresponding sequence of polynomials

$$(**) \qquad L_{t_{n,1} t_{n,2} \ldots t_{n,n}}(x, t).$$

THEOREM 2. For every choice of (*) there is an $x \in C[0, 1]$ for which (**) does not converge uniformly to x.

Proof. Denote by $l^{(i)}_{t_1 \ldots t_n}(t)$ the fraction appearing in the ith term of $L_{t_1 \ldots t_n}(x, t)$. Let

$$\lambda_{t_1 \ldots t_n}(t) = \sum_{i=1}^{n} |l^{(i)}_{t_1 \ldots t_n}(t)|.$$

The mapping $f \to L_{t_1 \ldots t_n}(f)$ is a linear transformation of $C[0, 1]$ into itself and

$$|L_{t_1 \ldots t_n}(x, t)| \leq \|f\| \cdot \lambda_{t_1 \ldots t_n}(t), \text{ for every } t \in [0, 1].$$

It follows that this transformation is bounded and that its norm does not exceed

$$\Lambda_{t_1 \ldots t_n} = \sup\,[\lambda_{t_1 \ldots t_n}(t) : t \in [0, 1]].$$

We show that its norm is precisely $\Lambda_{t_1 \ldots t_n}$. Since $\lambda_{t_1 \ldots t_n}$ is continuous on $[0, 1]$, there is $\xi \in [0, 1]$ such that

$$\Lambda_{t_1 \ldots t_n} = \lambda_{t_1 \ldots t_n}(\xi).$$

Let

$$x(t_i) = \operatorname{sgn} l^{(i)}_{t_1 \ldots t_n}(\xi), \quad i = 1, \ldots, n,$$

and let x be linear on the intervals $[t_i, t_{i+1}]$ and constant on $[0, t_1]$ and $[t_n, 1]$. Then $x \in C\,[0, 1]$, $\|x\| = 1$, and

$$L_{t_1 \ldots t_n}(x, \xi) = \sum_{i=1}^{n} \operatorname{sgn} l^{(i)}_{t_1 \ldots t_n}(\xi) \cdot l^{(i)}_{t_1 \ldots t_n}(\xi)$$

$$= \lambda_{t_1 \ldots t_n}(\xi) = \Lambda_{t_1 \ldots t_n}.$$

This proves the norm of the transformation is $\Lambda_{t_1 \ldots t_n}$. Corresponding to $(*)$ we have the sequence of norms $\Lambda_{t_{1,1}}, \Lambda_{t_{2,1} t_{2,2}}, \ldots$.

It is a fact, which in this case is not easy to prove (for a proof see Natanson [1], p. 370) that

$$\Lambda_{t_{k,1} \ldots t_{k,k}} > \frac{\log k}{8\pi}.$$

Thus our sequence of operators is unbounded and the theorem follows by theorem 1′.

Closely related to the principle of uniform boundedness is the so-called principle of the **condensation of singularities.** In its classical form, this says that under certain conditions, if for every point t_0 in a set S, there is a function x, in a given set X of functions, which has a singularity at $t = t_0$, then there is an $x \in X$ which has the singularity at every $t \in S$. The easiest case is that for which S is countable. This principle has been made precise in terms of linear operators on a Banach space for this case.

THEOREM 3. Let X be a Banach space and Y a normed vector space. For every $m = 1, 2, \ldots$, let $\{T_n(x, m)\}$ be a sequence of continuous linear operators on X with values in Y such that for every m there is an $x_m \in X$ for which $\overline{\lim_n}\, \|T_n(x_m, m)\| = +\infty$. Then there is an $x \in X$ for which $\overline{\lim_n}\, \|T_n(x, m)\| = +\infty$, $m = 1, 2, \ldots$.

Proof. We show that for every m, n and k, the set X_{mnk} of points $x \in X$ for which $\|T_\nu(x, m)\| > k$, for some $\nu > n$, is a dense open set. It is open since $X_{mnk} = \bigcup [Y_{m\nu k} : \nu > n]$, where $Y_{m\nu k}$ is the set for which $\|T_\nu(x, m)\| > k$, and every $Y_{m\nu k}$ is evidently open. That X_{mnk} is dense in X follows since for every real a and $x \in X$,

$$\|T_\nu(x + ax_m, m)\| \geq |a| \, \|T_\nu(x_m, m)\| - \|T_\nu(x, m)\|$$

so that either $\|T_\nu(x, m)\| > k$ or else $\|T_\nu(x + ax_m, m)\| > k$ for some $\nu > n$. Hence X_{mnk}, as a dense open set, is residual (i.e., its complement is of the first category in X). But the set

$$\bigcap [X_{mnk} : m, n, k = 1, 2, \ldots]$$

consists of those $x \in X$ for which $\overline{\lim_n} \|T_n(x, m)\| = +\infty$ for every m. Since this set is also residual and X is of the second category, the theorem is proved.

We prove a variant of this theorem, but first obtain the

LEMMA 1. If X is a Banach space and Y is a subspace of X which is a Borel set, then Y is either of the first category in X or is identical with X.

Proof. Suppose Y is of the second category in X. By a theorem of Banach and Blumberg (see Chapter 1, exercise 10.2), there is an open sphere S in X such that Y meets every subsphere of S in a set of the second category. Since Y is a subspace of X, the sphere S may be taken with center 0 and radius, say, $r > 0$. Moreover, since Y is a Borel set, it has the property of Baire (see Kuratowski [1]), which means here that it is residual in S. Since $0 \in S$, it follows that $x \in S$ implies $x \in x + S$, so that $S \cap (x + S)$ is non-empty. Let T be a sphere, center x, such that $T \subset S \cap (x + S)$. But then both Y and $x + Y$ are residual relative to T, so that $Y \cap (x + Y)$ is non-empty. There is thus a $y \in Y$ such that $y - x \in Y$. Since Y is a vector space, this implies $x \in Y$. Hence $S \subset Y$ and, since Y is a vector space,

$$X = \bigcup_{n=1}^{\infty} nS \subset Y.$$

We now prove

THEOREM 3'. Let X be a Banach space and Y a normed vector space. For every $m = 1, 2, \ldots$, let $\{T_n(x, m)\}$ be a sequence of continuous linear operators on X with values in Y such that, for every m, there is an $x_m \in X$ for which $\{T_n(x_m, m)\}$ does not converge. Then there is an $x \in X$ for which $\{T_n(x, m)\}$ does not converge for any $m = 1, 2, \ldots$.

Proof. The set X_m of points $x \in X$ for which $\{T_n(x, m)\}$ converges is easily seen to be a proper subspace and a Borel set, so that, by lemma 1, and the hypothesis, it is of the first category. But then there is an $x \in X$, $x \notin \bigcup_{m=1}^{\infty} X_m$ and the theorem is proved.

The above two principles are usually used together. Thus, by the principle of uniform boundedness, we showed that there are continuous functions whose Fourier series diverge at a single point. By then applying the principle of the condensation of singularities, we can show the existence of continuous functions whose Fourier expansions diverge on a dense (countable) set in $[0, 2\pi]$. The details of this and other similar applications are now easy to supply. It is worth mentioning that it is now not hard to show that there is a continuous function whose Fourier series diverges except at the points of a subset of $[0, 2\pi]$ of the first category.

2.14 Lemma of F. Riesz, Applications

If X is a finite dimensional normed vector space and Y is a proper subspace of X, then there is a point x_1 on the unit sphere of X, i.e., $\|x_1\| = 1$, such that the distance of x_1 from Y is one;

$$\inf [\|x - x_1\| : x \in Y] = 1.$$

This observation is intuitively evident and is, indeed, easy to prove. However, if the assumption of finite dimensionality is dropped, the conclusion no longer holds in general, as the following example shows.

Let X be the subspace of $C[0, 1]$ which consists of those functions x for which $x(0) = 0$. Let Y be the subspace of X whose elements satisfy

$$\int_0^1 x(t)\, dt = 0.$$

Clearly, Y is a closed subspace of X. Now, if $x_1 \in X$ and $\|x_1\| = \sup [|x_1(t)| : t \in [0, 1]] = 1$, then necessarily

$$\left| \int_0^1 x_1(t)\, dt \right| < 1,$$

because x_1 is continuous and $x_1(0) = 0$. We shall show that this conclusion is contradicted if it is assumed that $\|x - x_1\| \geq 1$ for all $x \in Y$. In fact for any $y \in X$, with $\|y\| = 1$, we would have

$$\left| \int_0^1 y(t)\, dt \right| \leq \left| \int_0^1 x_1(t)\, dt \right|.$$

This is obvious for $y \in Y$, and if $y \notin Y$, then we can form

$$x_1 - \frac{\int_0^1 x_1(t)\, dt}{\int_0^1 y(t)\, dt}\, y,$$

which is in Y, and so

$$1 \leq \left\| x_1 - \left(x_1 - \frac{\int_0^1 x_1(t)\, dt}{\int_0^1 y(t)\, dt}\, y \right) \right\| = \frac{\left| \int_0^1 x_1(t)\, dt \right|}{\left| \int_0^1 y(t)\, dt \right|}.$$

The contradiction follows from the observation that y may be chosen in such a way that $\left|\int_0^1 y(t)\,dt\right|$ is arbitrarily close to 1.

A result which holds in a general normed vector space is the

PROPOSITION 1. If Y is a closed proper subspace of X and $\epsilon > 0$, there is an x_ϵ on the unit sphere such that

$$\inf\,[\|x_0 - x_\epsilon\| : x_0 \in Y] > 1 - \epsilon.$$

Proof. Let $x \in X \sim Y$ and put

$$d = \inf\,[\|x - x_0\| : x_0 \in Y].$$

Since Y is closed, $d > 0$. Choose $y_0 \in Y$ so that

$$d \leq \|y_0 - x\| < d(1 + \epsilon)$$

and put

$$x_\epsilon = -\frac{y_0 - x}{\|y_0 - x\|}.$$

It is clear that $\|x_\epsilon\| = 1$. If $x_0 \in Y$,

$$\begin{aligned}\|x_0 - x_\epsilon\| &= \left\| x_0 + \frac{y_0 - x}{\|y_0 - x\|} \right\| \\ &= \frac{1}{\|y_0 - x\|}\,\|(y_0 + \|y_0 - x\|\,x_0) - x\| \\ &\geq \frac{d}{\|y_0 - x\|} > \frac{d}{d(1 + \epsilon)} = 1 - \frac{\epsilon}{1 + \epsilon} > 1 - \epsilon.\end{aligned}$$

This proposition has some interesting applications. If X is a normed vector space with the property that closed bounded sets are compact, then X is finite dimensional. For, if $x_1 \in X$ and x_1 does not generate X, there is x_2 with $\|x_2\| = 1$, so that $\|x_2 - x_1\| > \frac{1}{2}$. If x_1 and x_2 do not generate X, there is x_3 with $\|x_3\| = 1$, so that $\|x_3 - x_1\| > \frac{1}{2}$ and $\|x_3 - x_2\| > \frac{1}{2}$. This process must terminate, since otherwise we would obtain a sequence of points on the unit sphere which has no convergent subsequence, contrary to the assumed compactness of this set.

It is not hard to show that, conversely, the closed bounded sets in any finite dimensional normed vector space are compact. We may thus state the

PROPOSITION 2. A normed vector space is finite dimensional if and only if the closed bounded sets are compact.

A second application occurs in connection with the Baire classes of bounded functions, B_α, $\alpha < \omega_1$. We have noted that each B_α is a closed subspace of the space M of bounded functions and, hence, is a Banach space (see Sec. 1.7).

We consider now the class

$$B = \bigcup [B_\alpha : \alpha < \omega_0]$$

of all bounded functions of finite Baire type. Since each B_α, $\alpha < \omega_0$, is a closed proper subspace of B, hence is nowhere dense in B, it follows that B is of the first category in itself and so is not complete. Since the norm in M is the uniform norm, this conclusion may be stated as the

PROPOSITION 3. There is a uniformly convergent sequence of functions of finite Baire type whose limit is not of finite Baire type.

The limit is in $B_{\omega_0} \sim B$. The question arises whether every function in $B_{\omega_0} \sim B$ is the uniform limit of some sequence from B. The negative answer is obtained as follows.

Let $0 = a_0 < a_1 < \ldots < a_n < \ldots < 1$ and $\lim_{n\to\infty} a_n = 1$. Consider the Baire classes on $[a_{n-1}, a_n)$, say $B_\alpha^{(n)}$, $\alpha < \omega_1$. Each $B_\alpha^{(n)}$ is a closed proper subspace of the succeeding class $B_{\alpha+1}^{(n)}$. In particular, therefore, there exists a function f_n on $[a_{n-1}, a_n)$, of norm 1 in $B_n^{(n)}$, whose distance from $B_{n-1}^{(n)}$ exceeds $\frac{1}{2}$.

Let g_m, $m = 1, 2, 3, \ldots$, be defined on $[0, 1]$ by $g_m = f_1$ on $[a_0, a_1)$, $g_m = f_2$ on $[a_1, a_2), \ldots, g_m = f_m$ on $[a_{m-1}, a_m)$, and $g_m = 0$ on $[a_m, 1]$. The sequence $\{g_m\}$ converges on $[0, 1]$ to a function g.

If $f \in B$ then, for some $n, f \in B_n$, so that

$$\|g - f\| \geq \sup [|g(t) - f(t)| : t \in [a_n, a_{n+1})] > \tfrac{1}{2}.$$

Thus g is not the uniform limit of any sequence in B.

That $g \in B_{\omega_0}$ is a consequence of the fact that each g_m is of finite Baire type, a fact which is not hard to verify.

2.15 Application to Compact Transformations

We give some applications of proposition 1 of the last section to the theory of compact (completely continuous) operators.

Let X and Y be Banach spaces. A linear transformation T on X into Y is called **compact** if every bounded set in X is taken by T into a set whose closure is compact.

We give an example of a compact operator. Let $X = C[a, b]$. Let K be a continuous real function on $[a, b] \times [a, b]$. Define T by means of

$$Tf(x) = \int_a^b K(x, y) f(y)\, dy.$$

Then T is a linear operator on X into X.

Let $\epsilon > 0$. There is a $\delta > 0$ such that $|x_1 - x_2| < \delta$, $y \in [a, b]$ implies $|K(x_1, y) - K(x_2, y)| < \dfrac{\epsilon}{b - a}$, so that

$$\int_a^b |K(x_1, y) - K(x_2, y)|\, dy < \epsilon.$$

Thus

$$|Tf(x_1) - Tf(x_2)| \leq \int_a^b |K(x_1, y) - K(x_2, y)|\, |f(y)|\, dy$$
$$\leq \epsilon \, \|f\|$$

whenever $|x_1 - x_2| < \delta$. This means that if B is any bounded set in $C[0, 1]$, the set $TB = [Tf : f \in B]$ is equicontinuous. It then follows that T is a compact operator.

We have already discussed the Banach space $B(X, Y)$ of bounded transformations of a Banach space X into a Banach space Y. It is clear that a compact transformation is bounded. We consider the subset $C(X, Y) \subset B(X, Y)$ of compact transformations. $C(X, Y)$ is easily seen to be a vector space. We show that it is a closed subspace of $B(X, Y)$, whereby we have

PROPOSITION 1. $C(X, Y)$ is a Banach space.

Proof. Let T_n, $n = 1, 2, \ldots$ be a sequence of compact transformations which converges in $B(X, Y)$ to a transformation T, let $\{x_n\}$ be such that $x_n \in X$, $\|x_n\| \leq M$, $n = 1, 2, \ldots$. Then $\{x_n\}$ has a subsequence $\{x_{1,n}\}$ such that $\{T_1 x_{1,n}\}$ is Cauchy; $\{x_{1,n}\}$ has a subsequence $\{x_{2,n}\}$ such that $\{T_2 x_{2,n}\}$ is Cauchy. Continuing in this way and taking the diagonal sequence $y_n = x_{n,n}$, we obtain a subsequence $\{y_n\}$ of $\{x_n\}$ such that $\{T_m y_n\}$, $m = 1, 2, \ldots$ are all Cauchy sequences. Let $\epsilon > 0$. There is an m such that $\|T - T_m\| < \epsilon/4M$. Let N be such that $n, k > N$ implies $\|T_m y_n - T_m y_k\| < \epsilon/2$. Then

$$\|Ty_n - Ty_k\| \leq \|T - T_m\| \, \|y_n\| + \|T - T_m\| \, \|y_k\|$$
$$+ \|T_m y_n - T_m y_k\| < \epsilon,$$

so that $\{Ty_n\}$ is Cauchy.

We next define the adjoint of a transformation of X into Y. Let X' be the dual of X and Y' the dual of Y. Then X' and Y' are Banach spaces. Let $T \in B(X, Y)$. The **adjoint** T^* of T is a transformation $T^* \in B(Y', X')$ defined as follows: T^*y', $y' \in Y'$, is the element of X' which for each $x \in X$ has the value

$$(T^*y')(x) = y'(Tx).$$

It is easy to see that $T^* \in B(Y', X')$ and $\|T^*\| = \|T\|$. Moreover, if $T \in C(X, Y)$, it can be shown that $T^* \in C(Y', X')$ but we do not give the proof.

We now give two applications to compact operators of the proposition 1, Sec. 2.14.

PROPOSITION 2. If $T \in C(X, X)$, X a Banach space, the vector space of solutions of $x - Tx = 0$ is finite dimensional.

Proof. Suppose, on the contrary, that the equation has an infinite set x_n, $n = 1, 2, \ldots$ of linearly independent solutions. For every n, let X_n be the vector space spanned by $x_1, \ldots, x_n$.

For every $n > 1$, there is, by proposition 1, Sec. 14, a $y_n \in X_n$ such that $\|y_n\| = 1$ and $\|y_n - x\| > \frac{1}{2}$ for every $x \in X_{n-1}$. Now, $y_n = Ty_n$, $n = 1, 2, \ldots$, implies that $\{y_n\}$ has a Cauchy subsequence, contradicting the assertion that $\|y_m - y_n\| > \frac{1}{2}$, $m \neq n$.

We next consider the equation

$$x - \lambda Tx = 0.$$

A real number λ_0 is called a **characteristic number** of the equation if $x - \lambda_0 Tx = 0$ has a non-zero solution.

PROPOSITION 3. If $T \in C(X, X)$, the set of characteristic numbers of

$$x - \lambda Tx = 0$$

is an isolated set.

Proof. Let $\{\lambda_n\}$ be a sequence of distinct characteristic numbers and let $x_n \neq 0$, $x_n = \lambda_n Tx_n$, $n = 1, 2, \ldots$.

We show the x_n, $n = 1, 2, \ldots$, are linearly independent. Suppose

$$x_{n+1} = \sum_{i=1}^{n} a_i x_i.$$

Then

$$x_{n+1} = \lambda_{n+1} Tx_{n+1} = \sum_{i=1}^{n} \lambda_{n+1} a_i Tx_i$$

so that

$$x_{n+1} = \sum_{i=1}^{n} a_i \frac{\lambda_{n+1}}{\lambda_i} x_i.$$

It follows that

$$\sum_{i=1}^{n} a_i \left(1 - \frac{\lambda_{n+1}}{\lambda_i}\right) x_i = 0,$$

so that, since each $\lambda_{n+1}/\lambda_i \neq 1$, the $x_1, \ldots, x_n$ are linearly dependent. Consequently, x_1 is linearly dependent, which is a contradiction.

Now, for every $n = 1, 2, \ldots$, let X_n be the vector space spanned by $x_1, \ldots, x_n$. We note that $X_{n+1} \supset X_n$, where the inclusion is proper and X_n is closed. There is $y_{n+1} \in X_{n+1}$, $\|y_{n+1}\| = 1$ and $\|y_{n+1} - x\| > \frac{1}{2}$ for every $x \in X_n$, $n = 1, 2, \ldots$. Suppose $\{\lambda_n\}$ converges. Then since $\{\lambda_n y_n\}$ is bounded, $\{T(\lambda_n y_n)\}$ has a convergent subsequence.

But

$$\|T(\lambda_n y_n) - T(\lambda_m y_m)\| = \|y_n - (y_n - T(\lambda_n y_n) + T(\lambda_m y_m))\|$$

and $y_n \in X_n$ implies

$$y_n - \lambda_n T(y_n) = \sum_{i=1}^{n} a_i x_i - \sum_{i=1}^{n} \lambda_n \frac{a_i}{\lambda_i} x_i$$

$$= \sum_{i=1}^{n-1} a_i \left(1 - \frac{\lambda_n}{\lambda_i}\right) x_i \in X_{n-1},$$

and $T(\lambda_m y_m) \in X_m$. Hence, if $m < n$, $y_n - T(\lambda_n y_n) + T(\lambda_m y_m) \in X_{n-1}$, so that $\|T(\lambda_n y_n) - T(\lambda_m y_m)\| > \frac{1}{2}$. This contradicts the assertion that $\{T(\lambda_n y_n)\}$ has a convergent subsequence.

2.16 Applications, Weak Convergence, Summability Methods, Approximate Integration

The point-wise convergence of functionals is a type of convergence which is important in many applications. Various such notions of "weak" convergence and "weak" topology arise naturally in vector space theory and have extensive application. In this section we shall give a characterization of weak convergence for bounded linear functionals on a Banach space X and indicate some applications.

If X is a normed vector space and $x'_n \in X'$, $n = 1, 2, 3, \ldots$, we say that x'_n **converges weakly** to $x' \in X'$ if for each $x \in X$, $\lim_{n\to\infty} x'_n(x) = x'(x)$.

We shall call a subset of X **total** if its closed linear span is X.

PROPOSITION 1. If X is a Banach space, then $\{x'_n\}$ converges weakly to x' in X' if and only if the sequence $\{\|x'_n\|\}$ of norms is bounded and $\lim_n x'_n(x) = x'(x)$ for every x in some total subset of X.

Proof. (Sufficiency) If $x \in X$ and $\epsilon > 0$, we may choose a finite linear combination x_0 of elements in the total subset so that

$$\|x - x_0\| < \frac{\epsilon}{2M}$$

where M is an upper bound for the numbers $\|x'_n\| + \|x'\|$, $n = 1, 2, 3, \ldots$. For n sufficiently large

$$|x'_n(x_0) - x'(x_0)| < \frac{\epsilon}{2}.$$

For such indices n, we then obtain

$$\begin{aligned}|x'_n(x) - x'(x)| &\le |x'_n(x) - x'_n(x_0)| + |x'_n(x_0) - x'(x_0)| \\ &\quad + |x'(x_0) - x'(x)| \\ &\le \|x'_n\|\ \|x - x_0\| + \frac{\epsilon}{2} + \|x'\|\ \|x - x_0\| < \epsilon.\end{aligned}$$

(Necessity) If x'_n converges weakly to x', then in particular $[x'_n(x) : n = 1, 2, \ldots]$ is a bounded set of numbers for each $x \in X$. By the principle of uniform boundedness the sequence $\{\|x'_n\|\}$ is bounded.

We should remark that in the necessity, as we stated it, we have assumed the continuity of the limiting linear functional x'. But this is in fact a consequence of the existence of $\lim_n x'_n(x)$ for each $x \in X$.

Applying the above result we may obtain a basic theorem in the theory of divergent sequences. A matrix method of summability is characterized by a doubly infinite matrix

$$\mathcal{A} = \begin{pmatrix} a_{11} & a_{12} & \ldots & a_{1n} & \ldots \\ a_{21} & a_{22} & \ldots & a_{2n} & \ldots \\ & & \ldots & & \\ a_{m1} & a_{m2} & \ldots & a_{mn} & \ldots \\ & & \ldots & & \end{pmatrix}.$$

A sequence $x = (x_1, x_2, \ldots)$ is said to have $\mathcal{A}$-limit $= A(x)$ if for each $i = 1, 2, 3, \ldots,$ the sum

$$A_i(x) = \sum_{k=1}^{\infty} a_{ik} x_k$$

is convergent and $\lim_i A_i(x) = A(x)$.

The method $\mathcal{A}$ is said to be **regular** if for every convergent sequence x, the $\mathcal{A}$-limit exists and equals the ordinary limit of the sequence x. We shall prove a characterization, due to O. Toeplitz, of regular methods.

THEOREM 1. $\mathcal{A}$ is regular if and only if

(i) $\sum_{k=1}^{\infty} |a_{ik}| \le M,\ i = 1, 2, \ldots,$ for some M,

(ii) $\lim_i a_{ik} = 0,\ k = 1, 2, \ldots$

(iii) $\lim_i \sum_{k=1}^{\infty} a_{ik} = 1.$

Proof. The class of all convergent sequences is a Banach space c under $\|x\| = \sup |x_i|$. Suppose $\mathcal{A}$ is regular. Then the series $\sum_{k=1}^{\infty} a_{ik} x_k$ converges for each $x \in c$, $i = 1, 2, \ldots$. The partial sums $\sum_{k=1}^{n} a_{ik}x_k$ are clearly bounded linear functionals on c and the last remark amounts to the weak convergence in c' of the partial sums. Thus the limits A_i are in c' and have uniformly bounded norms, $\|A_i\| \leq M$ for all $i = 1, 2, \ldots$, by proposition 1. Taking $x_k = \operatorname{sgn} a_{ik}$ for $k = 1, \ldots, N$ and $x_k = 0$, $k > N$ gives, for any N,

$$\sum_{k=1}^{N} |a_{ik}| \leq \|A_i\|, \quad i = 1, 2, \ldots.$$

Thus $\sum_{k=1}^{\infty} |a_{ik}| \leq M$, 1, 2, . . . , proving the necessity of (i).

The choice $x = e_k = (0, 0, \ldots, 0, 1, 0, \ldots)$, with the 1 in the kth place, yields $A_i(x) = a_{ik}$ and the necessity of (ii) follows. Similarly, by choosing $x = e_0 = (1, 1, \ldots)$ we find $A_i(x) = \sum_{k=1}^{\infty} a_{ik}$ converges to 1, which completes the proof of necessity.

Suppose now that $\mathcal{A}$ has the three properties and, for a fixed i, consider the partial sum functionals

$$B_N(x) = \sum_{k=1}^{N} a_{ik}x_k, \quad N = 1, 2, 3, \ldots.$$

These functionals belong to c' and their norms are bounded by $\sum_{k=1}^{\infty} |a_{ik}|$. Whenever x is one of the points $e_0, e_1, e_2, \ldots$, the functionals converge as N increases. But these points form a total set in c, so that the B_N converge weakly and the limit functional $A_i(x) = \sum_{k=1}^{\infty} a_{ik}x_k$ is in c'.

The norms of the functionals A_i, $i = 1, 2, \ldots$, are bounded by M, by (i), and (ii), and (iii) express the fact that $\lim_i A_i(x)$ exists and equals the limit of the sequence x in case x is one of the points $e_0, e_1, e_2, \ldots$. Then this condition holds for any $x \in c$, which proves that $\mathcal{A}$ is regular.

We make a slight extension of the above result; $\mathcal{A}$ will be called **almost regular** if, for every convergent sequence $x = (x_1, x_2, \ldots)$, there is an N (which depends on x) such that, for every $m > N$, $\sum_{n=1}^{\infty} a_{mn}x_n$ converges and $\lim_{m\to\infty} \sum_{n=1}^{\infty} a_{mn}x_n = \lim_{n\to\infty} x_n$. We show that $\mathcal{A}$ is almost regular if and only if there is a fixed m_0 such that

(i) $\sum_{m=1}^{\infty} |a_{mn}| < M$ for fixed M and every $m > m_0$,

(ii) $\lim_m a_{mn} = 0$ for every n,

(iii) $\lim_m \sum_{n=1}^{\infty} a_{mn} = 1$.

This has the surprising implication that the above N may be taken to be independent of x. The proof depends upon the following lemma, whose proof we omit.

LEMMA 1. If $\{s_n^{(m)}\}$ is a sequence of sequences such that $\sum_{n=1}^{\infty} |s_n^{(m)}| = \infty$, $m = 1, 2, \ldots$, there is a sequence $\{t_n\}$, converging to zero, such that all the series $\sum_{n=1}^{\infty} s_n^{(m)} t_n$, $m = 1, 2, \ldots$, diverge.

The lemma implies there is an N such that $\sum_{n=1}^{\infty} |a_{mn}| < \infty$ for every $m > N$. Otherwise, there would be an $\{x_n\}$ converging to zero with infinitely many of the series $\sum_{n=1}^{\infty} a_{mn}x_n$ diverging, contrary to hypothesis.

For each $m > N$, let $E_m \subset c$ be the set of convergent sequences $x = (x_1, x_2, \ldots)$ for which $\sum_{n=1}^{\infty} a_{mn}x_n$ converges. Since $\sum_{n=1}^{\infty} |a_{mn}| < \infty$, it follows that E_m is a closed subspace of c. Let $S_m = \bigcap_{i=m}^{\infty} E_i$, $m = N + 1, N + 2 \ldots$. The S_m are closed subspaces of c such that

$$S_{N+1} \subset S_{N+2} \subset \ldots \subset S_m \subset \ldots .$$

Now, a closed subspace T of a Banach space X is either nowhere dense in X or else $T = X$. By hypothesis, $c = \bigcup_{m=N+1}^{\infty} S_m$. Since c is of the second category, there is an m_0 such that $c = S_{m_0}$.

It follows that $\mathcal{A}$, with its first $m_0 - 1$ rows removed, is regular and the result follows by theorem 1.

We make one further remark regarding matrix summability. Using theorem 1, it is not hard to show that if $\mathcal{A}$ is regular then there is a bounded sequence $x = \{x_n\}$ whose $\mathcal{A}$-limit does not exist. Let m be the vector space of bounded sequences with $\|x\| = \sup_n |x_n|$. It is easy to show that the subspace of m of sequences which are transformed by $\mathcal{A}$ into convergent sequences is a closed vector space. It is nowhere dense. Hence, there is no countable set of regular $\mathcal{A}_n$, $n = 1, 2, \ldots$ such that every bounded sequence is taken into a convergent one by at least one of the $\mathcal{A}_n$.

We give a second application of proposition 1.

Let $\alpha(t)$, $t \in [0, 1]$, be a non-decreasing function. The problem is to approximate $\int_0^1 x(t)\, d\alpha(t)$ by a sequence of finite sums of the form

$$S_n(x) = \sum_{k=1}^{k_n} a_{nk}x(t_{nk}), \quad n = 1, 2, \ldots,$$

where, for each n, $t_{nk} \in [0, 1]$, $k = 1, \ldots, k_n$, and $x \in C[0, 1]$.

A selection of points t_{nk} and coefficients a_{nk} determines a prospective method of approximate integration. If the choice yields the result

$$\lim_n S_n(x) = \int_0^1 x \, d\alpha$$

for every $x \in C[0, 1]$, then we have in fact a method of approximate integration. In other words, we require weak convergence on $C[0, 1]$ of the functionals S_n to the functional I given by

$$I(x) = \int_0^1 x \, d\alpha.$$

We shall say in this case that the method is convergent.

Various methods which suggest themselves, e.g., by integrating certain interpolating or approximating polynomials, have the property that $S_n(x) = I(x)$ whenever x is a polynomial of degree at most n, $n = 1, 2, 3, \ldots$. Such a condition serves to relate the numbers a_{nk} and the points t_{nk} by the system of equations

$$\int_0^1 t^i \, da = \sum_{k=1}^{k_n} a_{nk} t_{nk}^i, \quad i = 0, 1, \ldots, n.$$

Under the above condition, it makes sense to characterize the convergent methods by conditions on the coefficients a_{nk}. We have the

PROPOSITION 2. The method $\{S_n\}$ satisfying $S_n(x) = I(x)$, whenever x is a polynomial whose degree does not exceed n, is convergent if and only if the sums

$$\sum_{k=1}^{k_n} |a_{nk}|, \quad n = 1, 2, 3, \ldots,$$

are bounded.

Proof. We note that

$$|S_n(x)| \leq \|x\| \sum_{k=1}^{k_n} |a_{nk}|.$$

It follows that S_n is in the dual of $C[0, 1]$, and $\|S_n\|$ is no greater than the above sum. A proper choice of x shows that $\|S_n\|$ equals the sum.

The result follows from proposition 1 since the polynomials are dense in $C[0, 1]$.

COROLLARY. If each $a_{nk} \geq 0$, the method is convergent.

Proof. Since $S_n(x) = I(x)$ for all n, if x is the function $x(t) = 1$, $t \in [0, 1]$, we have

$$S_n(1) = \sum_{k=1}^{k_n} a_{nk} = I(1), \quad n = 1, 2, 3, \ldots,$$

so that the sums in proposition 2 are bounded.

2.17 Second Dual Space

We discuss the second dual of a normed vector space briefly.

Let X be a normed vector space, and let X' be the dual of X. For every $x_0 \neq 0$ in X, consider the functional $\tilde{x}_0$ on X' defined by

$$\tilde{x}_0(x') = x'(x_0).$$

Then $\tilde{x}_0$ is a linear functional, since

$$\begin{aligned}\tilde{x}_0(ax' + by') &= (ax' + by')(x_0) = ax'(x_0) + by'(x_0)\\ &= a\tilde{x}_0(x') + b\tilde{x}_0(y').\end{aligned}$$

Moreover, $\tilde{x}_0$ is bounded, since $|\tilde{x}_0(x')| = |x'(x_0)| \leq \|x'\|\,\|x_0\|$. The norm of $\tilde{x}_0$ satisfies

$$\|\tilde{x}_0\| \leq \|x_0\|.$$

By proposition 2, Sec. 2.12, there is an $x' \in X'$ such that $x'(x_0) = \|x_0\|$ and $\|x'\| = 1$. Hence

$$|\tilde{x}_0(x')| = |x'(x_0)| = \|x_0\|.$$

It follows that

$$\|\tilde{x}_0\| = \|x_0\|.$$

The mapping

$$x \to \tilde{x}$$

of X into X'' is thus norm preserving.

X is said to be **reflexive** if $X = X''$, where equality is understood in the sense of isomorphism under the canonical mapping. Thus reflexivity occurs if and only if the above mapping carries X onto X''.

We shall see that there exist reflexive spaces—necessarily Banach spaces, since X'' is complete. Any incomplete space is of course non-reflexive, but there are also non-reflexive Banach spaces.

Examples of these possibilities will arise as we discuss cases of the following general question: Given a particular normed vector space, find a concrete representation of its dual (i.e., a specific Banach space which can be mapped isomorphically onto the dual space in question).

Before doing this, let us prove another useful consequence of the Hahn-Banach theorem.

PROPOSITION 1. If Y is a proper subspace of a normed vector space X and x_0 is at positive distance $d > 0$ from Y, i.e.,

$$d = \inf\,[\|x - x_0\| : x \in Y] > 0,$$

then there is an $x' \in X'$ such that $\|x'\| = 1$, x' vanishes at all points of Y and $x'(x_0) = d$. The hypotheses are fulfilled, in particular, if Y is closed and $x_0 \notin Y$.

Proof. Define f on $(Y \cup \{x_0\})$ by

$$f(x + ax_0) = ad, \; x \in Y.$$

Clearly, f is a linear functional on $(Y \cup \{x_0\})$ and the result will follow if we show that f is bounded with $\|f\| = 1$.

First, f is bounded with $\|f\| \leq 1$, i.e., $|f(x + ax_0)| = |a|\, d \leq \|x + ax_0\|$ for all $x \in Y$ and all a. This is obvious for $a = 0$ and, for $a \neq 0$, it follows from

$$\left\|\frac{x}{a} + x_0\right\| = \left\|-\frac{x}{a} - x_0\right\| \geq \inf\,[\|z - x_0\| : z \in Y] = d,$$

since $-x/a \in Y$.

Finally, $\|f\| \geq 1$. For if $\{x_n\}$ is a sequence in Y such that $\lim_{n\to\infty} \|x_n - x_0\| = d$, then

$$\frac{|f(x_n - x_0)|}{\|x_n - x_0\|} = \frac{d}{\|x_n - x_0\|},$$

which converges to 1 as n goes to infinity. Thus

$$\|f\| = \sup\left[\frac{|f(z)|}{\|z\|} : z \in (Y \cup \{x_0\})\right] \geq 1.$$

A consequence of this result is that a Banach space X is either reflexive or its successive second dual spaces X'', X'''', . . . are all distinct. Namely, one can show that $X \neq X''$ implies $X' \neq X'''$ and then use an induction argument.

Another consequence of proposition 1 is the

PROPOSITION 2. If X' is separable, then X is separable.

(*Remark.* We shall see that the dual of l_1 is l_∞ and so, in view of the present proposition, l_1 is not reflexive. This example also shows that the converse of the proposition is false.)

Proof. Let $\{x_n'\}$ be a countable dense subset of the unit sphere in $X' : \|x_n'\| = 1$, $n = 1, 2, \ldots$.

For each n, choose an $x_n \in X$ with $\|x_n\| = 1$ and $|x_n'(x_n)| \geq \frac{3}{4}$. This is possible since $\|x_n'\| = 1$.

Let Y be the closed subspace of X generated by $x_1, x_2, \ldots$. We shall show that $Y = X$, so that X is separable, since then the linear combinations, with rational coefficients, of the x_n form a dense set.

Suppose $Y \neq X$. Then with $x_0 \in X \sim Y$, there is an x' on the unit sphere of X' which vanishes on Y but not at x_0. Thus for any $n = 1, 2, \ldots$

$$\begin{aligned} \tfrac{3}{4} \leq |x_n'(x_n)| &= |x_n'(x_n) - x'(x_n)| \\ &= |(x_n' - x')(x_n)| \leq \|x_n' - x'\|, \end{aligned}$$

which is impossible, since $\{x_n'\}$ is dense on the unit sphere in X'.

2.18 Dual of l_p

We now describe the duals of the l_p spaces, $1 \leq p < \infty$.

THEOREM 1. If $1 \leq p < \infty$, and q is conjugate to p, then $(l_p)' = l_q$.

(*Remark*. It follows that if $1 < p < \infty$, then l_p is reflexive, while l_1 and l_∞ are non-reflexive.)

Proof. Let

$$e_n = (0, \ldots, 0,1, 0, \ldots)$$

be the vector with 1 in the nth place and 0 in every other place. Then for all p and n, $e_n \in l_p$ and $\|e_n\|_p = 1$.

If $f \in (l_p)'$, for $1 \leq p < \infty$, we define

$$a_n = f(e_n), \quad n = 1, 2, \ldots,$$

and consider the mapping

$$\tau : f \rightarrow \{a_n\}.$$

We shall show that

(a) $f \in (l_p)'$ implies $\tau f = \{a_n\} \in l_q$ and

$$\|\tau f\|_q = \|\{a_n\}\|_q \leq \|f\|,$$

and

(b) τ maps $(l_p)'$ onto l_q and

$$\|\tau f\|_q \geq \|f\|.$$

Since τ is obviously linear, this will establish the isomorphism of the normed vector spaces $(l_p)'$ and l_q.

For (a), observe that

$$|a_n| = |f(e_n)| \leq \|f\| \, \|e_n\| = \|f\|, \; n = 1, 2, \ldots.$$

Hence $\{a_n\}$ is a bounded sequence and

$$\|\{a_n\}\|_\infty \leq \|f\|,$$

proving (a) in the case $p = 1$.

For the case $1 < p < \infty$, we must estimate the partial sums $\sum_{n=1}^{k} |a_n|^q$, $k = 1, 2, \ldots$. We do this by finding elements $x_k \in l_p$ such that $f(x_k)$ is the kth partial sum.

Let $x_k = (\xi_1, \xi_2, \ldots, \xi_k, 0, \ldots)$, where

$$\xi_i = |a_i|^{q-1} \operatorname{sgn} a_i, \; i = 1, 2, \ldots, k.$$

(sgn z means $z/|z|$ if $z \neq 0$ and 0 if $z = 0$.)

Clearly, $x_k \in l_p$ for all k, and we find

$$f(x_k) = f\left(\sum_{n=1}^{k} \xi_n e_n\right) = \sum_{n=1}^{k} \xi_n f(e_n) = \sum_{n=1}^{k} \xi_n a_n$$

$$= \sum_{n=1}^{k} |a_n|^{q-1} a_n \operatorname{sgn} a_n = \sum_{n=1}^{k} |a_n|^q.$$

Hence,

$$\sum_{n=1}^{k} |a_n|^q = |f(x_k)| \le \|f\| \; \|x_k\|_p.$$

But,

$$\|x_k\|_p = \left(\sum_{n=1}^{k} |a_n|^{(q-1)p}\right)^{1/p} = \left(\sum_{n=1}^{k} |a_n|^q\right)^{1/p}.$$

We obtain, by dividing by $\|x_k\|_p$, if this is not 0, and by inspection otherwise, the inequality

$$\left(\sum_{n=1}^{k} |a_n|^q\right)^{1/q} \le \|f\|, \quad k = 1, 2, \ldots,$$

from which (a) follows.

For (b), choose any $\{a_n\} \in l_q$ and define a linear functional f as follows (to obtain $\tau f = \{a_n\}$). Let

$$f(e_n) = a_n, \; n = 1, 2, \ldots$$

(this is permissible since the e_n's form a linearly independent set) and extend by linearity to all of l_p. Thus, if $x = (x_1, x_2, \ldots) \in l_p$, we put

$$f(x) = \sum_{n=1}^{\infty} a_n x_n.$$

The convergence of this series results from Hölder's inequality

$$|f(x)| \le \sum_{n=1}^{\infty} |a_n x_n| \le \|\{a_n\}\|_q \|x\|_p,$$

which also shows that f is a bounded linear functional on l_p and that

$$\|f\| \le \|\{a_n\}\|_q.$$

This completes the proof.

2.19 Dual of $C[a, b]$, Riesz Representation Theorem

The dual space of the space $C[a, b]$ can be identified with a subspace of the space $BV[a, b]$ of functions of bounded variation on $[a, b]$ (see example (x), Sec. 1.1).

The major step in the identification is the Riesz representation theorem, which asserts that any bounded linear functional F on $C[a, b]$ is a Riemann-Stieltjes integral with respect to a function $g \in BV[a, b]$:

$$F(f) = \int_a^b f \, dg, \qquad f \in C[a, b].$$

We begin by recalling the definition of the Riemann-Stieltjes integral. Let $f \in C[a, b]$ and $g \in BV[a, b]$. For every partition $a = x_0 < x_1 < \ldots < x_{n-1} < x_n = b$ of $[a, b]$ and $x_{i-1} \leq t_i \leq x_i$, $i = 1, 2, \ldots, n$, consider the sum

$$\sum_{i=1}^{n} f(t_i)[g(x_i) - g(x_{i-1})].$$

There is a real number I such that for every $\epsilon > 0$ there is a $\delta > 0$ such that $\max\,[|x_i - x_{i-1}| : i = 1, \ldots, n] < \delta$ implies

$$\left| I - \sum_{i=1}^{n} f(t_i)[g(x_i) - g(x_{i-1})] \right| < \epsilon.$$

The number I is the Riemann-Stieltjes integral of f with respect to g and is denoted $\int_a^b f\,dg$.

It is easy to show that the integral depends linearly on $f \in C[a, b]$ and on $g \in BV[a, b]$, so that

$$\int_a^b (\alpha f_1 + \beta f_2)\,dg = \alpha \int_a^b f_1\,dg + \beta \int_a^b f_2\,dg,$$

and

$$\int_a^b f\,d(\alpha g_1 + \beta g_2) = \alpha \int_a^b f\,dg_1 + \beta \int_a^b f\,dg_2.$$

Moreover, the inequality

$$\left| \int_a^b f\,dg \right| \leq \max\,[|f(x)| : x \in [a, b]] \cdot V(g)$$

holds.

In view of the definitions of the norms in C and BV (we use $\|g\| = |g(a)| + V(g)$ as norm in BV), we see that any $g \in BV$ determines a bounded linear functional F on C by the formula,

$$F(f) = \int_a^b f\,dg, \quad f \in C$$

and that

$$\|F\| \leq V(g).$$

Clearly, any two choices of g which differ by an additive constant define the same functional F. Thus, in order to specify a unique g determining F it is necessary to impose a normalizing condition, e.g., $g(a) = 0$. There remains a further ambiguity in g. It is easily seen that F may be the same for distinct functions in BV which vanish at a. A normalizing condition which removes this ambiguity consists of requiring that g be continuous from the right. The justification for this normalization lies in the fact that corresponding to any $g \in BV[a, b]$ there is exactly one $\bar{g} \in BV[a, b]$, continuous from the right, and such that

$$\int_a^b f\,d\bar{g} = \int_a^b f\,dg, \quad f \in C.$$

Moreover, $V(\bar{g}) \leq V(g)$.

The functions satisfying these normalizing conditions form a subspace BVN of $BV[a, b]$, and our observations so far show that the mapping

$$g \to \int_a^b f\,dg$$

is a one-one, linear, norm decreasing transformation of BVN into C'. We now show that the mapping is onto C' and preserves norms.

THEOREM 1. If $F \in C'[a, b]$, there exists a $g \in BVN[a, b]$ such that

$$F(f) = \int_a^b f\,dg, \quad f \in C[a, b]$$

and

$$\|g\| = V(g) = \|F\|.$$

Proof. We consider $C[a, b]$ as a subspace of the space $M[a, b]$ of bounded functions on $[a, b]$ and apply the Hahn-Banach theorem to extend F to all of $M[a, b]$, preserving the norm $\|F\|$. We may now apply F to the functions

$$\varphi_x(t) = \begin{cases} 1, & a \leq t < x \\ 0, & x \leq t \leq b \end{cases}, \quad a < x \leq b,$$

$$\varphi_a(t) = 0, \; a \leq t \leq b,$$

and define

$$g(x) = F(\varphi_x), \; a \leq x \leq b.$$

To compute $V(g)$, we choose

$$a = x_0 < x_1 < \ldots < x_{n-1} < x_n = b,$$

and evaluate

$$\begin{aligned}\sum_{i=1}^n |g(x_i) - g(x_{i-1})| &= \sum_{i=1}^n \{g(x_i) - g(x_{i-1})\} \operatorname{sgn}(g(x_i) - g(x_{i-1})) \\ &= F\left(\sum_{i=1}^n \{\varphi_{x_i} - \varphi_{x_{i-1}}\} \operatorname{sgn}(g(x_i) - g(x_{i-1}))\right) \\ &\leq \|F\| \cdot \left\| \sum_{i=1}^n \{\varphi_{x_i} - \varphi_{x_{i-1}}\} \operatorname{sgn}(g(x_i) - g(x_{i-1})) \right\| \\ &= \|F\|.\end{aligned}$$

We see that $g \in BV$ and $V(g) \leq \|F\|$. Since $g(a) = F(\varphi_a) = F(0) = 0$, this means $\|g\| \leq \|F\|$. The reverse inequality has been shown earlier. Hence $\|g\| = \|F\|$.

Now, if $f \in C$, we define a sequence of functions $f_n \in M$ such that $\{f_n\}$ converges to f in the space M and the values $F(f_n)$ are approximating sums for the integral $\int_a^b f\,dg$. Namely, writing s for $b - a$,

$$f_n(t) = \sum_{k=1}^n f\left(\frac{ks}{n}\right) [\varphi_{ks/n}(t) - \varphi_{(k-1)s/n}(t)].$$

Notice that f_n is a step function having the value $f(ks/n)$ in the interval $(k-1)s/n < t \leq ks/n$, $k = 1, 2, \ldots, n$.

It is evident that f_n converges uniformly to f.

Now

$$F(f_n) = \sum_{k=1}^{n} f\left(\frac{ks}{n}\right)[F(\varphi_{ks/n}) - F(\varphi_{(k-1)s/n})]$$

$$= \sum_{k=1}^{n} f\left(\frac{ks}{n}\right)\left[g\left(\frac{ks}{n}\right) - g\left(\frac{(k-1)s}{n}\right)\right],$$

so that $\{F(f_n)\}$ converges to $\int_a^b f\,dg$.

On the other hand, $F(f_n)$ converges to $F(f)$ by the continuity of F. Hence

$$F(f) = \int_a^b f\,dg.$$

By a remark above, g may be modified so as to belong to BVN and the proof is complete.

2.20 Open Mapping and Closed Graph Theorems

Differential operators provide important examples of linear transformations. Let us consider the simplest differential operator d/dt as an illustration of some general facts.

If X and Y are both taken to be the space $C[0, 1]$, then the subspace of X consisting of all continuously differentiable functions constitutes a natural domain on which d/dt defines a linear transformation from X into Y. Call this subspace C_1 and put

$$Tx = \frac{dx}{dt} \quad \text{for } x \in C_1.$$

The linear transformation T is not bounded since, for example, the functions

$$x_n(t) = t^n, \; n = 1, 2, \ldots$$

all have norm 1 in X while their images

$$(T x_n)(t) = nt^{n-1}$$

have norms arbitrarily large.

In general, linear transformations defined by differential operators are not bounded. They do, however, enjoy a related property.

A linear transformation T from a Banach space X into a Banach space Y, with domain $D(T) \subset X$, is said to be **closed** if $x_n \in D(T)$, $\lim_n x_n = x$ and $\lim_n Tx_n = y$ imply $x \in D(T)$ and $y = Tx$.

PROPOSITION 1. If $D(T)$ is closed in X and T is bounded, then T is closed.

We verify that the first order differential operator discussed above yields a closed transformation.

Let $x_n \in C_1$, $\lim_n x_n = x$ and $\lim_n \dfrac{dx_n}{dt} = y$, where the convergence is uniform. Then

$$\int_0^t y(s)\,ds = \int_0^t \lim_n \frac{dx_n}{dt}\,ds = \lim_n \int_0^t \frac{dx_n}{dt}\,ds$$

$$= \lim_n\,[x_n(t) - x_n(0)] = x(t) - x(0).$$

Thus

$$x(t) = x(0) + \int_0^t y(s)\,ds,$$

and it follows that $x \in C_1$ and $dx/dt = y$.

The property of being closed has a simple geometric interpretation in terms of the concept of the graph of the transformation.

The product $X \times Y$ of normed vector spaces X and Y is a normed vector space under the norm

$$\|(x, y)\| = \|x\| + \|y\|,$$

and $X \times Y$ is a Banach space if X and Y are such.

The **graph** of a mapping $T : X \rightarrow Y$ is the set of points $(x, Tx) \in X \times Y$, with $x \in D(T)$. It is designated as $G(T)$.

If T is linear, $G(T)$ is a vector subspace of $X \times Y$. The following proposition is immediate.

PROPOSITION 2. *T is a closed linear transformation if and only if $G(T)$ is a closed vector subspace of $X \times Y$.*

THEOREM 1 (Closed Graph Theorem). *If X and Y are Banach spaces and T is a linear transformation from X into Y, then $D(T)$ closed and $G(T)$ closed imply T is bounded.*

The proof is accomplished by an application of the

THEOREM 2 (Open Mapping Theorem). *If X and Y are Banach spaces and T is a bounded linear transformation which maps X onto all of Y, then T is an open mapping (i.e., the image under T of an open set in X is open in Y).*

Thus, if the mapping is also one-one, then T^{-1} is continuous.

Proof of Closed Graph Theorem. Since $D(T)$ is closed, it is itself a Banach space. Again $G(T)$, as a closed subspace of the Banach space $X \times Y$, is a Banach space. The mapping

$$(x, Tx) \rightarrow x$$

is a linear transformation of $G(T)$ onto $D(T)$. It is bounded because

$$\|x\| \leq \|x\| + \|Tx\| = \|(x, Tx)\|.$$

Hence, by the open mapping theorem, the inverse mapping

$$x \to (x, Tx)$$

is bounded. Thus, there is an M for which

$$\|Tx\| \leq \|Tx\| + \|x\| = \|(x, Tx)\| \leq M \|x\|,$$

for all $x \in D(T)$. This proves the boundedness of T.

Proof of Open Mapping Theorem. We first prove the

LEMMA 1. If X and Y are complete metric spaces and F is a continuous mapping of X onto Y such that for every $r > 0$, there is a $k > 0$ such that for every $x \in X$ the closure of the image of the sphere, $\sigma(x, r)$, in X, of center x and radius r, contains the sphere $\sigma(F(x), k)$ in Y, then for every $\rho > r$ the image of $\sigma(x, \rho)$ contains $\sigma(F(x), k)$.

Proof. Let $\sum_{n=1}^{\infty} r_n$ be a convergent series of positive terms such that $r_1 = r$ and $\sum_{n=1}^{\infty} r_n = \rho$. For every n, there is a $k_n > 0$ such that the closure of the image of $\sigma(x, r_n)$ contains $\sigma(F(x), k_n)$. The k_n may be chosen so that $\lim_n k_n = 0$. For $x \in X$, let $y \in \sigma(F(x), k)$. We define a sequence $\{x_n\}$ such that, for every n, $x_n \in \sigma(x_{n-1}, r_n)$ and $F(x_n) \in \sigma(y, k_{n+1})$. Having chosen $x_1 = x$, $x_2, \ldots, x_{n-1}$ so that $F(x_i) \in \sigma(y, k_{i+1})$, $i = 1, \ldots, n-1$, we have $y \in \sigma(F(x_{n-1}), k_n)$. Since the closure of the image of $\sigma(x_{n-1}, r_n)$ contains $\sigma(F(x_{n-1}), k_n)$, there is an $x_n \in \sigma(x_{n-1}, r_n)$ such that $F(x_n) \in \sigma(y, k_{n+1})$. The sequence $\{x_n\}$ is evidently a Cauchy sequence and so it converges to a point x_0 which is contained in $\sigma(x, \rho)$. But $\{F(x_n)\}$ converges to $F(x_0)$ since F is continuous. Moreover, since $\lim_n k_n = 0$, it follows that $\lim_n F(x_n) = y$. Hence $y = F(x_0)$.

To prove the theorem, we need only prove that the closure of the image of every sphere in X, with center 0, contains a sphere in Y with center 0. For this, let $r > 0$ and let $S \subset X$ be the sphere of center 0 and radius r. For every $x \in X$, the sequence $\left\{\frac{1}{n} x\right\}$ converges to 0, so there is an n with $x \in nS$. Then $Y = \bigcup_{n=1}^{\infty} nT(S)$. Since Y is of the second category, there is an n for which the closure of $nT(S)$ contains a sphere. It follows that the closure of $T(S)$ contains a sphere, whose center y may be taken in $T(S)$. Let this sphere, $\sigma(y, k)$, have radius k. Now, let $y = T(x)$ and $y' = T(x')$ for $y' \in \sigma(y, k)$. Then $T(x - x') = y - y'$. As y' varies over the closure of $T(S)$, $y - y'$ covers the sphere $\sigma(0, k)$. But a set of these y', dense in $\sigma(y, k)$, are of the form $y' = T(x')$, $x' \in S$. Since x and x' are in S, $x - x' \in \sigma(0, 2r)$. Hence the closure of the image of $\sigma(0, 2r)$ contains $\sigma(0, k)$.

2.21 Application, Projections

In this section, we shall discuss the decomposition of a Banach space as the sum of two subspaces and the related question of the existence of operators which are projections onto subspaces. These ideas are essential to the analysis of the structure of a linear transformation (see, e.g., Lorch [1]). Although we shall not discuss spectral theory here, these results are of geometric interest, and also provide an illustration of the use of the closed graph theorem.

A vector space X is the **direct sum** of two of its subspaces M and N,

$$X = M \dotplus N,$$

if every $x \in X$ has a unique decomposition

$$x = y + z$$

with $y \in M$ and $z \in N$.

Observe that if $X = M \dotplus N$ then $M \cap N = (0)$.

If $X = M \dotplus N$ and we define a mapping P of X into itself by $Px = y$, where for every $x \in X$, $x = y + z$ with $y \in M$, $z \in N$, then P is a linear transformation and

(i) $Px = x$ if and only if $x \in M$,
(ii) $Px = 0$ if and only if $x \in N$,
(iii) $P^2 = P$, i.e., $P(Px) = Px$ for all $x \in X$.

It is reasonable to think of P as the operation of projecting X onto M along N. Conversely, if we have a linear transformation P which is idempotent, $P^2 = P$, then X is the direct sum of the subspaces

$$M = \text{range of } P = [x \in X : Px = x]$$

and

$$N = \text{nullspace of } P = [x \in X : Px = 0].$$

For, if $x \in X$, then

$$x = Px + (I - P)x,$$

where I denotes the identity operator in X, given by $Ix = x$. It is clear that $Px \in M$ and $(I - P)x \in N$, so that this equation gives a decomposition of x according to the subspaces M and N. But the decomposition is unique, since $x = y + z, y \in M, z \in N$ gives $Px = Py = y$ and $(I - P)x = x - Px = x - y = z$.

Thus in the purely algebraic setting, we see the connection between direct sum decompositions of a space and idempotent linear operators in that space.

If X is a Banach space, we are particularly interested in closed subspaces, on the one hand, and in continuous linear operators on the other. The term **projection** will be used to refer to a bounded, linear, idempotent operator in a Banach space X, while by a **direct summand** of X we shall mean a closed linear subspace M such that for some closed linear subspace N, $X = M \dotplus N$.

If $M \neq (0)$, $M \neq X$, is a direct summand, there are infinitely many closed subspaces N for which $X = M \dotplus N$. This is clear if X is the plane and M is a line, for then N may be chosen as any line, distinct from M, through the origin. The general case is left as an exercise.

If M is a finite dimensional subspace of X, then M is a direct summand of X. We shall see in Chapter 4 that if X is a Hilbert space (e.g., l_2), then any closed subspace is a direct summand. However, it has been shown in a variety of particular cases that a Banach space may contain closed subspaces which are not direct summands.

It is easy to see that a linear operator T is continuous only if its null space $[x : Tx = 0]$ is closed. Application of this to P and $I - P$ gives, in view of earlier remarks,

PROPOSITION 1. If P is a projection in X, and M and N are its range and null space, respectively, then M and N are closed. Since $X = M \dotplus N$, M and N are direct summands.

The converse follows from the closed graph theorem.

PROPOSITION 2. If M is a direct summand of X and N is a closed subspace with $X = M \dotplus N$, then P is a projection, where $Px = y$ for $x = y + z$, $y \in M$, $z \in N$.

Proof. Only the continuity of P requires proof. Since P has domain X and range M, both Banach spaces, we need only show that P is closed. Suppose therefore that $\lim_n x_n = x$ and $\lim_n Px_n = y$. We have

$$x_n = Px_n + (I - P)x_n$$

converges to x, and since Px_n converges to y, $(I - P)x_n$ must converge and its limit z must satisfy $x = y + z$. But $y \in M$ and $z \in N$ since these subspaces are closed. Hence $y = Px$.

2.22 Application, Schauder Expansion

We now give an application of the open mapping theorem. Let X be an infinite dimensional separable Banach space. By a **Schauder basis** in X, we mean a set $x_n \in X$, $\|x_n\| = 1$, $n = 1, 2, \ldots$, such that for every $x \in X$ there is a unique series $\sum_{n=1}^{\infty} a_n x_n$ such that

$$\lim_n \left\| x - \sum_{i=1}^{n} a_i x_i \right\| = 0.$$

For example, a Schauder basis is given, in l_2, by $x_n = (0, \ldots, 0, 1, 0, \ldots)$, $n = 1, 2, \ldots$, where 1 appears in the nth place.

We may write the Schauder expansion of x as $\sum_{n=1}^{\infty} a_n(x)x_n$, and this emphasizes that the coefficients $a_n(x)$ are functionals on X. That they are linear is obvious. We show that they are bounded.

We introduce the vector space Y of all sequences $(a_1, \ldots, a_n, \ldots)$ for which $\sum_{n=1}^{\infty} a_n x_n$ converges in X. The norm

$$\|y\| = \sup_n \left\| \sum_{i=1}^{n} a_i x_i \right\|$$

converts Y into a normed vector space. We show it is a Banach space. Let $y_m = \{a_n^{(m)}\}$, $m = 1, 2, \ldots$, be a Cauchy sequence in Y. Let $\epsilon > 0$. There is an N such that $m, k > N$ implies

$$\|y_m - y_k\| = \sup_n \left\| \sum_{i=1}^{n} (a_i^{(m)} - a_i^{(k)}) x_i \right\| < \epsilon.$$

But this implies $|a_n^{(m)} - a_n^{(k)}| < 2\epsilon$ for every n. Hence, for every n, $\lim_m a_n^{(m)} = a_n$ exists.

It remains only to show that $y = (a_1, \ldots, a_n, \ldots) \in Y$ and $\lim_n y_n = y$ in Y. But these are computations of a standard type used in other completeness arguments in the text and the proof is left to the reader.

Consider the transformation $T : Y \to X$ for which $y = (a_1, a_2, \ldots) \in Y$ is taken into $T(y) = \sum_{n=1}^{\infty} a_n x_n \in X$. T is evidently linear, one-one, and onto. It is bounded since $\sup_n \left\| \sum_{i=1}^{n} a_i x_i \right\| \geq \lim_n \left\| \sum_{i=1}^{n} a_i x_i \right\|$.

By the open mapping theorem, the inverse T^{-1} of T is bounded. Now,

$$\begin{aligned} |a_n(x)| = |a_n| = \|a_n x_n\| &= \left\| \sum_{i=1}^{n} a_i x_i - \sum_{i=1}^{n-1} a_i x_i \right\| \\ &\leq 2 \sup_n \left\| \sum_{i=1}^{n} a_i x_i \right\| = 2\,\|y\| = 2\,\|T^{-1}x\| \leq 2\,\|T^{-1}\|\,\|x\|. \end{aligned}$$

This proves the boundedness of the $a_n(x)$.

2.23 A Theorem on Operators in $C[0, 1]$

In the final section, we show that certain sequences $\{F_n\}$ of operators on $C[0, 1]$ into itself have the property that $\lim_n F_n f = f$ for every $f \in C$.

An operator F is called **positive** if $f \geq 0$ implies $Ff \geq 0$. We show that the sequence $\{F_n\}$ has the desired property if each F_n is linear, positive and if $\lim_n F_n f = f$ for the three functions $1, x, x^2$, where $1(t) = 1$, $x(t) = t$, and $x^2(t) = t^2$ for each $t \in [0, 1]$.

We need the

LEMMA 1. If $\{F_n\}$ are linear, $\lim_n F_n 1 = 1$, $\lim_n F_n x = x$, and $\lim_n F_n x^2 = x^2$, then if x_u^2 is the function $x_u^2(t) = (t - u)^2$, it follows that $\lim_n F_n x_u^2(u) = 0$ uniformly in $u \in [0, 1]$.

The proof is left to the reader.

We now prove the

THEOREM 1. If $\{F_n\}$ is a sequence of linear positive operators on $C[0, 1]$ into itself such that $\lim_n F_n f = f$, in the C norm for $f = 1, x, x^2$, then $\lim_n F_n f = f$ for every $f \in C$.

Proof. Let $f \in C$ and $\epsilon > 0$. There is a $\delta > 0$ such that $|t - u| < \delta$ implies $|f(t) - f(u)| < \epsilon$, and an M such that $\|f\| = \max |f(t)| = M$. For every $t \in [0, 1]$,

$$-\epsilon - \frac{2M}{\delta^2}(t - u)^2 \leq f(t) - f(u) \leq \epsilon + \frac{2M}{\delta^2}(t - u)^2.$$

For, if $|t - u| < \delta$, we have

$$-\epsilon \leq f(t) - f(u) \leq \epsilon$$

and if $|t - u| \geq \delta$ we have

$$-\frac{2M}{\delta^2}(t - u)^2 \leq -2M \leq f(t) - f(u) \leq 2M \leq \frac{2M}{\delta^2}(t - u)^2.$$

Consider u fixed. Then, for every n,

$$-\epsilon F_n 1 - \frac{2M}{\delta^2} F_n x_u^2 \leq F_n f - f(u) F_n 1 \leq \epsilon F_n 1 + \frac{2M}{\delta^2} F_n x_u^2.$$

By lemma 1 and the fact that $\lim_n F_n 1 = 1$, we have for sufficiently large n,

$$-2\epsilon \cdot 1 \leq F_n f - f \leq 2\epsilon \cdot 1.$$

It follows, since $\epsilon > 0$ is arbitrary, that $\lim_n F_n f = f$ in C.

EXERCISES

1.1 Show that $0 \cdot x = 0$ and $(-1)x = -x$ in a vector space.

1.2 Show that $\Sigma[X_\alpha : \alpha \in A]$ is a vector space and $\Sigma[X_\alpha : \alpha \in A] = \Pi[X_\alpha : \alpha \in A]$ if and only if A is finite.

1.3 Show that the functions on $[0, 1]$ for which

$$\int_0^1 (e^{|f(x)|} - 1)\, dx < \infty$$

do not form a vector space.

2.1 Show that (S) is the set of all finite linear combinations of points in S.

3.1 Give the details of the proof that X/Y is a vector space.

4.1 Show that a set S is linearly independent if and only if for each finite subset $x_1, x_2, \ldots, x_n \in S$, the only linear combination

$$a_1x_1 + \ldots + a_nx_n$$

giving 0 is the one in which $a_1 = a_2 = \ldots = a_n = 0$.

4.2 Prove proposition 3.

4.3 Prove proposition 4.

4.4 Show that if Y is a subspace of X then X has a Hamel basis S which has a subset T which is a Hamel basis of Y.

5.1 Prove that for every vector space X with algebraic dual X^*, $\dim X^* \geq \dim X$.

5.2 Suppose $\dim X = \aleph_0$. What is $\dim X^*$?

5.3 Show that if X is infinite dimensional, then X^{**} is not isomorphic to X.

5.4 Show that if X is finite dimensional, the natural isomorphism of X into X^{**} is onto.

6.1 If S is convex, show that $x + S$ and aS are convex.

7.1 Show that the group G/H associated with the preordered group G is an ordered group.

7.2 Consider the abelian group of all real functions on $(-\infty, \infty)$, with f in the positive cone if $f(x) > 0$ for all x in some interval. Discuss the resulting preordered group and its associated ordered group.

7.3 Give details of the proof that every archimedean totally ordered group is isomorphic to a subgroup of the reals.

7.4 Give an example of a non-archimedean totally ordered group; of such a field.

8.1 Show directly that the set S of the example following theorem 1 does not have an interior point.

8.2 Complete the proof of theorem 2.

8.3 Prove theorem 3 using theorem 2.

9.1 Show that theorem 3, section 8, follows from the theorem of section 9.

10.1 Show that $p'(x) = \sup_j \varliminf \frac{1}{k} \sum_{i=1}^{k} x_{n_i+j}$.

10.2 Prove that for $x \in m$

$$\varliminf x_n \leq p'(x) \leq p(x) \leq \varlimsup x_n.$$

10.3 Show that the functional $\varlimsup x_n$ is subadditive on m.

10.4 If $\{x_n\}$ converges, show that $\left\{\frac{1}{n}(x_1 + \ldots + x_n)\right\}$ converges and has the same limit.

10.5 In the proof of proposition 5 give details of the proof that

$$p(x) = \inf_{y \in c} [p(y + x) - l(y)].$$

10.6 Show that the functional q does not satisfy proposition 3, by giving an example of a bounded sequence for which $q(x) < p(x)$.

11.1 Show that the norm is continuous on the metric space associated with a normed vector space.

11.2 Define the operations and the norm in the completion of a normed vector space so that it is a normed vector space.

11.3 Consider the class of all polynomials with

$$\|p\| = |a_0| + |a_1| + \ldots + |a_n|$$

if

$$p(t) = a_0 + a_1 t + \ldots + a_n t^n.$$

Show that this is a normed vector space but is not a Banach space. Discuss the completion.

11.4 Compare $P[0, 1]$ with the above norm and the norm of example (*b*).

11.5 Consider any norm in the plane. Show that the unit sphere $[x : \|x\| \leq 1]$ is a closed convex set which is symmetric with respect to the origin and contains the origin as an interior point. Conversely, if S is any set with these properties, then a norm is obtained by specifying

$$\|x\| = 1 \text{ if } x \text{ is on the boundary of } S,$$

$$\|x\| = a \text{ if } \frac{1}{a} x \text{ is on the boundary of } S,$$

$$\|0\| = 0.$$

Sketch the unit spheres for the norms $\|x\|_p$ for several values of p, $1 \leq p \leq \infty$.

11.6 Show that the sequences $x = \{x_n\}$ with $\sum_{n=1}^{\infty} |x_{n+1} - x_n| < \infty$ form a Banach space with norm

$$\|x\| = |x_1| + \sum_{n=1}^{\infty} |x_{n+1} - x_n|.$$

11.7 Show that c is a nowhere dense subspace of m.

11.8 Show that the space of all real sequences can be normed so as to become a Banach space.

11.9 Show that no norm in the space of all sequences has the property

$$s_n \geq t_n,\ n = 1, 2, \ldots, \text{ implies } \|s\| \geq \|t\|.$$

11.10 Show that

$$\|T\| = \sup\,[\|Tx\| : \|x\| \leq 1]$$
$$= \sup\,[\|Tx\| : \|x\| = 1].$$

11.11 A series $\sum_{n=1}^{\infty} x_n$ in a normed vector space is summable to x if

$$\lim_n \left\| x - \sum_{i=1}^{n} x_i \right\| = 0.$$

It is absolutely summable if $\sum_{n=1}^{\infty} \|x_n\| < \infty$. Show that a normed vector space is complete if and only if every absolutely summable series is summable.

12.1 If X is a Banach space and Y is a closed subspace, show that X/Y is a Banach space with the norm

$$\|\xi\| = \min\,[\|x\| : x \in \xi].$$

12.2 If X is a normed vector space and $Y \subset X$ is a Banach space, then if X/Y is a Banach space, show that X itself is a Banach space.

12.3 Let X be a complex vector space. Show that if F_1 is a linear functional on X considered as a real space, then

$$F(x) = F_1(x) - iF_1(ix)$$

is a linear functional on the complex space X.

12.4 Show directly that no countable set is a Hamel basis for $C[0, 1]$.

12.5 Consider the sequences of successive powers of a fixed $\xi \in (0, 1)$, $(\xi, \xi^2, \ldots, \xi^n, \ldots)$. Show that all such vectors form a linearly independent set in m.

12.6 Show that a finite dimensional subspace of a normed vector space is closed.

12.7 Show that a Banach space is finite dimensional if and only if every subspace is closed.

12.8 Show that a closed proper subspace of a normed vector space is nowhere dense.

12.9 Give a simple proof that a Hamel basis of an infinite dimensional Banach space is uncountable (without actually showing that its cardinality is at least c as in proposition 5).

13.1 Prove theorem 1′.

13.2 Prove that there exists a continuous function whose Fourier series diverges at all points except for those belonging to a set of the first category.

14.1 Prove that if X is finite dimensional and Y is a proper subspace, then there is a point on the unit sphere of X at distance 1 from Y.

14.2 Show that all norms on an n-dimensional vector space yield homeomorphic metric spaces.

14.3 Show that an infinite dimensional vector space has infinitely many norms, no two of which yield homeomorphic metric spaces.

14.4 Show that if f is of finite Baire type on $[0, a)$ and g is of finite Baire type on $[a, 1]$, then h is of finite Baire type on $[0, 1]$, where

$$h(x) = \begin{cases} f(x), & 0 \le x < a \\ g(x), & a \le x \le 1. \end{cases}$$

14.5 Give an example of a Banach space X and a decreasing sequence of non-empty bounded closed convex sets in X whose intersection is empty.

15.1 Show that $T^* \in B(Y', X')$ and $\|T^*\| = \|T\|$.

15.2 A Banach space X is called uniformly convex if $\|x_n\| = \|y_n\| = 1$, $n = 1, 2, \ldots$ and $\lim_n \|x_n + y_n\| = 2$ implies $\lim_n \|x_n - y_n\| = 0$. If X is uniformly convex and $K \subset X$ is a closed convex set, show that $\|x\|$ has a uniquely assumed minimum on K.

15.3 Show that a finite dimensional Banach space need not be uniformly convex.

16.1 Show that if X is a Banach space, $x_n' \in X'$, $n = 1, 2, \ldots$ and $\lim_n x_n'(x) = y(x)$ for each $x \in X$, then $y \in X'$.

16.2 Extend the result of the previous exercise to sequences of bounded linear transformations of one Banach space into another.

16.3 Prove that if $\{s_n^{(m)}\}$ is a sequence of sequences such that $\sum_{n=1}^{\infty} |s_n^{(m)}| = \infty$, $m = 1, 2, \ldots$, there is a $\{t_n\}$, converging to zero, such that all the series $\sum_{n=1}^{\infty} s_n^{(m)} t_n$, $m = 1, 2, \ldots$, diverge.

16.4 Prove that the spaces E_m, defined after lemma 1, are closed.

16.5 Apply theorem 1 to show that if $\mathcal{A}$ is regular, then there is a bounded sequence $x = \{x_n\}$ whose $\mathcal{A}$-limit does not exist.

17.1 Prove that a Banach space is either reflexive or its successive second duals X'', $X^{\text{iv}}, \ldots$ are all distinct.

17.2 If X is reflexive and $Y \subset X$ is a closed subspace, show that Y is reflexive.

18.1 Consider the plane under any norm. Describe the dual space and its norm.

18.2 Show that the dual of c is l_1, where c is the space of convergent sequences.

18.3 Identify the dual of the closed subspace c_0 of c, of sequences which converge to zero.

18.4 Let S be the vector space of all real sequences and let $x^m = (x_1^{(m)}, x_2^{(m)}, \ldots)$ converge to $x = (x_1, x_2, \ldots)$ if $\lim_m x_i^{(m)} = x_i$ for each $i = 1, 2, \ldots$. Show that every continuous linear functional on S is given by a sequence, all but a finite number of whose terms are zero.

18.5 Give an example of a Banach space which is not a dual. (Hint: Find example in the literature.)

19.1 Prove the inequality

$$\left|\int_a^b f\, dg\right| \leq \max\,[|f(x)| : x \in [a, b]] \cdot V(g).$$

19.2 Show that for any $g \in BV$ there is a unique $\bar{g} \in BV$, continuous from the right, such that

$$\int_a^b f\, d\bar{g} = \int_a^b f\, dg \text{ for all } f \in C \text{ and } V(\bar{g}) \leq V(g).$$

20.1 Prove that if $D(T) \subset X$ is closed and T is bounded, then T is closed.

20.2 Find a necessary and sufficient condition that a subspace G of $X \times Y$ be a graph of a linear transformation from X into Y.

20.3 If X and Y are Banach spaces and $T : X \to Y$ is a bounded linear transformation, show that either $T(X) = Y$ or is of the first category in Y.

20.4 Let X, Y and Z be Banach spaces. Let $T : X \to Z$ and $U : Y \to Z$ be bounded linear transformations. Suppose for every $x \in X$ the equation $Tx = Uy$ has a unique solution y. Show that the mapping $y = Vx$ so determined is linear and bounded.

20.5 Let $T : X \to Z$ be a bounded linear transformation. Suppose T is a composite $Tx = VUx$ for all $x \in X$, where $U : X \to Y$ is linear and $V : Y \to Z$ is one-one bounded and linear. Show that U is bounded.

20.6 Show that a linear operator T in a Banach space X is continuous only if its null space is closed.

21.1 Show that every finite dimensional subspace of a Banach space is a direct summand.

21.2 If X is a Banach space, M is a direct summand of X, $M \neq (0)$, $M \neq X$, then there are infinitely many distinct choices of the closed subspace N for which $M \dotplus N = X$.

22.1 Finish the proof of the completeness of Y.

22.2 Exhibit a Schauder basis in $C[0, 1]$.

22.3 Exhibit a Schauder basis in c.

23.1 Let C be the Banach space of continuous functions on $[0, 2\pi]$ with $f(0) = f(2\pi)$ and $\|f\| = \max [|f(t)| : t \in [0, 2\pi]]$. With 1, s, c defined by $1(t) = 1$, $s(t) = \sin t$, $c(t) = \cos t$, $t \in [0, 2\pi]$, show that if $\{F_n\}$ is a sequence of positive linear operators on C into itself such that $\lim_n F_n f = f$ for $f = 1, f = s$, and $f = c$, then $\lim_n F_n f = f$ for every $f \in C$.

23.2 Show that the Lagrange interpolation polynomials do not satisfy the conditions of theorem 1.

23.3 Show that the linear interpolation functions satisfy the conditions of theorem 1.

23.4 Let $F_n f$, $n = 1, 2, \ldots$ be the nth partial sum of the Fourier series of f. Show that the $\{F_n\}$ do not satisfy the conditions of exercise 23.1.

23.5 Let $\Phi_n = \frac{1}{n}(F_1 + \ldots + F_n)$, $n = 1, 2, \ldots$ where the F_n are those of exercise 23.4. Show that the $\{\Phi_n\}$ satisfy the conditions of exercise 23.1.

CHAPTER 3

MEASURE AND INTEGRATION

L_p SPACES

3.1 Lebesgue Measure for Bounded Sets in E_n

Let E_n be euclidean n-space. An **open interval** I in E_n, given by $a_i < b_i$, $i = 1, 2, \ldots, n$, is the set of all $x = (x_1, \ldots, x_n) \in E_n$ for which $a_i < x_i < b_i$, $i = 1, \ldots, n$. By the (n-dimensional) **measure** of I, we understand the number

$$m(I) = \prod_{i=1}^{n} (b_i - a_i).$$

A **closed interval** is the set of all $x = (x_1, \ldots, x_n)$ for which $a_i \leq x_i \leq b_i$, $i = 1, \ldots, n$.

If $I_1, \ldots, I_k$ are pair-wise disjoint open intervals, and if $J = \bigcup_{j=1}^{k} I_j$ then by the **measure** of J we understand the number

$$m(J) = \sum_{j=1}^{k} m(I_j).$$

We shall designate as **elementary sets** those sets which are unions of finite numbers of pair-wise disjoint open intervals. Let U_n be the open unit n-cube which is composed of all $x = (x_1, \ldots, x_n)$ for which $0 < x_i < 1$, $i = 1, \ldots, n$. For convenience, we consider, for the time being, only subsets of U_n.

We may attach a number, called the exterior measure of S, to every subset $S \subset U_n$. For this purpose, we consider a covering of S by means of a countable set of open intervals

$$\mathfrak{J} = \{I_1, I_2, \ldots\},$$

where the I_n, $n = 1, 2, \ldots$, may overlap. Let

$$\psi(S, \mathfrak{J}) = \sum_{n=1}^{\infty} m(I_n).$$

By the **exterior measure** of S we mean the number

$$m_e(S) = \inf \psi(S, \mathfrak{J}),$$

where the infimum is taken for all coverings $\mathfrak{J}$ of S.

Let $\{S_n\}$ be a sequence of sets, and let $S = \bigcup_{n=1}^{\infty} S_n$. Then

$$m_e(S) \leq \sum_{n=1}^{\infty} m_e(S_n).$$

In order to prove this, let $\epsilon > 0$. For every $n = 1, 2, \ldots$, let $\mathfrak{J}_n$ be a covering of S_n such that

$$\psi(S_n, \mathfrak{J}_n) < m_e(S_n) + \frac{\epsilon}{2^n}\,.$$

Let $\mathfrak{J} = \bigcup_{n=1}^{\infty} \mathfrak{J}_n$. Then $\mathfrak{J}$ is a covering of S, and

$$m_e(S) \leq \psi(S, \mathfrak{J}) = \sum_{n=1}^{\infty} \psi(S_n, \mathfrak{J}_n) \leq \sum_{n=1}^{\infty} m_e(S_n) + \epsilon.$$

Since this holds for every $\epsilon > 0$, the statement is proved.

By an easy application of the Borel covering theorem, it can be shown that if J is an elementary set, then

$$m_e(J) = m(J).$$

For any $i = 1, \ldots, n$, let H be a hyperplane obtained by fixing the ith coordinate, so that H is the set of $x = \{x_1, \ldots, x_n\}$ for which $x_i = x_i^{(0)}$. It is then easy to show that $m_e(H) = 0$.

We leave these proofs, as well as the proof that $T \subset S$ implies $m_e(T) \leq m_e(S)$, to the reader.

We are now ready to give the main definition. A set $S \subset U_n$ will be called **measurable** if, for every $\epsilon > 0$ there are sets J, T_1 and T_2 in U_n such that

(i) J is an elementary set,

(ii) $T_1 \subset J$, $T_2 \subset U_n \sim J$,

(iii) $m_e(T_1) < \epsilon$, $m_e(T_2) < \epsilon$,

and

(iv) $S = (J \sim T_1) \cup T_2$.

PROPOSITION 1. If S is measurable, then $U_n \sim S$ is measurable.

Proof. Let $\epsilon > 0$. Then

$$S = (J \sim T_1) \cup T_2,$$

where J is an elementary set,

$$T_1 \subset J, T_2 \subset U_n \sim J, m_e(T_1) < \epsilon \quad \text{and} \quad m_e(T_2) < \epsilon.$$

Now

$$U_n \sim S = [(K \cup T_3) \sim (T_2 \cup T_3')] \cup T_1,$$

where K is an elementary set, T_3 is a subset of a finite set of hyperplanes such that $K \cup T_3 = U_n \sim J$, and $T_3' \subset T_3$. So,

$$U_n \sim S = (K \sim T_4) \cup T_5$$

where

$$T_4 \subset T_2 \cup T_3 \quad \text{and} \quad T_5 \subset T_1 \cup T_3.$$

But $m_e(T_2 \cup T_3) < \epsilon$ and $m_e(T_1 \cup T_3) < \epsilon$, so that the proposition is proved.

PROPOSITION 2. Every open set is measurable.

Proof. We use the fact that every open set is the union of a sequence of closed intervals whose interiors are pair-wise disjoint. Let G be open. Let $J_1, J_2, \ldots$ be the interiors of closed intervals, with pair-wise disjoint interiors, whose union is G. Since $G \subset U_n$, $\sum_{k=1}^{n} m(J_k) \leq 1$, for every n, so that $\sum_{k=1}^{\infty} m(J_k) \leq 1$. Let $\epsilon > 0$. There is an N such that $\sum_{n=1}^{N} m(J_n) > \sum_{n=1}^{\infty} m(J_n) - \epsilon$. Now

$$G = \left(\bigcup_{n=1}^{N} J_n\right) \cup \left(\bigcup_{n=N+1}^{\infty} J_n \cup E\right),$$

where E is contained in a countable set of hyperplanes, so that $m_e(E) = 0$. It follows that $m_e\left(\bigcup_{n=N+1}^{\infty} J_n \cup E\right) < \epsilon$ and the proposition is proved.

PROPOSITION 3. Every closed set is measurable.

Proof. By propositions 1 and 2.

PROPOSITION 4. S is measurable if and only if, for every $\epsilon > 0$, there is an open set G such that $S \subset G$ and $m_e(G \sim S) < \epsilon$.

Proof. Suppose S is measurable. Let $\epsilon > 0$. There are sets J, T_1 and T_2 such that J is elementary, $m_e(T_1) < \epsilon/2$, $m_e(T_2) < \epsilon/2$, and

$$S = (J \sim T_1) \cup T_2.$$

There is an open set H such that $T_2 \subset H$ and $m_e(H) < \epsilon/2$. Let $G = J \cup H$. Then $S \subset G$ and $G \sim S \subset T_1 \cup H$. But $m_e(T_1 \cup H) < \epsilon$.

Suppose, for every $\epsilon > 0$, there is an open set G such that $S \subset G$ and $m_e(G \sim S) < \epsilon$. Now, $G = J \cup E$, where J is elementary and $m_e(E) < \epsilon$. Then

$$S = [J \sim (G \sim S) \cap J] \cup (S \cap E).$$

But $m_e[(G \sim S) \cap J] < \epsilon$ and $m_e(S \cap E) < \epsilon$. Hence S is measurable.

COROLLARY. S is measurable if and only if for every $\epsilon > 0$ there is a compact set F such that $F \subset S$ and $m_e(S \sim F) < \epsilon$.

The proof is immediate if "compact" is replaced by "closed." The required simple modification needed to complete the proof is left to the reader.

PROPOSITION 5. If $\{S_n\}$ is a sequence of measurable sets, then $S = \bigcup_{n=1}^{\infty} S_n$ is measurable.

Proof. Let $\epsilon > 0$. For every $n = 1, 2, \ldots$, there is an open set G_n such that $S_n \subset G_n$ and $m_e(G_n \sim S_n) < \epsilon/2^n$. Let $G = \bigcup_{n=1}^{\infty} G_n$. Now $G \sim S \subset \bigcup_{n=1}^{\infty} (G_n \sim S_n)$, so that $m_e(G \sim S) \leq \sum_{n=1}^{\infty} m_e(G_n \sim S_n) < \epsilon$. Hence, S is measurable.

PROPOSITION 6. If F_1 and F_2 are compact sets such that $F_1 \cap F_2 = 0$, then $m_e(F_1 \cup F_2) = m_e(F_1) + m_e(F_2)$.

Proof. We first observe that if I is an open interval and $\epsilon > 0$, then I may be covered by a finite set $I_1, \ldots, I_k$ of open intervals, each of diameter less than ϵ, such that $\sum_{j=1}^{k} m(I_j) < m(I) + \epsilon$. Let $d > 0$ be the distance between F_1 and F_2. Let $\{I_n\}$ be a covering of $F_1 \cup F_2$ such that $\sum_{n=1}^{\infty} m(I_n) < m_e(F_1 \cup F_2) + \epsilon$. By the above remark, there is a covering $\{J_n\}$ of $F_1 \cup F_2$ such that the diameter of each J_n is less than $d/2$ and $\sum_{n=1}^{\infty} m(J_n) < m_e(F_1 \cup F_2) + 2\epsilon$. Let $\{K_n\}$ be those intervals from among the J_n which contain points of F_1 and $\{L_n\}$ those which contain points of F_2. No J_n is both a K_i and an L_j. Then

$$m_e(F_1) + m_e(F_2) \leq \sum_{n=1}^{\infty} m(K_n) + \sum_{n=1}^{\infty} m(L_n) \leq \sum_{n=1}^{\infty} m(J_n) < m_e(F_1 \cup F_2) + 2\epsilon.$$

Thus, $m_e(F_1) + m_e(F_2) \leq m_e(F_1 \cup F_2)$. Since $m_e(F_1) + m_e(F_2) \geq m_e(F_1 \cup F_2)$, the proposition is proved.

PROPOSITION 7. If S_1 and S_2 are measurable sets and $S_1 \cap S_2 = 0$, then $m_e(S_1 \cup S_2) = m_e(S_1) + m_e(S_2)$.

Proof. Let $\epsilon > 0$. There are compact sets $F_1 \subset S_1$ and $F_2 \subset S_2$ such that $m_e(S_1 \sim F_1) < \epsilon$ and $m_e(S_2 \sim F_2) < \epsilon$. So,

$$\begin{aligned} m_e(S_1) + m_e(S_2) &< m_e(F_1) + m_e(F_2) + 2\epsilon \\ &= m_e(F_1 \cup F_2) + 2\epsilon \leq m_e(S_1 \cup S_2) + 2\epsilon. \end{aligned}$$

Thus, $m_e(S_1) + m_e(S_2) \leq m_e(S_1 \cup S_2)$. But, $m_e(S_1) + m_e(S_2) \geq m_e(S_1 \cup S_2)$, so that the proposition is proved.

PROPOSITION 8. If $\{S_n\}$ is a sequence of pair-wise disjoint measurable sets and $S = \bigcup_{n=1}^{\infty} S_n$, then $m_e(S) = \sum_{n=1}^{\infty} m_e(S_n)$.

Proof. By proposition 7,

$$m_e(S) \geq m_e\left(\bigcup_{k=1}^{n} S_k\right) = \sum_{k=1}^{n} m_e(S_k), \text{ for every } n = 1, 2, \ldots.$$

Hence,

$$m_e(S) \geq \sum_{n=1}^{\infty} m_e(S_n).$$

But,

$$m_e(S) \leq \sum_{n=1}^{\infty} m_e(S_n),$$

so that the proposition is proved.

We have now proved

THEOREM 1. The collection $\mathcal{M}$ of measurable sets is a closed system with respect to countable unions and complementation. On $\mathcal{M}$, the exterior measure is a countably additive non-negative set function, i.e., if $S_n \in \mathcal{M}$, $n = 1, 2, \ldots$ and $S_i \cap S_j = 0$ for $i \neq j$, then

$$m_e\left(\bigcup_{n=1}^{\infty} S_n\right) = \sum_{n=1}^{\infty} m_e(S_n).$$

For sets in $\mathcal{M}$, the exterior measure will be referred to as the **measure** and will be written as $m(S)$ rather than $m_e(S)$.

Finally, the term **almost everywhere** is used to mean "except for a set of measure zero."

3.2 Lebesgue Measure for Unbounded Sets

We now extend the definition of measure from subsets of U_n to subsets of E_n, including unbounded as well as bounded sets.

For every n-tuple $K = (k_1, \ldots, k_n)$ where the k_i, $i = 1, \ldots, n$, are integers, we consider the cube $U_n^{(K)}$, consisting of the points $x = (x_1, \ldots, x_n)$ for which $k_i < x_i < k_i + 1$, $i = 1, \ldots, n$. We may then define the set $\mathcal{M}_K$ of measurable subsets of $U_n^{(K)}$, and the measure on them, in a way analogous to that in which we defined $\mathcal{M}$. (Indeed, the sets in $\mathcal{M}_K$ are simply translations of the sets in $\mathcal{M}$.)

Now, we define a set $S \subset E_n$ to be measurable if, for every K, we have $S \cap U_n^{(K)} \in \mathcal{M}_K$. Let $\mathcal{S}$ be the collection of measurable subsets of E_n. It is an easy matter to see that the complement of a measurable set is measurable and that the union of a countable set of measurable sets is measurable.

For every $S \in \mathcal{S}$, we define the measure of S by

$$m(S) = \sum_K m(S \cap U_n^{(K)}).$$

Since all summands are non-negative, the order in which the series is summed does not matter. Moreover, there are sets $S \in \mathcal{S}$ for which $m(S) = +\infty$.

With the convention that for every $0 \leq a \leq +\infty$, $+\infty + a = +\infty$, $m(S)$ is a completely additive set function on $\mathcal{S}$.

3.3 Totally σ Finite Measures

We now say a few words about general measure theory. For this purpose, let S be a set and $\mathcal{S}$ a collection of subsets of S such that

(a) $S \in \mathcal{S}$,
(b) if $A \subset S$ and $A \in \mathcal{S}$ then $S \sim A \in \mathcal{S}$,
(c) if $A_n \subset S$ and $A_n \in \mathcal{S}$, $n = 1, 2, \ldots$, then $\bigcup_{n=1}^{\infty} A_n \in \mathcal{S}$.

Such a system of sets is called a **σ algebra.** Suppose now that there is a real-valued function μ on $\mathcal{S}$ such that

(α) $\mu(A) \geq 0$ for every $A \in \mathcal{S}$,
(β) $A_n \in \mathcal{S}$, $n = 1, 2, \ldots$ and $A_i \cap A_k = 0$, $i \neq k$, implies

$$\mu\left(\bigcup_{n=1}^{\infty} A_n\right) = \sum_{n=1}^{\infty} \mu(A_n).$$

Then μ is called a **measure** on $\mathcal{S}$ and the system $(S, \mathcal{S}, \mu)$ is called a **totally finite measure space.** Note that $S \in \mathcal{S}$ and $\mu(S) < \infty$.

Sometimes it is useful to make the further assumption

(γ) if $A \in \mathcal{S}$ and $\mu(A) = 0$ then, for every $B \subset A$, $B \in \mathcal{S}$.
It then follows, of course, that $\mu(B) = 0$.

If (γ) holds, the measure space is called **complete** (a work horse of a word which is perhaps overworked in this case).

The measure defined in Sec. 3.1 is totally finite and complete.

Now, let S be a σ algebra of subsets of a set S and let μ satisfy

(a′) $0 \leq \mu(A) \leq +\infty$ for every $A \in \mathcal{S}$,
(b′) if $A_n \in \mathcal{S}$, $n = 1, 2, \ldots$ and $A_i \cap A_j = 0$, $i \neq j$, then

$$\mu\left(\bigcup_{n=1}^{\infty} A_n\right) = \sum_{n=1}^{\infty} \mu(A_n),$$

(c′) there is a sequence T_n, $n = 1, 2, \ldots$ in $\mathcal{S}$ such that $S = \bigcup_{n=1}^{\infty} T_n$, and $\mu(T_n) < \infty$, $n = 1, 2, \ldots$.

Then μ is still called a measure on $\mathcal{S}$ but the measure space $(S, \mathcal{S}, \mu)$ is called **totally σ finite.**

The measure space discussed in Sec. 3.2 is totally σ finite and complete. It is called **Lebesgue measure.**

We now consider a special class of measure spaces. In each case, the set S will be the real line E_1.

Let f be a monotonically non-decreasing real-valued function on E_1. We define a measure μ on E_1 by means of the function f. For every $a \in E_1$, let $I_a = (-\infty, a]$. Define

$$\mu(I_a) = f(a).$$

Let $\{a_n\}$ be an increasing sequence which converges to a. Let

$$\mu(\{a\}) = \lim_{n \to \infty} [f(a) - f(a_n)].$$

It is clear that $\mu(\{a\})$ is independent of the choice of $\{a_n\}$. Then, for any open interval $I = (a, b)$, define

$$\mu(I) = \mu(I_b) - \mu(I_a) - \mu(\{b\}).$$

Now, let $J_n = (n, n + 1)$. Just as in the discussion of Sec. 3.1, $\mu(I)$ defined on the open subintervals of J_n determines an exterior measure for all sets in J_n. The measurable subsets of J_n are determined as before and on them, the exterior measure has the properties of a totally finite measure.

Now, a subset $A \subset E_1$ will be called measurable if $A \cap J_n$ is measurable for every n. Then the measure of A is defined as

$$\mu(A) = \sum_{n=-\infty}^{\infty} \mu(A \cap J_n) + \sum_{n=-\infty}^{\infty} \mu(A \cap \{n\}).$$

The measure space obtained in this way is totally σ finite and complete. It is totally finite if and only if $\lim_{x \to -\infty} f(x) > -\infty$ and $\lim_{x \to +\infty} f(x) < +\infty$.

General measure theory also deals with measures that are not totally σ finite. Such measures will not be of interest to us.

3.4 Measurable Functions, Egoroff Theorem

Let $(S, \mathcal{S}, \mu)$ be a totally σ finite measure space, which we assume to be complete for the sake of convenience.

A real function f on S is said to be **measurable** if, for every open set $G \subset E_1$, the set $f^{-1}(G)$ is measurable, i.e., $f^{-1}(G) \in \mathcal{S}$.

It is easy to see that f is measurable if and only if the inverse images of all Borel sets are measurable or, on the other hand, if and only if the inverse images of the open intervals (a, ∞) are measurable. It should also be clear that if f is measurable and $g(x) = f(x)$ almost everywhere, then g is measurable. We leave the easy proofs to the reader. If f and g are such that $f(x) = g(x)$ almost everywhere, we say that f is **equivalent** to g. The set of measurable functions equivalent to a given function will be called an **equivalence class** of measurable functions. We usually shall not distinguish between a function and its equivalence class, since there is little danger of confusion. Often we shall speak of an equivalence class as a function.

We shall also consider extended real-valued functions on S, i.e., functions which may assume the values $+\infty$ and $-\infty$, as well as real values. In this case, a measurable function f may be defined as one such that, for every real a, the inverse image of the interval (a, ∞) is measurable, and the set for which $f(x) = +\infty$ is measurable.

PROPOSITION 1. If f and g are extended real-valued measurable functions on a totally σ finite measure space $(S, \mathcal{S}, \mu)$, then the set $E = [x : f(x) > g(x)]$ is measurable.

Proof. For every rational r, let $E_r = [x : f(x) < r < g(x)]$. Now E_r, as the intersection of two measurable sets, is measurable. Morever,

$$E = \bigcup E_r$$

where the union is for all the rationals, and so E, as the union of a countable set of measurable sets, is measurable.

PROPOSITION 2. If f is measurable and a is a real number, then $f + a$ and af are measurable.

The easy proof is omitted.

In the next propositions, we assume that f is real-valued.

PROPOSITION 3. If f and g are measurable, then $f + g$ is measurable.

Proof. Let a be a real number. The set for which $f(x) + g(x) > a$ is the same as the set for which $f(x) > -g(x) + a$. But the function $-g + a$ is measurable, so that this set is measurable.

PROPOSITION 4. If f is measurable and a is real, then $|f|^a$ is measurable.

Proof. For $a = 0$, the proposition is trivially true. Let $a \neq 0$. The set for which $|f|^a$ exceeds a positive real number b is the union of the sets for which $f(x) > b^{1/a}$ and $f(x) < -b^{1/a}$ and so is measurable. This completes the proof.

PROPOSITION 5. If f and g are measurable, then fg is measurable.

Proof. $fg = \frac{1}{4}(f + g)^2 - \frac{1}{4}(f - g)^2$.

PROPOSITION 6. If f and g are measurable, then $|f|$, $\max(f, g)$ and $\min(f, g)$ are measurable.

Proof. That $|f|$ is measurable follows by proposition 4. Since

$$\max(f, g) = \tfrac{1}{2}(f+g) + \tfrac{1}{2}|f-g|$$

and

$$\min(f, g) = \tfrac{1}{2}(f+g) - \tfrac{1}{2}|f-g|,$$

these functions are measurable.

The set of measurable functions is thus closed with respect to addition, multiplication, and scalar multiplication. Moreover, every pair of measurable functions has a measurable greatest lower bound, the point-wise minimum, and a measurable least upper bound, the point-wise maximum.

We now consider sequences of measurable functions.

We first note that if $\{f_n\}$ is a monotonically non-decreasing sequence of extended real-valued functions, each of which is measurable, then the function f, defined by $f(x) = \lim_{n\to\infty} f_n(x)$, is also measurable. For, let a be any real number. Then $f(x) > a$ if and only if $f_n(x) > a$ for some n. So, the set of points for which $f(x) > a$, as the union of countably many measurable sets, is measurable. Thus, f is measurable.

We make the remark that if, for every n, g_n is equivalent to f_n, then $\{g_n\}$ converges almost everywhere to the function f. Indeed, in these considerations and others which follow, functions need only be defined almost everywhere.

PROPOSITION 7. If $\{f_n\}$ is a sequence of extended real-valued measurable functions, then $\limsup_n f_n$ and $\liminf_n f_n$ are measurable.

Proof. If $\{k_n\}$ is a sequence of extended real-valued measurable functions and, for every n, $g_n = \max(k_1, \ldots, k_n)$, then $\{g_n\}$ is monotonically non-decreasing and $\lim_n g_n$ is measurable. But $\lim_n g_n = \sup_n k_n$, so that we have shown that the supremum of a sequence of measurable functions is measurable. Now, for every n, let $h_n = \sup(f_n, f_{n+1}, \ldots)$. Then $\{h_n\}$ is a monotonically non-increasing sequence of measurable functions and $\lim_n h_n = \limsup_n f_n$. Thus $\limsup_n f_n$ is measurable. Similarly, $\liminf_n f_n$ is measurable.

COROLLARY. The limit of a convergent sequence of measurable functions is measurable.

We now consider only totally finite measure spaces. We show that convergent sequences of measurable functions on such spaces are "approximately" uniformly convergent.

THEOREM 1 (Egoroff). If $\{f_n\}$ is a sequence of measurable real functions defined on a totally finite measure space $(S, \mathcal{S}, \mu)$, and $\{f_n\}$ converges almost everywhere to a function f on S, then, for every $\epsilon > 0$, there is a set $E \in \mathcal{S}$ such that $\mu(E) > \mu(S) - \epsilon$ and $\{f_n\}$ converges uniformly to f on E.

Proof. Let $\epsilon > 0$. For every n, let E_n be the set of points $x \in S$ such that $|f(x) - f_k(x)| < \epsilon$ for every $k \geq n$. Then E_n, as the intersection of countably many measurable sets, is measurable. Moreover,

$$E_1 \subset E_2 \subset \ldots \subset E_n \subset \ldots$$

and, since $\{f_n\}$ converges to f almost everywhere,

$$\bigcup_{n=1}^{\infty} E_n = S \sim Z, \quad \text{where } \mu(Z) = 0.$$

Now,

$$\mu(S) = \mu(E_1) + \mu(E_2 \sim E_1) + \ldots + \mu(E_{n+1} \sim E_n) + \ldots,$$

so that there is an N such that

$$\mu(E_N) > \mu(S) - \epsilon.$$

There is a measurable set T_1, with $\mu(T_1) > \mu(S) - \epsilon/2$, and an n_1 such that, for all $k > n_1$ and all $x \in T_1$, $|f(x) - f_k(x)| < \epsilon/2$. There is a measurable set T_2, with $\mu(T_2) > \mu(S) - \epsilon/2^2$, and an $n_2 > n_1$ such that, for all $k > n_2$ and all $x \in T_2$, $|f(x) - f_k(x)| < \epsilon/2^2$. Continuing in this way, there is a sequence

$$n_1 < n_2 < \ldots < n_i < \ldots$$

and a sequence $\{T_i\}$ of measurable sets such that, for each i, for all $k > n_i$ and all $x \in T_i$, $|f(x) - f_k(x)| < \epsilon/2^i$. Let $E = \bigcap_{n=1}^{\infty} T_n$. Then it is easy to see that $\mu(E) > \mu(S) - \epsilon$, and $\{f_n\}$ converges uniformly to f on E.

This theorem is not always true for totally σ finite measure spaces.

3.5 Convergence in Measure

We now present a kind of convergence for sequences of measurable functions which is weaker than almost everywhere convergence and is of considerable importance in analysis and probability theory.

Let $(S, \mathcal{S}, \mu)$ be a totally finite measure space. A sequence $\{f_n\}$ of measurable functions on S is said to **converge in measure** to a function f if, for every $\epsilon > 0$, there is an N such that, for every $n > N$, the set E_n of points $x \in S$ for which $|f(x) - f_n(x)| > \epsilon$ has measure $\mu(E_n) < \epsilon$.

PROPOSITION 1. If $(S, \mathcal{S}, \mu)$ is a totally finite measure space and $\{f_n\}$ is a sequence of measurable functions on S which converges almost everywhere to a function f, then $\{f_n\}$ converges in measure to f.

Proof. Let $\epsilon > 0$. By theorem 1, Sec. 3.4, there is a set E such that $m(E) > m(S) - \epsilon$ and $\{f_n\}$ converges uniformly to f on E. There is an N such that $n > N$ and $x \in E$ implies $|f(x) - f_n(x)| < \epsilon$. This proves that $\{f_n\}$ converges in measure to f.

The converse to proposition 1 is false. For example, let S be the interval $(0, 1)$ and let μ be the measure defined in Sec. 3.1, i.e., Lebesgue measure. We consider the following sequence of functions:

$$f_1(x) = 1, \quad x \in (0, 1),$$

$$f_2(x) = \begin{cases} 1, & x \in (0, \frac{1}{2}) \\ 0, & x \in [\frac{1}{2}, 1), \end{cases}$$

$$f_3(x) = \begin{cases} 1, & x \in (\frac{1}{2}, 1) \\ 0, & x \in (0, \frac{1}{2}]. \end{cases}$$

Define the next four functions by

$$f_{3+i}(x) = \begin{cases} 1, & x \in \left(\dfrac{i-1}{4}, \dfrac{i}{4}\right), \quad i = 1, \ldots, 4. \\ 0, & \text{elsewhere on } (0, 1) \end{cases}$$

We continue in a manner which should be clear. Now, for every $x \in (0, 1)$, except for diadic points (which form a set of measure zero), there is an infinite set of values of n for which $f_n(x) = 1$ and an infinite set of values of n for which $f_n(x) = 0$. Thus $\{f_n\}$ does not converge almost everywhere.

However, let $\epsilon > 0$. There is then an N such that, for every $n > N$, $f_n(x) = 0$ except on a set of measure less than ϵ. Hence, $\{f_n\}$ converges in measure to 0.

On the other hand, we have

PROPOSITION 2. If $(S, \mathcal{S}, \mu)$ is a totally finite measure space, and $\{f_n\}$ converges in measure to f, then $\{f_n\}$ has a subsequence which converges almost everywhere to f.

Proof. There is an n_1 such that the set E_1 of points $x \in S$ for which

$$|f(x) - f_{n_1}(x)| > 1$$

has measure $\mu(E_1) < 1/2$. There is an $n_2 > n_1$ such that the set E_2 of points $x \in S$ for which $|f(x) - f_{n_2}(x)| > 1/2$ has measure $\mu(E_2) < 1/2^2$. Proceeding in this way, there is a sequence

$$n_1 < n_2 < \ldots < n_i < \ldots$$

such that for every $i = 1, 2, \ldots$ the set E_i of points $x \in S$ for which

$$|f(x) - f_{n_i}(x)| > 1/i$$

has measure $\mu(E_i) < 1/2^i$. We show that the sequence $\{f_{n_i}\}$ converges almost everywhere to f.

Let $\epsilon > 0$. There is an N such that $\sum_{i=N+1}^{\infty} \frac{1}{2^i} < \epsilon$. Hence

$$\mu(E_{N+1} \cup E_{N+2} \cup \ldots) < \epsilon.$$

Let

$$S_N = S \sim \bigcup_{i=N+1}^{\infty} E_i.$$

Then $\{f_{n_i}\}$ converges uniformly to f on S_N. It follows that $\{f_{n_i}\}$ converges to f on $T = \bigcup_{N=1}^{\infty} S_N$ and $\mu(S \sim T) = 0$.

It is now easy to prove the following: If $(S, \mathcal{S}, \mu)$ is a totally finite measure space, then a sequence $\{f_n\}$ of measurable real functions on S converges in measure to a function f if and only if every subsequence of $\{f_n\}$ has a subsequence which converges almost everywhere to f. We leave the proof to the reader.

3.6. Summable Functions

We now consider an important subset of the set of measurable functions, the so-called summable functions.

Let $(S, \mathcal{S}, \mu)$ be a totally σ finite measure space. Let $A_1, \ldots, A_n$ be a finite set of pair-wise disjoint sets belonging to $\mathcal{S}$, each of finite measure. We use the notation

$$s = (a_1, A_1; a_2, A_2; \ldots; a_n, A_n)$$

to represent the function which assumes the value a_i for all $x \in A_i$, $i = 1, \ldots, n$, and the value 0 on $S \sim \bigcup_{i=1}^{n} A_i$. This function is measurable and is called a **step function.** We define the **integral** of the step function s as

$$\int_S s\, d\mu = \sum_{i=1}^{n} a_i \mu(A_i).$$

The following simple facts are easy to prove:

If s is a step function and a is a real number, then as is a step function and $\int_S as\, d\mu = a \int_S s\, d\mu$.

If s and t are step functions and $s(x) \geq t(x)$, for every $x \in S$, then $\int_S s\, d\mu \geq \int_S t\, d\mu$. In particular, if s is non-negative, then $\int_S s\, d\mu \geq 0$.

The proofs will be omitted. We do prove

PROPOSITION 1. If s and t are step functions then $s + t$ is a step function and

$$\int_S (s + t)\, d\mu = \int_S s\, d\mu + \int_S t\, d\mu.$$

Proof. Let $s = (a_1, A_1; a_2, A_2; \ldots; a_n, A_n)$ and $t = (b_1, B_1; b_2, B_2; \ldots; b_m, B_m)$. Then

$$s + t = \left(a_1 + b_1, A_1 \cap B_1; \ldots; a_i + b_j, A_i \cap B_j; \ldots; \right.$$
$$\left. a_n + b_m, A_n \cap B_m; \ldots; a_i, A_i \sim \bigcup_{j=1}^{m} B_j; \ldots; b_j, B_j \sim \bigcup_{i=1}^{n} A_i \ldots\right).$$

Then

$$\begin{aligned}
\int_S (s + t)\, d\mu &= \sum_{i=1}^{n} \sum_{j=1}^{m} (a_i + b_j)\mu(A_i \cap B_j) \\
&\quad + \sum_{i=1}^{n} a_i\mu\left(A_i \sim \bigcup_{j=1}^{m} B_j\right) + \sum_{j=1}^{m} b_j\left(B_j \sim \bigcup_{i=1}^{n} A_i\right) \\
&= \sum_{i=1}^{n} \sum_{j=1}^{m} a_i\mu(A_i \cap B_j) + \sum_{i=1}^{n} a_i\mu\left(A_i \sim \bigcup_{j=1}^{m} B_j\right) \\
&\quad + \sum_{j=1}^{m} \sum_{i=1}^{n} b_j\mu(A_i \cap B_j) + \sum_{j=1}^{m} b_j\mu\left(B_j \sim \bigcup_{i=1}^{n} A_i\right) \\
&= \sum_{i=1}^{n} a_i\mu(A_i) + \sum_{j=1}^{m} b_j\mu(B_j) \\
&= \int_S s\, d\mu + \int_S t\, d\mu.
\end{aligned}$$

We proceed to extend the definition of integral to measurable functions which are not step functions. For this purpose, we need

PROPOSITION 2. If f is non-negative and measurable, there is a non-decreasing sequence $\{s_n\}$ of step functions which converges everywhere to f.

Proof. Let $\{A_n\}$ be an increasing sequence of sets of finite measure such that $S = \bigcup_{n=1}^{\infty} A_n$. Let

$$s_n(x) = \begin{cases} \dfrac{k}{2^n} \text{ if } x \in A_n \text{ and } \dfrac{k}{2^n} \leq f(x) < \dfrac{k+1}{2^n}, \ k = 0, 1, \ldots, n \cdot 2^n - 1, \\ n \text{ if } x \in A_n, f(x) \geq n \\ 0 \text{ if } x \notin A_n. \end{cases}$$

The sequence $\{s_n\}$ is easily seen to have the desired property.

We observe that for every non-decreasing sequence $\{s_n\}$ of step functions, the sequence

$$\left\{\int_S s_n \, d\mu\right\}$$

is a non-decreasing sequence of real numbers, so that it converges to a real number or to $+\infty$. The next fact is crucial to our discussion.

PROPOSITION 3. If f is a non-negative measurable function and $\{s_n\}$ and $\{t_n\}$ are non-decreasing sequences of step functions converging to f, then

$$\lim_{n\to\infty} \int_S s_n \, d\mu = \lim_{n\to\infty} \int_S t_n \, d\mu.$$

Proof. Fix N. Let $s_N = (a_1, A_1; a_2, A_2; \ldots; a_k, A_k)$. Let $\epsilon > 0$ and $M = \max(a_1, \ldots, a_k)$. Now,

$$\lim_{n\to\infty} t_n(x) = f(x) \geq s_N(x)$$

for every x. It follows that the sets

$$E_n = \left[x : t_n(x) < s_N(x) - \frac{\epsilon}{2\sum_{i=1}^{k} \mu(A_i)}\right]$$

satisfy $\lim_{n\to\infty} \mu(E_n) = 0$. Choose n so that $\mu(E_n) < \dfrac{\epsilon}{2M}$. Then

$$\int_S t_n \, d\mu > \int_S s_N \, d\mu - \left(\frac{\epsilon}{2\sum_{i=1}^{k} \mu(A_i)} \sum_{i=1}^{k} \mu(A_i) + \frac{\epsilon}{2M} M\right)$$

$$= \int_S s_N \, d\mu - \epsilon.$$

Thus, $\lim_{n\to\infty} \int_S t_n \, d\mu \geq \int_S s_N \, d\mu$. Since this holds for every N, we have $\lim_{n\to\infty} \int_S t_n \, d\mu \geq \lim_{n\to\infty} \int_S s_n \, d\mu$.

Propositions 2 and 3 allow us to define the integral of any non-negative measurable function on S. Thus, if f is measurable and non-negative on S, let $\{s_n\}$ be a non-decreasing sequence of step functions which converges to f. Define the **integral** of f as

$$\int_S f \, d\mu = \lim_{n\to\infty} \int_S s_n \, d\mu,$$

where the value $+\infty$ is allowed for the integral.

We now prove

PROPOSITION 4. If f and g are non-negative measurable functions on S, then

$$\int_S (f+g)\,d\mu = \int_S f\,d\mu + \int_S g\,d\mu.$$

Proof. Suppose that $\int_S f\,d\mu$ and $\int_S g\,d\mu$ are both finite. Let $\{s_n\}$ be a non-decreasing sequence of step functions which converges to f and $\{t_n\}$ a non-decreasing sequence of step functions which converges to g. The result follows from the facts that

$$\int_S f\,d\mu = \lim_{n\to\infty} \int_S s_n\,d\mu, \quad \int_S g\,d\mu = \lim_{n\to\infty} \int_S t_n\,d\mu,$$

and

$$\int_S (s_n + t_n)\,d\mu = \int_S s_n\,d\mu + \int_S t_n\,d\mu \quad \text{for } n = 1, 2, \ldots.$$

The case where one of the integrals is infinite is trivial.

The following facts, whose proofs are easy and will be omitted, also hold:

If f and g are non-negative and $f(x) \geq g(x)$ for every $x \in S$, then $\int_S f\,d\mu \geq \int_S g\,d\mu$.

If f is a non-negative measurable function and $a \geq 0$, then

$$\int_S af\,d\mu = a\int_S f\,d\mu.$$

We now extend the definition of integral to functions which assume both positive and negative values. Let f, accordingly, be a measurable real function on S. We consider the positive and negative parts of f defined as

$$f^+(x) = \begin{cases} f(x) & \text{if } f(x) > 0 \\ 0 & \text{if } f(x) \leq 0, \end{cases}$$

and

$$f^-(x) = \begin{cases} -f(x) & \text{if } f(x) < 0 \\ 0 & \text{if } f(x) \geq 0. \end{cases}$$

It is clear that f^+ and f^- are non-negative measurable functions and that

$$f = f^+ - f^-.$$

Moreover,

$$|f| = f^+ + f^-.$$

A measurable function f is said to be **summable** if both $\int_S f^+\,d\mu < \infty$ and $\int_S f^-\,d\mu < \infty$.

If f is summable, then the integral of f is defined as

$$\int_S f\,d\mu = \int_S f^+\,d\mu - \int_S f^-\,d\mu.$$

Otherwise, f is said to be non-summable. Sometimes functions are called **integrable** if at most one of the two integrals $\int_S f^+ \, d\mu$ and $\int_S f^- \, d\mu$ is infinite.

If $\int_S f^+ \, d\mu = +\infty$ and $\int_S f^- \, d\mu < +\infty$, then we say

$$\int_S f \, d\mu = +\infty.$$

If $\int_S f^+ \, d\mu < +\infty$ and $\int_S f^- \, d\mu = +\infty$, then we say

$$\int_S f \, d\mu = -\infty.$$

If both $\int_S f^+ \, d\mu = +\infty$ and $\int_S f^- \, d\mu = +\infty$, then f is non-integrable as well as non-summable.

Now, let f and g both be summable. Then

$$f + g = (f^+ + g^+) - (f^- + g^-).$$

Clearly, the function $f^+ + g^+ \geq f + g$. But $(f + g)^+$, as $\max(f + g, 0)$, is the smallest non-negative function which exceeds $f + g$. Therefore,

$$f^+ + g^+ = (f + g)^+ + h,$$

where h is non-negative. So

$$\int_S (f^+ + g^+) \, d\mu = \int_S (f + g)^+ \, d\mu + \int_S h \, d\mu.$$

Also,

$$f^- + g^- = (f + g)^- + h,$$

since

$$(f^+ + g^+) - (f^- + g^-) = (f + g)^+ - (f + g)^-,$$

so that

$$\int_S (f^- + g^-) \, d\mu = \int_S (f + g)^- \, d\mu + \int_S h \, d\mu.$$

Thus,

$$\int_S (f + g)^+ \, d\mu < \infty \quad \text{and} \quad \int_S (f + g)^- \, d\mu < \infty,$$

so that $f + g$ is summable. Moreover,

$$\begin{aligned} \int_S (f + g) \, d\mu &= \int_S (f + g)^+ \, d\mu - \int_S (f + g)^- \, d\mu \\ &= \int_S (f^+ + g^+) \, d\mu - \int_S h \, d\mu - \int_S (f^- + g^-) \, d\mu + \int_S h \, d\mu \\ &= \int_S f^+ \, d\mu - \int_S f^- \, d\mu + \int_S g^+ \, d\mu - \int_S g^- \, d\mu = \int_S f \, d\mu + \int_S g \, d\mu. \end{aligned}$$

We have thus proved

THEOREM 1. If f and g are summable, then $f + g$ is summable and

$$\int_S (f + g) \, d\mu = \int_S f \, d\mu + \int_S g \, d\mu.$$

It is also true that if f is summable and c is a real number, then cf is summable and

$$\int_S cf\,d\mu = c\int_S f\,d\mu.$$

3.7 Fatou and Lebesgue Dominated Convergence Theorems

We now prove some facts about sequences of summable functions.

PROPOSITION 1. If $\{f_n\}$ is a non-decreasing sequence of non-negative summable functions and f is the point-wise limit of $\{f_n\}$, then

$$\int_S f\,d\mu = \lim_{n\to\infty}\int_S f_n\,d\mu,$$

where possibly $\int_S f\,d\mu = +\infty$.

Proof. For every n, let $\{s_{nm}\}$ be a non-decreasing sequence of step functions which converges to f_n, and let

$$t_n = \max\,[s_{ij} : i, j \leq n].$$

Then $t_n(x) \leq f_n(x)$, for every $x \in S$, so that

$$\int_S t_n\,d\mu \leq \int_S f_n\,d\mu, \text{ and } \lim_{n\to\infty}\int_S t_n\,d\mu \leq \lim_{n\to\infty}\int_S f_n\,d\mu,$$

where the limits exist since $\{t_n\}$ and $\{f_n\}$ are non-decreasing, so that $\left\{\int_S t_n\,d\mu\right\}$ and $\left\{\int_S f_n\,d\mu\right\}$ are non-decreasing.

For every n, m for all $k \geq \max\,(m, n)$, $t_k(x) \geq s_{nm}(x)$, for every $x \in S$, so that

$$\lim_{k\to\infty}\int_S t_k\,d\mu \geq \lim_{m\to\infty}\int_S s_{nm}\,d\mu = \int_S f_n\,d\mu.$$

Hence,

$$\lim_{n\to\infty}\int_S t_n\,d\mu = \lim_{n\to\infty}\int_S f_n\,d\mu.$$

We have thus shown that

$$\lim_{n\to\infty}\int_S t_n\,d\mu = \lim_{n\to\infty}\int_S f_n\,d\mu.$$

Now, $\lim_{k\to\infty} t_k(x) \geq f_n(x)$ for every n and $x \in S$, and $t_k(x) \leq f(x)$ for every k and $x \in S$. Hence $f(x) = \lim_{k\to\infty} t_k(x)$ for every x. But, by definition, $\int_S f\,d\mu = \lim_{n\to\infty}\int_S t_n\,d\mu$. Hence $\int_S f\,d\mu = \lim_{n\to\infty}\int_S f_n\,d\mu$.

COROLLARY. If $\{f_n\}$ is a non-decreasing sequence of summable functions which converges to f, then

$$\lim_{n\to\infty}\int_S f_n\,d\mu = \int_S f\,d\mu.$$

Proof. Apply proposition 1 to the sequence $\{f_n - f_1\}$ and the limit function $f - f_1$.

It is possible for a sequence $\{f_n\}$ to converge to a function f, but for the integrals of the functions not to converge to the integral of f. We give examples for which this fails in different ways.

EXAMPLE 1. Let S be the open interval $(0, 1)$ and let μ be Lebesgue measure. (In this case, and in others, we shall write $\int f\,dx$, or $\int_0^1 f(x)\,dx$, or $\int f$, or other variants, instead of $\int_S f\,d\mu$.) For every $n = 1, 2, \ldots$, let

$$f_n(x) = \begin{cases} 2k(k+1) & \text{if } x \in \left(\dfrac{1}{k+1}, \dfrac{2k+1}{2k(k+1)}\right) \\ -2k(k+1) & \text{if } x \in \left(\dfrac{2k+1}{2k(k+1)}, \dfrac{1}{k}\right) \qquad k = 1, \ldots, n \\ 0 & \text{elsewhere.} \end{cases}$$

Then f_n is summable, for every $n = 1, 2, \ldots$. Indeed, $\int f_n = 0$. But $\{f_n\}$ converges to the function f on $(0, 1)$ defined by

$$(x) = \begin{cases} 2k(k+1) & \text{if } x \in \left(\dfrac{1}{k+1}, \dfrac{2k+1}{2k(k+1)}\right) \\ -2k(k+1) & \text{if } x \in \left(\dfrac{2k+1}{2k(k+1)}, \dfrac{1}{k}\right) \\ & k = 1, 2, \ldots. \end{cases}$$

But $\int f^+ = +\infty$ and $\int f^- = +\infty$, so that f is not summable.

EXAMPLE 2. Let S be the open interval $(0, 1)$ and let μ be Lebesgue measure. For every $n = 1, 2, \ldots$, let

$$f_n(x) = \begin{cases} n(n+1), & x \in \left(\dfrac{1}{n+1}, \dfrac{1}{n}\right) \\ 0 & \text{elsewhere.} \end{cases}$$

Then $\{f_n\}$ converges on $(0, 1)$ to the function f which is identically zero. But $\int f_n = 1$, $n = 1, 2, \ldots$, and $\int f = 0$.

On the positive side, we already have proposition 1. We prove the basic

THEOREM 1 (Fatou). If $(S, \mathcal{S}, \mu)$ is a totally σ finite measure space and $\{f_n\}$ is a sequence of non-negative summable functions on S, then

$$\int_S \left[\liminf_{n\to\infty} f_n\right] d\mu \leq \liminf_{n\to\infty} \int_S f_n\,d\mu.$$

Proof. For every $n = 1, 2, \ldots$, let

$$g_n = \inf [f_n, f_{n+1}, \ldots].$$

Then $\{g_n\}$ is a non-decreasing sequence of summable functions, and

$$\lim_{n\to\infty} g_n(x) = \liminf_{n\to\infty} f_n(x), \quad \text{for every } x \in S.$$

Hence,

$$\int_S \left[\liminf_{n\to\infty} f_n\right] d\mu = \int_S \left[\lim_{n\to\infty} g_n\right] d\mu$$

$$= \lim_{n\to\infty} \int_S g_n \, d\mu.$$

But $g_n(x) \le f_n(x)$ for every $n = 1, 2, \ldots$, and $x \in S$, so that

$$\int_S g_n \, d\mu \le \int_S f_n(x) \, d\mu \quad \text{for every } n = 1, 2, \ldots.$$

Thus,

$$\int_S \left[\liminf_{n\to\infty} f_n\right] d\mu = \lim_{n\to\infty} \int_S g_n \, d\mu \le \liminf_{n\to\infty} \int_S f_n \, d\mu.$$

COROLLARY. If $\{f_n\}$ is a sequence of non-positive summable functions then

$$\int_S \left[\limsup_{n\to\infty} f_n\right] d\mu \ge \limsup_{n\to\infty} \int_S f_n \, d\mu.$$

The proof is left to the reader.

We now prove the so-called dominated convergence theorem.

THEOREM 2 (Lebesgue). If $(S, \mathcal{S}, \mu)$ is a totally σ finite measure space and $\{f_n\}$ is a convergent sequence of summable functions on S such that there is a summable g for which $|f_n(x)| \le g(x)$, for every $n = 1, 2, \ldots$ and $x \in S$, then

$$\lim_{n\to\infty} \int_S f_n \, d\mu = \int_S \left[\lim_{n\to\infty} f_n\right] d\mu.$$

Proof. The functions in the sequence $\{g + f_n\}$ are non-negative and summable, so that, by theorem 1,

$$\int_S \left[\liminf_{n\to\infty} (g + f_n)\right] d\mu \le \liminf_{n\to\infty} \int_S (g + f_n) \, d\mu.$$

But,

$$\int_S \left[\liminf_{n\to\infty} (g + f_n)\right] d\mu = \int_S g \, d\mu + \int_S \left[\liminf_{n\to\infty} f_n\right] d\mu$$

and

$$\liminf_{n\to\infty} \int_S (g + f_n) \, d\mu = \int_S g \, d\mu + \liminf_{n\to\infty} \int_S f_n \, d\mu.$$

Hence,

$$\int_S \left[\liminf_{n\to\infty} f_n\right] d\mu \leq \liminf_{n\to\infty} \int_S f_n \, d\mu.$$

Similarly, by using the corollary to theorem 1,

$$\int_S \left[\limsup_{n\to\infty} f_n\right] d\mu \geq \limsup_{n\to\infty} \int_S f_n \, d\mu.$$

Accordingly,

$$\liminf_{n\to\infty} \int_S f_n \, d\mu \geq \int_S \left[\lim_{n\to\infty} f_n\right] d\mu \geq \limsup_{n\to\infty} \int_S f_n \, d\mu,$$

so that $\lim_{n\to\infty} \int_S f_n \, d\mu$ exists and equals $\int_S \left[\lim_{n\to\infty} f_n\right] d\mu$.

3.8 Integral as a Set Function

Let $(S, \mathcal{S}, \mu)$ be a totally σ finite measure space, and let f be a non-negative summable function on S. We show, in this section, that f determines a measure on S which is totally finite.

For every $E \in \mathcal{S}$, we define the integral of f on E as

$$\int_E f \, d\mu = \int_S f \cdot \chi_E \, d\mu,$$

where $\chi_E = \begin{cases} 1, & x \in E \\ 0, & x \notin E \end{cases}$ is the characteristic function of E.

Since, for every $x \in S$,

$$0 \leq (f \cdot \chi_E)(x) \leq f(x),$$

it follows that $f \cdot \chi_E$ is summable and

$$0 \leq \int_E f \, d\mu \leq \int_S f \, d\mu.$$

Let

$$\nu(E) = \int_E f \, d\mu.$$

Then ν is a non-negative function on $\mathcal{S}$. We show it is a measure.

Suppose $E_1, E_2 \in \mathcal{S}$ and $E_1 \cap E_2 = 0$. Then,

$$\begin{aligned} \nu(E_1 \cup E_2) &= \int_{E_1 \cup E_2} f \, d\mu = \int_S f \cdot \chi_{E_1 \cup E_2} \, d\mu \\ &= \int_S f \cdot (\chi_{E_1} + \chi_{E_2}) \, d\mu = \int_S f\chi_{E_1} \, d\mu + \int_S f\chi_{E_2} \, d\mu \\ &= \int_{E_1} f \, d\mu + \int_{E_2} f \, d\mu = \nu(E_1) + \nu(E_2). \end{aligned}$$

This shows that the set function ν is additive. Now, we show it is completely additive. For this purpose, let $E_n \in \mathcal{S}$, $n = 1, 2, \ldots$, where $E_i \cap E_j = 0$, if $i \neq j$, and let $E = \bigcup_{n=1}^{\infty} E_n$. Moreover, let $F_n = \bigcup_{k=1}^{n} E_k$, $n = 1, 2, \ldots$.

Then $\{f \cdot \chi_{F_n}\}$ is a non-decreasing sequence of non-negative functions which converges to the function $f \cdot \chi_E$. It follows that

$$\int_S f\chi_E \, d\mu = \lim_{n\to\infty} \int_S f\chi_{F_n} \, d\mu$$
$$= \sum_{n=1}^{\infty} \int_S f\chi_{E_n} \, d\mu.$$

In other words, $\nu(E) = \sum_{n=1}^{\infty} \nu(E_n)$.

This proves that ν is a measure. Since $\nu(S) = \int_S f\,d\mu < \infty$, the measure $(S, \mathcal{S}, \nu)$ is totally finite.

Suppose, however, that f is summable but not non-negative. We shall see that it again generates a set function ν, which is no longer non-negative. This leads to the notion of a signed measure.

3.9 Signed Measure, Decomposition into Measures

Let S be a set and $\mathcal{S}$ a σ algebra of subsets of S. Then we have $S \in \mathcal{S}$; the empty set $0 \in \mathcal{S}$; for every $E \in \mathcal{S}$, the complement $S \sim E \in \mathcal{S}$; and for every $E_n \in \mathcal{S}$, $n = 1, 2, \ldots$, the union $\bigcup_{n=1}^{\infty} E_n \in \mathcal{S}$.

Let μ be a real valued function on $\mathcal{S}$ such that if $E_n \in \mathcal{S}$, $n = 1, 2, \ldots$, and $E_i \cap E_j = 0$, whenever $i \neq j$, then

$$\mu\left(\bigcup_{n=1}^{\infty} E_n\right) = \sum_{n=1}^{\infty} \mu(E_n).$$

Such a function μ is called a **finite signed measure** on $\mathcal{S}$, and we write $(S, \mathcal{S}, \mu)$ for the signed measure.

PROPOSITION 1. If $(S, \mathcal{S}, \mu)$ is a finite signed measure, there is an M such that $|\mu(E)| < M$ for every $E \in \mathcal{S}$.

Proof. Suppose the assertion is false. Then

$$\sup\,[|\mu(E)| : E \in \mathcal{S}] = +\infty.$$

Since $\mu(S)$ is finite, there is then a $T \in \mathcal{S}$ such that

$$|\mu(T)| \geq |\mu(S)| + 1.$$

If

$$\sup\,[|\mu(E)| : E \in \mathcal{S}, E \subset T] = +\infty,$$

let $S_1 = T$; otherwise, let $S_1 = S \sim T$. In either case, $|\mu(S_1)| \geq 1$ and

$$\sup\,[|\mu(E)| : E \in \mathcal{S}, E \subset S_1] = +\infty.$$

By induction, we obtain a decreasing sequence

$$S_1 \supset S_2 \supset \ldots \supset S_n \supset \ldots$$

such that $|\mu(S_n)| \geq n$, and

$$\sup [|\mu(E)| : E \in \mathcal{S}, E \subset S_n] = +\infty,$$

for every $n = 1, 2, \ldots$. Let

$$R = \bigcap_{n=1}^{\infty} S_n.$$

It is easily seen from the countable additivity of μ that

$$\mu(R) = \lim_{n \to \infty} \mu(S_n).$$

Hence

$$|\mu(R)| = +\infty$$

in contradiction with the hypothesis.

In particular, let $(S, \mathcal{S}, \mu)$ be a totally σ finite measure space and let f be a real valued summable function on S. For every $E \in \mathcal{S}$, let

$$\nu(E) = \int_E f \, d\mu = \int_S f \cdot \chi_E \, d\mu.$$

Since f is summable, the functions $f^+ = \max(f, 0)$ and $f^- = -\min(f, 0)$ are non-negative summable functions, and the corresponding set functions ν^+ and ν^-, defined by

$$\nu^+(E) = \int_E f^+ \, d\mu$$

and

$$\nu^-(E) = \int_E f^- \, d\mu,$$

are completely additive non-negative set functions, so that they are measures. Moreover, it is clear that

$$\nu(E) = \nu^+(E) - \nu^-(E)$$

and it readily follows that ν is a finite signed measure.

Every signed measure behaves essentially like the one obtained above, and this is the meaning of the next few results.

Let $(S, \mathcal{S}, \mu)$ be a finite signed measure. We shall show that there are non-negative totally finite measures μ^+, μ^-, and $|\mu|$ such that, for every $E \in \mathcal{S}$,

$$\mu(E) = \mu^+(E) - \mu^-(E)$$

and

$$|\mu|(E) = \mu^+(E) + \mu^-(E).$$

Indeed, we define μ^+ and μ^- by

$$\mu^+(E) = \sup [\mu(T) : T \in \mathcal{S}, T \subset E]$$

and

$$\mu^-(E) = -\inf [\mu(T) : T \in \mathcal{S}, T \subset E].$$

It is almost immediately evident that μ^+ and μ^- are completely additive and non-negative. We show that for every $E \in \mathcal{S}$,

$$\mu(E) = \mu^+(E) - \mu^-(E).$$

Let $T \subset E$, $T \in \mathcal{S}$. Then

$$\mu(T) = \mu(E) - \mu(E \sim T).$$

Since $\mu^+(E) \geq \mu(E \sim T)$, it follows that

$$\mu(T) \geq \mu(E) - \mu^+(E).$$

This holds for every $T \subset E$, $T \in \mathcal{S}$, so that

$$-\mu^-(E) \geq \mu(E) - \mu^+(E),$$

and

$$\mu^+(E) - \mu^-(E) \geq \mu(E).$$

A similar computation shows that

$$\mu^+(E) - \mu^-(E) \leq \mu(E).$$

Since μ^+ and μ^- are measures, $|\mu| = \mu^+ + \mu^-$ is also a measure. It is called the **total variation** of the measure μ. Indeed, it is easy to show that, for every $E \in \mathcal{S}$,

$$|\mu|\,(E) = \sup \sum_{i=1}^{k} |\mu(E_i)|,$$

where the supremum is taken for all partitions $E = \bigcup_{i=1}^{k} E_i$, $E_i \cap E_j = 0$, $i \neq j$. We leave the proof for the reader.

Now let $(S, \mathcal{S}, \mu)$ be a finite signed measure. We show that there are sets $A \in \mathcal{S}$, $B \in \mathcal{S}$ such that $A \cup B = S$, $A \cap B = 0$, and such that $\mu(E) \geq 0$ whenever $E \subset A$, $E \in \mathcal{S}$, and $\mu(E) \leq 0$, whenever $E \subset B$, $E \in \mathcal{S}$. This follows from

LEMMA 1. If $(S, \mathcal{S}, \mu)$ is a finite signed measure, there is an $E \in \mathcal{S}$ such that

$$\mu(E) = \sup\,[\mu(T) : T \in \mathcal{S}].$$

Proof. Let $m = \sup\,[\mu(T) : T \in \mathcal{S}]$. There is a set $E_1 \in \mathcal{S}$ such that $\mu(E_1) > m - 1/2$. There is a set $E_2 \in \mathcal{S}$ such that $E_2 = (E_1 \sim R_1) \cup S_1$, where $|\mu(R_1)| < 1/2$, $|\mu(S_1)| < 1/2$ and $\mu(E_2) > m - 1/2^2$. Proceeding in this way, we obtain a sequence $E_n \in \mathcal{S}$, $n = 1, 2, \ldots$, such that

$$E_{n+1} = (E_n \sim R_n) \cup S_n,$$

where $|\mu(R_n)| < 1/2^n$, $|\mu(S_n)| < 1/2^n$, and $\mu(E_n) > m - 1/2^n$, for every $n = 1, 2, \ldots$. Let

$$E = \bigcup_{k=1}^{\infty} \bigcap_{n=k}^{\infty} E_n.$$

It is easily seen that

$$\mu(E) > \mu(E_n) - \frac{1}{2^{n-1}} > m - \frac{1}{2^{n-2}},$$

for every n, so that $\mu(E) = m$.

Let $A = E$, $B = S \sim E$. If A has a subset $T \in \mathcal{S}$ such that $\mu(T) < 0$, then $\mu(A \sim T) = \mu(A) - \mu(T) > m$, which is impossible. Also, if B had a subset $T \in \mathcal{S}$ such that $\mu(T) > 0$, then $\mu(A \cup T) = m + \mu(T) > m$, which is impossible. We may now state

THEOREM 1. If $(S, \mathcal{S}, \mu)$ is a finite signed measure, there are sets $A \in \mathcal{S}$, $B \in \mathcal{S}$ such that $A \cup B = S$, $A \cap B = 0$, $\mu(T) \geq 0$ if $T \in \mathcal{S}$, $T \subset A$, and $\mu(T) \leq 0$ if $T \in \mathcal{S}$, $T \subset B$.

The set functions $\mu^+(E) = \mu(E \cap A)$ and $\mu^-(E) = -\mu(E \cap B)$ are measures, and $\mu = \mu^+ - \mu^-$, $|\mu| = \mu^+ + \mu^-$.

3.10 Absolute Continuity and Singularity of Measures

Let $(S, \mathcal{S}, \mu)$ be a totally finite measure space, and f a summable function on S. Then the set function ν given by

$$\nu(E) = \int_E f \, d\mu$$

defines a finite signed measure $(S, \mathcal{S}, \nu)$.

Let $\epsilon > 0$. Then, by the definition of summability, there is an n such that if

$$A = [x : f(x) > n] \quad \text{and} \quad B = [x : f(x) < -n]$$

then

$$\int_A f \, d\mu < \frac{\epsilon}{4} \quad \text{and} \quad \int_B -f \, d\mu < \frac{\epsilon}{4}.$$

Now, let $C = S \sim (A \cup B)$, let $\delta = \epsilon/2n$, and let $E \in \mathcal{S}$ with $\mu(E) < \delta$. Then

$$|\nu|(E) = \int_E |f| \, d\mu = \int_{C \cap E} |f| \, d\mu + \int_{A \cap E} |f| \, d\mu + \int_{B \cap E} |f| \, d\mu$$
$$\leq \delta \cdot n + \frac{\epsilon}{4} + \frac{\epsilon}{4} = \epsilon.$$

We consider the following definition. If $(S, \mathcal{S}, \mu)$ and $(S, \mathcal{S}, \nu)$ are finite signed measures, ν is said to be **absolutely continuous** with respect to μ if $|\mu|(E) = 0$ implies $|\nu|(E) = 0$.

The main fact regarding absolute continuity is the Radon-Nikodym theorem below, which by the above analysis implies that if ν is absolutely continuous with respect to μ, where μ is a measure, then for every $\epsilon > 0$ there is a $\delta > 0$ such that $\mu(E) < \delta$ implies $|\nu|(E) < \epsilon$.

THEOREM 1. If $(S, \mathcal{S}, \mu)$ is a totally finite measure space and $(S, \mathcal{S}, \nu)$ is a finite signed measure, then if ν is absolutely continuous with respect to μ, there is a summable f on S such that, for every $E \in \mathcal{S}$,

$$\nu(E) = \int_E f \, d\mu.$$

Proof. We suppose first that ν is non-negative and obtain a non-negative summable function with the desired property. In order to obtain such a function, we let $\mathcal{M}$ be the set of non-negative summable functions on S such that, for every $E \in \mathcal{S}$,

$$\int_E f \, d\mu \leq \nu(E).$$

The set $\mathcal{M}$ is non-empty since $0 \in \mathcal{M}$. We note that if $f \in \mathcal{M}$ and $g \in \mathcal{M}$, then $\max(f, g) \in \mathcal{M}$. For, if $E \in \mathcal{S}$, let

$$E_1 = [x : x \in E, f(x) \geq g(x)] \quad \text{and} \quad E_2 = E \sim E_1.$$

Then

$$\begin{aligned} \int_E \max(f, g) \, d\mu &= \int_{E_1} f \, d\mu + \int_{E_2} g \, d\mu \\ &\leq \nu(E_1) + \nu(E_2) = \nu(E). \end{aligned}$$

Now, let

$$m = \sup \left[\int_S f \, d\mu : f \in \mathcal{M} \right].$$

There is a sequence $f_n \in \mathcal{M}$, $n = 1, 2, \ldots$ such that

$$\lim_{n \to \infty} \int_S f_n \, d\mu = m.$$

For every n, let

$$g_n = \max(f_1, f_2, \ldots, f_n).$$

Then $g_n \in \mathcal{M}$, $n = 1, 2, \ldots$, is a non-decreasing sequence, and

$$\lim_{n \to \infty} \int_S g_n \, d\mu = m.$$

Let $g = \lim_{n \to \infty} g_n$. Since

$$\int_E g \, d\mu = \lim_{n \to \infty} \int_E g_n \, d\mu,$$

for every $E \in \mathcal{S}$, it follows that $g \in \mathcal{M}$ and $\int_S g \, d\mu = m$.

We now define a measure ω by

$$\omega(E) = \nu(E) - \int_E g \, d\mu.$$

Since $g \in \mathcal{M}$, ω is non-negative. We show it is identically zero. Suppose then that there is an $E \in \mathcal{S}$ with $\omega(E) > 0$. There is then a $k > 0$ such that $\omega(E) > k\mu(E)$, i.e.,

$$(\omega - k\mu)(E) > 0.$$

There is then a subset $A \subset E$, $A \in \mathcal{S}$, such that

$$(\omega - k\mu)(A) > 0,$$

and for every $T \subset A$, $T \in \mathcal{S}$,

$$(\omega - k\mu)(T) \geq 0.$$

It is evident that $\omega - k\mu$ is absolutely continuous with respect to μ, and it follows that $\mu(A) > 0$.

We next notc that

$$g + k\chi_A \in \mathcal{M}.$$

For every $E \in \mathcal{S}$,

$$\begin{aligned}\int_E (g + k\chi_A)\, d\mu &= \int_{E \sim A} g\, d\mu + \int_{E \cap A} (g + k\chi_A)\, d\mu \\ &\leq \nu(E \sim A) + \nu(E \cap A) = \nu(E).\end{aligned}$$

But,

$$\int_S (g + k\chi_A)\, d\mu = m + k\mu(A) > m,$$

which is a contradiction. Thus, for every $E \in \mathcal{S}$,

$$\nu(E) = \int_E g\, d\mu.$$

The general case follows by letting $\nu = \nu^+ - \nu^-$ and finding the measurable functions g and h which correspond to ν^+ and ν^-, respectively. Then for every $E \in \mathcal{S}$,

$$\nu(E) = \int_E (g - h)\, d\mu.$$

A companion notion to absolute continuity is that of singularity. If $(S, \mathcal{S}, \mu)$ and $(S, \mathcal{S}, \nu)$ are finite signed measures, then ν is **singular** with respect to μ if there are $A \in \mathcal{S}$, $B \in \mathcal{S}$ such that $A \cup B = S$ and $A \cap B = 0$, and $\mu(A) = 0$, $\nu(B) = 0$. Clearly, ν is singular with respect to μ if and only if μ is singular with respect to ν. The next theorem will be stated and proved only for the case of totally finite measures.

THEOREM 2. If $(S, \mathcal{S}, \mu)$ and $(S, \mathcal{S}, \nu)$ are totally finite measures, there are measures α and β, which are singular with respect to each other, such that α is absolutely continuous with respect to μ, β is singular with respect to μ, and $\nu = \alpha + \beta$. There is then a summable f, with respect to μ, such that, for every $E \in \mathcal{S}$,

$$\nu(E) = \int_E f\, d\mu + \beta(E).$$

Proof. Let

$$\mathcal{A} = [E : E \in \mathcal{S},\ \mu(E) = 0],$$

and let

$$m = \sup\, [\nu(E) : E \in \mathcal{A}].$$

For every n, there is an $E_n \in \mathcal{A}$ such that $\nu(E_n) > m - 1/n$. Let

$$B = \bigcup_{n=1}^{\infty} E_n.$$

Then $\mu(B) = 0$, so that $B \in \mathcal{A}$, and $\nu(B) = m$. Let $A = S \sim B$. We now let

$$\alpha(E) = \nu(E \cap A) \quad \text{and} \quad \beta(E) = \nu(E \cap B),$$

for every $E \in \mathcal{S}$. Then it is clear that α and β are measures which are singular with respect to each other and that β is singular with respect to μ. Moreover, $\nu = \alpha + \beta$. We show that α is absolutely continuous with respect to μ. Otherwise, there would be an $E \subset A$, $E \in \mathcal{S}$, such that $\mu(E) = 0$ and

$$\nu(E) > 0.$$

Then $\mu(E \cup B) = 0$ and $\nu(E \cup B) = \nu(E) + m > m$, which is impossible.

3.11 The L_p Spaces, Completeness

In this section, we introduce an important class of Banach spaces, the L_p spaces. Let $(S, \mathcal{S}, \mu)$ be a totally σ finite measure space, and let $p \geq 1$. A measurable real function f on S is said to be in $L_p = L_p(S, \mathcal{S}, \mu)$ if

$$\int_S |f|^p \, d\mu < \infty.$$

We first note that L_p is a vector space. If $f \in L_p$, $a \in R$, then

$$\int_S |af|^p \, d\mu = |a|^p \int_S |f|^p \, d\mu < \infty,$$

so that $af \in L_p$. If $f \in L_p$, $g \in L_p$,

$$A = [x : |f(x)| \geq |g(x)|] \quad \text{and} \quad B = S \sim A,$$

then

$$\begin{aligned} \int_S |f + g|^p \, d\mu &= \int_A |f + g|^p \, d\mu + \int_B |f + g|^p \, d\mu \\ &\leq 2^p \int_A |f|^p \, d\mu + 2^p \int_B |g|^p \, d\mu \\ &\leq 2^p \Big[\int_S |f|^p \, d\mu + \int_S |g|^p \, d\mu\Big] < \infty. \end{aligned}$$

Hence, L_p is a vector space.

The vector space L_p has a "natural" norm for which it is a Banach space. We consider the case $p = 1$ first, and let

$$\|f\| = \int_S |f| \, d\mu$$

for $f \in L_1$. Then $\|f\| \geq 0$, and $\|f\| = 0$ if and only if f is zero almost everywhere. It is accordingly clear that the objects we are considering are not functions but are equivalence classes of functions. The objects of L_p, $p > 1$, will also be equivalence classes of functions. We use the function notation for these objects. That we have a norm in L_1 now follows since

$$\|af\| = \int_S |af|\, d\mu = |a| \int_S |f|\, d\mu = |a|\, \|f\|,$$

and

$$\|f + g\| = \int_S |f + g|\, d\mu \leq \int_S |f|\, d\mu + \int_S |g|\, d\mu$$
$$= \|f\| + \|g\|.$$

Thus, with this norm, L_1 is a normed vector space.

For $p > 1$, the norm is taken as

$$\|f\| = \left[\int_S |f|^p\, d\mu\right]^{1/p}.$$

Again, $\|f\| \geq 0$, and $\|f\| = 0$ if and only if f is zero almost everywhere. Moreover, it is trivial that

$$\|af\| = |a|\, \|f\|.$$

The proof that $\|f + g\| \leq \|f\| + \|g\|$ involves the Hölder and Minkowski inequalities. Since the proofs differ little from those given in Chapter 1, we leave the minor changes needed for the proofs to the reader. Having conceded this much, we know that L_p, $p > 1$ is a normed vector space. We now prove

THEOREM 1. If $(S, \mathcal{S}, \mu)$ is a totally σ finite measure space and $p \geq 1$, then the normed vector space $L_p = L_p(S, \mathcal{S}, \mu)$ is complete.

Proof. Let $\{f_n\}$ be a Cauchy sequence in L_p. This implies that, for every $\epsilon > 0$, there is an N such that $m, n > N$ implies

$$\int_S |f_n - f_m|^p\, d\mu < \epsilon.$$

Now, $\{f_n\}$ converges in measure since $\int_S |f_n - f_m|^p\, d\mu < \epsilon$ implies

$$\mu[x : |f_n(x) - f_m(x)| > \epsilon^{1/2p}] < \epsilon^{1/2}.$$

It follows that $\{f_n\}$ has a subsequence $\{f_{n_k}\}$ which converges almost everywhere to a function f.

Now, $\{|f_{n_k}|^p\}$ converges almost everywhere to $|f|^p$, and it follows that

$$\int_S |f|^p\, d\mu \leq \liminf_{k \to \infty} \int_S |f_{n_k}|^p\, d\mu < \infty,$$

whereby $f \in L_p$.

Since, for every k, the sequence $\{|f_{n_r} - f_{n_k}|^p\}$ converges almost everywhere to $|f - f_{n_k}|^p$, we have

$$\int_S |f - f_{n_k}|^p\, d\mu \leq \liminf_{r \to \infty} \int_S |f_{n_r} - f_{n_k}|^p\, d\mu.$$

It follows that, for every $\epsilon > 0$, there is an N, such that $k > N$ implies $\|f - f_{n_k}\| < \epsilon$. Thus $\{f_{n_k}\}$ converges in L_p to f and, since $\{f_n\}$ is a Cauchy sequence, $\{f_n\}$ converges in L_p to f.

The spaces $L_p = L_p(S, \mathcal{S}, \mu)$, $p \geq 1$, are thus Banach spaces.

We now consider the case where $(S, \mathcal{S}, \mu)$ is one dimensional Lebesgue measure and $S = [0, 1]$. We show that in this case the L_p spaces are separable.

LEMMA 1. Let S_i, $i = 1, \ldots, n$, be a partition of S into measurable sets which are pair-wise disjoint and such that $S = \bigcup_{i=1}^{n} S_i$. Then, for every $\epsilon > 0$, there are elementary sets J_i, $i = 1, \ldots, n$, which are pair-wise disjoint and such that

$$S \sim \left(\bigcup_{i=1}^{n} J_i\right)$$

is finite, and

$$\mu(S_i \sim J_i) < \epsilon, \quad i = 1, \ldots, n.$$

Proof. This is a routine application of the definition of measurable set, and the proof is left to the reader.

We shall call a function a **stair** function if there is a partition $0 = a_0 < a_1 < \ldots < a_n = 1$, on each interval of which the function is constant.

LEMMA 2. If f is a bounded measurable function on S, then for every $\epsilon > 0$ there is a stair function g, with the same bound as f, and such that

$$|f(x) - g(x)| < \epsilon$$

except on a set of measure less than ϵ.

Proof. This is an easy application of lemma 1 and the details are left to the reader.

COROLLARY. If f is a bounded measurable function on S, then for every $\epsilon > 0$, there is a continuous function h, with the same bound as f, and such that

$$|f(x) - h(x)| < \epsilon$$

except on a set of measure less than ϵ.

Proof. By a modification of the function g of lemma 2.

PROPOSITION 1. If $p \geq 1$, the continuous functions are dense in $L_p = L_p(S, \mathcal{S}, \mu)$.

Proof. Let $f \in L_p$, and $\epsilon > 0$. There is a bounded g such that $\|f - g\| < \epsilon/2$. But, by the corollary to lemma 2, there is a continuous h such that $\|g - h\| < \epsilon/2$.

COROLLARY. The spaces $L_p = L_p(S, \mathcal{S}, \mu)$ are separable.

Proof. Let $f \in L_p$ and $\epsilon > 0$. There is a continuous g such that $\|f - g\| < \epsilon/2$. By the Weierstrass approximation theorem, there is a polynomial p, with rational coefficients, such that $\|g - p\| < \epsilon/2$. Since the set of polynomials with rational coefficients is countable, the result follows.

3.12 Approximation and Smoothing Operations

We next discuss the approximation of functions in L_p by means of special functions in a more systematic way.

As a first example, suppose for every $n = 1, 2, \ldots,$ we divide $S = [0, 1]$ into n equal intervals

$$S_1 = \left[0, \frac{1}{n}\right], \quad S_i = \left(\frac{i-1}{n}, \frac{i}{n}\right], \quad i = 2, \ldots, n.$$

Let $p > 1$. For every n, and $f \in L_p$, let

$$f_n(x) = n \int_{S_i} f(x)\, dx, \quad x \in S_i, \quad i = 1, \ldots, n.$$

(We shall write $\int f\, d\mu$ or $\int f(x)\, dx$ for the integral.)

Then f_n is constant on each S_i, the constant being the mean value of f on S_i. We first show that the norm of f_n, in the space L_p, is not greater than that of f.

$$\begin{aligned}
\int_S |f_n|^p\, d\mu &= \sum_{i=1}^{n} \int_{S_i} |f_n|^p\, d\mu \\
&= \sum_{i=1}^{n} \int_{S_i} \left| n \int_{S_i} f(x)\, dx \right|^p dt \\
&= \sum_{i=1}^{n} n^p \int_{S_i} \left| \int_{S_i} f(x)\, dx \right|^p dt \\
&\le \sum_{i=1}^{n} n^p \int_{S_i} \left\{ \int_{S_i} |f(x)|\, dx \right\}^p dt \\
&\le \sum_{i=1}^{n} n^p \int_{S_i} \left(\int_{S_i} |f(x)|^p\, dx \right) \left(\int_{S_i} |1|^q\, dx \right)^{p/q} dt,
\end{aligned}$$

where $1/p + 1/q = 1$, and the Hölder inequality is applied in each term to the function pair f, 1. Then,

$$\begin{aligned}
\int_S |f_n|^p\, d\mu &\le \sum_{i=1}^{n} n^p \int_{S_i} |f|^p\, d\mu \cdot \left(\frac{1}{n}\right)^{1+p/q} \\
&= \sum_{i=1}^{n} \int_{S_i} |f|^p\, d\mu = \int_S |f|^p\, d\mu.
\end{aligned}$$

It follows that $\|f_n\| \le \|f\|$.

The proof for the case $p = 1$, which is easier, is left to the reader. We now prove

PROPOSITION 1. For every $p \geq 1$, and $f \in L_p$, the sequence $\{f_n\}$ converges to f in the L_p norm.

Proof. Let $\epsilon > 0$. There is a continuous g such that

$$\|f - g\| < \frac{\epsilon}{3}.$$

We show that $\{g_n\}$ converges to g uniformly; hence, in the L_p norm. Let $\eta > 0$. There is a $\delta > 0$ such that $|x - y| < \delta$ implies $|g(x) - g(y)| < \eta$. Choose N so that $1/N < \delta$. Then $n > N$ implies $|g(x) - g_n(x)| < \eta$ for all $x \in S$.

Now, suppose η chosen so small that $n > N$ implies

$$\|g - g_n\| < \frac{\epsilon}{3}.$$

Then,

$$\|f - f_n\| \leq \|f - g\| + \|g - g_n\| + \|g_n - f_n\|.$$

But,

$$\|g_n - f_n\| = \|(g - f)_n\| \leq \|g - f\|,$$

so that

$$\|f - f_n\| \leq \frac{\epsilon}{3} + \frac{\epsilon}{3} + \frac{\epsilon}{3} = \epsilon, \quad \text{for every } n > N.$$

Hence $\{f_n\}$ converges to f in the L_p norm.

Instead of taking the mean values on intervals, as above, we may take a sliding average. Let $S = [0, 1]$ and let μ be Lebesgue measure. Then, for every $f \in L_p$, extend the domain of f so that it is defined for all reals and is zero outside of S. For every $h > 0$, let

$$f_h(x) = \frac{1}{2h} \int_{x-h}^{x+h} f(t)\, dt, \quad 0 \leq x \leq 1.$$

We shall need

LEMMA 3. For every totally finite measure space $(S, \mathcal{S}, \mu)$, and every $p \geq 1$, $L_1 \supset L_p$.

Proof. Let $f \in L_p$, and let

$$E = [x : |f(x)| > 1], \quad \text{and} \quad F = S \sim E.$$

Then,

$$\begin{aligned} \int_S |f|\, d\mu &= \int_F |f|\, d\mu + \int_E |f|\, d\mu \\ &\leq \int_F |f|\, d\mu + \int_E |f|^p\, d\mu < \infty. \end{aligned}$$

Hence, $f \in L_1$.

PROPOSITION 2. For every $f \in L_1$, and $h > 0$, the function f_h is continuous.

Proof. For every $\epsilon > 0$, there is a $\delta > 0$, such that $\mu(E) < \delta$ implies $\int_E |f(t)|\,dt < \epsilon h$. Now, $|x - y| < \delta$ implies

$$|f_h(x) - f_h(y)| = \frac{1}{2h}\left|\int_{x-h}^{x+h} f(t)\,dt - \int_{y-h}^{y+h} f(t)\,dt\right|$$
$$= \frac{1}{2h}\left|\int_{x-h}^{y-h} f(t)\,dt - \int_{y+h}^{x+h} f(t)\,dt\right| < \epsilon.$$

PROPOSITION 3. If $p \geq 1, f \in L_p$, and $h > 0$, then

$$\|f_h\| \leq \|f\|.$$

Proof. We first suppose $p = 1$. Suppose f is non-negative. Then

$$\int_0^1 f_h(x)\,dx = \int_0^1 \left\{\frac{1}{2h}\int_{x-h}^{x+h} f(t)\,dt\right\}dx$$
$$= \frac{1}{2h}\int_{-h}^{h} dx \int_x^{1+x} f(t)\,dt \leq \int_0^1 f(t)\,dt,$$

since

$$\int_x^{1+x} f(t)\,dt \leq \int_0^1 f(t)\,dt \text{ for every } x \in S.$$

(We have used Fubini's theorem, whose proof has been relegated to the exercises.)

Now, consider any $f \in L_1$. Then, for every $x \in S$,

$$|f_h(x)| = \left|\frac{1}{2h}\int_{x-h}^{x+h} f(t)\,dt\right| \leq \frac{1}{2h}\int_{x-h}^{x+h} |f(t)|\,dt$$
$$= |f|_h(x).$$

Then,

$$\int_0^1 |f_h(t)|\,dt \leq \int_0^1 |f|_h(t)\,dt \leq \int_0^1 |f(t)|\,dt.$$

Suppose $p > 1$. Let $f \in L_p$ and $h > 0$. By the Hölder inequality,

$$\left|\int_{x-h}^{x+h} f(t)\,dt\right| \leq \left(\int_{x-h}^{x+h} |f|^p\,d\mu\right)^{1/p}\left(\int_{x-h}^{x+h} d\mu\right)^{1/q},$$

whereby

$$\left|\int_{x-h}^{x+h} f(t)\,dt\right|^p \leq \int_{x-h}^{x+h} |f|^p\,d\mu \cdot (2h)^{p/q},$$

and it follows that

$$|f_h(x)|^p \leq \frac{1}{2h}\int_{x-h}^{x+h} |f|^p\,d\mu = \{|f|^p\}_h(x).$$

So,

$$\int_0^1 |f_h|^p\,d\mu \leq \int_0^1 \{|f|^p\}_h\,d\mu.$$

But,

$$\int_0^1 \{|f|^p\}_h \, d\mu \leq \int_0^1 |f|^p \, d\mu,$$

so that

$$\|f_h\| \leq \|f\|.$$

THEOREM 1. For every $f \in L_p$, $p \geq 1$,

$$\lim_{h \to 0} \|f - f_h\| = 0.$$

Proof. Suppose g is continuous. Then it is easy to see that the g_h are uniformly bounded and converge uniformly to g on every compact subset of $(0, 1)$. Hence,

$$\lim_{h \to 0} \|g - g_n\| = 0.$$

Let $f \in L_p$ and $\epsilon > 0$. There is a continuous g such that $\|f - g\| < \epsilon/3$. Let $\delta > 0$ be such that $0 < h < \delta$ implies $\|g - g_h\| < \epsilon/3$. Then,

$$\|f - f_h\| \leq \|f - g\| + \|g - g_h\| + \|g_h - f_h\|$$

and

$$\|g_h - f_h\| = \|(g - f)_h\| \leq \|g - f\|$$

imply that $\|f - f_h\| < \epsilon$, and the theorem is proved.

There are other interesting ways in which this theorem may be proved, and various generalizations which we leave as exercises.

Using the approximating functions f_h, we may obtain an interesting characterization of compact sets in L_p, $p > 1$.

THEOREM 2. If $S = [0, 1]$ and μ is Lebesgue measure, then for every $p > 1$, a set $A \subset L_p$ is compact if and only if it is closed, bounded, and

$$\lim_{h \to 0} \|f - f_h\| = 0$$

uniformly on A.

Proof. We first show that if $\{f_n\}$ is a Cauchy sequence in L_p, then

$$\lim_{h \to 0} \|f_n - f_{n,h}\| = 0$$

and the convergence is uniform in n.

Let $\epsilon > 0$. There is an N such that $n > N$ implies $\|f_n - f_N\| < \epsilon/3$. There is a $\delta_1 > 0$ such that $0 < h < \delta_1$ implies $\|f_N - f_{N,h}\| < \epsilon/3$. Then, for every $n > N$,

$$\|f_n - f_{n,h}\| \leq \|f_n - f_N\| + \|f_N - f_{N,h}\| + \|f_{N,h} - f_{n,h}\|.$$

But,

$$\|f_{N,h} - f_{n,h}\| = \|(f_N - f_n)_h\| \leq \|f_N - f_n\|,$$

so that

$$\|f_n - f_{n,h}\| < \epsilon.$$

There is a $\delta_2 > 0$ such that $0 < h < \delta_2$ and $n \leq N$ implies $\|f_n - f_{n,h}\| < \epsilon$. Let $\delta = \min(\delta_1, \delta_2)$. Then $0 < h < \delta$ implies

$$\|f_n - f_{n,h}\| < \epsilon, \text{ for all } n.$$

We assume $A \subset L_p$ is compact. Then A is closed and bounded. Suppose $\|f - f_h\|$ does not converge uniformly to 0 on A. Then there is a sequence $\{h_n\}$ converging to 0 and $\{f_n\}, f_n \in A$, and a $k > 0$ such that

$$\|f_n - f_{n,h_n}\| > k, \quad n = 1, 2, \ldots.$$

Since A is compact, the sequence $\{f_n\}$ may be assumed to be convergent. But this contradicts the assertion proved above. Thus, the conditions are necessary.

Suppose now that $A \subset L_p$ is closed, bounded, and that $\lim_{h \to 0} \|f - f_h\| = 0$ uniformly on A. In order to show that A is compact, we need only show that it is totally bounded. We note that we will know this as soon as we know that for every $h > 0$ the set

$$[f_h : f \in A]$$

is totally bounded, for if $\epsilon > 0$ is prescribed, there is a $\delta > 0$ so that $0 < h < \delta$ implies

$$\|f - f_h\| < \epsilon$$

for every $f \in A$. Hence, if we choose a finite set $g_1, \ldots, g_n$ such that for every f_h there is a g_i with $\|f_h - g_i\| < \epsilon$, then $\|f - g_i\| < 2\epsilon$.

But the functions

$$[f_h : f \in A]$$

are continuous and their total boundedness follows from uniform boundedness and equicontinuity, which we now prove.

There is an M such that $\|f\| \leq M$ for every $f \in A$. Let $h > 0$ and $f \in A$. Then for every $x \in [0, 1]$,

$$|f_h(x)| \leq \frac{1}{2h}\left(\int_{x-h}^{x+h} |f|^p \, d\mu\right)^{1/p}\left(\int_{x-h}^{x+h} d\mu\right)^{1/q} \leq \frac{M}{(2h)^{1/p}}.$$

This shows that $[f_h : f \in A]$ is uniformly bounded.

Finally, if $0 \leq x < y \leq 1$, the Hölder inequality implies

$$|f_h(y) - f_h(x)| < \frac{M}{h}|y - x|,$$

so that $[f_h : f \in A]$ is equicontinuous. This completes the proof.

3.13. The Dual of L_p, $p > 1$

We now show that if $p > 1$, the dual of L_p is L_q, where $1/p + 1/q = 1$. It is convenient, in certain parts of the proof, to assume that $(S, \mathcal{S}, \mu)$ is totally finite, and we make this assumption. The necessary changes for the totally

σ finite case are cumbersome but minor. We first consider some preliminary material.

PROPOSITION 1. If $f \in L_p$, $p > 1$, then

$$\|f\| = \max \left[\int_S |fg| \, d\mu : g \in L_q, \|g\| = 1\right].$$

Proof. By Hölder's inequality,

$$\int_S |fg| \, d\mu \leq \|f\| \, \|g\| \leq \|f\|, \text{ if } \|g\| = 1.$$

Hence,

$$\|f\| \geq \sup \left[\int_S |fg| \, d\mu : g \in L_q, \|g\| = 1\right].$$

Let

$$g = \operatorname{sgn} f \cdot |f|^{p-1} \|f\|^{-p/q},$$

where

$$\operatorname{sgn} f(x) = \begin{cases} +1 & \text{if } f(x) \geq 0 \\ -1 & \text{if } f(x) < 0. \end{cases}$$

Then,

$$\int_S |g|^q \, d\mu = \int_S |f|^{(p-1)q} \|f\|^{-p} \, d\mu = 1,$$

so that $\|g\| = 1$. But

$$\int_S |fg| \, d\mu = \int_S |f|^p \, d\mu \cdot \|f\|^{-p/q} = \|f\|^{p-p/q} = \|f\|.$$

PROPOSITION 2. If $f \notin L_p$, $p > 1$, then

$$\sup \left[\int_S |fg| \, d\mu : g \in L_q, \|g\| = 1\right] = \infty.$$

Proof. Let

$$f_n(x) = \begin{cases} n & \text{if } f(x) > n \\ f(x) & \text{if } |f(x)| \leq n \\ -n & \text{if } f(x) < -n. \end{cases}$$

Then $\{\|f_n\|\}$ is an unbounded increasing sequence. Now,

$$\begin{aligned} \sup &\left[\int_S |fg| \, d\mu : g \in L_q, \|g\| = 1\right] \\ &\geq \lim_{n\to\infty} \max \left[\int_S |f_n g| \, d\mu : g \in L_q, \|g\| = 1\right] \\ &= \lim_{n\to\infty} \|f_n\| = \infty. \end{aligned}$$

COROLLARY. If $f \notin L_p$, there is $g \in L_q$ with $\int_S fg \, d\mu = \infty$.

Proof. Let $\epsilon_n > 0$, $n = 1, 2, \ldots$ and $\sum_{n=1}^{\infty} \epsilon_n < \infty$. By proposition 2, there are $g_n \in L_q$, $n = 1, 2, \ldots$ such that fg_n is non-negative, $\|g_n\| = \epsilon_n$ and

$\int_S fg_n \, d\mu > 1$. The function $g = \sum_{n=1}^{\infty} g_n$ is easily seen to have the desired property.

Let $(S, \mathcal{S}, \mu)$ be a totally finite measure space, and let $g \in L_q$, $q > 1$.

PROPOSITION 3. The functional

$$\varphi(f) = \int_S fg \, d\mu$$

is linear and continuous on L_p, and

$$\|\varphi\| = \|g\|.$$

Proof. The functional is linear since for $f_1, f_2 \in L_p$ and $a_1, a_2 \in R$,

$$\varphi(a_1 f_1 + a_2 f_2) = \int_S (a_1 f_1 + a_2 f_2) g \, d\mu = a_1 \int_S f_1 g \, d\mu + a_2 \int_S f_2 g \, d\mu$$
$$= a_1 \varphi(f_1) + a_2 \varphi(f_2).$$

By proposition 1, with the roles of p and q interchanged,

$$\max\,[|\varphi(f)| : f \in L_p, \quad \|f\| = 1] = \|g\|.$$

This completes the proof.

We shall prove that for every continuous linear functional φ on L_p, $p > 1$, there is a $g \in L_q$ such that

$$\varphi(f) = \int_S fg \, d\mu$$

for every $f \in L_p$. Thus, let φ be a continuous linear functional on L_p. We define a function g as follows:

For every $E \in \mathcal{S}$, let

$$\nu(E) = \varphi(\chi_E).$$

Then ν is a signed measure. Indeed, that it is finite and additive follows from the linearity of φ. Let $\{E_n\}$ be a sequence of pair-wise disjoint sets in $\mathcal{S}$. Let $F_n = \bigcup_{k=1}^{n} E_k$, $n = 1, 2, \ldots$, and let $E = \bigcup_{n=1}^{\infty} E_n$. Then χ_E is the limit of $\{\chi_{F_n}\}$ in the L_p norm so that, since φ is continuous, $\nu(E) = \lim_{n \to \infty} \nu(F_n)$. Hence, ν is a signed measure. Moreover, ν is absolutely continuous with respect to μ. By the Radon-Nikodym theorem, there is a summable g such that

$$\nu(E) = \int_E g \, d\mu$$

for every $E \in \mathcal{S}$. In other words,

$$\varphi(\chi_E) = \nu(E) = \int_E g \, d\mu = \int_S \chi_E g \, d\mu.$$

It follows from the linearity of φ and of the integral that, for every step function f,

$$\varphi(f) = \int_S fg \, d\mu.$$

Let f be a bounded measurable function. There is a sequence $\{f_n\}$ of step functions, which converges uniformly to f, so that $\{f_n g\}$ converges boundedly to fg. Hence,

$$\varphi(f) = \lim_{n\to\infty} \varphi(f_n) = \lim_{n\to\infty} \int_S f_n g \, d\mu = \int_S fg \, d\mu.$$

We next show that the function g is in the space L_q. Suppose $f \in L_p$ and, for every n, let

$$f_n(x) = \begin{cases} |f(x)| \operatorname{sgn} g(x) & \text{if } |f(x)| \leq n \\ 0 & \text{if } |f(x)| > n. \end{cases}$$

Since $\|f_n\| \leq \|f\|$, we have

$$|\varphi(f_n)| \leq \|\varphi\| \, \|f_n\| \leq \|\varphi\| \, \|f\|.$$

Since $\{|f_n g|\}$ converges to $|fg|$,

$$\begin{aligned} \int_S |fg| \, d\mu &\leq \liminf_{n\to\infty} \int_S |f_n g| \, d\mu \\ &= \liminf_{n\to\infty} \int_S f_n g \, d\mu = \liminf_{n\to\infty} |\varphi(f_n)| \\ &\leq \|\varphi\| \, \|f\|. \end{aligned}$$

Thus, $\int_S fg \, d\mu < \infty$, for every $f \in L_p$. By the corollary to proposition 2, it follows that $g \in L_q$.

Now, let $f \in L_p$ and, for every n, let

$$f_n(x) = \begin{cases} n & \text{if } f(x) > n \\ f(x) & \text{if } |f(x)| \leq n \\ -n & \text{if } f(x) < -n. \end{cases}$$

Then $\{f_n\}$ converges to f in the L_p norm, so that $\varphi(f) = \lim_{n\to\infty} \varphi(f_n)$. Moreover, $\{f_n g\}$ converges to fg, and $\{|f_n g|\}$ is dominated by $|fg|$, so that

$$\int_S fg \, d\mu = \lim_{n\to\infty} \int_S f_n g \, d\mu.$$

Since $\varphi(f_n) = \int_S f_n g \, d\mu$, we have shown that $\varphi(f) = \int_S fg \, d\mu$. This completes the proof of

THEOREM 1. If $(S, \mathcal{S}, \mu)$ is a totally finite measure space and φ is a continuous linear functional on $L_p = L_p(S, \mathcal{S}, \mu)$, $p > 1$, there is a $g \in L_q$ such that

$$\varphi(f) = \int_S fg \, d\mu$$

for every $f \in L_p$, and $\|\varphi\| = \|g\|$.

Thus, the dual $(L_p)'$ of L_p is isomorphic, as a Banach space, with L_q.

3.14 The Dual of L_1

We now consider the case $p = 1$. We show that the dual of $L_1 = L_1(S, \mathcal{S}, \mu)$, where $(S, \mathcal{S}, \mu)$ is totally finite, is the space M of bounded measurable functions with

$$\|f\| = \text{ess sup}\,[|f(x)| : x \in S],$$

where

$$\text{ess sup}\,[|f(x)| : x \in S] = \inf\,[y : \mu\{x : |f(x)| > y\} = 0].$$

We obtain

(a) If $f \in L_1$, then

$$\|f\| = \max\left[\int_S |fg|\, d\mu : g \in M,\ \|g\| = 1\right].$$

The proof of (a) is easy, and is left to the reader.

(b) If $f \notin L_1$, then

$$\sup\left[\int_S |fg|\, d\mu : g \in M,\ \|g\| = 1\right] = \infty.$$

The proof of (b) is also easy.

(c) If $f \in M$, then

$$\|f\| = \sup\left[\int_S |fg|\, d\mu : g \in L_1,\ \|g\| = 1\right].$$

That $\|f\| \geq \int_S |fg|\, d\mu$, for every $g \in L_1$, $\|g\| = 1$, is obvious. Suppose $\|f\| > 0$. Let $\epsilon > 0$. Let $E \in \mathcal{S}$, $\mu(E) > 0$, be such that $|f(x)| > \|f\| - \epsilon$ for every $x \in E$. Define g so that

$$f(x) = \begin{cases} \dfrac{1}{\mu(E)} \operatorname{sgn} f(x) & \text{if } x \in E \\ 0 & \text{if } x \notin E. \end{cases}$$

Then,

$$\int_S fg\, d\mu \geq \|f\| - \epsilon$$

and $\|g\| = 1$ in the L_1 space.

(d) If $f \notin M$, then

$$\sup\left[\int_S |fg|\, d\mu : g \in L_1,\ \|g\| = 1\right] = \infty.$$

The proof is like that of proposition 2, Sec. 3.13.

(e) If $f \notin M$, there is $g \in L_1$ with $\int_S fg\, d\mu = \infty$. This follows from (d) in the same way as the corollary follows from proposition 2, Sec. 3.13.

(f) If $f \notin L_1$, there is $g \in M$ with $\int_S fg\, d\mu = \infty$. The function $g = \operatorname{sgn} f$ does the job.

PROPOSITION 1. If $g \in M$, then the functional

$$\varphi(f) = \int_S fg \, d\mu$$

is continuous and linear on $L_1 = L_1(S, \mathcal{S}, \mu)$ and $\|\varphi\| = \|g\|$. $(S, \mathcal{S}, \mu)$ is again restricted to being totally finite.

Proof. The proof differs little from that of proposition 3, Sec. 3.13. The details are left to the reader.

Conversely, for every continuous linear φ on L_1, there is a $g \in M$ such that

$$\varphi(f) = \int_S fg \, d\mu,$$

for every $f \in L_1$.

However, there exist continuous linear functionals φ on M such that there is no $g \in L_1$ for which $\varphi(f) = \int_S fg \, d\mu$ for all $f \in M$. We discuss these facts.

Let φ be a continuous linear functional on L_1. Then

$$\nu(E) = \varphi(\chi_E)$$

is easily seen to be a finite signed measure on $\mathcal{S}$.

(The proof breaks down at this point in the case of a continuous linear functional on M.)

ν is absolutely continuous with respect to μ. There is, accordingly, a summable g such that

$$\nu(E) = \varphi(\chi_E) = \int_S \chi_E \cdot g \, d\mu.$$

The extension of this equality to the bounded measurable functions, the proof that g is essentially bounded, and the proof that

$$\varphi(f) = \int_S fg \, d\mu$$

for all $f \in L_1$, all proceed in much the same way as for $L_p, p > 1$. The details are omitted. We then obtain

THEOREM 1. If $(S, \mathcal{S}, \mu)$ is a totally finite measure space, and φ is a continuous linear functional on $L_1 = L_1(S, \mathcal{S}, \mu)$, there is a $g \in M$ such that

$$\varphi(f) = \int_S fg \, d\mu$$

for every $f \in L_1$ and such that $\|\varphi\| = \|g\|$. Thus, the dual $(L_1)'$ of L_1 is isomorphic with M as a Banach space.

Let S be the interval $[0, 1]$ and μ Lebesgue measure. We determine a continuous linear functional φ on $M = M(S, \mathcal{S}, \mu)$ as follows: For every continuous f on $[0, 1]$, let $\varphi(f) = f(1)$. Then φ is continuous and linear on the subspace of M composed of continuous functions. By the Hahn-Banach theorem, φ may be extended so as to be continuous and linear on M. For every $n = 1, 2, \ldots$, let

$$f_n(x) = \begin{cases} 0 & \text{if } 0 \le x \le 1 - \dfrac{1}{n} \\ nx - n + 1 & \text{if } 1 - \dfrac{1}{n} \le x \le 1. \end{cases}$$

Then $\varphi(f_n) = 1$. But for every summable g,

$$\lim_{n \to \infty} \int_S f_n g \, d\mu = 0.$$

Thus, there is no summable g such that

$$\varphi(f) = \int_S fg \, d\mu$$

for every $f \in M$.

Another way of proving this is by noting that M is not separable. It is shown in Chapter 2 that if a Banach space is not separable, then its dual is also not separable. Since L_1 is separable, it cannot be the dual of M.

We have also proved that the spaces L_p, $p > 1$ are reflexive and that L_1 is not reflexive.

3.15 The Individual Ergodic Theorem

In the remaining sections, we present a sampling of the many interesting further developments of the material of this chapter.

The ergodic theorem is concerned with measure preserving transformations, and is considered to be important because of an old theorem of Liouville which says that certain transformations which arise in the space of sets of coordinates and velocities of a finite system of particles are measure preserving. There are two types of ergodic theorems, individual and mean. We consider the individual ergodic theorem here. Many mathematicians have written proofs of this theorem since the original one given by G. D. Birkhoff in 1933. The proof presented here, which is relatively simple, is due to F. Riesz. We start with

LEMMA 1. If $a_1, a_2, \ldots, a_n$ is an ordered set of real numbers, if $m < n$, and if $a_i \in S_m$ whenever $a_i + a_{i+1} + \ldots + a_{i+p-1} \ge 0$ for some $p \le m$, then

$$\sum [a_i : a_i \in S_m] \ge 0.$$

The proof is left as an exercise for the reader. For convenience, the members of S_m are called m-favorable.

Let $(S, \mathcal{S}, \mu)$ be a totally finite measure space, and let T be a one-one measure preserving transformation of S onto itself. Then

$$\mu(T(E)) = \mu(E), \quad \text{for every } E \in \mathcal{S}.$$

For every $x \in S$, we consider the sequence

$$x,\ Tx,\ T^2x = T(Tx), \ldots, T^nx = T(T^{n-1}x), \ldots .$$

PROPOSITION 1. If $f \in L_1$ and E is the set of points x for which

$$f(x) + f(Tx) + \ldots + f(T^nx) \geq 0$$

for some $n = 0, 1, 2, \ldots$, then

$$\int_E f\,d\mu \geq 0.$$

Proof. For every m, let E_m be the set of points x such that

$$f(x) + f(Tx) + \ldots + f(T^kx) \geq 0$$

for some $k \leq m$. It is enough to show that $\int_{E_m} f\,d\mu \geq 0$, for every m.
For every x, consider the ordered set

$$f(x), f(Tx), \ldots, f(T^{n+m-1}x)$$

where n is fixed but arbitrary. Let F_k be the set of points x for which $f(T^kx)$ is m-favorable. By lemma 1,

$$\sum_{k=0}^{n+m-1} \int_{F_k} f(T^kx)\,d\mu \geq 0.$$

But, it is evident that $TF_k = F_{k-1}$ for every $k \leq n - 1$ and that $F_0 = E_m$. Hence, the last inequality reduces to

$$n \int_{E_m} f\,d\mu + \sum_{k=n}^{n+m-1} \int_{F_k} f(T^kx)\,d\mu \geq 0.$$

This yields,

$$n \int_{E_m} f\,d\mu + m \int_S |f|\,d\mu \geq 0.$$

Since this holds for every n, it follows that $\int_{E_m} f\,d\mu \geq 0$ and the proposition is proved.

We now prove the individual ergodic theorem for totally finite measures.

THEOREM 1. If T is measure preserving and $f \in L_1$, then if

$$\varphi_n(x) = \frac{1}{n}\{f(x) + f(Tx) + \ldots + f(T^{n-1}x)\},$$

$\lim_{n\to\infty} \varphi_n(x)$ exists for almost every x.

Proof. Let $a < b$ be rational numbers, and let A be the set of points x for which

$$\liminf_{n\to\infty} \varphi_n(x) < a < b < \limsup_{n\to\infty} \varphi_n(x).$$

It is clearly sufficient to show that $\mu(A) = 0$. The definition of A readily implies that $TA = A$. Using A as our space, which is permissible since it is invariant under T, proposition 1 yields

$$\int_A (f - b)\, d\mu \geq 0 \text{ and } \int_A (a - f)\, d\mu \geq 0.$$

It follows that $\int_A (a - b)\, d\mu \geq 0$. Since $a - b < 0$, this implies $\mu(A) = 0$ and the theorem is proved.

3.16 L_p Convergence of Fourier Series

In the last section we gave a proof of the ergodic theorem due to F. Riesz. In this section, we present a theorem on Fourier series due to his brother M. Riesz. Not all details in the proof will be given; the reader is referred to the books of Bari [1] and Zygmund [1].

Let $f \in L_p(-\pi, \pi)$, $p > 1$, extend f by periodicity, of period 2π to the real line, and let

$$\frac{a_0}{2} + \sum_{n=1}^{\infty} (a_n \cos nx + b_n \sin nx)$$

be the Fourier series of f.

There is a function $\bar{f}$ defined by

$$\begin{aligned} \bar{f}(x) &= -\frac{1}{\pi} \int_0^{\pi} \frac{f(x + t) - f(x - t)}{2 \tan t/2}\, dt \\ &= \lim_{\varepsilon \to 0} \left\{ -\frac{1}{\pi} \int_{\varepsilon}^{\pi} \frac{f(x + t) - f(x - t)}{2 \tan t/2}\, dt \right\}. \end{aligned}$$

We omit the proof of the existence, and also that f and $\bar{f}$ are boundary functions of conjugate harmonic functions. To be precise, there is an analytic $F(z)$, $|z| < 1$, with

$$\begin{aligned} F(z) &= u(z) + iv(z) \\ &= u(re^{i\theta}) + iv(re^{i\theta}) \end{aligned}$$

and

$$\lim_{r \to 1} u(re^{i\theta}) = f(\theta), \quad \lim_{r \to 1} v(re^{i\theta}) = \bar{f}(\theta)$$

almost everywhere. Moreover, $0 < r < 1$ implies

$$\|u(re^{i\theta})\| \leq \|f\|,$$

and $f(\theta) \geq 0$, not identically zero, implies $u(z) > 0$, $|z| < 1$.

The Fourier series of $\bar{f}$ is

$$\sum_{n=1}^{\infty} (-b_n \cos nx + a_n \sin nx).$$

(These are the only major facts whose proofs are omitted.)

We shall show that there is a constant A, depending only on p, such that $\|\tilde{f}\| \leq A \|f\|$. In other words, the linear operator which takes each function in L_p into its conjugate function is a bounded operator on L_p to L_p. We need

LEMMA 1. If $x \in [-\pi/2, \pi/2]$, $1 < p \leq 2$, then

$$|\sin x|^p \leq a |\cos x|^p - b \cos px,$$

where a and b depend only on p.

This is an elementary computation and is left to the reader.

PROPOSITION 1. With the above notation, and the additional hypothesis that $u(z) > 0$ for $|z| < 1$, $v(0) = 0$, we have

$$\|v(re^{i\theta})\| \leq A\|u(re^{i\theta})\|,$$

where A depends only on p, and $0 \leq r < 1$.

Proof. Let $u = R\cos\varphi$, $v = R\sin\varphi$. We first show that

$$\int_0^{2\pi} R^p \cos p\varphi \, dx > 0.$$

Since $u(z) > 0$, $F(z) \neq 0$, so that F^p is analytic for $|z| < 1$. Then

$$\frac{1}{2\pi}\int_0^{2\pi} R^p \cos p\varphi \, dx = \operatorname{Re}\left[\frac{1}{2\pi i}\int \frac{F(z)^p}{z}\, dz\right] = \operatorname{Re}\,[F(0)^p]$$

$$= u(0)^p > 0.$$

Multiplying the inequality of lemma 1 by R^p, and integrating,

$$\int_0^{2\pi} |v(re^{i\theta})|^p \, d\theta \leq a \int_0^{2\pi} |u(re^{i\theta})|^p \, d\theta - b \int_0^{2\pi} R^p \cos p\varphi \, d\theta$$

and so

$$\|v(re^{i\theta})\| \leq A \|u(re^{i\theta})\|,$$

where $1 < p \leq 2$.

We next show how to obtain this inequality for q, having it for p, where $1/p + 1/q = 1$.

For every trigonometric polynomial g, with p norm $\|g\| \leq 1$, and, considering the q norms of u and v, we have

$$\int_0^{2\pi} |u\bar{g}| \, dx \leq \|u\| \, \|\bar{g}\| \leq \|u\| \, A \, \|g\| \leq A \, \|u\|,$$

and

$$\int_0^{2\pi} |vg| \, dx \leq A \, \|u\|, \text{ since } \int_0^{2\pi} vg \, dx = -\int_0^{2\pi} u\bar{g} \, dx,$$

by the Parseval equality (see Chapter 4), which applies since all functions involved are in L_2. But

$$\sup\left[\int_0^{2\pi} |vg| \, dx : \|g\| = 1\right] = \|v\|,$$

since the trigonometric polynomials are dense in L_p, whereby

$$\|v\| \le A \|u\|.$$

We now prove

THEOREM 1. For every $f \in L_p$, $p > 1$,

$$\|\tilde{f}\| \le A \|f\|.$$

Proof. Suppose $f \ge 0$. Then by the Fatou theorem,

$$\|\tilde{f}\| \le \limsup_{r\to 1} \|v(re^{i\theta})\|.$$

But

$$\limsup_{r\to 1} \|u(re^{i\theta})\| \le \|f\|,$$

so that

$$\|\tilde{f}\| \le A \|f\|.$$

For f arbitrary, let

$$f_1 = \max(f, 0)$$
$$f_2 = f - f_1.$$

Then,

$$\begin{aligned}\|\tilde{f}\| \le \|\tilde{f}_1\| + \|\tilde{f}_2\| &\le A[\|f_1\| + \|f_2\|] \\ &\le 2A \|f\|.\end{aligned}$$

We shall use this theorem to show that if $f \in L_p$, $p > 1$, then the Fourier series of f converges to f in the L_p norm.

We introduce the sequence $\{s_n^*\}$ defined by

$$s_n^*(x) = \frac{1}{\pi}\int_{-\pi}^{\pi} \frac{f(x+t)\sin nt}{2\tan t/2}\,dt.$$

Then

$$\begin{aligned}s_n^*(x) &= \cos nx \cdot \frac{1}{\pi}\int_{-\pi}^{\pi} \frac{f(x+t)\sin n(t-x)}{2\tan t/2}\,dt \\ &\quad - \sin nx \cdot \frac{1}{\pi}\int_{-\pi}^{\pi} \frac{f(x+t)\cos n(t-x)}{2\tan t/2}\,dt.\end{aligned}$$

Let

$$f_1(x) = f(x)\sin nx, \quad f_2(x) = f(x)\cos nx,$$

then

$$s_n^*(x) = \cos nx\,\tilde{f}_1(x) - \sin nx\,\tilde{f}_2(x),$$

and it follows that

$$\|s_n^*\| \le \|\tilde{f}_1\| + \|\tilde{f}_2\| \le A[\|f_1\| + \|f_2\|] \le 2A \|f\|.$$

Since the partial sum s_n of the Fourier series of f satisfies

$$s_n(x) = s_n^*(x) + \frac{1}{2\pi}\int_{-\pi}^{\pi} f(x+t)\cos nt\,dt,$$

we have

$$\|s_n\| \le \|s_n^*\| + \|f\| \le 2(A+1)\,\|f\|.$$

Thus the sequence of linear operators $\{s_n\}$ on L_p to L_p is uniformly bounded.

Let T be a trigonometric polynomial such that, for a prescribed $\epsilon > 0$, $\|f - T\| < \epsilon$. $\left(\text{The averages } \dfrac{1}{n+1}(s_0 + s_1 + \ldots + s_n) \text{ of the partial sums of the Fourier series of } f \in L_p,\ p > 1, \text{ converge to } f \text{ in the } L_p \text{ norm.}\right)$ Now,

$$\begin{aligned} f - s_n(f) &= (f - T) + (T - s_n(f)) \\ &= f - T + s_n(T - f), \end{aligned}$$

where n is greater than the degree of T. But then

$$\|f - s_n\| < \epsilon + 2(A+1)\epsilon.$$

This proves

THEOREM 2. If $f \in L_p$, then the Fourier series of f converges to f in the L_p norm.

3.17 Functions Whose Fourier Series Diverge Almost Everywhere

Let X be a Banach space, and let M be the space of equivalence classes of measurable functions on a subset S of the real line, of finite measure. Then M is a vector space. We shall show in Chapter 5 that M is a metric space with distance

$$d(f, g) = \int_S \frac{|f(t) - g(t)|}{1 + |f(t) - g(t)|}\, dt.$$

We prove

LEMMA 1. Let X be a Banach space, and $\{U_n\}$ a sequence of continuous linear operators on X into M. Let $H \subset X$ be of the second category. Suppose that, for every $x \in H$, there is a set $A_x \subset S$, of measure greater than $\alpha > 0$, such that

$$\limsup_{n \to \infty} |U_n(x)(t)| < \infty$$

for every $t \in A_x$. Let $0 < \epsilon < \alpha$. Then there is a set $A \subset S$ of measure greater than $\alpha - \epsilon$ such that, for every $x \in X$,

$$\limsup_{n \to \infty} |U_n(x)(t)| < \infty$$

almost everywhere on A.

Proof. We first recall that we have shown in exercise 3.3 that the equivalence classes of measurable subsets of S form a metric space with distance

$$d(E, F) = \mu(E \sim F) + \mu(F \sim E).$$

This metric space is separable in the present case. Let $\mathcal{A}$ be a countable dense set in this metric space. Let $A_1, A_2, \ldots$ be those sets in $\mathcal{A}$ whose measure exceeds $\alpha - \epsilon/2$. Let $H_{km} \subset X$ consist of those x for which there is an $R(x)$ such that

(a) $|U_n(x)(t)| \leq m$, for every n and $t \in R(x)$,

and

(b) $\mu(A_k \sim R(x)) < \epsilon/2$.

Then

$$H \subset \bigcup H_{km}$$

and at least one of the sets in the union, say $H_{k_1 m_1}$, is of the second category in X. Moreover, since the operators U_n are continuous, $H_{k_1 m_1}$ is closed, and so it contains a sphere K.

By linearity,

$$\limsup_{n \to \infty} |U_n(x)(t)| < \infty,$$

for every $x \in X$ and $t \in A_{k_1} \cap R(x)$.

Designate A_{k_1} by $A^{(1)}$ and $A_{k_1} \cap R(x)$ by $R^{(1)}(x)$. By applying the above argument to $A^{(1)}$, we obtain $R^{(2)}(x) \subset A^{(1)}$, for every $x \in X$, such that

$$\limsup_{n \to \infty} |U_n(x)(t)| < \infty$$

for every $t \in R^{(2)}(x)$, where $m(A^{(2)} \sim R^{(2)}(x)) < \epsilon/4$ and $A^{(2)} \subset A^{(1)}$, $m(A^{(2)}) \geq \alpha - \frac{3}{4}\epsilon$.

By continuing in this way, we obtain a sequence

$$A^{(1)} \supset A^{(2)} \supset \ldots,$$

and for every $x \in X$, $R^{(k)}(x) \subset A^{(k)}$, with $m(A^{(k)} \sim R^{(k)}(x)) < \epsilon/2^k$ and $m(A^{(k)}) > \alpha - (1 - 1/2^k)\epsilon$.

Let $A = \bigcap_{k=1}^{\infty} A^{(k)}$. It is an easy matter to verify that A satisfies the requirements of the lemma.

PROPOSITION 1. Let $\{u_n(x, t)\} = \{U_n(x)(t)\}$. There is an $A \subset S$ such that

$$\limsup_{n \to \infty} |u_n(x, t)| < \infty$$

for almost all $t \in A$, for every $x \in X$, and

$$\limsup_{n \to \infty} |u_n(x, t)| = \infty$$

for almost all $t \notin A$, for every $x \in X$, except for a set of the first category.

Proof. Let β be the supremum of all α for which there is an $H(\alpha) \subset X$, of the second category, such that, for every $x \in H(\alpha)$,

$$\limsup_{n\to\infty} |u_n(x, t)| < \infty$$

on a set of measure α. If $\beta = 0$, the proposition holds. Suppose $\beta > 0$.

By lemma 1, for every p, there is an $A_p \subset S$, with $m(A_p) \geq \beta - 1/p$, such that, for every $x \in X$,

$$\limsup_{n\to\infty} |u_n(x)(t)| < \infty.$$

The set $A = \bigcap A_p$ is then seen to have the desired property without too much difficulty.

This proposition has an interesting application to Fourier series.

Let $X = L_1[-\pi, \pi]$, and for every n, $x \in L_1$, $t \in [-\pi, \pi]$, let $u_n(x, t)$ be the nth partial sum of the Fourier series of x, evaluated at t. It is a deep fact, proved by Kolmogoroff, that there is an $x \in L_1$ such that

$$\limsup_{n\to\infty} |u_n(x, t)| = \infty$$

for almost all t. (The proof of this theorem is given in Zygmund [1] and Hardy and Rogosinski [1].)

It follows that, for this case, the set A of proposition 1 is empty. In other words, the set of functions in L_1 whose Fourier series diverge almost everywhere is all of L_1 except for a set of the first category.

3.18 Continuous Functions Which Differ from All Those Having a Given Modulus

In the last section, we gave an example in which we showed how the existence of one function of a certain type implied, with the help of functional analysis, the existence of many functions of this type. In most applications of functional analysis to real variable theory, the existence itself is proved and the analysis is not nearly so deep.

We give one example which seems interesting. We shall need the Lebesgue density theorem, which we have not discussed in this text. We shall simply define **metric density** and state the theorem. The proof can be found in many places; in particular, in Goffman [1].

Let $S \subset [0, 1]$ be measurable. For every interval $I \subset [0, 1]$ we define the relative measure of S in I by

$$\frac{\mu(S \cap I)}{\mu(I)}.$$

The metric density is said to exist at a point $x \in (0, 1)$ and equal d if for every $\epsilon > 0$ there is a $\delta > 0$ such that if I is an interval, with $x \in I$ and

$\mu(I) < \delta$, then

$$\left| \frac{\mu(S \cap I)}{\mu(I)} - d \right| < \epsilon.$$

Clearly, if the metric density exists at a point, its value is between 0 and 1.

The **Lebesgue density theorem** asserts that if S is measurable, then the metric density of S exists and is equal to 1 almost everywhere on S.

We shall deal with the space $C[0, 1]$ of continuous functions on $[0, 1]$. We recall that this is a Banach space with norm

$$\|f\| = \max \{|f(x)| : x \in [0, 1]\}.$$

A function σ will be called a **modulus of continuity** if σ is defined for positive reals, is increasing, and

$$\lim_{x \to 0} \sigma(x) = 0.$$

A function $f \in C[0, 1]$ will be said to have σ as its modulus of continuity, and **belong to $C(\sigma)$,** if $x, y \in [0, 1]$ implies

$$|f(x) - f(y)| \leq \sigma(|x - y|).$$

It follows readily from the definition of continuity that every $f \in C[0, 1]$ is in $C(\sigma)$ for some σ.

What we shall show is that for every modulus of continuity σ there is an $f \in C[0, 1]$ such that for every $g \in C(\sigma)$

$$\mu\{x : f(x) = g(x)\} = 0.$$

We start with

LEMMA 1. Let ϵ and n be given, where $1 > \epsilon > 0$ and n is a positive integer. There are δ and η, where $1/n > \delta > 0$ and $\eta > 0$, and a continuous function f, with $\|f\| \leq \epsilon$, such that for every $x \in [0, 1 - 1/n]$ and every g, with $\|f - g\| < \eta$, we have either

$$|g(x) - g(y)| > \sigma(|x - y|) \text{ for every } y \in \left[x - \frac{\delta}{2}, x - \frac{\delta}{4}\right]$$

or

$$|g(x) - g(y)| > \sigma(|x - y|) \text{ for every } y \in \left[x + \frac{\delta}{4}, x + \frac{\delta}{2}\right].$$

Proof. Choose $\delta < 1/n$, so that

$$\sigma(\delta) < \frac{\epsilon}{8}.$$

Let f be continuous, with graph composed of a finite set of straight line segments, such that

$$f(0) = 0$$
$$0 \leq f(x) \leq \epsilon \text{ for all } x \in [0, 1],$$

and such that the line segments which form the graph of f have slopes $\pm\epsilon/\delta$, and all but the last has length equal to $(\epsilon^2 + \delta^2)^{1/2}$. There is a positive integer k such that $k\delta < 1 \leq (k + 1)\delta$. Then $[0, k\delta] \supset [0, 1 - 1/n]$.

For every $x \in [0, k\delta]$, it is true that $(y, f(y))$ is on the same line segment of the graph of f as is $(x, f(x))$ either for all $y \in \left[x - \frac{\delta}{2}, x - \frac{\delta}{4}\right]$ or for all $y \in \left[x + \frac{\delta}{4}, x + \frac{\delta}{2}\right]$. For all such y,

$$\begin{aligned} |f(x) - f(y)| &= \frac{\epsilon}{\delta}|x - y| \\ &> 8\sigma(|x - y|) \cdot \frac{|x - y|}{\delta} \\ &\geq 2\sigma(|x - y|), \end{aligned}$$

since $\sigma(|x - y|) < \sigma(\delta) < \epsilon/8$ and $|x - y| \geq \delta/4$. Let

$$\eta = \frac{1}{2}\sigma\left(\frac{\delta}{4}\right).$$

Suppose $\|f - g\| < \eta$. Then for x and y as above

$$\begin{aligned} |g(x) - g(y)| &> |f(x) - f(y)| - 2\eta \\ &> 2\sigma(|x - y|) - \sigma\left(\frac{\delta}{4}\right) \\ &\geq 2\sigma(|x - y|) - \sigma(|x - y|) \\ &= \sigma(|x - y|), \end{aligned}$$

since $\sigma(|x - y|) \geq \sigma(\delta/4)$.

LEMMA 2. For every positive integer n, the set $E_n \subset C[0, 1]$, of functions f for which there is a δ, with $0 < \delta < 1/n$, depending on f, such that, for every $x \in \left[0, 1 - \frac{1}{n}\right]$,

$$|f(x) - f(y)| > \sigma(|x - y|)$$

either for every $y \in \left[x - \frac{\delta}{2}, x - \frac{\delta}{4}\right]$ or for every $y \in \left[x + \frac{\delta}{4}, x + \frac{\delta}{2}\right]$, contains a dense open set in $C[0, 1]$.

Proof. Let $g \in C[0, 1]$ and ζ such that $0 < \zeta < 1$. There is a polynomial p such that

$$\|p - g\| < \frac{\zeta}{2}.$$

Let

$$M = \max\{|p'(x)| : x \in [0, 1]\}.$$

Letting the roles of ϵ and σ in lemma 1 be played by $\zeta/2$ and the function ω defined by

$$\omega(x) = \sigma(x) + Mx,$$

there are δ and η, where $0 < \delta < 1/n$ and $\eta > 0$, and a continuous function h, with $\|h\| \leq \zeta/2$, such that $\|k - h\| < \eta$ implies that, for every $x \in \left[0, 1 - \frac{1}{n}\right]$,

$$|k(x) - k(y)| > \omega(|x - y|)$$

either for every $y \in \left[x - \frac{\delta}{2}, x - \frac{\delta}{4}\right]$ or for every $y \in \left[x + \frac{\delta}{4}, x + \frac{\delta}{2}\right]$. Then, for all such x and y,

$$\begin{aligned} |(p + k)(x) - (p + k)(y)| &\geq |k(x) - k(y)| - |p(x) - p(y)| \\ &> \omega(|x - y|) - M\,|x - y| \\ &= \sigma(|x - y|). \end{aligned}$$

Since

$$\|g - (p + h)\| \leq \|g - p\| + \|h\| < \zeta,$$

the sphere of center g and radius ζ contains an open subset of E_n. Since $g \in C[0, 1]$ and $\zeta > 0$ are arbitrary, E_n contains a dense, open subset of $C[0, 1]$.

We may now prove

THEOREM 1. For every modulus of continuity σ, there exist functions $f \in C[0, 1]$ such that, for every $g \in C(\sigma)$,

$$\mu\{x : f(x) = g(x)\} = 0.$$

Moreover, the set of functions which do not have this property is of the first category.

Proof. Let E_n, $n = 1, 2, \ldots$, be as defined in lemma 2, and let

$$E = \bigcap_{n=1}^{\infty} E_n.$$

Then the set E is residual in $C[0, 1]$, i.e., its complement is of the first category in $C[0, 1]$.

Let $f \in E$. Then for every n, there is a δ_n, with $0 < \delta_n < 1/n$, such that for every $x \in \left[0, 1 - \frac{1}{n}\right]$, we have

$$|f(x) - f(y)| > \sigma(|x - y|),$$

either for every $y \in \left[x - \frac{\delta_n}{2}, x - \frac{\delta_n}{4}\right]$ or for every $y \in \left[x + \frac{\delta_n}{4}, x + \frac{\delta_n}{2}\right]$.

Now, let $g \in C(\sigma)$, and suppose the set

$$A = \{x : f(x) = g(x)\}$$

has positive measure. Let $x \in (0, 1)$ be a point in A at which the metric density of A is 1. Then the set

$$B = \{y : |f(x) - f(y)| > \sigma(|x - y|)\}$$

has metric density 0 at x. Choose n_0 so that $x < 1 - \dfrac{1}{n_0}$. Then, for every $n \geq n_0$, since $f \in E_n$,

$$|f(x) - f(y)| > \sigma(|x - y|)$$

on a subset of $\left[x - \dfrac{\delta_n}{2}, x + \dfrac{\delta_n}{2}\right]$ of relative measure $\frac{1}{4}$. Since $\lim_{n\to\infty} \delta_n = 0$, this contradicts the assertion that B has metric density 0 at x.

EXERCISES

1.1 Show that $S \subset T$ implies $m_e(S) \leq m_e(T)$.

1.2 If J is an elementary set, show that $m_e(J) = m(J)$.

1.3 Show that if $H \subset E_n$ is a hyperplane, then $m_e(H) = 0$.

1.4 Show that every open set in E_n is the union of a countable set of intervals whose interiors are pair-wise disjoint.

1.5 Give the details of the proof of the corollary to proposition 4 which says that a subset $S \subset U$ is measurable if and only if, for every $\epsilon > 0$, there is a compact $F \subset S$ such that $m_e(S \sim F) < \epsilon$.

1.6 Prove that there is a nowhere dense measurable set in [0, 1] whose intersection with every open set is either empty or of positive measure.

1.7 Prove that there is a measurable set S in [0, 1] such that, for every open set $G \subset [0, 1]$, both $m(S \cap G) > 0$ and $m(G \cap CS) > 0$.

1.8 Let S be a measurable subset of [0, 1] with the property that if $x \in S$ and the decimal expansion of x is changed in finitely many places to yield y, then $y \in S$. Show that $m(S)$ is 0 or 1.

1.9 Discuss the measurability and exterior measure of the subset of [0, 1] of points whose decimal expansions do not contain the digit 5.

1.10 Show that $S \subset U_n$ is measurable if and only if, for every $A \subset U_n$,

$$m_e(S) = m_e(S \cap A) + m_e(S \cap CA).$$

1.11 Given any set Z of measure 0, show that there is a continuous monotonically increasing function whose derivative is $+\infty$ at every point of Z.

2.1 Give an example of an unbounded connected open set of finite measure in E_2.

2.2 Give an example of an unbounded everywhere dense connected open set of finite measure in E_2.

2.3 Show that for every measurable S in E_n there is a set A, which is the union of an increasing sequence of compact sets, such that $A \subset S$ and $m(S \sim A) = 0$.

2.4 Give an example of a homeomorphic image of the unit circle $x^2 + y^2 = 1$ in E_2 whose measure in E_2 is positive.

3.1 Define a class of measures on E_n analogous to the ones defined in the text on E_1.

3.2 Give an example of a measure space which is not σ finite.

3.3 If $(S, \mathcal{S}, \mu)$ is a totally finite measure space, let

$$d(A, B) = \mu(A \sim B) + \mu(B \sim A),$$

for $A, B \in \mathcal{S}$. Show that the equivalence classes of sets in $\mathcal{S}$ form a metric space, where A is equivalent to B if $d(A, B) = 0$.

3.4 Show that the metric space of exercise 3.3 is complete.

3.5 A totally finite measure space is called separable if its associated metric space is separable. Show that the measure space of Section 3.1 is separable.

3.6 A totally finite measure space $(S, \mathcal{S}, \mu)$ is called non-atomic if for every $A \in \mathcal{S}$, $\mu(A) > 0$, there is $B \in \mathcal{S}$, $B \subset A$, $0 < \mu(B) < \mu(A)$. It is called normal if $\mu(S) = 1$. Show that if $(S, \mathcal{S}, \mu)$ and $(T, \mathcal{T}, \nu)$ are separable, non-atomic, and normal, there is a one-one mapping φ between their equivalence classes of sets such that $\nu(\varphi(A)) = \mu(A)$, $\varphi(A \cup B) = \varphi(A) \cup \varphi(B)$ and $\varphi(S \sim A) = T \sim \varphi(A)$.

4.1 Using the definition in the text, show that a function f is measurable if and only if the inverse image of every Borel set is measurable.

4.2 Using the definition in the text, show that a function f is measurable if and only if the inverse image of $[a, \infty)$ is measurable for every a.

4.3 Show, by example, that the measurability of the inverse image of (a, ∞) for every a does not imply the measurability of f.

4.4 If f is measurable on the unit interval $[0, 1]$, show that for every $\epsilon > 0$ there is a continuous g on $[0, 1]$ such that $|f(x) - g(x)| < \epsilon$, except on a set of measure less than ϵ.

4.5 Show, accordingly, that there is a sequence of continuous functions which converges almost everywhere to f.

4.6 However, using exercise 1.7, show that there is a measurable function which is not equivalent to any Baire 1 function.

4.7 Nevertheless, show that every measurable function is equivalent to a Baire 2 function.

4.8 If f is measurable on $[0, 1]$, show that, for every $\epsilon > 0$, there is a continuous g on $[0, 1]$ such that $f(x) = g(x)$, except on a set of measure less than ϵ.

4.9 If $(S, \mathcal{S}, \mu)$ is a totally finite measure space and we order equivalence classes of measurable sets by $S \geq T$ if $\mu(T \sim S) = 0$, then every collection of equivalence classes has a least upper bound.

4.10 If equivalence classes of measurable functions on a totally finite measure space are ordered so that $f \geq g$ if $f(x) \geq g(x)$ almost everywhere, show that every set of equivalence classes which has an upper bound has a least upper bound.

4.11 Give an example of a totally σ finite measure space in which theorem 1 fails.

5.1 If $\{f_n\}$ converges in measure to f and $\{g_n\}$ converges in measure to g, show that $\{f_n + g_n\}$ converges in measure to $f + g$.

5.2 If $\{f_n\}$ converges in measure to f and $\{g_n\}$ converges in measure to g, show that $\{f_n g_n\}$ converges in measure to fg.

5.3 If $(S, \mathcal{S}, \mu)$ is a totally finite measure space, show that a sequence $\{f_n\}$ of measurable real functions on S converges in measure to f if and only if every subsequence of $\{f_n\}$ has a subsequence which converges almost everywhere to f.

6.1 Let $(S, \mathcal{S}, \mu)$ be a totally σ finite measure space, and let A be the set of step functions on S. Show that with

$$\|s\| = \int_S |s| \, d\mu,$$

A is a normed vector space.

6.2 Show that A is dense in the space of summable functions with this norm.

6.3 Show that if $E \subset [0, 1] \times [0, 1]$ is measurable, and if, for every $x_0 \in [0, 1]$,

$$I_{x_0} = \{(x_0, y) : y \in [0, 1]\},$$

then $E \cap I_x$ is measurable for almost all x, and

$$\mu_2(E) = \int_0^1 \mu_1(E \cap I_x) \, dx;$$

where μ_k indicates k-dimensional Lebesgue measure.

6.4 Use the result of exercise 6.3 to prove that if f is summable on E_2, then $f(x, y)$ is summable as a function of each variable for almost all values of the other variable, and that

$$\int_{E_2} f \, d\mu_2 = \int_{-\infty}^{\infty} dy \left\{ \int_{-\infty}^{\infty} f(x, y) \, dx \right\} = \int_{-\infty}^{\infty} dx \left\{ \int_{-\infty}^{\infty} f(x, y) \, dy \right\}.$$

This is called Fubini's theorem.

6.5 Generalize Fubini's theorem to E_k.

7.1 Prove the corollary to theorem 1.

7.2 Give an example where f, g are summable, but fg is not summable.

7.3 For f, g summable on E_1, with respect to Lebesgue measure, define

$$(f * g)(t) = \int_{-\infty}^{\infty} f(t - u) g(u) \, du.$$

Show that $f * g$ is summable.

7.4 If f is summable on E_1, with respect to Lebesgue measure, show that

$$\lim_{t \to 0} \int_{-\infty}^{\infty} |f(x + t) - f(x)| \, dt = 0.$$

7.5 Show that if $(S, \mathcal{S}, \mu)$ is totally finite and $\{f_n\}$ converges to f in the sense that

$$\lim_{n \to \infty} \int_S |f_n - f| \, d\mu = 0,$$

then $\{f_n\}$ converges in measure to f.

7.6 Show that convergence in the sense of exercise 7.5 does not always imply almost everywhere convergence.

7.7 Show that almost everywhere convergence does not always imply convergence in the sense of exercise 7.5.

7.8 If $\{a_n\}$ is a sequence of real numbers, and $b_n = \inf (a_n, a_{n+1}, \ldots)$ for every n, show that

$$\lim_{n \to \infty} b_n = \liminf_{n \to \infty} a_n.$$

9.1 If $(S, \mathcal{S}, \mu)$ is a finite signed measure and

$$\mu^+(E) = \sup [\mu(T) : T \in \mathcal{S}, T \subset E],$$

give details of the proof that μ^+ is completely additive and non-negative.

9.2 Having defined μ^+ as above and μ^- in similar fashion, as in the text, and $|\mu| = \mu^+ + \mu^-$, show that

$$|\mu| (E) = \sup \sum_{i=1}^{k} |\mu(E_i)|$$

for all partitions of E, for every $E \in \mathcal{S}$.

10.1 Prove the Radon-Nikodym theorem for the totally σ finite case.

10.2 Construct an example of a measure space (not totally σ finite) in which the Radon-Nikodym theorem fails to hold.

11.1 Prove the Hölder and Minkowski inequalities in the context of this chapter.

11.2 Prove that if $(S, \mathcal{S}, \mu)$ is a separable measure space, then $L_p = L_p(S, \mathcal{S}, \mu)$ is a separable metric space for every $p > 1$.

11.3 Show that for every $p_1 \geq 1$, $p_2 \geq 1$, and totally σ finite $(S, \mathcal{S}, \mu)$, the set $L_{p_1} \cap L_{p_2}$ is dense in the space L_{p_1}.

11.4 If $(S, \mathcal{S}, \mu)$ is a measure space, not necessarily totally σ finite, show that, for every $p \geq 1$, every function in L_p vanishes outside of a totally σ finite subspace.

11.5 For $(S, \mathcal{S}, \mu)$ totally σ finite, but not totally finite, $p_1 \geq 1$, $p_2 \geq 1$, $p_1 \neq p_2$, show that neither L_{p_1} nor L_{p_2} is a subset of the other.

11.6 If $p_1 > p_2 \geq 1$ and $(S, \mathcal{S}, \mu)$ is totally finite, show that $L_{p_1} \subset L_{p_2}$.

11.7 If $(S, \mathcal{S}, \mu)$ is totally finite and $p > 1$, show that convergence in the L_p norm implies convergence in measure.

11.8 If $(S, \mathcal{S}, \mu)$ is totally finite, we say that $\{f_n\}$ converges asymptotically in the L_p norm to f if for every $\epsilon > 0$ there is an $E \in \mathcal{S}$, $\mu(S \sim E) < \epsilon$ such that $\{f_n\}$ converges to f in the L_p norm on E. Is asymptotic convergence in the L_p norm implied by convergence in measure? Prove or give a counter example.

12.1 Prove that the mapping which takes each $f \in L_1[0, 1]$ into the stair function f_n is norm decreasing.

12.2 Give an example of a space, not an L_p space, for which proposition 1 holds.

12.3 If $(S, \mathcal{S}, \mu)$ is a separable, non-atomic measure space of norm 1, show that an analog of proposition 1 holds.

12.4 If $(S, \mathcal{S}, \mu)$ is non-separable, show that an analog of proposition 1 does not hold.

12.5 If M is the space of equivalence classes of bounded measurable functions on [0, 1] with

$$\|f\| = \inf \{y : \mu[x : f(x) > y] = 0\},$$

show that proposition 1 does not hold for M.

12.6 Let $\{f_n\}$ be a sequence of continuous, non-negative functions on E_1 such that, for every n, $f_n(x) = 0$ if $|x| \geq 1/n$ and $\int_{E_1} f_n(x)\, dx = 1$. For every $f \in L_1$, which vanishes off a compact set, let

$$F_n(x) = \int_{-\infty}^{\infty} f(x - t) f_n(t)\, dt.$$

Show that the functions F_n are continuous and that

$$\lim_{n \to \infty} \|f - F_n\| = 0,$$

where we are using the L_1 norm.

12.7 If $f \in L_p$, $p > 1$, show that

$$\lim_{n \to \infty} \|f - F_n\| = 0$$

for the L_p norm.

12.8 Give an example of a Banach function space X, not an L_p space, such that $f \in X$ implies

$$\lim_{n \to \infty} \|f - F_n\| = 0$$

in the norm of X.

12.9 Give an example of a Banach function space, such that the $\{F_n\}$ as defined above do not converge to f in the space.

12.10 Show that if the $\{f_n\}$ of exercise 12.6 are of class C^∞, then so are the $\{F_n\}$.

14.1 Give the details of the proof that the dual of L_1 is M for the totally finite case.

14.2 Prove that the dual of L_1 is M also for the totally σ finite case. Is this restriction on the measure space necessary?

14.3 Prove the existence of a continuous linear functional on M for which the associated set function

$$\nu(E) = \varphi(\chi_E)$$

is not completely additive.

14.4 Give the details of the construction which proves (*e*).

15.1 Prove that if $a_1, a_2, \ldots, a_n$ is an ordered set of real numbers, if $m \leq n$, and if $a_i \in S_m$ whenever $a_i + a_{i+1} + \ldots + a_{i+p-1} \geq 0$, for some $p \leq m$, then

$$\sum [a_i : a_i \in S_m] \geq 0.$$

16.1 Prove the lemma that if $x \in [-\pi/2, \pi/2]$, $1 < p \leq 2$, then

$$|\sin x|^p \leq a\, |\cos x|^p - b \cos px$$

where a and b depend only on p.

16.2 Prove that the trigonometric polynomials are dense in $L_p[0, 1]$ from the fact that the algebraic polynomials have this property.

16.3 Prove the fact that the trigonometric polynomials are dense in L_p by considering the averages of the partial sums of Fourier series and then applying a modified, but still valid, version of exercise 12.6.

18.1 If σ and τ are moduli of continuity, and $\lim\limits_{n \to 0} \dfrac{\sigma(x)}{\tau(x)} = \infty$, show that there is a function in $C(\sigma)$ which is not in $C(\tau)$ on any interval.

18.2 Given any set Z of measure 0, show that there is a measurable set S whose metric density does not exist at any point of Z.

18.3 Give an example of a measurable set whose metric density exists and is equal to $\frac{1}{2}$ at a point x.

CHAPTER 4

HILBERT SPACE

4.1 Inner Product, Hilbert Space

A Hilbert space is a Banach space in which the norm is derived from a scalar-valued function called an inner product. We shall define this concept and show that linear spaces with an inner product retain many geometric properties of finite dimensional euclidean spaces.

Let X be a vector space over the field F (real or complex). A mapping of $X \times X$ into F which takes each ordered pair $\{x, y\} \in X \times X$ into the number $(x, y) \in F$ is called an **inner product** in X if

(i) $(x, y) = \overline{(y, x)}$,
(ii) $(ax_1 + bx_2, y) = a(x_1, y) + b(x_2, y)$,
(iii) $(x, x) \geq 0$ and $(x, x) = 0$ if and only if $x = 0$.

Condition (i) clearly reduces to $(x, y) = (y, x)$ if X is a real vector space. In general, we confine our discussion to the complex case. However, the real case will also be considered when natural.

From (i) and (ii), we obtain

(iv) $(x, cy_1 + dy_2) = \bar{c}(x, y_1) + \bar{d}(x, y_2)$.

We shall show that the quantity

$$\|x\| = (x, x)^{1/2}$$

is a norm in the space X. The only property which is not obvious is the triangle inequality and this follows directly from the fundamental

PROPOSITION 1 (Cauchy-Schwarz inequality). For every x, $y \in X$, we have

$$|(x, y)| \leq \|x\| \, \|y\|.$$

Proof. We may suppose that $x \neq 0$, $y \neq 0$. For any a, $\|x + ay\|^2 \geq 0$ by (iii). Thus, using (ii) and (iv), and then (i),

$$\begin{aligned} 0 \leq (x + ay, x + ay) &= \|x\|^2 + a(y, x) + \bar{a}(x, y) + |a|^2 \, \|y\|^2 \\ &= \|x\|^2 + 2 \operatorname{Re}(a(y, x)) + |a|^2 \, \|y\|^2. \end{aligned}$$

Choosing the argument of a in such a way that $a(y, x) \leq 0$, we may write,

$$2\,|a|\,|(x, y)| \leq \|x\|^2 + |a|^2\,\|y\|^2,$$

and the choice

$$|a| = \frac{\|x\|}{\|y\|}$$

reduces this to the desired inequality.

The proof of the triangle inequality is now obtained by

$$\begin{aligned}\|x + y\|^2 &= \|x\|^2 + (x, y) + (y, x) + \|y\|^2 \\ &= \|x\|^2 + 2\,\mathrm{Re}\,(x, y) + \|y\|^2 \\ &\leq \|x\|^2 + 2\,|(x, y)| + \|y\|^2 \\ &\leq \|x\|^2 + 2\,\|x\|\,\|y\| + \|y\|^2 \\ &= (\|x\| + \|y\|)^2.\end{aligned}$$

A vector space X, together with an inner product in X, is called an **inner product space** and is, as we have seen, a special kind of normed vector space. If X is complete under the norm obtained from its inner product, then X is called a **Hilbert space.**

The reader will verify without difficulty that

(a) n-space under

$$(x, y) = \sum_{i=1}^{n} x_i \bar{y}_i$$

is a Hilbert space. (In the complex case this space is called n dimensional **unitary** space; in the real case, euclidean n-space.)

(b) l_2 is a Hilbert space with

$$(x, y) = \sum_{i=1}^{\infty} x_i \bar{y}_i.$$

(c) $L_2[0, 1]$ is a Hilbert space with

$$(x, y) = \int_0^1 x(t)\,\overline{y(t)}\,dt.$$

(d) More generally, if $(S, \mathcal{S}, \mu)$ is any measure space, then $L_2(S, \mathcal{S}, \mu)$ is a Hilbert space with

$$(x, y) = \int_S x\bar{y}\,d\mu.$$

(e) An inner product space, not complete, is the vector space $P[0, 1]$ of real valued polynomials on $[0, 1]$ with

$$(p, q) = \int_0^1 p(t)q(t)\,dt.$$

That this is not complete follows since its Hamel basis is countable.

A more interesting example, along these lines, is the set X of real-valued functions on $R \cup C$, where R is a domain in the uv-plane and its boundary C is a smooth simple closed curve, which are continuous on $R \cup C$ and have continuous second partial derivatives in R. Define (x, y) by

$$(x, y) = \int_R (x_u y_u + x_v y_v)\, du\, dv + \int_C xy\, ds.$$

The Cauchy-Schwarz inequality shows that the inner product is a continuous function on $X \times X$ into F, where the norm of $\{x, y\} \in X \times X$ is $\|x\| + \|y\|$. That is,

$$\lim_n x_n = x \text{ and } \lim_n y_n = y \text{ implies } \lim_n (x_n, y_n) = (x, y).$$

A consequence of this fact is that the completion of an inner product space is again an inner product space, hence a Hilbert space.

In euclidean 3 space, the inner product

$$(x, y) = x_1 y_1 + x_2 y_2 + x_3 y_3$$

is used to express the angle between two non-zero vectors by means of the formula

$$\cos \theta = \frac{(x, y)}{\|x\|\ \|y\|}, \quad 0 \leq \theta \leq \pi.$$

In particular, orthogonality ($\theta = \pi/2$) of x and y is characterized by $(x, y) = 0$, if we accept the convention that the vector 0 is orthogonal to every vector. In a general inner product space we define x and y to be **orthogonal,** written $x \perp y$, if $(x, y) = 0$.

The **Pythagorean theorem** can be stated as

$$x \perp y \text{ implies } \|x\|^2 + \|y\|^2 = \|x + y\|^2$$

and is seen to be valid in any inner product space, by direct computation.

Similarly, it is easy to verify directly that the **parallelogram law** holds in an inner product space (the sum of the squares of the diagonals of a parallelogram equals the sum of the squares of the sides):

$$\|x + y\|^2 + \|x - y\|^2 = 2\,\|x\|^2 + 2\,\|y\|^2.$$

The converse is also valid. Thus, a Banach space in which the norm satisfies the parallelogram law is necessarily a Hilbert space. We consider the real case, and suppose the parallelogram law holds in a Banach space X. We define an inner product in X by

$$(x, y) = \frac{1}{4}\left[\|x + y\|^2 - \|x - y\|^2\right].$$

Then $(x, x) \geq 0$ and $(x, x) = 0$ if and only if $x = 0$. Moreover, $(x, x) = \|x\|^2$ and $(x, y) = (y, x)$. It is only necessary to show that

$$(x_1 + x_2, y) = (x_1, y) + (x_2, y), \text{ and } (ax, y) = a(x, y).$$

First,

$$\|u + v + w\|^2 + \|u + v - w\|^2 = 2\,\|u + v\|^2 + 2\,\|w\|^2$$

and

$$\|u - v + w\|^2 + \|u - v - w\|^2 = 2\,\|u - v\|^2 + 2\,\|w\|^2$$

so that

$$\|u + v + w\|^2 + \|u + v - w\|^2 - \|u - v + w\|^2 - \|u - v - w\|^2 = 2\,\|u + v\|^2 - 2\,\|u - v\|^2$$

or

$$(u + w, v) + (u - w, v) = 2\,(u, v).$$

Setting $u = w$, this implies immediately that

$$(2u, v) = 2(u, v).$$

Now, let $x_1 = u + w$, $x_2 = u - w$, and $y = v$ to obtain

$$(x_1, y) + (x_2, y) = (x_1 + x_2, y).$$

We leave it to the reader to show that $(ax, y) = a(x, y)$.
We thus have

PROPOSITION 2. A Banach space is a Hilbert space if and only if the parallelogram law holds.

It follows that a real Banach space is a Hilbert space if and only if every two-dimensional subspace is euclidean.

The identity

$$(x, y) = \frac{1}{4}\,[\|x + y\|^2 - \|x - y\|^2]$$

used in the proof of proposition 2 is called the **polarization identity.** In complex Hilbert space it has the form

$$(x, y) = \frac{1}{4}\,[\|x + y\|^2 - \|x - y\|^2 + i\,\|x + iy\|^2 - i\,\|x - iy\|^2].$$

It may be established by direct computation.

The next section presents further special aspects of Hilbert spaces which set them apart from general Banach spaces.

4.2 Basic Lemma, Projection Theorem, Dual

We shall show that any closed subspace M of a Hilbert space X is a direct summand of X. Indeed we let

$$M^{\perp} = [z \in X : z \perp y \text{ for all } y \in M].$$

It is immediate that $M^\perp$ is a closed subspace of X. We shall show that $X = M \dotplus M^\perp$. We first show that $M \neq X$ implies $M^\perp \neq (0)$. For this purpose, we prove

LEMMA 1. If K is a closed, convex set in X and $x_0 \in X \sim K$, there is a unique $y_0 \in K$ such that

$$\|x_0 - y_0\| = \inf [\|x_0 - y\| : y \in K].$$

Proof. Let $d = \inf [\|x_0 - y\| : y \in K]$. Choose $y_1, y_2, \ldots$ in K so that

$$\lim_{n\to\infty} \|x_0 - y_n\| = d.$$

By the parallelogram law,

$$\|y_m - y_n\|^2 = 2\,\|y_m - x_0\|^2 + 2\,\|y_n - x_0\|^2 - \|(y_m - x_0) + (y_n - x_0)\|^2$$
$$= 2\,\|y_m - x_0\|^2 + 2\,\|y_n - x_0\|^2 - 4\Big\|\frac{y_m + y_n}{2} - x_0\Big\|^2.$$

Since K is convex, $\left\|\dfrac{y_m + y_n}{2} - x_0\right\| \geq d$. Hence,

$$\|y_m - y_n\|^2 \leq 2\,\|y_m - x_0\|^2 + 2\,\|y_n - x_0\|^2 - 4\,d^2$$

from which $\{y_n\}$ is a Cauchy sequence. Let $y_0 = \lim_n y_n$. Then $y_0 \in K$ and $\|x_0 - y_0\| = d$.

For the uniqueness, suppose $z_0 \in K$ and $\|x_0 - z_0\| = d$. Then

$$\|y_0 - z_0\|^2 = 2\,\|y_0 - x_0\|^2 + 2\,\|x_0 - z_0\|^2 - \|y_0 + z_0 - 2x_0\|^2$$
$$= 4d^2 - 4\left\|\frac{y_0 + z_0}{2} - x_0\right\|^2 \leq 0.$$

Hence, $z_0 = y_0$.

Apply lemma 1 to the closed convex set M. It associates a unique $y_0 \in M$ with every $x_0 \in X \sim M$. We show that $x_0 - y_0 \in M^\perp$.

If $y \in M$ and a is a scalar, then $y_0 + ay \in M$, so that

$$\|x_0 - (y_0 + ay)\|^2 \geq \|y_0 - x_0\|^2$$

and

$$-\bar{a}(x_0 - y_0, y) - a(y, x_0 - y_0) + |a|^2\,\|y\|^2 \geq 0.$$

With $a = \epsilon$, $\epsilon > 0$, and then $a = -\epsilon$, this yields

$$2\,|\mathrm{Re}\,(x_0 - y_0, y)| \leq \epsilon\,\|y\|^2$$

and with $a = i\epsilon$ and then $a = -i\epsilon$, it gives

$$2\,|\mathrm{Im}\,(x_0 - y_0, y)| \leq \epsilon\,\|y\|^2.$$

Since $\epsilon > 0$ is arbitrary, it follows that $(x_0 - y_0, y) = 0$ and so $x_0 - y_0 \in M^\perp$.

Now since $M \cap M^\perp = (0)$ and any $x_0 \in X$ can be written as

$$x_0 = y_0 + (x_0 - y_0),$$

where y_0 is the vector in M at minimum distance from x_0, we see that

$$X = M \dotplus M^{\perp}.$$

The mapping P of X onto M given by $Px_0 = y_0$ is a projection operator with the special property that its range and null space are mutually orthogonal subspaces of X. For this reason it is called the orthogonal projection on M.

Our results so far are summarized in

THEOREM 1 (Projection theorem). If M is a closed subspace of a Hilbert space X, then every $x \in X$ has a unique decomposition

$$x = y + z, \qquad y \in M, z \in M^{\perp}.$$

The operator $x \to Px = y$ is a bounded linear idempotent (a projection), such that $M = PX$ and $M^{\perp} = (I - P)X$ are mutually orthogonal closed subspaces with

$$X = M \dotplus M^{\perp}.$$

For each $x \in X$, Px is the unique element of M which minimizes the distance from x to M.

We may use this result to identify the dual space of a Hilbert space, as follows.

If $y_0 \in X$, then (x, y_0) is a linear functional (in x) on X. The Cauchy-Schwarz inequality gives

$$\frac{|(x, y_0)|}{\|x\|} \leq \|y_0\|, \quad x \neq 0$$

and leads immediately to the observation that for each $y_0 \in X$,

$$x \to (x, y_0)$$

is a bounded linear functional on X with norm $\|y_0\|$.

We thus have a natural, norm preserving mapping φ on X into X'. It is not linear (unless X is real) but has the property

$$\varphi(ax + by) = \bar{a}\varphi(x) + \bar{b}\varphi(y).$$

We shall show that φ is onto, thus establishing a natural identification of a Hilbert space with its own dual.

THEOREM 2. If x' is a bounded linear functional on a Hilbert space X, there is a unique $y_0 \in X$ such that

$$x'(x) = (x, y_0), \; x \in X.$$

Proof. If $x' = 0$, take $y_0 = 0$. Suppose $x' \neq 0$. The null space N of x' is a closed subspace of X and $N \neq X$. Consequently $N^{\perp} \neq (0)$. Let $x_0 \in N^{\perp}$, $x_0 \neq 0$.

We choose $y_0 = ax_0$, where a is determined so that $x'(x_0) = (x_0, y_0)$, that is, $x'(x_0) = a \|x_0\|^2$, giving

$$a = \frac{\overline{x'(x_0)}}{\|x_0\|^2}.$$

To verify that y_0 has the desired property, let $x \in X$ and put $b = x'(x)/x'(x_0)$. Then $x - bx_0 \in N$, since $x'(x - bx_0) = 0$. Hence

$$(x, y_0) = (x - bx_0, y_0) + (bx_0, y_0) = b(x_0, y_0)$$

since $y_0 \perp N$. But

$$b(x_0, y_0) = bx'(x_0) = x'(x).$$

Thus, for any $x \in X$,

$$(x, y_0) = x'(x)$$

and the proof is complete.

4.3 Application, Mean Ergodic Theorem

The lemma of the preceding section is a form of variational principle which has several applications in analysis. We present the mean ergodic theorem as an example.

If $(S, \mathcal{S}, \mu)$ is a totally σ finite measure space and T is a one-one measure preserving transformation of S into itself, then for any function f defined on S, we consider the means φ_n defined by

$$\varphi_n(x) = \frac{1}{n}[f(x) + f(Tx) + \ldots + f(T^{n-1}x)],\ x \in S, \quad n = 1, 2, \ldots,$$

and seek a limiting function for $\{\varphi_n\}$.

The individual ergodic theorem establishes the existence of an almost everywhere limit in case $f \in L_1$ (see Sec. 3.15). The mean ergodic theorem establishes the existence of a limit in the mean, i.e., in the sense of L_2 convergence, in case $f \in L_2(S)$.

The transition to a problem in abstract Hilbert space comes about through the observation that if $f \in L_2$ then $f \to Vf$, where

$$Vf(x) = f(Tx),\ x \in S,$$

defines a linear transformation of $L_2(S)$ into itself, which is norm preserving;

$$\|Vf\| = \|f\|,\ f \in L_2.$$

The proof of this fact is left to the reader.

The mean ergodic theorem is thus a consequence of

PROPOSITION 1. Let V be a bounded linear operator on a Hilbert space X with $\|V\| \leq 1$. Let $f \in X$ and put

$$\varphi_n = \frac{1}{n}[f + Vf + \ldots + V^{n-1}f], \quad n = 1, 2, \ldots.$$

There is then a $\varphi \in X$ such that $\lim_n \varphi_n = \varphi$ in X.

Proof. Let K be the set of all finite convex combinations of the elements

$$f, Vf, V^2f, \ldots,$$

i.e., all g of the form (taking $V^0 f = f$)

$$g = \sum_{i=0}^{m-1} C_i V^i f, \quad C_i \geq 0, \quad \sum_{i=0}^{m-1} C_i = 1.$$

Clearly K is convex. Its closure $\bar{K}$ is convex and lemma 1, Sec. 4.2, assures the existence of a unique $\varphi \in \bar{K}$ realizing the distance from 0 to $\bar{K}$. Then

$$\|\varphi\| = \inf[\|g\| : g \in \bar{K}].$$

To show that $\lim_n \varphi_n = \varphi$, it suffices to show that $\{\varphi_n\}$ is a minimizing sequence, i.e., for every $\epsilon > 0$ there is an N such that $n > N$ implies

$$\|\varphi_n\| < \|\varphi\| + \epsilon.$$

We may choose $g \in K$ so that

$$\|g\| < \|\varphi\| + \frac{\epsilon}{2}.$$

Putting

$$\psi_n = \frac{1}{n}[g + Vg + \ldots + V^{n-1}g],$$

$$\begin{aligned}
\|\varphi_n\| &\leq \|\varphi_n - \psi_n\| + \|\psi_n\| \\
&\leq \|\varphi_n - \psi_n\| + \frac{1}{n}\sum_{i=0}^{n-1} \|V^i g\| \\
&\leq \|\varphi_n - \psi_n\| + \frac{1}{n}\sum_{i=0}^{n-1} \|g\| = \|\varphi_n - \psi_n\| + \|g\| \\
&< \|\varphi_n - \psi_n\| + \|\varphi\| + \frac{\epsilon}{2},
\end{aligned}$$

where we have used the fact that $\|V\| \leq 1$.

Now, we must show that $\|\varphi_n - \psi_n\| < \epsilon/2$ for all n sufficiently large. Since $g \in K$, we have

$$g = \sum_{j=0}^{m-1} C_j V^j f, \quad C_j \geq 0, \quad \sum_{j=0}^{m-1} C_j = 1.$$

Thus, for $n \geq m$,

$$\psi_n = \frac{1}{n} \sum_{i=0}^{n-1} \sum_{j=0}^{m-1} C_j V^{i+j} f$$

$$= \frac{1}{n} \sum_{j=0}^{m-1} C_j \sum_{i=0}^{n-1} V^{i+j} f.$$

Also,

$$\varphi_n = \frac{1}{n} \sum_{i=0}^{n-1} V^i f = \frac{1}{n} \sum_{j=0}^{m-1} C_j \sum_{i=0}^{n-1} V^i f$$

and so

$$\psi_n - \varphi_n = \frac{1}{n} \sum_{j=0}^{m-1} C_j \sum_{i=0}^{n-1} (V^{i+j} f - V^i f).$$

Cancellations occur in the inner sum, leaving $2j$ terms of the form $\pm V^k f$. Hence, applying the triangle inequality and the assumption on V,

$$\|\psi_n - \varphi_n\| \leq \frac{1}{n} \sum_{j=0}^{m-1} C_j 2j \, \|f\|$$

$$\leq \frac{2(m-1)}{n} \|f\| \sum_{j=0}^{m-1} C_j = \frac{2(m-1)}{n} \|f\|,$$

and the result follows.

4.4 Orthonormal Sets, Fourier Expansion

A set S of elements of an inner product space X is an **orthogonal set** if $x \perp y$ whenever $x \in S$, $y \in S$ and $x \neq y$. If, in addition, each $x \in S$ has norm one, S is called **orthonormal.**

$$(x, y) = \begin{cases} 0 \text{ if } x \neq y \\ 1 \text{ if } x = y \end{cases}, \quad x, y \in S.$$

A set S is **complete** in X if no non-zero $x \in X$ is orthogonal to S (i.e., to every element of S). Since the orthonormal sets in an $X \neq (0)$ form a non-empty class partially ordered by inclusion, and the complete orthonormal sets are the maximal elements, it follows by Zorn's lemma that complete orthonormal sets exist in X.

We shall show that every $x \in X$ has a unique series expansion in terms of a complete orthonormal set in X, but first we discuss a few concrete examples.

(i) The set $e_1 = (1, 0, 0, \ldots)$, $e_2 = (0, 1, 0, \ldots)$, $\ldots$ is clearly a complete orthonormal set in l_2. It is also clear that any $x = (x_1, x_2, \ldots) \in l_2$ is a sum $x = \sum_{i=1}^{\infty} x_i e_i$, where the partial sums converge in the l_2 norm and $x_i = (x, e_i)$.

(ii) In the complex space $L_2(0, 2\pi)$, the sequence

$$\frac{1}{\sqrt{2\pi}} e^{ikt}, \quad k = 0, \pm 1, \pm 2, \ldots$$

is immediately seen to be orthonormal. We show it is complete. Indeed, we show that for any $f \in L_1 \supset L_2$,

$$\int_0^{2\pi} f(t)e^{ikt}\, dt = 0, \quad k = 0, \pm 1, \pm 2, \ldots$$

implies $f = 0$.

Put

$$F(t) = \int_0^t f(u)\, du.$$

For any constant a, by partial integration,

$$\int_0^{2\pi} [F(t) - a]e^{ikt}\, dt = 0, \quad k = \pm 1, \pm 2, \ldots,$$

and a may be chosen so that this holds also for $k = 0$.

Since $F - a$ is periodic of period 2π (because $\int_0^{2\pi} f(t)\, dt = 0$), there is a trigonometric polynomial

$$T(t) = \sum_{k=-n}^{n} c_k e^{ikt},$$

given $\epsilon > 0$, such that

$$\max\,[|F(t) - a - T(t)| : t \in [0, 2\pi]] < \epsilon.$$

This follows from the fact that $F(t)$ is continuous, by the Weierstrass theorem. Now,

$$\begin{aligned}\int_0^{2\pi} |F(t) - a|^2\, dt &= \int_0^{2\pi} (F(t) - a)(\overline{F(t)} - \bar{a} - T(t))\, dt \\ &\leq \epsilon \int_0^{2\pi} |F(t) - a|\, dt \\ &\leq \epsilon \left[\int_0^{2\pi} |F(t) - a|^2\, dt\right]^{1/2} \left[\int_0^{2\pi} dt\right]^{1/2}.\end{aligned}$$

Hence

$$\int_0^{2\pi} |F(t) - a|^2\, dt \leq 2\pi\epsilon^2,$$

so that $F(t)$ is constant. Thus $f(t)$ is zero almost everywhere.

The corresponding system for real $L_2(0, 2\pi)$ is the sequence

$$\frac{1}{\sqrt{2\pi}}, \frac{1}{\sqrt{\pi}} \cos nt, \frac{1}{\sqrt{\pi}} \sin nt, \quad n = 1, 2, 3, \ldots,$$

and the series for these systems are the usual Fourier series.

(iii) Consider the collection X of functions on $-\infty < t < \infty$ representable in the form

$$x(t) = \sum_{k=1}^{n} a_k e^{i\lambda_k t},$$

for arbitrary n, real numbers $\lambda_1, \lambda_2, \ldots, \lambda_n$, and complex coefficients $a_1, a_2, \ldots, a_n$.

X is a vector space and an inner product is defined in X by

$$(x, y) = \lim_{T\to\infty} \frac{1}{2T}\int_{-T}^{T} x(t)\overline{y(t)}\, dt.$$

In this space, the uncountable set $[e^{i\lambda t} : -\infty < \lambda < \infty]$ forms a complete orthonormal set, as direct computation shows.

(iv) The functions holomorphic on the unit disk $|z| < 1$ in the complex z plane and having finite Dirichlet integral

$$\iint_{|z|<1} |f(z)|^2\, dx\, dy < \infty$$

form an inner product space under

$$(f, g) = \iint_{|z|<1} f\bar{g}\, dx\, dy.$$

An orthonormal sequence is given by

$$\varphi_n(z) = \sqrt{\frac{n}{\pi}}\, z^{n-1}, \quad n = 1, 2, 3, \ldots.$$

Some additional examples appear in the exercises and in later sections.

We are interested in expansions of the form $x = \Sigma\, a_\alpha x_\alpha$, for $x \in X$ and $\{x_\alpha\}$ an orthonormal family in X, possibly uncountable. The meaning to be given to such a sum will be described below. First we note formally that

$$(x, x_\beta) = \sum a_\alpha(x_\alpha, x_\beta) = a_\beta,$$

so that the coefficients of interest are the numbers (x, x_β), called the **Fourier coefficients** of x with respect to $\{x_\alpha\}$.

PROPOSITION 1. If $\{x_1, x_2, \ldots, x_n\}$ is an orthonormal set, then for any $x \in X$,

$$\sum_{i=1}^{n} |(x, x_i)|^2 \le \|x\|^2.$$

Proof.

$$\left\| x - \sum_{i=1}^{n}(x, x_i)x_i \right\|^2 = \|x\|^2 - \left(x, \sum_{i=1}^{n}(x, x_i)x_i\right) - \left(\sum_{i=1}^{n}(x, x_i)x_i, x\right) + \left\| \sum_{i=1}^{n}(x, x_i)x_i \right\|^2 \ge 0.$$

But

$$\left(x, \sum_{i=1}^{n} (x, x_i)x_i\right) = \sum_{i=1}^{n} \overline{(x, x_i)}(x, x_i) = \sum_{i=1}^{n} |(x, x_i)|^2,$$

$$\left(\sum_{i=1}^{n} (x, x_i)x_i, x\right) = \sum_{i=1}^{n} (x, x_i)\overline{(x, x_i)} = \sum_{i=1}^{n} |(x, x_i)|^2$$

and, by the pythagorean relation,

$$\left\| \sum_{i=1}^{n} (x, x_i)x_i \right\|^2 = \sum_{i=1}^{n} |(x, x_i)|^2 \, \|x_i\|^2 = \sum_{i=1}^{n} |x, x_i|^2.$$

It follows that

$$\|x\|^2 \geq \sum_{i=1}^{n} |(x, x_i)|^2.$$

Now we let $S = [x_\alpha : \alpha \in A]$ be an orthonormal set in X and define the sum

$$\sum_{\alpha \in A} (x, x_\alpha)x_\alpha$$

for any $x \in X$.

First, by proposition 1, the number of x_α for which $|(x, x_\alpha)| > 1/k$ is finite for each $k = 1, 2, \ldots$. Thus, for each $x \in X$ the set of Fourier coefficients of x which do not vanish is countable. The sum

$$\sum_{\alpha \in A} (x, x_\alpha)x_\alpha$$

is thus a countable sum.

With x fixed, arrange the countable set of x_α for which $(x, x_\alpha) \neq 0$ as a sequence $x_1, x_2, \ldots$.

By proposition 1, the series $\sum_{i=1}^{\infty} |(x, x_i)|^2$ converges, since its partial sums do not exceed $\|x\|^2$. Since

$$\left\| \sum_{i=n}^{n+k} (x, x_i)x_i \right\|^2 = \left(\sum_{i=n}^{n+k} (x, x_i)x_i, \sum_{j=n}^{n+k} (x, x_j)x_j \right)$$

$$= \sum_{i=n}^{n+k} \sum_{j=n}^{n+k} (x, x_i)\overline{(x, x_j)}(x_i, x_j) = \sum_{i=n}^{n+k} |(x, x_i)|^2,$$

it follows that the series $\sum_{i=1}^{\infty} (x, x_i)x_i$ converges. Moreover, from a similar computation the sum is independent of the arrangement of the $\{x_i\}$. This value is the sum of the series

$$\sum_{\alpha \in A} (x, x_\alpha)x_\alpha.$$

We have also proved

PROPOSITION 2 (Bessel's inequality). If X is a Hilbert space, $S = [x_\alpha : \alpha \in A]$ is an orthonormal set in X and $x \in X$, then

$$\sum_{\alpha \in A} |(x, x_\alpha)|^2 \leq \|x\|^2.$$

Fixing the Hilbert space X and the orthonormal set S as above, we say that S is a **basis** (orthonormal basis; also called a closed orthonormal set) in X if the Fourier series representation

$$x = \sum_{\alpha \in A} (x, x_\alpha) x_\alpha$$

is valid for each $x \in X$.

We have two theorems.

THEOREM 1. S is a basis if and only if it is complete in X.

THEOREM 2 (Parseval's formula). S is a basis if and only if equality holds in Bessel's inequality

$$\|x\|^2 = \sum_{\alpha \in A} |(x, x_\alpha)|^2, \quad x \in X$$

or, equivalently,

$$(x, y) = \sum_{\alpha \in A} (x, x_\alpha)\overline{(y, x_\alpha)}, \quad x, y \in X.$$

Proofs. Suppose S is a basis, then if $x \in X$ satisfies $(x, x_\alpha) = 0$, $\alpha \in A$, the definition of sum gives

$$x = \sum_{\alpha \in A} (x, x_\alpha) x_\alpha = 0.$$

Thus, S is complete.

Moreover,

$$\|x\|^2 = \left(\sum_{\alpha \in A} (x, x_\alpha) x_\alpha, \quad \sum_{\beta \in A} (x, x_\beta) x_\beta \right)$$
$$= \sum_{\alpha \in A} |(x, x_\alpha)|^2.$$

The equivalence with the second form of Parseval's formula is left as an exercise.

That Parseval's formula implies completeness is obvious, so we need only show that the completeness of S implies S is a basis. Let $\beta \in A$. Then for any $x \in X$,

$$\left(x - \sum_{\alpha \in A} (x, x_\alpha) x_\alpha, x_\beta \right) = (x, x_\beta) - \sum_{\alpha \in A} (x, x_\alpha)(x_\alpha, x_\beta)$$
$$= (x, x_\beta) - (x, x_\beta) = 0,$$

and completeness yields

$$x = \sum_{\alpha \in A} (x, x_\alpha) x_\alpha.$$

To conclude this section, we describe a constructive procedure, the **Gramm-Schmidt process,** for orthonormalizing a given countable linearly independent set of vectors U in an inner product space X. The process is inductive. It will suffice to indicate a few steps.

Choose any $u_1 \in U$. The first member of the proposed orthonormal set $\{x_n\}$ is $x_1 = u_1/\|u_1\|$. Choose $u_2 \in U \sim \{u_1\}$. Then

$$v_2 = u_2 - (u_2, x_1)\, x_1$$

is non-zero, by linear independence, and is orthogonal to x_1. Put $x_2 = v_2/\|v_2\|$. Choose $u_3 \in U \sim \{u_1, u_2\}$ and put

$$v_3 = u_3 - [(u_3, x_1)x_1 + (u_3, x_2)x_2] \quad \text{and} \quad x_3 = v_3/\|v_3\|.$$

Continue in this way.

A linearly independent set of elements of X is a complete set in X if and only if the set of finite linear combinations of its elements is dense in X. We leave the proof to the reader.

4.5 Application, Isoperimetric Theorem

The Parseval formula for the real trigonometric system is

$$a_0^2 + \sum_{n=1}^{\infty} (a_n^2 + b_n^2) = \int_0^{2\pi} |f(t)|^2\, dt,$$

where

$$\frac{a_0}{2} + \sum_{n=1}^{\infty} (a_n \cos nt + b_n \sin nt)$$

is the Fourier series of $f \in L_2\,(0, 2\pi)$. This may be applied to prove the isoperimetric theorem, in the following form:

THEOREM 1. Among all simple, closed, piecewise smooth curves of length L in the plane, the circle encloses the maximum area.

Proof. Let $x = x(s)$, $y = y(s)$, $0 \le s \le L$, be a parametric representation of such a curve with arc length as parameter. Let A be the area enclosed by this curve. Introduce the parameter $t = 2\pi s/L$, $0 \le t \le 2\pi$, and let the Fourier coefficients of $x(t)$ be a_n and b_n, while denoting those of $y(t)$ by c_n and d_n. Then the Fourier coefficients of dx/dt and dy/dt are, respectively, nb_n, $-na_n$ and nd_n, $-nc_n$.

Now,

$$\left(\frac{dx}{dt}\right)^2 + \left(\frac{dy}{dt}\right)^2 = \left(\frac{L}{2\pi}\right)^2 \left[\left(\frac{dx}{ds}\right)^2 + \left(\frac{dy}{ds}\right)^2\right] = \left(\frac{L}{2\pi}\right)^2,$$

so that

$$2\pi \left(\frac{L}{2\pi}\right)^2 = \int_0^{2\pi} \left[\left(\frac{dx}{dt}\right)^2 + \left(\frac{dy}{dt}\right)^2\right] dt = \sum_{n=1}^{\infty} n^2(a_n^2 + b_n^2 + c_n^2 + d_n^2).$$

Moreover, by Green's theorem,

$$A = \int_0^{2\pi} x \frac{dy}{dt}\, dt = \sum_{n=1}^{\infty} n(a_n d_n - b_n c_n).$$

Thus,

$$L^2 - 4\pi A = 2\pi \sum_{n=1}^{\infty} [(na_n - d_n)^2 + (nb_n + c_n)^2 + (n^2 - 1)(c_n^2 + d_n^2)] \geq 0.$$

This establishes the inequality

$$L^2 \geq 4\pi A$$

with equality only if

$$a_1 = d_1,\ b_1 = -c_1 \quad \text{and} \quad a_n = b_n = c_n = d_n = 0 \quad \text{for } n = 2, 3, \ldots.$$

Since the maximum enclosed area occurs in the case of equality, the curve which realizes it is given by the circle

$$x(t) = \frac{a_0}{2} + a_1 \cos t + b_1 \sin t$$

$$y(t) = \frac{b_0}{2} - b_1 \cos t + a_1 \sin t.$$

4.6 Müntz Theorem

Let $x_1, x_2, \ldots, x_n \in L_2[0, 1]$. The **Gramm determinant** of $x_1, x_2, \ldots, x_n$ is

$$G = G(x_1, x_2, \ldots, x_n) = \begin{vmatrix} (x_1, x_1) & (x_1, x_2) & \ldots & (x_1, x_n) \\ (x_2, x_1) & (x_2, x_2) & \ldots & (x_2, x_n) \\ \cdot & \cdot & & \cdot \\ \cdot & \cdot & & \cdot \\ \cdot & \cdot & & \cdot \\ (x_n, x_1) & (x_n, x_2) & \ldots & (x_n, x_n) \end{vmatrix}.$$

We leave to the reader the proof of

PROPOSITION 1. A necessary and sufficient condition for $(x_1, x_2, \ldots, x_n)$ to be linearly independent is that $G \neq 0$.

Now if $x_1, x_2, \ldots, x_n$ are linearly independent and if $y \in L_2[0, 1]$, then the minimum distance d from y to the subspace generated by $x_1, x_2, \ldots, x_n$ is realized by the projection

$$Py = \sum_{i=1}^{n} a_i x_i$$

of y on that subspace. We wish to evaluate the coefficients a_i and the minimal distance d.

Since $y - Py \perp x_j$, $j = 1, 2, \ldots, n$, we obtain a system of equations for the coefficients:

$$(y, x_j) - \sum_{i=1}^{n} a_i(x_i, x_j) = 0, \quad j = 1, 2, \ldots, n.$$

The determinant of coefficients is just G, so a unique solution exists and is given by

$$a_i = \frac{G^{(i)}}{G}, \quad i = 1, \ldots, n,$$

where $G^{(i)}$ is obtained from G by replacing the ith column by the constants (y, x_j).

Now,

$$\begin{aligned} d^2 = \|y - Py\|^2 &= (y, y - Py) - (Py, y - Py) \\ &= (y, y - Py) \\ &= \|y\|^2 - \sum_{i=1}^{n} a_i(y, x_i). \end{aligned}$$

Let G^+ be the Gramm determinant of the set $y, x_1, x_2, \ldots, x_n$. We show that

$$\frac{G^+}{G} = \|y\|^2 - \sum_{i=1}^{n} a_i(y, x_i) = d^2.$$

Indeed,

$$\frac{G^+}{G} = \frac{1}{G}\begin{vmatrix} (y, y) & (y, x_1) & \ldots & (y, x_n) \\ (x_1, y) & (x_1, x_1) & \ldots & (x_1, x_n) \\ \cdot & \cdot & & \cdot \\ \cdot & \cdot & & \cdot \\ \cdot & \cdot & & \cdot \\ (x_n, y) & (x_n, x_1) & \ldots & (x_n, x_n) \end{vmatrix}$$

$$= \frac{(y, y)G - (y, x_1)M_1 + \ldots + (-1)^n(y, x_n)M_n}{G},$$

where M_i is the ith minor in the expansion of the determinant by its first row. But, it is clear that $M_i = (-1)^{i-1}G^{(i)}$. Hence,

$$\begin{aligned} \frac{G^+}{G} &= \|y\|^2 \\ &\quad + \frac{(-1)(y, x_1)G^{(1)} + (-1)^3(y, x_2)G^{(2)} + \ldots + (-1)^{2n-1}(y, x_n)G^{(n)}}{G} \\ &= \|y\|^2 - \sum_{i=1}^{n}\left[\frac{G^{(i)}}{G}\right](y, x_i) = d^2. \end{aligned}$$

We shall make use of these observations to prove a remarkable result known as Müntz's theorem.

The set of functions

$$1, t, t^2, t^3, \ldots$$

is a total set in $C[0, 1]$ by the Weierstrass theorem. It is also total (complete) in $L_2[0, 1]$, since the polynomials are dense in that space as well. Müntz's theorem gives necessary and sufficient conditions that the set

$$1, t^{n_1}, t^{n_2}, \ldots, 1 \le n_1 < n_2 < \ldots$$

enjoy the same properties.

Let $x \in C[0, 1]$. The inequality

$$\left[\int_0^1 \left| x(t) - \sum_{i=1}^n a_i t^{n_i} \right|^2 dt\right]^{1/2} \le \max \left| x(t) - \sum_{i=1}^n a_i t^{n_i} \right|$$

shows that the sequence is complete in L_2 if it is total in $C[0, 1]$. Conversely, to show that the sequence is total in $C[0, 1]$ if it is complete in L_2, it is enough to show that an inequality in the reverse direction holds for the functions t^m, $m = 1, 2, \ldots$. We have

$$\begin{aligned}\left| t^m - \sum_{i=1}^n a_i t^{n_i} \right| &= m \left| \int_0^t \left[t^{m-1} - \sum_{i=1}^n b_i t^{n_i - 1} \right] dt \right| \\ &\le m \int_0^1 \left| t^{m-1} - \sum_{i=1}^n b_i t^{n_i - 1} \right| dt \\ &\le m \left[\int_0^1 \left| t^{m-1} - \sum_{i=1}^n b_i t^{n_i - 1} \right|^2 dt\right]^{1/2},\end{aligned}$$

fulfilling the requirement.

Thus the theorem, which we now state, may be regarded as a generalization of the Weierstrass theorem as well as an L_2 theorem.

THEOREM 1. A necessary and sufficient condition for the set

$$t^{n_1}, t^{n_2}, \ldots, 1 \le n_1 < n_2 < \ldots$$

to be complete in L_2 is that

$$\sum_{i=1}^{\infty} \frac{1}{n_i} = \infty.$$

REMARK. The function 1 must be added in the case of $C[0, 1]$ but is redundant for L_2.

Proof. The completeness holds if and only if for each $m \ge 1$, with $m \ne n_i$, $i = 1, 2, \ldots$, the minimal distance d_k of t^m from the subspace generated by $t^{m_1}, \ldots, t^{m_k}$ approaches zero as k goes to infinity.

In view of the previous observations, this condition may be stated as

$$\lim_{k \to \infty} \frac{G(t^m, t^{n_1}, \ldots, t^{n_k})}{G(t^{n_1}, \ldots, t^{n_k})} = \lim_{k \to \infty} d_k^2 = 0.$$

Computing a typical element of such a determinant requires only the formula

$$(t^p, t^q) = \int_0^1 t^{p+q}\, dt = \frac{1}{p + q + 1}.$$

The determinants themselves can then be explicitly evaluated (see the exercises for an outline of the computation). The result is

$$d_k^2 = \frac{\prod_{i=1}^{k} (n_i - m)^2}{(2m+1)^2 \prod_{i=1}^{k} (m + n_i + 1)^2}.$$

In view of this, the condition becomes

$$\lim_{k\to\infty} \left[\sum_{i=1}^{k} \left(\log \left(1 - \frac{m}{n_i}\right) - \log\left(1 + \frac{m+1}{n_i}\right)\right)\right] = -\infty.$$

Now the series $\sum_{i=1}^{\infty} \log (1 + a_i)$ and the series $\sum_{i=1}^{\infty} a_i$ converge or diverge simultaneously, since

$$\lim_{x\to 0} \frac{\log(1+x)}{x} = \lim_{x\to 0} \frac{1}{1+x} = 1,$$

and so, for any $\epsilon > 0$,

$$(1-\epsilon)a_i < \log(1 + a_i) < (1+\epsilon)a_i$$

if i is sufficiently large.

Thus the convergence of

$$\sum_{i=1}^{\infty} \left(\log \left(1 - \frac{m}{n_i}\right) - \log\left(1 + \frac{m+1}{n_i}\right)\right)$$

occurs simultaneously with the convergence of

$$\sum_{i=1}^{\infty} \frac{1}{n_i}$$

which establishes the theorem.

4.7 Dimension, Riesz-Fischer Theorem

Let $S = [x_\alpha : \alpha \in A]$ and $T = [y_\beta : \beta \in B]$ be two bases for the Hilbert space X, and denote by $|A|$ and $|B|$ their cardinal numbers. Since, for each $\alpha \in A$,

$$x_\alpha = \sum_{\beta\in B} (x_\alpha, y_\beta) y_\beta,$$

the subset $B_\alpha \subset B$ of indices for which $(x_\alpha, y_\beta) \neq 0$ is countable. Moreover,

$$B = \bigcup_{\alpha\in A} B_\alpha,$$

since otherwise some y_β would be orthogonal to the complete set S, which is impossible. Hence, $|B| \leq \aleph_0 |A|$ and, by symmetry, $|A| \leq \aleph_0 |B|$.

If A and B are infinite sets, this proves the following proposition, which is well known from linear algebra in case A or B is finite.

PROPOSITION 1. The cardinality is the same for each basis of a given Hilbert space.

In view of this fact, we define the (orthogonal) **dimension** of a Hilbert space to be the cardinality of its bases.

Now we show

PROPOSITION 2. X is separable if and only if there is a countable basis for X, i.e., the dimension of X does not exceed $\aleph_0$.

Proof. If X is separable and $(x_1, x_2, \ldots)$ is a dense set in S, then by an inductive procedure we can construct an orthonormal set of linear combinations of the x_i's. The procedure differs from the Gramm-Schmidt procedure only in that at each stage the next x_i may be linearly dependent on the preceding, in which case it is skipped. The resulting orthonormal set is complete since it is a total set in X.

Conversely, if $(x_1, x_2, \ldots)$ is a basis for X, then the linear combinations with rational coefficients form a countable dense set.

Now, suppose that X and Y are two Hilbert spaces of the same dimension. Let S be a basis for X and let T be a basis for Y. Let φ be a one-one map of S onto T. Extend φ to X as follows: If $x \in X$, $x = \sum_{\alpha \in A} (x, x_\alpha)x_\alpha$, where $S = [x_\alpha : \alpha \in A]$, put

$$\varphi(x) = \sum_{\alpha \in A} (x, x_\alpha)\varphi(x_\alpha).$$

The latter series converges because the former does and each convergence is equivalent to

$$\sum_{\alpha \in A} |(x, x_\alpha)|^2 < \infty.$$

Now φ is clearly a linear transformation, which is one-one since $\varphi(x) = 0$ implies $(x, x_\alpha) = 0$ for all $\alpha \in A$, so that $x = 0$. Moreover, φ maps X onto Y since, if $y \in Y$, then

$$y = \sum_{\alpha \in A} (y, \varphi(x_\alpha))\varphi(x_\alpha)$$

and we may put

$$x = \sum_{\alpha \in A} (y, \varphi(x_\alpha))x_\alpha$$

to obtain $(x, x_\alpha) = (y, \varphi(x_\alpha))$, $\alpha \in A$, and

$$\varphi(x) = \sum_{\alpha \in A} (x, x_\alpha)\varphi(x_\alpha) = \sum_{\alpha \in A} (y, \varphi(x_\alpha))\varphi(x_\alpha) = y.$$

Finally, the Parseval formula implies that φ is norm preserving and so preserves inner product:

$$\|\varphi(x)\| = \|x\| \text{ and } (\varphi(x), \varphi(x')) = (x, x').$$

We have exhibited a mapping φ which is an isomorphism of the Hilbert spaces X and Y in the sense of being one-one and preserving all the Hilbert space structure. Thus we have proved

PROPOSITION 3. Two Hilbert spaces are isomorphic if they have the same dimension.

The converse is obvious.

We have seen that both l_2 and $L_2(0, 2\pi)$ are separable. Hence they are isomorphic. The isomorphism φ is simply the mapping which associates a function f with its sequence of Fourier coefficients.

The deep part of this result is the classical Riesz-Fischer theorem in the theory of Fourier series:

THEOREM 1. If $\sum_{n=-\infty}^{\infty} |c_n|^2 < \infty$, then there is an $f \in L_2(0, 2\pi)$ such that $\sum_{n=-\infty}^{\infty} c_n e^{inx}$ is the Fourier series of f and converges to f in the sense of L_2.

Clearly, the essential part of this theorem is the completeness of L_2, and so the latter assertion is often called the Riesz-Fischer theorem.

The space of example (iii) in Sec. 4.4 generates, upon completion, a Hilbert space which is non-separable.

4.8 Reproducing Kernel

The introduction of a Banach or Hilbert space in a problem in analysis often involves the step of completing a class of functions on a set S with respect to a norm appropriate to the problem. The functions of the class are usually defined everywhere in S, and there is some relation between the norm and the functional values at points of S. In general, all this is lost in the abstract process of completion, i.e., by imbedding in a space of equivalence classes of Cauchy sequences. For this reason, a more concrete completion is to be preferred if it is available. In particular, a completion by the adjunction of functions is desired. This problem was studied by Aronszajn and Smith [1]. The simple case of completing the class of continuous functions on [0, 1] under the norm of $L_p[0, 1]$ already shows that one cannot expect, in general, to obtain functions defined everywhere in S for the completion, but rather classes of equivalent functions in the sense of being defined and equal except on certain exceptional sets in S (the sets of measure zero in the example, where $L_p[0, 1]$ is the completion).

Under adequate conditions relating the norm and the functional values, however, this phenomenon does not occur. It is this case which we shall discuss here.

A vector space X of functions on a set S, together with a norm in X, is called a **proper functional space** if for every $s \in S$ the evaluation functional at s is continuous, i.e., there is an M_s such that

$$|x(s)| \leq M_s \|x\|, \ x \in X.$$

A **proper functional completion** of X is a proper functional space $\overline{X}$ on the same basic set S such that $\overline{X}$ is complete and X is a dense subspace of $\overline{X}$.

Since the abstract completion is unique up to isomorphism and since if $\overline{X}$ exists then the function values of its elements are well determined by the continuous extension there of the evaluation functionals, it is clear that at most one proper functional completion of X can exist.

We have

PROPOSITION 1. A proper functional space X has a proper functional completion if and only if for each Cauchy sequence $\{x_n\}$ in X,

$$\lim_n \ x_n(s) = 0 \text{ for all } s \in S \text{ implies } \lim_n \|x_n\| = 0.$$

Proof. Suppose $\overline{X}$ exists. If $\{x_n\}$ is a Cauchy sequence, $\lim_n \|x_n\|$ exists and $\|\bar{x}\| = \lim_n \|x_n\|$, if $\bar{x}$ is the limit in X of $\{x_n\}$. Suppose $\|\bar{x}\| > 0$. Then $\bar{x}(s) \neq 0$ for some $s \in S$. But $\bar{x}(s) = \lim_n x_n(s) = 0$ by assumption and the continuity of evaluation at s. This contradiction establishes the condition.

Conversely, assuming the condition, let $\overline{X}$ consist of all functions on S defined by

$$\bar{x}(s) = \lim_n x_n(s), \quad s \in S,$$

where $\{x_n\}$ is any Cauchy sequence in X (the limit exists since $|x_n(s) - x_m(s)| \leq M_s \|x_n - x_m\|$). In addition, let

$$\|\bar{x}\| = \lim_n \|x_n\|.$$

This choice is independent of the sequence defining $\bar{x}$, since if

$$\bar{x}(s) = \lim_n y_n(s)$$

also, then $\|x_n - y_n\|$ has limit zero by the condition, and so

$$\lim_n | \ \|x_n\| - \|y_n\| \ | \leq \lim_n \|x_n - y_n\| = 0.$$

Since the constant sequence $(x, x, \ldots)$ defines $x \in X$ in the above way, we have $X \subset \overline{X}$ with agreement of norms.

Let $\bar{x} \in \overline{X}$ and $\{x_n\}$ define $\bar{x}$. Then

$$\lim_n \|\bar{x} - x_n\| = \lim_k \lim_n \|x_k - x_n\| = 0,$$

since $\{x_n\}$ is a Cauchy sequence. Hence any defining sequence for $\bar{x}$ converges to $\bar{x}$ in $\overline{X}$. In particular, X is dense in $\overline{X}$.

Now, for $s \in S$,

$$|\bar{x}(s)| = |\lim_n x_n(s)| = \lim_n |x_n(s)| \leq M_s \lim_n \|x_n\| = M_s \|\bar{x}\|.$$

Thus $\bar{X}$ is a proper functional space.

Finally, $\bar{X}$ is complete. Let $\{\bar{x}_k\}$ be a Cauchy sequence in $\bar{X}$ and, for each k, let $\{x_{k,n}\}$ define $\bar{x}_k$,

$$\bar{x}_k(s) = \lim_n x_{k,n}(s), \quad s \in S \quad \text{and} \quad \bar{x}_k = \lim_n x_{k,n} \text{ in } \bar{X}.$$

Choose $1 < N(1) < N(2) < \ldots$ so that

$$\|\bar{x}_k - x_{k,N(k)}\| < \frac{1}{2^k}.$$

The sequence $\{x_{k,N(k)}\}$ is Cauchy in X, so the limit

$$\bar{x}(s) = \lim_k x_{k,N(k)}(s), \quad s \in S$$

exists by the proper functional space property. That $\bar{x} = \lim_k \bar{x}_k$ in $\bar{X}$ follows from

$$\|\bar{x}_k - \bar{x}\| \leq \|\bar{x}_k - x_{k,N(k)}\| + \|x_{k,N(k)} - \bar{x}\|$$

and the fact that in $\bar{X}$

$$\bar{x} = \lim_k x_{k,N(k)}.$$

If X is a proper functional Hilbert space, then for each $t \in S$ the evaluation functional is represented by an element, K_t, of X. Then,

$$x(t) = (x, K_t), \qquad x \in X, t \in S.$$

Denote by $K(s, t)$ the function on $S \times S$ given by

$$K(s, t) = K_t(s) = (K_t, K_s)$$

and adopt the notation $K(\cdot, t)$ for the element $K_t \in X$.

A Hilbert space of functions on a set S is said to have a **reproducing kernel** if there is a function $K(s, t)$ on $S \times S$ with the properties

(i) $K(\cdot, t) \in X$, $t \in S$

(ii) (reproducing property) $x(t) = (x, K(\cdot, t))$, $x \in X$, $t \in S$.

Thus, any proper functional Hilbert space has a reproducing kernel. It is immediately seen to be unique. On the other hand, if X has a reproducing kernel, then

$$|x(t)| = |(x, K(\cdot, t))| \leq \|K(\cdot, t)\| \|x\|, \qquad x \in X, t \in S,$$

and X is a proper functional space.

In our next remarks, we shall characterize those functions on a set $S \times S$ which are reproducing kernels, and indicate how the proper functional space can be constructed from the kernel function.

A function $K(s, t)$ on $S \times S$ is a **positive matrix** if for each $n = 1, 2, \ldots$ and each choice of points $t_1, t_2, \ldots, t_n$ the quadratic form

$$\sum_{i,j=1}^{n} K(t_i, t_j)\xi_i\bar{\xi}_j$$

in $\xi_1, \xi_2, \ldots, \xi_n$ is non-negative.

If K is the reproducing kernel of a space X, this expression is easily seen to be just $\left\| \sum_{i=1}^{n} K(\cdot\,, t_i)\xi_i \right\|^2$. Thus each reproducing kernel is a positive matrix. Conversely, if K is a positive matrix, then the class of all functions on S representable in the form $\sum_{i=1}^{n} K(s, t_i)\xi_i$ for some $n, t_1, t_2, \ldots, t_n \in S$ and scalars $\xi_1, \xi_2, \ldots, \xi_n$, is a proper functional inner product space under the norm

$$\left\| \sum_{i=1}^{n} K(\cdot\,, t_i)\xi_i \right\|^2 = \sum_{i,j=1}^{n} K(t_i, t_j)\xi_i\bar{\xi}_j.$$

The proper functional completion X of this space has reproducing kernel K. (The condition for existence of a proper functional completion is clearly satisfied.)

If X is a proper functional Hilbert space with $K(s, t)$ as reproducing kernel and if $[x_\alpha : \alpha \in A]$ is a complete orthonormal set in X, then

$$K(\cdot\,, t) = \sum_{\alpha \in A} (K(\cdot\,, t), x_\alpha)x_\alpha = \sum_{\alpha \in A} \overline{x_\alpha(t)}x_\alpha, \quad t \in S,$$

or

$$K(s, t) = \sum_{\alpha \in A} x_\alpha(s)\overline{x_\alpha(t)}.$$

This expansion makes it possible to give an explicit representation of the reproducing kernel when a complete orthonormal set for X is known.

At first we only have the convergence of this expansion in the sense that, for each t, only countably many terms of $\sum_{\alpha \in A} \overline{x_\alpha(t)}x_\alpha$ are non-zero and this series converges in X to $K(\cdot\,, t)$.

However, the convergence is actually point-wise in S for each t since, if $\{x_n\}$ converges in X to x, then, for each $s \in S$,

$$|x_n(s) - x(s)| = |(x_n - x, K(\cdot\,, s))| \leq \|x_n - x\|\,\|K(\cdot\,, s)\|.$$

4.9 Application, Bergman Kernel

Let D be a bounded domain in the $z = x + iy$ plane whose boundary consists of a finite number of smooth simple closed curves.

Consider the class $L_2(D)$ of all holomorphic functions in D for which the Dirichlet integral exists

$$\iint_D |f(z)|^2\, dx\, dy < \infty.$$

This integral is to be understood in the sense of

$$\lim_n \iint_{K_n} |f(z)|^2 \, dx \, dy,$$

where K_n is a non-decreasing sequence of compact subsets of D whose union is D.

We shall show that $L_2(D)$, with $\|f\|^2$ taken as the Dirichlet integral, is a proper functional Hilbert space and illustrate the importance of its reproducing kernel K_D, called the Bergman kernel of D, by an application to conformal mapping.

We need the mean value theorem for holomorphic functions. If f is holomorphic in a disk $|z - z_0| < r$, then

$$f(z_0) = \frac{1}{\pi r^2} \iint_{|z-z_0|<r} f(z) \, dx \, dy.$$

This follows from the identity

$$\iint_D f(z)\overline{g'(z)} \, dx \, dy = \frac{1}{2i} \int_C f(z)\overline{g(z)} \, dz,$$

a consequence of Green's theorem in the plane, because

$$\begin{aligned} f(z_0) &= \frac{1}{2\pi i} \int_{|z-z_0|=r} \frac{f(z)}{z - z_0} \, dz \\ &= \frac{1}{2\pi i r^2} \int_{|z-z_0|=r} f(z)\overline{(z - z_0)} \, dz \\ &= \frac{1}{\pi r^2} \iint_{|z-z_0|<r} f(z) \, dx \, dy. \end{aligned}$$

If we apply the Cauchy-Schwarz inequality in the mean value theorem, we obtain

$$|f(z_0)| \leq \frac{1}{\sqrt{\pi} r} \|f\|, \quad z_0 \in D, \quad f \in L_2(D),$$

where r is chosen so that the disk $|z - z_0| < r$ is contained in D. Thus, $L_2(D)$ is a proper functional space.

Next, we show that $L_2(D)$ is complete. If K is any compact subset of D, then a single R may be found such that for any $z_0 \in K$

$$|f(z_0)| \leq \frac{1}{\sqrt{\pi} R} \|f\|, \quad f \in L_2(D).$$

Thus convergence in $L_2(D)$ of a sequence $\{f_n\}$ implies convergence in D, uniformly on compact subsets of D. By a standard theorem of Weierstrass, such convergence yields a holomorphic limit function. Thus, if $\{f_n\}$ is a Cauchy sequence in $L_2(D)$, it converges point-wise (uniformly on compact sets) to a holomorphic function f on D. We show that $f \in L_2(D)$.

Let $K \subset D$ be compact and choose N so that

$$|f(z) - f_N(z)| < 1, \; z \in K,$$

and

$$|\, \|f_N\| - \lim_n \|f_n\| \,| < 1.$$

Thus, if m denotes planar measure,

$$\begin{aligned}\left[\iint_K |f(z)|^2 \, dx \, dy\right]^{1/2} & \\ &\leq \left[\iint_K |f(z) - f_N(z)|^2 \, dx \, dy\right]^{1/2} + \left[\iint_K |f_N(z)|^2 \, dx \, dy\right]^{1/2} \\ &< \sqrt{m(K)} + \|f_N\| \\ &< \sqrt{m(D)} + \lim_n \|f_n\| + 1.\end{aligned}$$

Since the final bound is independent of K, we have $f \in L_2(D)$.

It remains for us to show that $\{f_n\}$ converges to f in $L_2(D)$. Let $\epsilon > 0$. Choose N so that $m, n \geq N$ implies $\|f_n - f_m\| < \epsilon$. Next choose a compact set $K \subset D$ so that both

$$\|f\chi_K - f\| < \epsilon \text{ and } \|f_N\chi_K - f_N\| < \epsilon,$$

where the norm sign here indicates the norm in the Hilbert space of all square summable functions in D, the characteristic function of K being non-analytic. That there exists such a compact set K follows since

$$\|f\|^2 = \lim_n \iint_{K_n} |f|^2 dx \, dy$$

and similarly for f_N.

Next, by the uniform convergence of f_n to f on K, we may choose $N' > N$ such that

$$|f_{N'}(z) - f(z)|^2 < \frac{\epsilon^2}{m(K)}, \quad z \in K,$$

and hence such that

$$\begin{aligned}\|(f_{N'} - f)\chi_K\| &= \left[\iint_K |f_{N'}(z) - f(z)|^2 \, dx \, dy\right]^{1/2} \\ &< \left[\frac{\epsilon^2}{m(K)} \cdot m(K)\right]^{1/2} = \epsilon.\end{aligned}$$

Now, for $n > N'$,

$$\begin{aligned}\|f_n - f\| \leq \|f_n - f_N\| + \|f_N - f_N\chi_K\| + \|(f_N - f_{N'})\chi_K\| \\ + \|(f_{N'} - f)\chi_K\| + \|f\chi_K - f\| < 5\epsilon.\end{aligned}$$

We recall the celebrated Riemann mapping theorem which asserts that if D is a simply connected domain having more than one boundary point, then there exists a holomorphic function in D which maps D one-one onto the unit disk. We assume, as above, that D is bounded. Moreover, if $\xi \in D$ there is a unique mapping function $f(z) = f(z, \xi)$ for which $f(\xi) = 0$ and $f'(\xi) > 0$.

If $K = K_D$ denotes the Bergman kernel of D, we have

THEOREM 1. The mapping function f is expressed in terms of K by the formula

$$f'(z) = \sqrt{\frac{\pi}{K(\xi, \xi)}}\, K(z, \xi).$$

Proof. Let D_r denote the subdomain of D which is mapped by f onto the disk $|w| < r$, where $r < 1$ and $w = f(z)$. Denote the boundary of D_r by C_r.

If $g \in L_2(D)$, then $g(z)/f(z)$ is meromorphic in D_r with a single pole at $z = \xi$. The residue at this pole is obtained from

$$\lim_{z \to \xi} \frac{(z - \xi)g(z)}{f(z)} = \frac{g(\xi)}{f'(\xi)}.$$

Thus, by the residue theorem,

$$\frac{g(\xi)}{f'(\xi)} = \frac{1}{2\pi i} \int_{C_r} \frac{g(z)}{f(z)}\, dz = \frac{1}{2\pi i r^2} \int_{C_r} \overline{f(z)} g(z)\, dz,$$

since $|f(z)|^2 = r^2$ for $z \in C_r$. Applying the identity (consequence of Green's theorem) mentioned at the beginning of this section, we obtain

$$\frac{g(\xi)}{f'(\xi)} = \frac{1}{\pi r^2} \iint_{D_r} \overline{f'(z)} g(z)\, dx\, dy.$$

Letting r converge to 1, we get

$$g(\xi) = \iint_D \frac{\overline{f'(z)}\,\overline{f'(\xi)}}{\pi}\, g(z)\, dx\, dy.$$

In other words, the function

$$K(z, \xi) = \frac{f'(z)\overline{f'(\xi)}}{\pi}$$

has the reproducing property for $L_2(D)$. Since reproducing kernels are unique, it follows that $K(z, \xi)$ is the Bergman kernel of D.

In case $\xi = z$,

$$[f'(\xi)]^2 = \pi K(\xi, \xi)$$

and substitution for $f'(\xi)$ in the formula for K gives the desired result.

We conclude with the remark that if a complete orthonormal set for $L_2(D)$ is known, K may be written in the form

$$K(z, \xi) = \sum_{n=1}^{\infty} \varphi_n(z)\overline{\varphi_n(\xi)}$$

and so the mapping function can be obtained.

We refer the reader to Nehari [1] for additional interesting applications of Bergman kernels.

4.10 Examples of Complete Orthonormal Sets

In this section, we shall discuss a few of the many important examples of complete orthonormal sets.

First, we consider the sequence

$$1, z, z^2, z^3, \ldots$$

in the space $L_2(D)$. This set is complete for a variety of domains D, each leading to a set of orthonormal polynomials. We shall give the proof only in case D is the unit disk $|z| < 1$.

In this case, the orthonormal set is $\varphi_n(z) = \sqrt{\frac{n}{\pi}}\, z^{n-1}$, $n = 1, 2, 3, \ldots$.

Thus, if $f \in L_2(D)$, the Fourier coefficients are given by

$$\begin{aligned} a_n &= \sqrt{\frac{n}{\pi}} \iint_{|z|<1} f(z)\bar{z}^{n-1}\, dx\, dy \\ &= \lim_{r\to 1} \sqrt{\frac{n}{\pi}} \iint_{|z|<r} f(z)\bar{z}^{n-1}\, dx\, dy \\ &= \lim_{r\to 1} \frac{1}{2i} \sqrt{\frac{n}{\pi}} \int_{|z|=r} f(z)\, \frac{\bar{z}^n}{n}\, dz, \end{aligned}$$

where we have applied the complex Green's formula of the preceding section.

Since $|z|^2 = r^2$, we have $\bar{z} = r^2 z^{-1}$ and so

$$a_n = \lim_{r\to 1} \frac{1}{\sqrt{\pi n}} \frac{r^{2n}}{2i} \int_{|z|=r} \frac{f(z)}{z^n}\, dz.$$

Now, if

$$f(z) = b_0 + b_1 z + b_2 z^2 + \ldots, \quad |z| < 1$$

is the power series expansion for f, then

$$b_{n-1} = \frac{1}{2\pi i} \int_{|z|=r} \frac{f(z)}{z^n}\, dz.$$

Hence,

$$a_n = \lim_{r\to 1} \sqrt{\frac{\pi}{n}}\, r^{2n} b_{n-1} = \sqrt{\frac{\pi}{n}}\, b_{n-1}, \quad n = 1, 2, \ldots.$$

We proceed to verify Parseval's formula,

$$\iint_{|z|<1} |f(z)|^2\,dx\,dy = \sum_{n=1}^{\infty} |a_n|^2 = \pi \sum_{n=1}^{\infty} \frac{|b_{n-1}|^2}{n},$$

thereby establishing the completeness. For this,

$$\begin{aligned}
\iint_{|z|<1} |f(z)|^2\,dx\,dy &= \int_0^1 \int_0^{2\pi} |f(re^{i\theta})|^2 r\,dr\,d\theta \\
&= \int_0^1 \int_0^{2\pi} \left(\sum_{n=0}^{\infty} b_n r^n e^{in\theta}\right)\left(\sum_{n=0}^{\infty} \bar{b}_n r^n e^{-in\theta}\right) r\,dr\,d\theta \\
&= \int_0^1 \int_0^{2\pi} \sum_{n=0}^{\infty} \sum_{m=0}^{\infty} b_n \bar{b}_m r^{n+m} e^{i(n-m)\theta} r\,dr\,d\theta \\
&= \int_0^1 \sum_{n=0}^{\infty} \sum_{m=0}^{\infty} b_n \bar{b}_m \int_0^{2\pi} e^{i(n-m)\theta}\,d\theta r^{n+m+1}\,dr,
\end{aligned}$$

where the term by term integration is valid because the power series converges uniformly on $|z| \leq r$ for $r < 1$. Then,

$$\begin{aligned}
\iint_{|z|<1} |f(z)|^2\,dx\,dy &= \int_0^1 2\pi \sum_{n=0}^{\infty} |b_n|^2 r^{2n+1}\,dr \\
&= \pi \sum_{n=0}^{\infty} \frac{|b_n|^2}{n+1} = \pi \sum_{n=1}^{\infty} \frac{|b_{n-1}|^2}{n}.
\end{aligned}$$

With this result as a basis, it is not hard to establish a complete orthonormal system for $L_2(D)$, in case D is a simply connected domain with at least two boundary points, in terms of any mapping function $w = f(z)$ which maps D conformally on the unit disk $|w| < 1$. Indeed,

$$\varphi_n(z) = \sqrt{\frac{n}{\pi}}\,[w(z)]^{n-1} w'(z), \quad n = 1, 2, \ldots$$

is such a system.

A complete orthonormal system for the space $L_2(0, \infty)$ is given by the Laguerre functions

$$\varphi_n(x) = \frac{1}{n!}\,e^{-x/2} L_n(x), \quad n = 0, 1, 2, \ldots,$$

where $L_n(x)$ is the Laguerre polynomial, defined as the coefficient of e^{-x} in the nth derivative of $x^n e^{-x}$:

$$\begin{aligned}
L_n(x) &= e^x \frac{d^n}{dx^n}(x^n e^{-x}) \\
&= \sum_{k=0}^{n} (-1)^k \binom{n}{k} n(n-1)\ldots(k+1)x^k.
\end{aligned}$$

For the orthogonality,

$$\int_0^{\infty} e^{-x} L_n(x) L_m(x)\,dx = 0, \quad n > m,$$

we compute, for $k < n$,

$$\int_0^\infty e^{-x}x^k L_n(x)\,dx = \int_0^\infty x^k \frac{d^n}{dx^n}(x^n e^{-x})\,dx$$

$$= (-1)^k k! \int_0^\infty \frac{d^{n-k}}{dx^{n-k}}(x^n e^{-x})\,dx = 0,$$

by repeated integration by parts. Also,

$$\int_0^\infty e^{-x}L_n^2(x)\,dx = \int_0^\infty \frac{d^n}{dx^n}(x^n e^{-x})\sum_{k=0}^{n}(-1)^k\binom{n}{k}n(n-1)\ldots(k+1)x^k\,dx$$

$$= \int_0^\infty (-1)^n x^n \frac{d^n}{dx^n}(x^n e^{-x})\,dx$$

$$= n!\int_0^\infty x^n e^{-x}\,dx = (n!)^2$$

which shows that $\{\varphi_n\}$ is orthonormal.

We show that the Laguerre functions are complete in $L_2(0, \infty)$. Thus, we show that if $f \in L_2(0, \infty)$ and

$$\int_0^\infty f(x)e^{-x/2}L_n(x)\,dx = 0, \quad n = 0, 1, 2, \ldots,$$

then f is zero almost everywhere. Let

$$g(x) = f(x)e^{-x/2}.$$

Then we need only show that

$$\int_0^\infty g(x)x^n\,dx = 0, \quad n = 0, 1, 2, \ldots$$

implies $g(x) = 0$ almost everywhere for $g \in L_1(0, \infty)$, since $f \in L_2$, $e^{-x/2} \in L_2$ implies $g \in L_1$.

Now consider

$$\Phi(s) = \int_0^\infty e^{-st}g(t)\,dt.$$

Since $g \in L_1$, Φ is defined for all $s = \sigma + i\tau$, $\sigma \geq 0$. Moreover, Φ is regular for all $s = \sigma + i\tau$, $\sigma > 0$ (see Widder, [1]).

We consider the series

$$\sum_{n=0}^{\infty} s^n \cdot \frac{1}{n!}\int_0^\infty g(t)t^n\,dt$$

for real s. An easy computation using the fact that

$$\int_0^\infty e^{-x/2}x^n\,dx = 2^{n+1}n!$$

shows that this series converges for $0 \leq s < \frac{1}{4}$ to the function Φ. Thus Φ is zero for all real $s \geq 0$.

But

$$\Phi(s) = \int_0^1 t^{s-1}g(-\log t)\,dt = 0, \quad s \geq 0$$

implies, as the reader may verify, that

$$g(-\log t) = 0 \text{ a.e. in } (0, 1)$$

so that

$$g(x) = 0 \text{ a.e. in } (0, \infty).$$

4.11 Systems of Haar, Rademacher, Walsh; Applications

We now consider the Haar system in $L_2[0, 1]$. Let

$$\varphi_0^{(0)}(x) = 1, \quad x \in [0, 1]$$

$$\varphi_1^{(0)}(x) = \begin{cases} 1, & x \in [0, \frac{1}{2}) \\ -1, & x \in (\frac{1}{2}, 1] \\ 0, & x = \frac{1}{2} \end{cases}$$

and for $m = 1, 2, \ldots; k = 1, \ldots, 2^m$

$$\varphi_m^{(k)}(x) = \begin{cases} \sqrt{2^m}, & x \in \left(\dfrac{k-1}{2^m}, \dfrac{k-1/2}{2^m}\right) \\ -\sqrt{2^m}, & x \in \left(\dfrac{k-1/2}{2^m}, \dfrac{k}{2^m}\right) \\ 0, & x \in [0, 1] \sim \left[\dfrac{k-1}{2^m}, \dfrac{k}{2^m}\right] \end{cases}$$

and at that finite set of points at which $\varphi_m^{(k)}(x)$ has not yet been defined, let $\varphi_m^{(k)}(x)$ be the average of the left and right limits of $\varphi_m^{(k)}(x)$ as x approaches the point in question. At 0 and 1, let $\varphi_m^{(k)}(x)$ assume the value of the one-sided limit. The set of functions

$$\{\varphi_0^{(0)}, \varphi_1^{(0)}, \varphi_m^{(k)}, m = 1, 2, \ldots; k = 1, \ldots, 2^m\}$$

is known as the **Haar system.**

PROPOSITION 1. The Haar system is orthonormal in $L_2[0, 1]$.

Proof. If $m = n$, then $\varphi_m^{(k)}(x) \cdot \varphi_n^{(l)}(x) = 0$, $k \neq l$, except possibly at one point, so that

$$\int_0^1 \varphi_m^{(k)}(x)\varphi_n^{(l)}(x)\, dx = 0.$$

If $m \neq n$, then

$$\int_0^1 \varphi_m^{(k)}(x) \cdot \varphi_n^{(l)}(x)\, dx = \int_{\frac{l-1}{2^n}}^{\frac{l}{2^n}} \varphi_m^{(k)}(x)\varphi_n^{(l)}(x)\, dx$$

$$= \sqrt{2^n} \int_{\frac{l-1}{2^n}}^{\frac{l-1/2}{2^n}} \varphi_m^{(k)}\, dx - \sqrt{2^n} \int_{\frac{l-1/2}{2^n}}^{\frac{l}{2^n}} \varphi_m^{(k)}\, dx = 0,$$

since both integrals vanish.

For normality,

$$\int_0^1 (\varphi_m^{(k)}(x))^2\,dx = 2\left((\sqrt{2^m})^2 \cdot \frac{1}{2^{m+1}}\right) = 1.$$

We leave it to the reader to show that the Haar functions are complete in $L_2[0, 1]$. We show that they have a deeper property.

The **Haar expansion** of a function $f \in L_1$ is the series

$$a_0^{(0)}\varphi_0^{(0)} + a_1^{(0)}\varphi_1^{(0)} + \sum_{m=1}^{\infty}\left(\sum_{k=1}^{2^m} a_m^{(k)}\varphi_m^{(k)}\right),$$

where $a_m^{(k)} = \int_0^1 \varphi_m^{(k)}(x)f(x)\,dx$, and the terms are added in the order indicated.

We shall need the fact that if f is summable and F is defined by $F(x) = \int_0^x f(t)\,dt$, then F' exists almost everywhere and $F'(x) = f(x)$ almost everywhere. For a proof see, for example, Goffman [1].

THEOREM 1. If $f \in L_1$, then the Haar expansion of f converges to f almost everywhere.

Proof. Note that $a_m^{(k)}\varphi_m^{(k)}(x) = \int_0^1 \varphi_m^{(k)}(x)\varphi_m^{(k)}(t)f(t)\,dt$. The partial sums

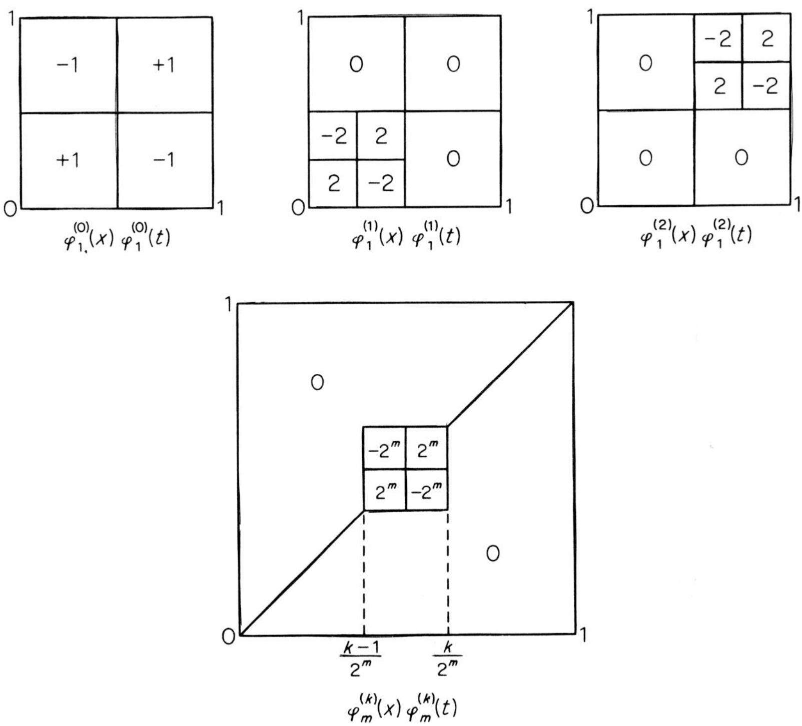

Figure 3

$$s_{m_0}^{(k_0)}(x) = a_0^{(0)}\varphi_0^{(0)}(x) + a_1^{(0)}\varphi_1^{(0)}(x) + \sum_{m=1}^{m_0-1}\left(\sum_{k=1}^{2^m} a_m^{(k)}\varphi_m^{(k)}(x)\right) + \sum_{k=1}^{k_0} a_{m_0}^{(k)}\varphi_{m_0}^{(k)}(x)$$

will be abbreviated by defining the kernel

$$K_{m_0}^{(k_0)}(x, t) = \varphi_0^{(0)}(x)\varphi_0^{(0)}(t) + \ldots + \varphi_m^{(k)}(x)\varphi_m^{(k)}(t) + \ldots + \ldots + \varphi_{m_0}^{(k_0)}(x)\varphi_{m_0}^{(k_0)}(t).$$

Thus,

$$s_{m_0}^{(k_0)}(x) = \int_0^1 f(t)K_{m_0}^{(k_0)}(x, t)\, dt.$$

Consider now the products $\varphi_m^{(k)}(x)\varphi_m^{(k)}(t)$ on the unit square.

$\varphi_0^{(0)}(x)\varphi_0^{(0)}(t) = 1$ on the unit square. The others will be given by diagram (Figure 3) for the sake of clarity. Since the values on the bounding lines do not matter in this discussion, they are omitted.

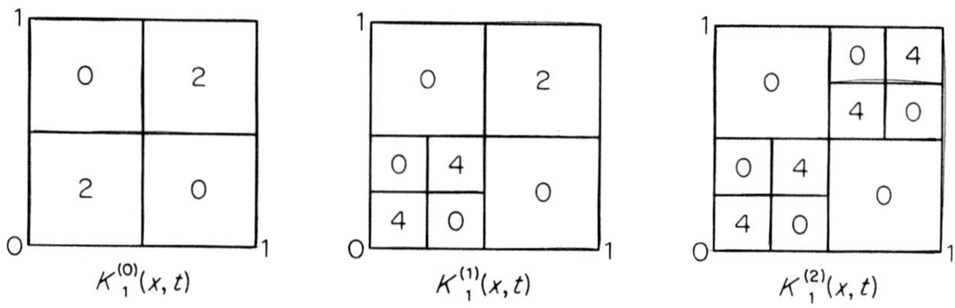

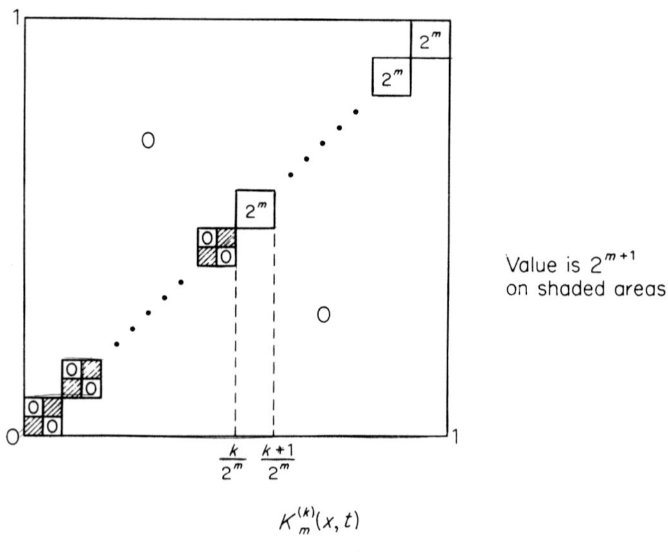

Figure 4

The kernels then take on the following values. $K_0^{(0)}(x, t) = 1$ on the unit square and the others are as in Figure 4.

Note that the value of the kernel on each diagonally situated square is the reciprocal of the length of the side of that square. Let $x \in [0, 1]$ not be of the form $p/2^n$ for any non-negative integers p and n. Then

$$\begin{aligned} s_m^{(k)}(x) &= \int_0^1 K_m^{(k)}(x, t) f(t)\, dt \\ &= \frac{1}{m(I)} \int_I f(t)\, dt, \end{aligned}$$

where I contains x and has length either $1/2^m$ or $1/2^{m+1}$. But if $F(x) = \int_0^x f(t)\, dt$, then clearly $\lim s_m^{(k)}(x) = F'(x) = f(x)$ almost everywhere.

A system closely related to the Haar system is the **Rademacher system:**

$$r_0(x) = \varphi_0^{(0)}(x)$$

$$r_1(x) = \varphi_1^{(0)}(x)$$

$$r_n(x) = \sqrt{2^{-n}} \sum_{k=1}^{2^n} \varphi_n^{(k)}(x), \quad n = 1, 2, \ldots .$$

It is clear that the Rademacher functions may be defined as follows: If $x \in [0, 1]$ has a unique binary expansion $0 \cdot a_1 a_2 \ldots$, then

$$r_n(x) = \begin{cases} 1, & a_n = 0 \\ -1, & a_n = 1. \end{cases}$$

If $x \in [0, 1]$ has two expansions, $0 \cdot a_1 a_2 \ldots$ and $0 \cdot b_1 b_2 \ldots$, then

$$r_n(x) = \begin{cases} 1 & \text{if } a_n = b_n = 0 \\ 0 & \text{if } a_n \neq b_n \\ -1 & \text{if } a_n = b_n = 1. \end{cases}$$

PROPOSITION 2. The Rademacher system is orthonormal in $L_2[0, 1]$. The proof is left to the reader.

Moreover, the Rademacher system is not complete. By symmetry considerations, $f(x) = \cos 2\pi x$ is orthogonal to all the Rademacher functions. The details are left to the reader.

PROPOSITION 3. If $\sum_{n=0}^{\infty} C_n^2 < \infty$, then $\sum_{n=0}^{\infty} C_n r_n(x)$ converges almost everywhere.

Proof. Let $C_n^{(k)} = C_n/\sqrt{2^n}$, $k = 1, \ldots, 2^n$; $n = 1, 2, \ldots$. So

$$\sum_{n=1}^{\infty} C_n r_n(x) = \sum_{n=1}^{\infty} \sum_{k=1}^{2^n} \sqrt{2^{-n}} C_n \varphi_n^{(k)}(x) = \sum_{n=1}^{\infty} \sum_{k=1}^{2^n} C_n^{(k)} \varphi_n^{(k)}(x),$$

which is the Haar expansion of some function in $L_1 \supset L_2$, by the Riesz-Fischer theorem, since

$$\sum_{n=1}^{\infty} \sum_{k=1}^{2^n} (C_n^{(k)})^2 = \sum_{n=1}^{\infty} C_n^2 < \infty.$$

Hence, by theorem 1, $\sum_{n=0}^{\infty} C_n r_n(x)$ converges almost everywhere.

COROLLARY 1. If $\sum_{n=1}^{\infty} C_n^2 < \infty$, then $\sum_{n=1}^{\infty} \pm C_n$ converges for almost all choices of signs.

Proof. "Almost all" is taken in the sense that to each sequence $s_1, s_2, \ldots$ of signs corresponds the point $0 \cdot a_1 a_2 \ldots \in [0, 1]$, where

$$a_i = 1 \quad \text{if } s_i \text{ is a } + \text{ sign,}$$
$$a_i = 0 \quad \text{if } s_i \text{ is a } - \text{ sign,}$$

and the measure of a set of sequences of signs is the Lebesgue measure of the corresponding set of points in [0, 1]. This mating is one-one, except for a countable set.

The result follows by considering $\sum_{n=1}^{\infty} C_n r_n(x)$ at each $x \in [0, 1]$.

COROLLARY 2. If $\{\varphi_n\}$ is any orthonormal system in $L_2[0, 1]$ and

$$\sum_{n=1}^{\infty} C_n^2 < \infty, \quad \text{then} \quad \sum_{n=1}^{\infty} \pm C_n \varphi_n$$

converges almost everywhere for almost all choices of signs.

Proof. Since the system is normal,

$$\sum_{n=1}^{\infty} C_n^2 = \sum_{n=1}^{\infty} \int_0^1 C_n^2 \varphi_n^2(x)\, dx < \infty.$$

So,

$$\left\{ \int_0^1 \sum_{n=1}^{m} C_n^2 \varphi_n^2(x)\, dx \right\}$$

is a bounded increasing sequence, and it follows that $\sum_{n=1}^{\infty} C_n^2\varphi_n^2(x) < \infty$ for almost all x. By proposition 3, it follows that

$$\sum_{n=1}^{\infty} C_n\varphi_n(x)r_n(t)$$

converges for almost all values of t for almost all choices of x.

But the set of pairs (x, t) for which this series converges is measurable. Thus its planar measure is 1. It follows that for almost all choices of t, $\sum_{n=1}^{\infty} C_n\varphi_n(x)r_n(t)$ converges almost everywhere in x. Then, just as in corollary 1, $\sum_{n=1}^{\infty} \pm C_n\varphi_n(x)$ converges almost everywhere in x for almost all choices of signs.

We shall need the following lemma, whose proof we leave to the reader.

LEMMA. If at least two of the integers $k_1, k_2, \ldots, k_n$ are distinct, then $\int_0^1 r_{k_1}(x)\ldots r_{k_n}(x)\,dx = 0$.

PROPOSITION 4. If $\sum_{n=0}^{\infty} C_n^2 = \infty$, then $\sum_{n=0}^{\infty} C_nr_n(x)$ diverges almost everywhere.

Proof. Suppose the series $\sum_{n=0}^{\infty} C_nr_n(x)$ converges on a set of positive measure. Then by Egoroff's theorem, it converges uniformly on a set E of positive measure. Let

$$S_n(x) = \sum_{k=0}^{n} C_kr_k(x), \quad n = 0, 1, 2, \ldots.$$

Let M be a positive integer and n such that $|S_m(x) - S_n(x)| < M$ for all $m > n$ and $x \in E$. Then, for every $m > n$,

$$M^2m(E) > \int_E \left[\sum_{k=n+1}^{m} C_kr_k(x)\right]^2 dx$$

$$= m(E)\sum_{k=n+1}^{m} C_k^2 + 2\sum_{k=n+1}^{m-1}\sum_{l=k+1}^{m} C_kC_l\int_E r_k(x)r_l(x)\,dx.$$

By the Cauchy-Schwarz inequality,

$$\left|\sum_{k=n+1}^{m-1}\sum_{l=k+1}^{m} C_kC_l\int_E r_k(x)r_l(x)\,dx\right|$$

$$\leq \left[\sum_{k=n+1}^{m}\sum_{l=n+1}^{m} C_k^2C_l^2\right]^{1/2}\left[\sum_{k=n+1}^{\infty}\sum_{l=k+1}^{\infty}\left(\int_E r_k(x)r_l(x)\,dx\right)^2\right]^{1/2}.$$

The functions $f_{kl} = r_k \cdot r_l$ form an orthonormal system, by the lemma, so we may expand in terms of them. The kl Fourier coefficient of χ_E is given by

$$\int_0 \chi_E(x) r_k(x) r_l(x)\, dx = \int_E r_k(x) r_l(x)\, dx.$$

By Bessel's inequality,

$$\sum_{k=0}^{\infty} \sum_{l=0}^{\infty} \left(\int_E r_k(x) r_l(x)\, dx\right)^2 \leq \int_0^1 \chi_E^2(x)\, dx = m(E).$$

Hence for sufficiently large n,

$$\sum_{k=n+1}^{\infty} \sum_{l=k+1}^{\infty} \left(\int_E r_k(x) r_l(x)\, dx\right)^2 < \frac{1}{9} m(E)^2.$$

Moreover,

$$\left[\sum_{k=n+1}^{m} \sum_{l=n+1}^{m} C_k^2 C_l^2\right]^{1/2} = \sum_{k=n+1}^{m} C_k^2.$$

Hence,

$$M^2 m(E) \geq m(E) \sum_{k=n+1}^{m} C_k^2 - \frac{2}{3} m(E) \sum_{k=n+1}^{\infty} C_k^2,$$

so that $\sum_{k=n+1}^{m} C_k^2 \leq 3M^2$ for every $m > n$. Therefore, $\sum_{k=0}^{\infty} C_k^2 < \infty$, a contradiction.

The Rademacher system has an extension to a complete orthonormal system, the **Walsh system,** given by

$$w_0 = 1$$

$$w_1 = r_1$$

$$\ldots$$

$$w_n = r_{\nu_1+1} \ldots r_{\nu_p+1}, \text{ for } n \geq 1,$$

where $\nu_1 < \nu_2 < \ldots < \nu_p$ are determined from the binary expansion of $n = 2^{\nu_1} + \ldots + 2^{\nu_p}$.

Observe that for $n = 2^p$, $w = r_n$, so that the Walsh functions include the Rademacher functions.

THEOREM 2. The Walsh system is a complete orthonormal set in $L_2\,[0, 1]$.

Proof. The orthogonality follows by the lemma. The normality is easy to verify.

To prove completeness, take any $f \in L_2[0, 1]$, or even $L_1[0, 1]$, and define $F(x) = \int_0^x f(t)\, dt$. Suppose $\int_0^1 f(x) w_n(x)\, dx = 0$, $n = 0, 1, 2, \ldots$. f will be shown equal to zero almost everywhere by proving $F(x) = 0$ on a dense set in [0, 1]. Since F is continuous, it must then be identically zero. Since $F'(x) = f(x)$ almost everywhere, it follows that $f(x) = 0$ almost everywhere.

Now, $F(0) = 0$.

$$0 = \int_0^1 f(x)w_0(x)\,dx = F(1).$$

$$0 = \int_0^1 f(x)w_1(x)dx = \left[F\left(\frac{1}{2}\right) - F(0)\right] - \left[F(1) - F\left(\frac{1}{2}\right)\right] = 2F\left(\frac{1}{2}\right).$$

$$\begin{aligned} 0 &= \int_0^1 f(x)w_2(x)\,dx \\ &= \left[F\left(\frac{1}{4}\right) - F(0)\right] - \left[F\left(\frac{1}{2}\right) - F\left(\frac{1}{4}\right)\right] \\ &\quad + \left[F\left(\frac{3}{4}\right) - F\left(\frac{1}{2}\right)\right] - \left[F(1) - F\left(\frac{3}{4}\right)\right] \\ &= 2\left[F\left(\frac{1}{4}\right) + F\left(\frac{3}{4}\right)\right] \end{aligned}$$

and

$$0 = \int_0^1 f(x)w_3(x)\,dx = 2\left[F\left(\frac{1}{4}\right) - F\left(\frac{3}{4}\right)\right]$$

so that

$$F\left(\frac{1}{4}\right) = F\left(\frac{3}{4}\right) = 0.$$

We leave it to the reader to show that $F(x) = 0$, $x = k/2^m$, for all $m = 0, 1, 2, \ldots, k = 1, \ldots, 2^m$.

EXERCISES

1.1 Show that equality holds in the Cauchy-Schwarz inequality if and only if x and y are proportional.

1.2 Derive conditions for equality in the triangle inequality.

1.3 Verify the pythagorean theorem and the parallelogram law.

1.4 Let $x_1, x_2, \ldots, x_n$ satisfy $x_i \neq 0$ and $x_i \perp x_j$ if $i \neq j$, $i, j = 1, 2, \ldots, n$. Show that the x_i's are linearly independent and extend the pythagorean theorem.

1.5 Verify the polarization identities in both the real and complex cases.

1.6 Give an example of a Schauder basis in a separable Hilbert space which is not a complete orthonormal set.

2.1 Give an example of a Banach space, not a Hilbert space, in which lemma 1 holds.

2.2 Give an example of a Banach space in which lemma 1 fails to hold.

2.3 Show that the minimizing point y_0 in lemma 1 is characterized by

$$\text{Re}\,(y_0 - x_0, y - y_0) \geq 0 \text{ for all } y \in K.$$

2.4 Let $f \in L_2(0, 1)$. Show that for each $n = 1, 2, \ldots$, there is a unique polynomial of degree n, or less, which is the best approximant to f in the sense of L_2.

2.5 Show that if M and N are closed subspaces of a Hilbert space X, then $M \dotplus N$ is closed provided $x \perp y$ for all $x \in M$, $y \in N$.

2.6 If S is a subset of a Hilbert space X, show that $S^\perp = [x \in X : x \perp y$ for all $y \in S]$ is a closed subspace of X and that $(S^\perp)^\perp$ is the closed linear span of X.

2.7 Show that a projection P is an orthogonal projection if and only if $(Px, y) = (x, Py)$, $x, y \in X$.

2.8 A sequence $\{x_n\}$ in a Hilbert space X is said to be weakly convergent if $\lim_n (x_n, y)$ exists for each $y \in X$. We say $\{x_n\}$ converges weakly to $x \in X$, written $x_n \rightharpoonup x$, if $\lim_n (x_n, y) = (x, y)$ for every $y \in X$. Show that if $\{x_n\}$ converges weakly, then there is an x for which $x_n \rightharpoonup x$.

2.9 If $\|x_n\| \le 1$, $n = 1, 2, \ldots$ in separable Hilbert space, show that a subsequence of $\{x_n\}$ converges weakly.

2.10 Show that $e_n = (0, \ldots, 0, 1, 0, \ldots)$, where 1 is in the nth place, converges weakly to $(0, 0, \ldots)$ in l_2.

2.11 Show that $\{x_n\}$ converges to x in the Hilbert space norm (called strong convergence) if and only if $x_n \rightharpoonup x$ and $\lim_n \|x_n\| = \|x\|$.

2.12 Let X be separable and let $\{\varphi_m\}$ be a complete orthonormal sequence in X. Show that $\{x_n\}$ converges weakly to x if and only if it is bounded and $\{(x_n, \varphi_m)\}$ converges to (x, φ_m) for each $m = 1, 2, \ldots$.

2.13 Show that if $x_n \rightharpoonup x$, there is a subsequence $\{x_{n_k}\}$ such that $\left\{\frac{1}{m}\sum_{i=1}^{m} x_{n_i}\right\}$ converges strongly to x.

3.1 Show that V is norm preserving, i.e.,

$$\int_S |f(x)|^2 \, d\mu = \int_S |f(Tx)|^2 \, d\mu, \quad f \in L_2.$$

4.1 Show that for the functions in example (iii),

$$(x, y) = \lim_{T \to \infty} \frac{1}{2T} \int_{-T}^{T} x(t) y(t) \, dt$$

exists. Show that $\{e^{i\lambda t} : -\infty < \lambda < \infty\}$ is a complete orthonormal set.

4.2 Prove that $\varphi_n(z) = \sqrt{\dfrac{n}{\pi}}\, z^{n-1}$, $n = 1, 2, \ldots$, is orthonormal with

$$(f, g) = \iint_{|z|<1} f\bar{g} \, dx \, dy.$$

4.3 Show that the two forms of Parseval's formula are equivalent.

4.4 Show that if $S = [x_\alpha : \alpha \in A]$ is orthonormal, then $\sum_{\alpha \in A} (x, x_\alpha) x_\alpha$ is the projection of x on the closed subspace spanned by S.

4.5 Let $x_1, x_2, \ldots, x_n$ be orthonormal in X and let $x \in X$. Show that for all sums of the form $\sum_{i=1}^{n} a_i x_i$, the distance (mean square error)

$$\|x - \sum_{i=1}^{n} a_i x_i\|$$

is a minimum when the a_i's are the Fourier coefficients of x.

4.6 Apply the preceding exercise to prove theorem 2.

4.7 Let $f \in L_2[-1, 1]$. Give an expression for the polynomial of degree n or less which best approximates to f in mean square.

4.8 Show that a linearly independent set in X is complete if and only if it is total.

4.9 Apply the Gramm-Schmidt process to the functions $1, t, t^2, \ldots$ in $L_2[-1, 1]$. Show that the resulting orthonormal set consists of $\sqrt{n + \frac{1}{2}}\, P_n(t)$, where

$$P_n(t) = \frac{1}{2^n n!} \frac{d^n}{dt^n} (t^2 - 1)^n$$

is the nth Legendre polynomial.

4.10 Prove completeness for the set in exercise 4.9.

4.11 Show that the trigonometric functions are complete in $L_2[0, 2\pi]$ by showing that if f is continuously differentiable, then its Fourier series converges uniformly to f.

4.12 Show that $\{\varphi_n\}$ is complete for $L_2[a, b]$ if and only if

$$\sum_{n=1}^{\infty} \left(\int_a^x \varphi_n(t)\, dt\right)^2 = x - a$$

for every $x \in [a, b]$.

4.13 Show that $\{\varphi_n\}$ is complete for $L_2[a, b]$ if and only if

$$\sum_{n=1}^{\infty} \int_a^b \left[\int_a^x \varphi_n(t)\, dt\right]^2 dx = \frac{(b - a)^2}{2}.$$

6.1 Let $x_1, x_2, \ldots, x_n \in X$ and let

$$\begin{aligned} y_1 &= c_{11}x_1 \\ y_2 &= c_{21}x_1 + c_{22}x_2 \\ &\ldots \\ y_n &= c_{n1}x_1 + c_{n2}x_2 + \ldots + c_{nn}x_n \end{aligned}$$

be obtained by the Gramm-Schmidt process from $x_1, x_2, \ldots, x_n$. Put

$$D_n = \begin{vmatrix} c_{11} & 0 & 0 & \ldots & 0 \\ c_{21} & c_{22} & 0 & \ldots & 0 \\ \cdot & \cdot & & & \cdot \\ \cdot & \cdot & & & \cdot \\ \cdot & \cdot & & & \cdot \\ c_{n1} & c_{n2} & & \ldots & c_{nn} \end{vmatrix}.$$

The determinant

$$G_n = \det (x_i, x_j)$$

is called the Gramm determinant of the set $x_1, x_2, \ldots, x_n$. Show that

$$D_n G_n \bar{D}_n = 1.$$

6.2 Prove that $x_1, x_2, \ldots, x_n$ are linearly independent if and only if $G_n \neq 0$.

6.3 Let

$$\Delta_n = \begin{vmatrix} \frac{1}{a_1 + b_1} & \frac{1}{a_1 + b_2} & \cdots & \frac{1}{a_1 + b_n} \\ \cdot & \cdot & & \cdot \\ \cdot & \cdot & & \cdot \\ \cdot & \cdot & & \cdot \\ \frac{1}{a_n + b_1} & \frac{1}{a_n + b_2} & \cdots & \frac{1}{a_n + b_n} \end{vmatrix}.$$

By subtracting the last row from each of the others, removing common factors, subtracting the last column and again removing common factors, show that

$$\Delta_n = \frac{\prod_{i=1}^{n-1} (a_n - a_i)(b_n - b_i)}{\prod_{i=1}^{n} (a_n + b_i) \prod_{i=1}^{n-1} (a_i + b_n)} \begin{vmatrix} & & & 1 \\ & \Delta_{n-1} & & 1 \\ & & & 1 \\ & & & \cdot \\ & & & \cdot \\ & & & \cdot \\ 0 & 0 & \cdots \; 0 & 1 \end{vmatrix}$$

and thus evaluate Δ_n.

6.4 Apply the above to verify the value of $d_{m,k}^2$ given in the text.

6.5 Show that if $\{\varphi_n\}$ is complete for $L_2[0, 1]$, then $1, \varphi_n, n = 1, 2, \ldots$ is total for $C[0, 1]$.

6.6 Show that Müntz's theorem is valid for every $L_p[0, 1]$, $1 \leq p < \infty$.

7.1 Let $\{x_k\}$ be a complete orthonormal system in X (separable). Suppose $\{y_k\}$ satisfies

$$\sum_{k=1}^{\infty} \|x_k - y_k\|^2 < 1.$$

Show that $\{y_k\}$ is complete.

7.2 Let $\{x_n\}$ be a complete sequence in a separable Hilbert space X. Find necessary and sufficient conditions on $y \in X$ in order that

$$y, x_2, x_3, \ldots$$

be a complete sequence.

7.3 Give the details of the proof of the Riesz-Fischer theorem.

8.1 Determine whether l_2 is a proper functional space. If it is, find its reproducing kernel.

8.2 Prove that a reproducing kernel, if it exists, is unique.

8.3 Let S be a metric space and X a proper functional Hilbert space of functions on S with reproducing kernel $K(s, t)$. Show that the mapping

$$t \to K(\cdot\,, t)$$

is continuous if and only if the spheres $[f \in X : \|f\| < M]$ in X are equicontinuous.

8.4 Suppose X, with reproducing kernel K, is a closed subspace of a Hilbert space Y. Show that if P is the orthogonal projection of Y on X, then $Py(t) = (y, K(\cdot, t)), y \in Y$.

8.5 If Y is a closed subspace of X, where X has reproducing kernel K, then Y has a reproducing kernel. The kernels K' and K'' of Y and $Y^{\perp}$ satisfy $K = K' + K''$.

8.6 Suppose $S_1 \subset S$ and X is a proper functional Hilbert space on X with reproducing kernel K. Show that the restriction of K to $S_1 \times S_1$ is the reproducing kernel K_1 of the space X_1, where X_1 consists of all restrictions to S_1 of functions in X, and the norm in X_1 is

$$\|x_1\| = \min \|x\|,$$

the minimum being taken over all $x \in X$ whose restriction to S_1 is x_1.

8.7 Give details for the construction of X given the positive matrix K on S.

9.1 Derive the identity

$$\iint_D f(z)\overline{g'(z)}\,dx\,dy = \frac{1}{2i}\int_C f(z)\overline{g(z)}\,dz$$

from Green's theorem.

9.2 Verify the conditions for proper functional completion in the case of $L_2(D)$.

9.3 Find the mapping function $f(z)$ which maps the unit circle onto itself and takes ζ into 0 with $f'(\zeta) > 0$. Use the fact that a complete orthonormal set in the unit circle is given by

$$\varphi_n(z) = \sqrt{\frac{n}{\pi}}\, z^{n-1}, \quad n = 1, 2, \ldots.$$

10.1 Prove that

$$\varphi_n(z) = \sqrt{\frac{n}{\pi}}\, [w(z)]^{n-1} w'(z), \quad n = 1, 2, \ldots$$

is a complete orthonormal set for $L_2(D)$, where D satisfies the conditions of the Riemann mapping theorem and w is a conformal mapping of D on the unit disk.

10.2 Prove the identity

$$\sum_{n=0}^{\infty} \frac{L_n(x)}{n!}\, t^n = \frac{e^{-xt/1-t}}{1-t}.$$

11.1 Show that the Haar functions are complete in $L_2[0, 1]$.

11.2 Show that the Haar functions are a Schauder basis for $L_p[0, 1]$, $p \geq 1$.

11.3 Show that the set of (x, t) for which $\sum_{n=1}^{\infty} C_n\varphi_n(x)r_n(t)$ converges is measurable.

CHAPTER 5

TOPOLOGICAL VECTOR SPACES

5.1 Topology

A **topological space** $(X, \mathcal{G})$ consists of a set X together with a collection $\mathcal{G} = [G]$ of subsets of X, called **open** sets, such that

(a) $X \in \mathcal{G}$, and the empty set $0 \in \mathcal{G}$,
(b) if $G_\alpha \in \mathcal{G}$, $\alpha \in A$, then $\bigcup [G_\alpha : \alpha \in A] \in \mathcal{G}$,
(c) if $G_1, G_2 \in \mathcal{G}$, then $G_1 \cap G_2 \in \mathcal{G}$.

If Y is a subset of X, $(X, \mathcal{G})$ is a topological space, and if $\mathcal{H} = [Y \cap G : G \in \mathcal{G}]$, then $(Y, \mathcal{H})$ is easily seen to be a topological space. This topology in Y is called the **topology induced by the topology in *X*.**

Every set X can be topologized in two trivial ways.

(i) Let $\mathcal{G}$ consist of the two sets X and 0.
(ii) Let $\mathcal{G}$ consist of all subsets of X.

If $(X, \mathcal{G})$ and $(X, \mathcal{H})$ are topological spaces, with the same set X, then the topology given by $\mathcal{G}$ is said to be **finer** than the one given by $\mathcal{H}$ if $\mathcal{G} \supset \mathcal{H}$. The topology given by $\mathcal{H}$ is then said to be **coarser** than that given by $\mathcal{G}$.

The topology (i) above is the coarsest among all the topologies possible for X; the topology (ii) is the finest.

If $(X, \mathcal{G})$ is a topological space and $\mathcal{B} \subset \mathcal{G}$, then $\mathcal{B}$ is called a **base** for the topology if every $G \in \mathcal{G}$ is the union of sets in $\mathcal{B}$. The reader may verify with ease that $\mathcal{B}$ is a base if and only if, for every $x \in X$ and $G \in \mathcal{G}$ such that $x \in G$, there is an $H \in \mathcal{B}$ such that $x \in H$ and $H \subset G$.

For fixed $x \in X$, a set $\mathcal{B}_x \subset \mathcal{G}$ is called a **base at *x*** for the topology if, for every $G \in \mathcal{G}$ such that $x \in G$, there is an $H \in \mathcal{B}_x$ such that $x \in H$ and $H \subset G$. Moreover, $x \in H$ for every $H \in \mathcal{B}_x$.

A subset $\mathcal{S} \subset \mathcal{G}$ is called a **subbase** for the topology if there is a base $\mathcal{B}$ for the topology such that, for every $G \in \mathcal{B}$, there is a finite set in $\mathcal{S}$, G_i, $i = 1, \ldots, n$, such that $G = \bigcap_{i=1}^{n} G_i$.

Finally, a set $\mathcal{S}_x \subset \mathcal{G}$ is called a **subbase at *x*,** $x \in X$, for the topology, if there is a base $\mathcal{B}_x$ at x such that, for every $G \in \mathcal{B}_x$, there is a finite set in $\mathcal{S}_x$, G_i, $i = 1, \ldots, n$, such that $G = \bigcap_{i=1}^{n} G_i$.

A topological space is associated in a unique way with every metric space. Let X be a metric space with metric d. Let $\mathcal{B}$ consist of all open spheres in X, and let $\mathcal{G}$ be the set of all unions of subsets of $\mathcal{B}$. It is a routine matter to show that $(X, \mathcal{G})$ is a topological space and that, for $x \in X$, $x_n \in X$, $n = 1, 2, \ldots$, $\lim_{n\to\infty} d(x, x_n) = 0$ if and only if, for every $G \in \mathcal{G}$ with $x \in G$, $x_n \in G$ for all but finitely many values of n.

A topological space $(X, \mathcal{G})$ is said to be **metrizable** if there is a metric d for X such that the topological space corresponding, as above, to this metric space, is $(X, \mathcal{G})$.

A topological space $(X, \mathcal{G})$ is said to be a **Hausdorff space** if, for every x, $y \in X$, $x \neq y$, there are G, $H \in \mathcal{G}$ such that $x \in G$, $y \in H$, and $G \cap H = 0$.

In particular, (ii) is a Hausdorff space, (i) is not a Hausdorff space. On the other hand, every metrizable space is a Hausdorff space.

Let $(X, \mathcal{G})$ and $(Y, \mathcal{H})$ be topological spaces and let f be a mapping of X into Y. Then f is said to be **continuous at** $\boldsymbol{x} \in X$ if, for every open set H with $f(x) \in H$, there is an open set G, with $x \in G$, such that $f(G) \subset H$. The mapping f is said to be **continuous** if it is continuous at every $x \in X$. We prove the following statement.

f is continuous if and only if the inverse image of every open set in Y is an open set in X.

Proof. Suppose the inverse image of every open set in Y is open. Let $x \in X$ and let H be an open set in Y with $f(x) \in H$. Then $f^{-1}(H)$ is open, $f[f^{-1}(H)] \subset H$, and $x \in f^{-1}(H)$, so that f is continuous.

Suppose f is continuous. Let H be open and let $x \in f^{-1}(H)$. Then H is an open set with $f(x) \in H$. There is, accordingly, an open set G, with $x \in G$ and $f(G) \subset H$. Then $G \subset f^{-1}(H)$, and it follows that $f^{-1}(H)$ is open.

As a special case, let X be a set with topologies $\mathcal{G}$ and $\mathcal{H}$, where $\mathcal{G}$ is finer than $\mathcal{H}$. The identity mapping

$$i : (X, \mathcal{G}) \to (X, \mathcal{H})$$

is continuous.

5.2 Tychonoff Theorem, Application in Banach Space

If X and Y are topological spaces, the space $X \times Y$ is defined as follows:

(a) $X \times Y = [(x, y) : x \in X, y \in Y]$,

(b) the open sets in $X \times Y$ are all the unions of sets of the form $G \times H$, where G is open in X and H is open in Y.

We may define the topological product of an arbitrary set X_α, $\alpha \in A$, of topological spaces. Let

$$X = \Pi[X_\alpha : \alpha \in A]$$

be the set of functions $[x_\alpha]$ such that for every $\alpha \in A$, $x_\alpha \in X_\alpha$. We topologize the space by defining a subbase for the topology. For each $\alpha \in A$, and open set $G_\alpha \subset X_\alpha$, let $U(\alpha, G_\alpha)$ be the set of $[x_\beta] \in X$ such that $x_\alpha \in G_\alpha$. We let the set $\mathcal{U}$ of all $U(\alpha, G_\alpha)$ be a subbase for the topology.

It is easy to see that a base for the topology is given by the sets

$$U(\alpha_1, \ldots, \alpha_n; G_{\alpha_1}, \ldots, G_{\alpha_n})$$

of elements $[x_\alpha] \in X$ for which $x_{\alpha_i} \in G_{\alpha_i}$, $i = 1, \ldots, n$, for all finite sets $\{\alpha_1, \ldots, \alpha_n\} \subset A$ and open sets $G_{\alpha_i} \subset X_{\alpha_i}$, $i = 1, \ldots, n$.

Certain properties of topological spaces are preserved by the product operation. We present the most important one of these properties.

We first recall that the closed sets in a topological space are merely the complements of the open sets. Thus, in a topological space X, the set X and the empty set are closed. The intersection of any collection of closed sets is closed. For every $S \subset X$, the intersection of all closed sets containing S is called the closure $\bar{S}$ of S. It is easy to see that $\bar{S}$ is closed and that $x \in \bar{S}$ if and only if, for every open set G such that $x \in G$, the set $G \cap S \neq 0$.

A topological space X is said to be **compact** if for every family $\mathcal{G}$ of open sets which covers X (i.e., for every $x \in X$ there is a $G \in \mathcal{G}$ such that $x \in G$), there is a finite subset of $\mathcal{G}$ which covers X. For example, a bounded closed set in the reals is a compact space, with the usual topology, but a non-empty open set is not compact.

It is an immediate consequence of the definition that a space X is compact if and only if every collection $\mathcal{F}$ of closed sets in X with the property that every finite subset of $\mathcal{F}$ has non-empty intersection is such that the intersection of all the sets in $\mathcal{F}$ is non-empty.

We have shown in Chapter 1 that a metric space is compact if and only if every infinite set in X has a limit point in X. Moreover, every closed subset of a compact Hausdorff space is easily seen to be compact with the induced topology.

We now state what is perhaps the main theorem of general topology.

THEOREM 1. If X_α, $\alpha \in A$, are compact spaces, then $X = \Pi[X_\alpha : \alpha \in A]$ is compact.

Proof. The proof is subtle but simple. A collection of sets in a topological space is said to have the finite intersection property if the intersection of any finite number of sets in the collection is non-empty. The compactness of X then follows from

(α) If $\mathcal{F}$ is a collection of sets in X, with the finite intersection property, then

$$\bigcap[\bar{F} : F \in \mathcal{F}] \neq 0.$$

In order to prove (α) we observe that if the family of collections of sets in X with the finite intersection property is ordered by inclusion, then every chain of such collections has an upper bound; indeed, the union of the members of the chain is in the family. Thus, by Zorn's lemma, every collection of sets in X with the finite intersection property is contained in a maximal collection of the same type. Thus, in order to prove the theorem, it remains only to prove

(β) If $\mathcal{F}$ is a maximal collection of sets in X, with the finite intersection property, then

$$\bigcap[\bar{F} : F \in \mathcal{F}] \neq 0.$$

Now, let $\mathcal{F}$ be a maximal collection of sets in X with the finite intersection property. For each $F \in \mathcal{F}$ and $\alpha \in A$, let F_α be the projection of F in X_α i.e.,

$$F_\alpha = \{x_\alpha : [x_\beta] \in F\}.$$

Then the collection $\{F_\alpha : F \in \mathcal{F}\}$ evidently has the finite intersection property. Since X_α is compact, there is an $x_\alpha \in \bigcap \{\bar{F}_\alpha : F \in \mathcal{F}\}$. Consider a point $[x_\beta] \in X$ such that for each $\alpha \in A$, $x_\alpha \in \bigcap \{\bar{F}_\alpha : F \in \mathcal{F}\}$. It remains only to show that, for every $F \in \mathcal{F}$, we have $[x_\alpha] \in \bar{F}$. For this purpose, let $\alpha \in A$ and let G_α be an open set in X_α, with $x_\alpha \in G_\alpha$. Then let G be the set of $[y_\beta] \in X$ such that $y_\alpha \in G_\alpha$. Then the union of $\mathcal{F}$ with the single set G continues to have the finite intersection property. Since $\mathcal{F}$ is maximal, $G \in \mathcal{F}$. Moreover, every finite intersection of such sets G also belongs to $\mathcal{F}$. But these finite intersections form a base for the topology in X at the point $[x_\alpha]$. This implies that every $F \in \mathcal{F}$ meets every open set in X to which $[x_\alpha]$ belongs. Thus, $[x_\alpha] \in \bar{F}$.

This theorem has many important applications. We now give one of them. Let X be a Banach space. The closed unit ball in X (i.e., the set for which $\|x\| \leq 1$) may not be compact. For example, consider the space l_2 of sequences $\{x_n\}$ of real numbers for which $\sum_{n=1}^{\infty} x_n^2 < \infty$. The closed unit ball in this space is not compact (see Chapter 1).

Now the dual X' of a Banach space is also a Banach space with the usual norm. But X' also has the **weak topology** defined as follows:

For each $x \in X$, and open $G \subset R$, let

$$U(x, G) = [x' \in X' : x'(x) \in G].$$

Let the sets $U(x, G)$ be a subbase for the topology in X'. This topology is called the weak topology in X' determined by X. (It is sometimes called the weak-star topology to distinguish it from the weak topology in X' determined by X''.)

For each $x \in X$, let R_x be the set of real numbers. The space $\Pi[R_x : x \in X]$ evidently contains X', and the weak topology in X' is the one induced by the topology in $\Pi[R_x : x \in X]$. Let K be the subset of $\Pi[R_x : x \in X]$ of points whose x-coordinate has absolute value not greater than $\|x\|$, for every $x \in X$. By theorem 1, K is compact. We show that the closed unit ball $B \subset X'$ is a weakly closed subset of K.

Let $\psi \in \bar{B}$, the weak closure of B in K. Then $\psi \in K$. We show that ψ is linear. For this, let $x, y, x + y \in X$. Let $\epsilon > 0$. By the definition of the topology $\Pi[R_x : x \in X]$, there is an $x' \in X'$ such that

$$|x'(x) - \psi(x)| < \epsilon, |x'(y) - \psi(y)| < \epsilon,$$

and $|x'(x + y) - \psi(x + y)| < \epsilon$. Since $x'(x + y) = x'(x) + x'(y)$, it follows that $|\psi(x + y) - \psi(x) - \psi(y)| < 3\epsilon$. Since this holds for every $\epsilon > 0$, $\psi(x + y) = \psi(x) + \psi(y)$. It is just as easy to show that for every $x \in X$ and $a \in R$, $\psi(ax) = a\psi(x)$. Since $|\psi(x)| \leq \|x\|$, for every $x \in X$, it follows that $\psi \in B$. Hence B is closed and we have proved

THEOREM 2. If X is a Banach space, the closed unit ball $B \subset X'$ is compact in the weak topology.

5.3 Topological Vector Space

A **topological vector space** is a set X which is:

(a) A vector space.

(b) A topological space.

(c) The operations $(x, y) \to x + y$ of $X \times X$ into X and $(a, x) \to ax$ of $R \times X$ into X are continuous.

We shall also suppose:

(d) If $x \neq 0$, there is an open set U such that $0 \in U$ and $x \notin U$.

The condition (d) is not always assumed by writers on the subject, but it is satisfied by all interesting examples.

PROPOSITION 1. If X is a topological vector space, then for every $x \in X$, the mapping $f_x : X \to X$, where $f_x(y) = x + y$, is a homeomorphism.

Proof. For every $y \in X$, $f_x(y - x) = (y - x) + x = y$, so that f_x is onto. If $y \neq z$, then $f_x(y) \neq f_x(z)$, since $y + x \neq z + x$. Hence f_x is one-one. The first part of condition (c) implies the continuity of f_x and of its inverse.

Proposition 1 implies that the open sets G_x containing x are the translations $x + G = [x + y : y \in G]$ of the open sets G containing 0. Thus, the topology in X is completely determined by a base for the open sets containing 0 (i.e., for the neighborhoods of 0). (Axioms for a topological vector space can be given in terms of a neighborhood base of 0, but we shall not do this here.)

5.4 Normable Space

If X is a Banach space, then its dual X', with the weak topology, is a topological vector space. In order to show this, one needs only make the easy verification that the vector space operations are continuous. We leave the details for the reader.

It is also easy to show that a normed vector space is a topological vector space. In this case, a base for the open sets containing 0 is given by the sets

$$U_a = [x : \|x\| < a],$$

for all $a > 0$.

A topological vector space is said to be **normable** if its topology may be given in terms of a norm.

It is of interest to know which topological vector spaces are normable.

For this purpose, we need the notion of a bounded set in a topological vector space. A set $A \subset X$ is said to be **bounded** if for every neighborhood U of 0 there is an a such that $A \subset aU$. In the particular case of a normed space, this simply means that the set $[\|x\| : x \in A]$ is bounded, which coincides with previous convention. We recall that a set $A \subset X$ is **convex** if for every $x, y \in A$ and $0 \leq a \leq 1$, $ax + (1 - a)y \in A$.

THEOREM 1. A topological vector space X is normable if and only if there is a non-empty $A \subset X$ which is open, bounded, and convex.

Proof. Let X be normable, and let $\| \ \|$ be a norm in X which yields the given topology. Let $U = [x : \|x\| < 1]$. Then U is non-empty and open. Let G be a neighborhood of 0. Then G contains the set $V = [x : \|x\| < a]$ for some $a > 0$. Then $\frac{1}{a} V \supset U$, and U is bounded. Finally, let $x, y \in U$ and $0 \leq a \leq 1$. Then $\|x\| < 1$, $\|y\| < 1$. It follows that $\|ax + (1 - a)y\| \leq a \|x\| + (1 - a) \|y\| < a + (1 - a) = 1$. Thus U is convex.

Conversely, suppose X contains a non-empty, open, bounded, convex set H. Let $x \in H$. Then the set $U = H - x = [y : y = z - x, z \in H]$ is also non-empty and open, and is easily seen to be convex and bounded. Moreover, the set $V = U \cap (-U)$ is also non-empty, open, bounded, convex and is also such that $x \in V$ implies $-x \in V$. We define a norm in X as follows:

$$\|x\| = 0 \quad \text{if } x = 0$$

and

$$\|x\| = \sup [a : a \geq 0, x \notin aV] \quad \text{if } x \neq 0.$$

Let $x \neq 0$. There is a neighborhood G of 0 such that $x \notin G$. There is an $a > 0$ such that $aV \subset G$, since V is bounded. Thus $x \notin aV$ and $\|x\| \geq a > 0$.

Let $b \neq 0$, $x \neq 0$. Then $x \in aV$ if and only if $bx \in |b|\, aV$, since $x \in V$ implies $-x \in V$. Hence $\|bx\| = |b|\, \|x\|$.

Finally, let $x \in aV$, $y \in bV$, so that $\|x\| \leq a$, $\|y\| \leq b$. Then $x + y \in (a + b)V$, so that $\|x + y\| \leq a + b$. Hence $\|x + y\| \leq \|x\| + \|y\|$. Thus, the defined $\|\ \|$ is indeed a norm.

We note that the topology defined by this norm agrees with the given topology in X. For, if G is any neighborhood of 0, there is an $a > 0$ such that $aV \subset G$, since G is bounded. This completes the proof.

5.5 Space of Measurable Functions

We now consider a special topological vector space. Let M be the vector space of equivalence classes of measurable functions on [0, 1]. We shall define a topology on this space in terms of a metric. For every $f, g \in M$, let

$$d(f, g) = \int_0^1 \frac{|f(t) - g(t)|}{1 + |f(t) - g(t)|}\, dt.$$

It is evident that $d(f, g) \geq 0$ and that $d(f, g) = 0$ if and only if $f(t) = g(t)$ almost everywhere. Moreover, $d(f, g) = d(g, f)$. In order to show that $d(f, g) + d(g, h) \geq d(f, h)$ it is only necessary to show that for any real numbers a, b

$$\frac{|a + b|}{1 + |a + b|} \leq \frac{|a|}{1 + |a|} + \frac{|b|}{1 + |b|}.$$

This is a routine calculation which we leave to the reader.

The metric d is an invariant metric, i.e., $d(f, g) = d(f + h, g + h)$ for every $f, g, h \in M$. Thus, in order to show that M is a topological vector space it is only necessary to show that if $\{f_n\}$ and $\{g_n\}$ converge to 0, then $\{f_n + g_n\}$ converges to 0, and if $\{f_n\}$ converges to 0 and $\{a_n\}$ converges, then $\{a_n f_n\}$ converges to 0.

In order to prove this, we consider an equivalent metric on M. Thus, for every $k > 0$, let

$$E_k = [x : |f(x)| \geq k],$$

and let

$$\delta(f, 0) = \inf\, [k : m(E_k) < k].$$

Now let

$$\delta(f, g) = \delta(f - g, 0).$$

We leave it as an exercise for the reader to show that $\lim\limits_{n \to \infty} \delta(f_n, 0) = 0$ if and only if $\lim\limits_{n \to \infty} d(f_n, 0) = 0$. Now, it is an easy matter to show that if $\delta(f_n, 0)$ and $\delta(g_n, 0)$ converge to 0, then $\delta(f_n + g_n, 0)$ converges to 0, and that if $\delta(f_n, 0)$ converges to 0 and $\{a_n\}$ is bounded, then $\delta(a_n f_n, 0)$ converges to 0. We leave these details to the reader as well.

The space M is separable. Let $f \in M$. Then, for every $\epsilon > 0$, there is a continuous g such that

$$|f(x) - g(x)| < \frac{\epsilon}{2}$$

except on a set of measure less than $\epsilon/2$. But there is a polynomial p, with rational coefficients, such that

$$|g(x) - p(x)| < \frac{\epsilon}{2}$$

everywhere. Hence, $\delta(f, p) < \epsilon$. Thus, the polynomials with rational coefficients are dense in M. But the set of such polynomials is countable, so that the separability of M is proved.

We remark that the topology in M is the topology of convergence in measure. Suppose that $\{f_n\}$ converges in measure to f. Then, for every $\epsilon > 0$, there is an N such that $n > N$ implies

$$|f_n(x) - f(x)| < \epsilon$$

except on a set of measure less than ϵ, so that $\delta(f, f_n) < \epsilon$. The converse, that convergence in M implies convergence in measure, is shown in analogous fashion.

In particular, it follows that every convergent sequence in M has a subsequence which converges almost everywhere.

Proposition 1. The only continuous linear functional φ on M is the trivial one, $\varphi(f) = 0$.

Proof. Let φ be a linear functional on M which is not identically zero. Then there is an $f \in M$ such that $\varphi(f) \neq 0$. Let

$$J_1 = \left[0, \frac{1}{2}\right], \quad J_2 = \left[\frac{1}{2}, 1\right],$$

and let

$$g_1 = f\chi_{J_1}, \qquad g_2 = f\chi_{J_2}.$$

Since $f = g_1 + g_2$, either $\varphi(g_1) \neq 0$ or $\varphi(g_2) \neq 0$. Let f_1 be one of the g_1, g_2 for which $\varphi(f_1) \neq 0$. Then $m[x : f_1(x) \neq 0] \leq \frac{1}{2}$. It should be clear that, by proceeding in this way, we obtain a sequence $\{f_n\}$ such that $\varphi(f_n) \neq 0$ and

$$m[x : f_n(x) \neq 0] \leq \frac{1}{2^n}, \quad n = 2, 3, \ldots.$$

For every n, let

$$F_n = \frac{1}{\varphi(f_n)} f_n.$$

Then $\varphi(F_n) = 1$, $n = 1, 2, \ldots$, and $\{F_n\}$ converges to 0 in M. Thus φ is not continuous on M.

PROPOSITION 2. Let $\{f_n\}$ be a countable set of functions in M such that the finite linear combinations of the $\{f_n\}$ are dense in M. Let X be the closure in M of the finite linear combinations of $\{f_2, f_3, \ldots\}$. Then every $f \in M$ is of the form $f = af_1 + g$, where $g \in X$ and $a \in R$.

Proof. There are $g_n \in X$, $a_n \in R$, $n = 1, 2, \ldots$ such that

$$f = \lim_{n\to\infty} (a_n f_1 + g_n).$$

Consider first the case where $\{a_n\}$ is unbounded. It suffices to assume that $\lim_{n\to\infty} a_n = +\infty$. Then

$$\lim_{n\to\infty} \left(f_1 + \frac{1}{a_n} g_n\right) = 0,$$

so that

$$f_1 = -\lim_{n\to\infty} \frac{1}{a_n} g_n$$

and $f_1 \in X$. Thus, $a_n f_1 + g_n \in X$, $n = 1, 2, \ldots$, and, since X is closed, $f \in X$.

Suppose then that $\{a_n\}$ is bounded. We may assume that $\{a_n\}$ converges. Let $a = \lim_{n\to\infty} a_n$. Now

$$f = \lim_{n\to\infty} (a_n f_1 + g_n) = af_1 + g,$$

where $g = \lim_{n\to\infty} g_n$ exists, since $\lim_{n\to\infty} a_n f_1$ exists.

The simple observations made in propositions 1 and 2 have some surprising consequences.

Let X be a separable metrizable topological vector space. We say that a countable set $A \subset X$ is **total** in X if the finite linear combinations of elements of A are dense in X. A total set A is said to be **minimal** if no proper subset of A is total. Many Banach spaces are known to have minimal total sets, and there is no known example of a separable Banach space which does not have a minimal total set. On the other hand, the space M has no minimal total set. Indeed, the following stronger fact holds.

THEOREM 1. If A is a total set for M and B is a subset of A obtained by discarding a finite subset of A, then B is a total set for M.

Proof. It is only necessary to show that, if $\{f_1, f_2, \ldots\}$ is total, then $\{f_2, f_3, \ldots\}$ is also total. Let X be the closure of the set of finite linear combinations of $\{f_2, f_3, \ldots\}$. Then $\{f_2, f_3, \ldots\}$ is total for X. Suppose $f_1 \notin X$. Define a linear functional φ on M by letting $\varphi(f) = a$, where

$$f = af_1 + g,\ g \in X.$$

Then φ is a non-trivial continuous linear functional on M, in contradiction with proposition 1. Hence, $f_1 \in M$ and $X = M$.

We have the following corollary to theorem 1.

COROLLARY. If A is total for M, there is an infinite subset $B \subset A$ such that $A \sim B$ is total for M.

Proof. Let $A = \{f_1, f_2, \ldots\}$ be total for M and $\{\epsilon_n\}$ a sequence of positive numbers which converges to 0. Delete f_1 from A. There is a finite linear combination

$$a_{2,2}f_2 + \ldots + a_{2,n_2}f_{n_2}$$

such that

$$d(f_1, a_{2,2}f_2 + \ldots + a_{2,n_2}f_{n_2}) < \epsilon_1.$$

Delete f_{n_2+1} from A. There are finite linear combinations

$$a_{3,i,n_2+2}f_{n_2+2} + \ldots + a_{3,i,n_3}f_{n_3}, \quad i = 1, 2$$

such that

$$d(f_i, a_{3,i,n_2+2}f_{n_2+2} + \ldots + a_{3,i,n_3}f_{n_3}) < \epsilon_2, \quad i = 1, 2.$$

Delete f_{n_3+1} from A. By continuing in this way, we delete an infinite subset $B \subset A$ from A such that the finite linear combinations of elements of $A \sim B$ are dense in B. Thus, they are dense in M.

A series $\sum_{n=1}^{\infty} a_n f_n$ is said to be **universal** in a space X if, for every $x \in X$, a subsequence of the sequence of partial sums of the series converges to x in the space X.

THEOREM 2. If $A = \{f_1, f_2, \ldots\}$ is total for M, there is a sequence $\{a_n\}$ such that $\sum_{n=1}^{\infty} a_n f_n$ is a universal series for M.

Proof. Let $\{g_n\}$ be a countable dense subset of M and $\{\epsilon_n\}$ a sequence of positive real numbers which converges to 0. There is a linear combination $a_1 f_1 + \ldots + a_{n_1} f_{n_1}$ such that

$$d(g_1, a_1 f_1 + \ldots + a_{n_1} f_{n_1}) < \epsilon_1.$$

There is then a linear combination $a_{n_1+1} f_{n_1+1} + \ldots + a_{n_2} f_{n_2}$ such that

$$d(g_1 - a_1 f_1 - \ldots - a_{n_1} f_{n_1}, a_{n_1+1} f_{n_1+1} + \ldots + a_{n_2} f_{n_2}) < \epsilon_2.$$

There is then a linear combination $a_{n_2+1} f_{n_2+1} + \ldots + a_{n_3} f_{n_3}$ such that

$$d(g_2 - a_1 f_1 - \ldots - a_{n_2} f_{n_2}, a_{n_2+1} f_{n_2+1} + \ldots + a_{n_3} f_{n_3}) < \epsilon_2.$$

By continuing in a way which should be clear to the reader, we obtain a series $\sum_{n=1}^{\infty} a_n f_n$ whose sequence of partial sums has, for every n, a subsequence which converges in M to g_n. It thus has a subsequence converging to any prescribed $f \in M$.

Consider now a Banach space X of equivalence classes of measurable functions defined on $[0, 1]$. Then $X \subset M$. For every measurable $G \subset [0, 1]$, with $m(G) > 0$, let X_G be the set of restrictions of the functions in X to the set G. Suppose X_G is normed so as to be a Banach space, and the following conditions hold.

(a) If $\{g_n\}$ converges to g in X, then $\{g_n\}$ converges to g in X_G.
(b) The set of bounded measurable functions is a dense subset of X.
(c) For every G, uniform convergence on G implies convergence in X_G, and convergence in X_G implies convergence in measure on G.

It follows from (a) and (b) that the set of bounded measurable functions is a dense subset of X_G for every G.

THEOREM 3. With the above set-up, if $A = \{f_1, f_2, \ldots\}$ is total for X, and $\epsilon > 0$, there is a measurable $G \subset [0, 1]$, with $m(G) > 1 - \epsilon$, such that $\{f_2, f_3, \ldots\}$ is total for X_G.

Proof. Let $\{g_1, g_2, \ldots\}$ be dense in X. Since $\{f_1, f_2, \ldots\}$ is total for X, every bounded measurable function is, by (b), a limit in X of finite linear combinations of $\{f_1, f_2, \ldots\}$; hence, by (c), in M. Since the bounded functions are dense in M, $\{f_1, f_2, \ldots\}$ is total for M. By theorem 1, $\{f_2, f_3, \ldots\}$ is total for M. Fix g_n. There is a sequence $\{\varphi_1, \varphi_2, \ldots\}$ of finite linear combinations of $\{f_2, f_3, \ldots\}$ which converges in measure to g_n. It has a subsequence $\{\psi_1, \psi_2, \ldots\}$ which converges uniformly to g_n on a measurable set G_n, with $m(G_n) > 1 - \dfrac{\epsilon}{2^n}$. Let $G = \bigcap_{n=1}^{\infty} G_n$.

Since uniform convergence on G implies convergence in X_G and since $\{g_1, g_2, \ldots\}$ is dense in X_G, it follows that $\{f_2, f_3, \ldots\}$ is total for X_G.

COROLLARY. If $\{f_1, f_2, \ldots\}$ is complete for $L_2[0, 1]$, and $\epsilon > 0$, there is a measurable $G \subset [0, 1]$, with $m(G) > 1 - \epsilon$, such that $\{f_2, f_3, \ldots\}$ is complete for $L_2(G)$.

As a companion to this corollary, we have a related phenomenon discovered by Boas and Pollard.

THEOREM 4. Let $\{f_1, f_2, \ldots\}$ be a complete orthonormal set for $L_2[a, b]$. There is a bounded g such that $\{gf_2, gf_3, \ldots\}$ is complete for $L_2[a, b]$.

Proof. Choose g so that $g(x) > 0$, for all x, g is bounded, and $f_1 g^{-1} \notin L_2$. Let $F \in L_2$. Suppose

$$\int_a^b g f_n F = 0, \quad n = 2, 3, \ldots .$$

Then $gF = cf_1$. Since $f_1 g^{-1} \notin L_2$ and $F \in L_2$, it follows that $c = 0$, and so $F = 0$. Thus $\{gf_2, \ldots, gf_n, \ldots\}$ is complete.

Indeed, Boas and Pollard proved the above theorem for the case where any finite number of functions is discarded from $\{f_1, f_2, \ldots\}$, but we shall not give the proof.

5.6 Locally Convex Space

An important class of topological vector spaces is the set of **locally convex** spaces. In this case, every open set containing 0 contains a convex open set containing 0. It then follows that the topology is given by a set of semi-norms.

Recall that a **semi-norm** p on a vector space X is a non-negative real function on X such that $p(x + y) \leq p(x) + p(y)$, for every $x, y \in X$ and $p(ax) = |a|\, p(x)$ for every $a \in R$ and $x \in X$.

Now, let U be a convex open set containing 0. Then $V = U \cap (-U)$ is also a convex open set containing 0. It is easy to see that for every $x \in X$ there is an $a \in R$ such that $x \in aV$. Moreover, $x \in aV$ if and only if $-x \in aV$. Let

$$p(x) = \sup\,[a : x \notin aV,\ a \geq 0], \text{ if } x \neq 0,$$
$$p(0) = 0.$$

It is a routine matter to verify that p is a semi-norm and that the sets

$$U_{p,r} = [x : p(x) < r],$$

for all p obtained in this way, and all $r > 0$, form a base for the topology in X at 0.

Thus, in a locally convex topological vector space, the topology is given by a set p_α, $\alpha \in A$, of semi-norms. The requirement that for every $x \neq 0$ there is an open set $G \subset X$ such that $0 \in G$ and $x \notin G$ is translated into the requirement that for every $x \neq 0$ there is an $\alpha \in A$ such that $p_\alpha(x) \neq 0$.

In contrast with the example of Sec. 5.5, in a locally convex space X, the dual X' is non-trivial. Indeed, the Hahn-Banach theorem holds here, and it assures the existence of many continuous linear functionals on X. The proof is quite the same as for normed spaces and will not be given here.

Let X^* be the algebraic dual of X, i.e., the set of all linear functionals on X, continuous or not. An $x' \in X^*$ is continuous on X if and only if there is an open set containing 0 on which it is bounded. It is only necessary to show that if there is such a set U, then x' is continuous. Let $\epsilon > 0$ and suppose $|x'(x)| < M$ if $x \in U$. Let $a > 0$ and $aM < \epsilon$. Then $|x'(x)| < \epsilon$ if $x \in aU$

and aU is an open set containing 0. Hence x' is continuous. This result implies that x' is continuous if and only if there is an $\alpha \in A$ and an M such that

$$|x'(x)| \leq Mp_\alpha(x), \text{ for every } x \in X.$$

This observation allows us to penetrate into the nature of the topological dual X' of a locally convex topological vector space X. Let X be a vector space, without topology. Let p be any semi-norm on X and $M > 0$. Let

$$S_{p,M} = [x' \in X^* : |x'(x)| \leq Mp(x) \text{ for all } x \in X].$$

Then, let

$$S_p = \bigcup [S_{p,M} : \text{for all } M > 0].$$

S_p will be called the **semi-norm set** of p. It is clearly a vector subspace of X^*.

Now, let X be a locally convex topological vector space where the topology is given by a collection p_α, $\alpha \in A$, of semi-norms in the sense that the sets

$$U_{p_\alpha,r} = [x : p_\alpha(x) < r]$$

for all $\alpha \in A$ and $r > 0$ form a base for the topology at 0. The dual X' of X is then given by

$$X' = \bigcup [S_{p_\alpha} : \alpha \in A].$$

We start over again, let X be a vector space, and observe that semi-norms in X may be defined by means of certain subsets $A \subset X^*$. For any $A \subset X^*$, let

$$p_A(x) = \sup [|x'(x)| : x' \in A],$$

for each $x \in X$. It is almost immediate that p_A has all the properties of a semi-norm except that it might take on the value $+\infty$. We thus restrict A so that it is point-wise bounded, i.e.,

$$\sup [|x'(x)| : x' \in A] < \infty$$

for each $x \in X$. These are precisely the sets A for which p_A is a semi-norm.

Now, X^* as a space of functions, may be given the product topology. Point-wise boundedness is the same as boundedness in this topology. We shall call a subset $A \subset X^*$ **distinguished** if it is bounded and closed in the product topology, if it is convex, and if $x' \in A$, $|a| \leq 1$, implies $ax' \in A$.

LEMMA 1. For every semi-norm p on a vector space X, the set $A_p \subset X^*$ of functionals x', such that $p(x) \leq 1$ implies $|x'(x)| \leq 1$ is distinguished. Conversely, if A is a distinguished subset of X^*, there is a semi-norm p such that $A = A_p$.

Proof. The first part of the lemma consists of a direct and simple verification. For the converse, let

$$p(x) = \sup [|x'(x)| : x' \in A].$$

Then p is a semi-norm. If $x' \in A$, $x \in X$, and $p(x) \leq 1$, then $|x'(x)| \leq 1$.

Suppose then that $x' \notin A$. By the Hahn-Banach theorem, applied to X^* with the weak topology induced by X, there is an $x \in X$ such that $|y'(x)| \leq 1$ for all $y' \in A$ and $x'(x) > 1$. Thus $p(x) \leq 1$, but $x'(x) > 1$.

Now, let $X' \subset X^*$ be a subspace of X^*. Using lemma 1, we are able to find semi-norms on X whose semi-norm sets are contained in X'. Indeed, the lemma tells us that every distinguished $A \subset X'$ produces such a semi-norm. We show that every locally convex topology on X, with respect to which X' is the dual, is given by means of such semi-norms.

Indeed, it is now obvious that if A_α is a collection of distinguished sets which are all subsets of X', and whose union is X', then X' is the dual of the locally convex space given by the semi-norms p_{A_α}. (Such a collection of distinguished sets are the sets $[ax' : |a| \leq 1]$ for all $x' \in X'$.)

Suppose, conversely, that we have a locally convex topology in X and that X' is the dual. Let p be one of the defining semi-norms for the topology. The set A_p of $x' \in X^*$ such that $p(x) \leq 1$ implies $|x'(x)| \leq 1$ is a distinguished subset of X^*. But A_p, as a subset of the semi-norm set of p, is a subset of X'. Moreover, $p = p_{A_p}$, by use of the Hahn-Banach theorem.

THEOREM 1. If X' is a subspace of X^*, every locally convex topology on X, for which X' is the dual, is determined by the semi-norms associated with a system of distinguished subsets of X^* whose union is X'. The finest topology exists and is given by the system of all distinguished subsets of X'.

If among all the locally convex topologies on X, for which X' is the dual of X, there happens to be one for which X is a Banach space, then it is the finest locally convex topology on X for which X' is the dual. We prove this by using the principle of uniform boundedness. Suppose, then, that there is a norm $\| \ \|$ for which X is a Banach space and X' is the dual. Let A be a distinguished set in X^* which is a subset of X'. Since A is point-wise bounded, it is norm bounded by the principle of uniform boundedness. Thus, there is an M such that $|x'(x)| \leq M \|x\|$ for every $x' \in A$ and $x \in X$. Let p be the semi-norm in X determined by A. Then $p(x) \leq M \|x\|$, for every $x \in X$. This means that if X is topologized in any way so as to be locally convex and have X' as its dual, then the Banach topology is finer. This proves

THEOREM 2. If X is a vector space and X' is a subspace of X^*, then X can be normed in at most one way (within topological equivalence) so as to be a Banach space and have X' as its dual.

5.7 Metrizable Space, Space of Entire Functions

An important class of locally convex topological vector spaces consists of those which are metrizable. Since a metrizable space is such that, at each point in the space, a base for the topology at the point is countable, it is evident that the topology for a metrizable locally convex topological vector space can be given in terms of countably many semi-norms.

We show that, conversely, if X is a locally convex topological vector space, in which the topology is given by countably many semi-norms, $\{p_n\}$, then X is metrizable. Indeed, we consider the number

$$d(x, y) = \sum_{n=1}^{\infty} \frac{1}{2^n} \frac{p_n(x-y)}{1+p_n(x-y)}.$$

This is easily seen to be an invariant metric. We show that the topology given by this metric is the same as the one given by the semi-norms $\{p_n\}$.

Let $\{x_n\}$ be a sequence which converges to 0 in the locally convex space. Then

$$\lim_{n\to\infty} p_m(x_n) = 0, \quad m = 1, 2, \ldots.$$

Let $\epsilon > 0$. Choose N' so that $\sum_{n=N'}^{\infty} \frac{1}{2^n} < \frac{\epsilon}{2}$. Choose N so that $n > N$, $m < N'$ implies $p_m(x_n) < \epsilon/2$. Now, for $n > N$,

$$\begin{aligned} d(x_n, 0) &= \sum_{m=1}^{N'-1} \frac{1}{2^m} \frac{p_m(x_n)}{1+p_m(x_n)} + \sum_{m=N'}^{\infty} \frac{1}{2^m} \frac{p_m(x_n)}{1+p_m(x_n)} \\ &\leq \sum_{m=1}^{N'-1} \frac{1}{2^m} \cdot \frac{\epsilon}{2} + \sum_{m=N'}^{\infty} \frac{1}{2^m} < \epsilon. \end{aligned}$$

Thus $\{x_n\}$ converges to 0 in the metric.

The converse is trivial.

An interesting example is given by the space E of entire functions. This is a vector space over the complex numbers. The functions in E are given by power series

$$f(z) = \sum_{n=0}^{\infty} a_n z^n,$$

for which

$$\lim_{n\to\infty} \sqrt[n]{|a_n|} = 0.$$

If

$$f(z) = \sum_{n=0}^{\infty} a_n z^n \quad \text{and} \quad g(z) = \sum_{n=0}^{\infty} b_n z^n$$

are entire functions, let

$$d(f, g) = \sup\,[|a_0 - b_0|, \sqrt[n]{|a_n - b_n|} : n = 1, 2, \ldots].$$

PROPOSITION 1. $\{f_n\}$ converges to f in the space E if and only if it converges uniformly to f on every compact set.

Proof. Suppose $\{f_n\}$ converges to f uniformly on every compact set. Let

$$f(z) = \sum_{m=0}^{\infty} a_m z^m, \quad f_n(z) = \sum_{m=0}^{\infty} a_m^{(n)} z^m, \quad n = 1, 2, \ldots.$$

Let $\epsilon > 0$. Choose $R > 1/\epsilon$. Then, there is an N such that $n > N$ implies

$$|a_0^{(n)} - a_0| < \epsilon$$

and, by the Cauchy inequalities applied to the circle of radius R,

$$|a_m^{(n)} - a_m| \, R^m \leq 1, \quad m = 1, 2, \ldots,$$

so that

$$|a_m^{(n)} - a_m|^{1/m} < \epsilon.$$

Hence, we have convergence with respect to the metric.

Conversely, suppose $\{f_n\}$ converges to f in the metric. Choose R and $\epsilon > 0$. There is an N such that $m > N$ implies

$$|a_0^{(m)} - a_0| < \epsilon, \quad |a_n^{(m)} - a_n|^{1/n} < \epsilon, \quad n = 1, 2, \ldots.$$

It follows that $|z| = R$ implies

$$\begin{aligned} |f(z) - f_m(z)| &\leq |a_0 - a_0^{(m)}| + \sum_{n=1}^{\infty} |a_n^{(m)} - a_n| \, R^n \\ &\leq \epsilon + \sum_{n=1}^{\infty} \epsilon^n R^n. \end{aligned}$$

Since, for fixed R, the right-hand side converges to 0 as ϵ converges to 0, the proposition follows by the maximum modulus principle.

We consider the following semi-norms in E. For every $n = 1, 2, \ldots$ let

$$p_n(f) = \max \, [|f(z)| : |z| \leq n].$$

The topology given E by this set of semi-norms is the same as that of uniform convergence on compact sets. In Chapter 1, we observed that the metric space E is complete.

We examine the dual of E. For this purpose we need

LEMMA 1. The series $\sum_{n=0}^{\infty} a_n b_n$ converges for every $\{a_n\}$ for which

$$\lim_{n \to \infty} \sqrt[n]{|a_n|} = 0$$

if and only if $\{\sqrt[n]{|b_n|}\}$ is bounded.

Proof. Suppose $|b_0| \leq M$, $|b_n| \leq M^n$, $n = 1, 2, \ldots$. Let $\{a_n\}$ be such that $\lim_{n\to\infty} \sqrt[n]{|a_n|} = 0$. There is an N such that $n > N$ implies $|a_n| \leq 1/(2^n M^n)$. Then, clearly $\sum_{n=0}^{\infty} |a_n b_n| < \infty$.

For the converse, suppose $\{|b_n|^{1/n}\}$ is not bounded. For every k, there is an n_k with $|b_{n_k}| > k^{n_k}$. Let $a_{n_k} = 1/k^{n_k}$, and $a_n = 0$ otherwise. Then $\lim_{n\to\infty} \sqrt[n]{|a_n|} = 0$, but $\sum_{n=0}^{\infty} a_n b_n$ does not converge.

THEOREM 1. Every continuous linear functional φ on E is of the form

$$\varphi(f) = \sum_{n=0}^{\infty} b_n a_n,$$

where $f(z) = \sum_{n=0}^{\infty} a_n z^n$ and $\{b_n\}$ is a fixed sequence for which $\{|b_n|^{1/n}\}$ is bounded, and for every such $\{b_n\}$ the corresponding φ, defined as above, is a continuous linear functional on E.

Proof. Suppose $\{|b_n|^{1/n}\}$ is bounded. Let

$$\varphi(f) = \sum_{n=0}^{\infty} a_n b_n.$$

Since the series on the right converges for every $f \in E$, φ is defined on E. It is clear that φ is linear. Suppose $\{f_m\}$ converges to 0 in E, where

$$f_m(z) = \sum_{n=0}^{\infty} a_n^{(m)} z^n, \quad m = 1, 2, \ldots .$$

Let

$$M = \sup \{|b_0|, |b_n|^{1/n} : n = 1, 2, \ldots\}.$$

Let $\epsilon > 0$. There is an N such that $m > N$ implies

$$|a_0^{(m)}| < \epsilon, \quad |a_n^{(m)}|^{1/n} < \epsilon, \quad n = 1, 2, \ldots .$$

Then

$$|\varphi(f_m)| = \left| \sum_{n=0}^{\infty} a_n^{(m)} b_n \right| \le \sum_{n=0}^{\infty} |a_n^{(m)} b_n|$$
$$\le \epsilon M + \sum_{n=1}^{\infty} (\epsilon M)^n.$$

But the right-hand side converges to 0 as ϵ converges to 0. Hence

$$\lim_{m \to \infty} \varphi(f_m) = 0.$$

Conversely, suppose φ is a continuous linear functional on E. Let

$$b_n = \varphi(z^n), \quad n = 0, 1, 2, \ldots .$$

Now, let

$$f(z) = \sum_{n=0}^{\infty} a_n z^n$$

be in E, and let

$$f_n(z) = \sum_{k=0}^{n} a_k z^k, \quad n = 0, 1, 2, \ldots .$$

Then $\{f_n\}$ converges in E to f, so that

$$\varphi(f) = \lim_{n \to \infty} \varphi(f_n).$$

But

$$\varphi(f_n) = \sum_{k=0}^{n} a_k b_k, \quad n = 1, 2, \ldots ,$$

so that

$$\varphi(f) = \sum_{n=0}^{\infty} a_n b_n.$$

Moreover, since for every $\{a_n\}$ for which $\lim_{n\to\infty} \sqrt[n]{|a_n|} = 0$, the series

$$\sum_{n=0}^{\infty} a_n b_n$$

converges, it follows that $\{\sqrt[n]{|b_n|}\}$ is bounded.

REMARK. A power series $\sum_{n=0}^{\infty} b_n z^n$ has positive radius of convergence if and only if $\{\sqrt[n]{|b_n|}\}$ is bounded. The dual of E may accordingly be identified with the function elements at 0 (or at any other point on the Riemann sphere).

We now use theorem 1 to obtain a simple but interesting condition for a set of functions to form a total set in the space E of entire functions. We first note a general criterion for a set to form a total set. Let $A \subset E$, and let X be the closure of the vector space generated by A. If $X \neq E$, by the Hahn-Banach theorem, for every $f \in E \sim X$, there is a continuous linear functional φ on E such that $\varphi(f) = 1$ and $\varphi(g) = 0$ for every $g \in X$. It follows that $X = E$ if and only if every continuous linear functional which vanishes on A is identically zero on E. In other words, a necessary and sufficient condition that A be total in E is that the zero functional is the only continuous linear functional on E which vanishes on A.

Now, suppose that

$$f(z) = \sum_{n=0}^{\infty} a_n z^n \in E$$

and that none of the coefficients a_n, $n = 0, 1, 2, \ldots$ are zero. Let S be a set of complex numbers which has at least one finite limit point. For every $\zeta \in S$, let

$$f_\zeta(z) = \sum_{n=0}^{\infty} a_n(\zeta z)^n.$$

Then $f_\zeta \in E$. Let

$$A = [f_\zeta : \zeta \in S].$$

PROPOSITION 2. A is total in E.

Proof. Suppose φ is a continuous linear functional on E which vanishes on A. There is a sequence $\{b_n\}$, with $\{|b_n|^{1/n}\}$ bounded, such that

$$\varphi(g) = \sum_{n=0}^{\infty} c_n b_n$$

for every $g(z) = \sum_{n=0}^{\infty} c_n z^n \in E$. Then, for every $\zeta \in S$,

$$\varphi(f_\zeta) = \sum_{n=0}^{\infty} a_n b_n \zeta^n = 0.$$

Consider the function given by

$$h(z) = \sum_{n=0}^{\infty} a_n b_n z^n.$$

Since $\lim_{n\to\infty} \sqrt[n]{|a_n b_n|} = 0$, $h \in E$. But $h(\zeta) = 0$ for every $\zeta \in S$, so that $h(z) = 0$, since S has a finite limit point. Thus $a_n b_n = 0$ for every $n = 0, 1, 2, \ldots$. Since no a_n is 0, it follows that $b_n = 0$ for $n = 0, 1, 2, \ldots$ and φ is identically zero.

5.8 FK Spaces

The space E discussed above is typical of a wide class of spaces, the *FK* spaces. An *FK* **space** X is:

(a) A vector space, over the real numbers (or complex numbers), whose elements are sequences of real numbers (or complex numbers).

(b) X is a locally convex metrizable topological vector space in which the topology is, accordingly, given by a countable set $\{p_n\}$ of semi-norms.

(c) The metric space X is complete.

(d) If we use the notation

$$x = (a_1(x), a_2(x), \ldots)$$

for $x \in X$, the linear functionals $a_n(x)$, $n = 1, 2, \ldots$, are continuous on X.

The space E, discussed in Sec. 5.7, may be interpreted as being composed of those sequences $\{a_n\}$ of complex numbers for which $\lim_{n\to\infty} \sqrt[n]{|a_n|} = 0$. This is a vector space, over the complex numbers, with semi-norms

$$p_n(f) = \sup \left[\left| \sum_{n=0}^{\infty} a_n(f) z^n \right| : |z| = n \right], \quad n = 0, 1, 2, \ldots .$$

E is a complete metric space. If

$$f(z) = \sum_{n=0}^{\infty} a_n(f) z^n, \quad f \in E,$$

the Cauchy inequalities imply that $a_n(f)$ is a continuous linear functional on E for every $n = 0, 1, 2, \ldots$.

A complete metrizable locally convex topological vector space is called a **Fréchet space.** Fréchet spaces have many properties in common with Banach spaces. In particular, the closed graph theorem holds, the proof being the same as for Banach spaces. We shall see that this implies the following uniqueness result:

Let X be a vector space whose elements are sequences of numbers. If X may be topologized in such a way as to be an *FK* space, then this may be done in only one way.

Indeed, a more general fact holds:

If X and Y are *FK* spaces, and $X \subset Y$, then the identity mapping of X into Y is continuous. In other words, the topology induced in X by the *FK* topology in Y is coarser than the *FK* topology in X.

PROPOSITION 1. Let X and Y be *FK* spaces, with $X \subset Y$. The identity mapping

$$I : X \to Y$$

is continuous.

Proof. We need only show that the graph of I is closed in $X \times Y$. Let $\{x_m, x_m\}$ be a Cauchy sequence in the graph of I. Let

$$x_m = (a_1^{(m)}, a_2^{(m)}, \ldots), \quad m = 1, 2, \ldots,$$

converge to

$$x = (a_1, a_2, \ldots)$$

in X, and to

$$y = (b_1, b_2, \ldots)$$

in Y. By continuity of the coordinate functions in an *FK* space, we have

$$a_n = \lim_{m \to \infty} a_n^{(m)} \quad \text{and} \quad b_n = \lim_{m \to \infty} a_n^{(m)}, \quad n = 1, 2, \ldots .$$

Hence, $a_n = b_n$, for every n, and $x = y$. Hence, $\lim\limits_{m \to \infty} (x_m, x_m) = (x, x)$ is on the graph of I, and the graph of I is closed. By the closed graph theorem, I is continuous.

COROLLARY. Let X be a vector space of sequences of real or complex numbers. If there are two sets of semi-norms $\{p_n\}$ and $\{q_n\}$, for each of which X is an *FK* space, then the corrresponding topologies are the same.

Proof. Let t_1 and t_2 be the corresponding topologies. By proposition 1, the identity mapping of X, with either topology, onto X, with the other topology, is continuous. Hence, the topologies are the same.

Thus, a vector space of sequences either is an *FK* space or it is not an *FK* space. If it is an *FK* space, the *FK* topology that it carries is unique.

Moreover, larger *FK* spaces have coarser topologies. Precisely, if $X \subset Y$ are *FK* spaces, the topology in X, induced by the *FK* topology in Y, is coarser than the *FK* topology in X.

We give some examples of *FK* spaces. In the first few examples, the sequences of numbers are either real or complex.

(a) The space of all sequences. This is an *FK* space with semi-norms

$$p_n(x) = |a_n(x)|, \quad n = 1, 2, \ldots,$$

where $x = (a_1(x), a_2(x), \ldots)$.

The p_n are evidently semi-norms. Convergence in this space is coordinatewise convergence, and it follows easily that the space is complete. Moreover, it is clear that the functionals $a_n(x)$, $n = 1, 2, \ldots$, are continuous and linear.

(b) The space of bounded sequences with the single semi-norm

$$p(x) = \sup_n |a_n(x)|.$$

The confirmation that this is an *FK* space is routine.

(c) For every $p \geq 1$, the space l_p of sequences

$$x = (a_1(x), a_2(x), \ldots),$$

for which

$$\sum_{n=1}^{\infty} |a_n(x)|^p < \infty,$$

with the semi-norm

$$p(x) = \left[\sum_{n=1}^{\infty} |a_n(x)|^p\right]^{1/p}.$$

We have only to show that the $a_n(x)$ are continuous. But

$$|a_n(x)|^p \leq \sum_{n=1}^{\infty} |a_n(x)|^p,$$

for every n, so that $|a_n(x)| \leq p(x)$.

(d) Let X be the vector space of sequences $\{a_n(x)\}$ such that

$$\sum_{n=1}^{\infty} |a_n(x)|^p < \infty$$

for all $p > 1$. We first note that $p > p'$ implies $l_p \supset l_{p'}$. Hence

$$X = \bigcap_{n=1}^{\infty} l_{1+1/n}, \quad n = 1, 2, \ldots$$

Thus, X is the intersection of a properly decreasing sequence of *FK* spaces. Let X have as its set of semi-norms, the semi-norms of the spaces $l_{1+1/n}$, $n = 1, 2, \ldots$, i.e., the semi-norms

$$q_n(x) = \left[\sum_{k=1}^{\infty} |a_k(x)|^{1+1/n}\right]^{n/n+1}, \quad n = 1, 2, \ldots .$$

We show that X is an *FK* space. It is necessary only to show that X is complete and that the $a_n(x)$ are continuous on X.

In order to show that X is complete, let $\{x_n\}$ be a Cauchy sequence in X. This implies that $\{q_m(x_n)\}$ is a Cauchy sequence of numbers for every $m = 1, 2, \ldots$. Accordingly, $\{x_n\}$ is a Cauchy sequence in each of the spaces $l_{1+1/m}$, $m = 1, 2, \ldots$. We next note that the limit of $\{x_n\}$, in each $l_{1+1/m}$, is the same sequence of numbers. This is true because the $l_{1+1/m}$ are all *FK* spaces so that convergence implies coordinate-wise convergence. Let x be this common limit. Then, since for every m,

$$q_m(x) = \lim_{n\to\infty} q_m(x_n),$$

x is also the limit of $\{x_n\}$ in X. Hence, X is complete.

It is implicit in the above argument that convergence in X implies coordinate-wise convergence.

We next note that there are vector spaces of sequences which are not *FK* spaces. Indeed, the union of any strictly increasing sequence of *FK* spaces is not an *FK* space. In order to prove this, we need a fact from the general theory of Fréchet spaces.

PROPOSITION 2. If X and Y are Fréchet spaces and f is a continuous linear mapping of X into Y, then $f(X)$ is either all of Y or is of the first category in Y.

Proof. We do not give all the details of the proof, but point out that it is similar to that of the open mapping theorem for Banach spaces. Just as in the proof of that theorem, if the image of the unit sphere in X is dense in a sphere in Y, then it contains a sphere in Y. Thus $f(X)$ is of the first category in Y or else $f(X) = Y$.

PROPOSITION 3. If $\{X_n\}$ is a sequence of *FK* spaces, $X_n \subset X_{n+1}$, $X_n \neq X_{n+1}$, $n = 1, 2, \ldots$, then $X = \bigcup_{n=1}^{\infty} X_n$ is not an *FK* space.

Proof. Suppose X were an *FK* space. Then, by proposition 1, for every n, the identity mapping of X_n into X would be continuous. Since $X \neq X_n$, X_n would be of the first category in X. Then X, as the union of a countable set of subsets of the first category, would be of the first category in itself. But an *FK* space is a complete metric space and so is of the second category.

In contrast with this fact, we have

PROPOSITION 4. If X_n, $n = 1, 2, \ldots$, is a sequence of *FK* spaces, then $X = \bigcap_{n=1}^{\infty} X_n$ is an *FK* space.

The proof of this theorem follows the lines of the special case given in example (*d*) above, and will be omitted.

5.9 Application to Summability Methods

We shall apply *FK* spaces to the theory of limitation of divergent sequences. In this regard, we consider sequence-to-sequence transformations given by infinite matrices. A more complete discussion than the one given in Chapter 2 will be given here. Accordingly, let

$$A = (a_{mn}) = \begin{pmatrix} a_{11} & a_{12} & \cdots & a_{1n} & \cdots \\ a_{21} & a_{22} & \cdots & a_{2n} & \cdots \\ & & \cdots & & \\ a_{m1} & a_{m2} & \cdots & a_{mn} & \cdots \\ & & \cdots & & \end{pmatrix}$$

be an infinite matrix.

Let $E(A)$ be the set of sequences $\{x_n\}$ for which all the series

$$\sum_{n=1}^{\infty} a_{mn}x_n, \quad m = 1, 2, \ldots$$

converge. Then A acts as a linear transformation on the set $E(A)$ to the vector space of all sequences, where the image of $\{x_n\} \in E(A)$ is the sequence $\left\{\sum_{n=1}^{\infty} a_{mn}x_n\right\}$. $E(A)$ is evidently a vector space.

We consider two vector subspaces of $E(A)$, the space $C(A)$ of all $\{x_n\} \in E(A)$ for which $\left\{\sum_{n=1}^{\infty} a_{mn}x_n\right\}$ converges, and the subspace $P(A)$ of $C(A)$ of sequences $\{x_n\}$ for which $\{x_n\}$ converges and

$$\lim_{m\to\infty} \sum_{n=1}^{\infty} a_{mn}x_n = \lim_{n\to\infty} x_n.$$

We recall that if $P(A)$ is precisely the set of convergent sequences, then A is said to be **regular.** The main fact in this subject is the following theorem which was proved in Chapter 2, and for which the following proof is also of interest.

THEOREM 1. *A is regular if and only if*

(a) $\sum_{n=1}^{\infty} |a_{mn}| < M$ *for some M and all m.*

(b) $\lim_{m\to\infty} a_{mn} = 0$ *for every n.*

(c) $\lim_{m\to\infty} \sum_{n=1}^{\infty} a_{mn} = 1.$

Proof. Suppose A is regular. Let c be the space of convergent sequences. Since $\sum_{n=1}^{\infty} a_{mn}x_n$ converges for every $\{x_n\} \in c$, for every m, it follows that

$\{a_{mn}\} \in l_1$, $m = 1, 2, \ldots$, since l_1 is the dual of c. But this sequence of linear functionals is bounded for every $\{x_n\} \in c$. By the principle of uniform boundedness, its sequence of norms is bounded in l_1. This proves that (a) must hold. The necessity of (b) and (c) is not as deep. For (b), consider the sequence which is 1 in the nth place and 0 everywhere else. This sequence is taken by the matrix A into the sequence $\{a_{mn}\}$. Thus we must have $\lim_{m\to\infty} a_{mn} = 0$ for every n. In order to prove that (c) holds, it is only necessary to consider the transform of the sequence $\{1, 1, \ldots, 1, \ldots\}$.

The sufficiency of the conditions (a), (b), (c) may be proved by a direct computation as follows:

Let $\{x_n\} \in c$ and suppose $\lim_{n\to\infty} x_n = s$. Then

$$\begin{aligned}\lim_{m\to\infty} \left| \sum_{n=1}^{\infty} a_{mn}x_n - s \right| &= \lim_{m\to\infty} \left| \sum_{n=1}^{\infty} a_{mn}(x_n - s) \right| \le \lim_{m\to\infty} \sum_{n=1}^{\infty} |a_{mn}|\,|x_n - s| \\ &= \lim_{m\to\infty} \left[\sum_{n=1}^{N} |a_{mn}|\,|x_n - s| + \sum_{n=N+1}^{\infty} |a_{mn}|\,|x_n - s| \right], \text{ for every } N.\end{aligned}$$

Now, let $\epsilon > 0$. Choose N so that $n > N$ implies $|x_n - s| < \epsilon/M$. Since $\lim_{m\to\infty} \sum_{n=1}^{N} |a_{mn}|\,|x_n - s| = 0$, we have $\lim_{m\to\infty} \left|\sum_{n=1}^{\infty} a_{mn}x_n - s\right| < \epsilon$, for every $\epsilon > 0$.

The following fact will be of central importance to the remaining considerations of this section.

PROPOSITION 1. Let X be an FK space with semi-norms $\{p_n\}$, Y an FK space with semi-norms $\{q_n\}$, and $\{f_n\}$ a sequence of continuous linear functionals on X. The set H of elements of X, for which $\{f_n(x)\} \in Y$, is an FK space with semi-norms

$$\{p_n(x)\}, \{q_n(\{f_m(x)\})\}.$$

Proof. For every $x \in X$, the sequence

$$\{f_1(x), f_2(x), \ldots\}$$

will be called the transform of x. H is clearly a vector space. H has countably many semi-norms. Since the topology in H is finer than that induced in H by X, the coordinate functionals are continuous. It remains only to prove that H is complete. Accordingly, let $\{x_m\}$ be a Cauchy sequence in H. Then, for every n, the sequence $\{p_n(x_m)\}$ converges, and the sequence $\{q_n(y_m)\}$ converges, where y_m is the transform of x_m. By the completeness of X and Y, the sequence $\{x_m\}$ converges to an $x \in X$ and the sequence $\{y_m\}$ converges to a $y \in Y$. Now, by coordinate-wise convergence in Y, and the continuity of the f_n, the nth coordinate of y is given by

$$\lim_{m\to\infty} f_n(x_m) = f_n\left(\lim_{m\to\infty} x_m\right) = f_n(x).$$

Thus, y is the transform of x. This means that $\{x_m\}$ converges to x in H so that H is complete.

COROLLARY 1. If $\{a_n\}$ is a sequence of real numbers, the set H of sequences $x = \{x_n\}$, for which $\sum_{n=1}^{\infty} a_n x_n$ converges, is an *FK* space with semi-norms

$$|x_n|, \quad n = 1, 2, \ldots$$

and

$$\sup_n \left| \sum_{k=1}^{n} a_k x_k \right|.$$

Proof. Let X be the *FK* space of all sequences. Then the determining semi-norms for X are

$$p_n(x) = |x_n|, \quad n = 1, 2, \ldots$$

for $x = (x_1, x_2, \ldots)$. Let

$$f_n(x) = \sum_{k=1}^{n} a_k x_k, \quad n = 1, 2, \ldots .$$

Let Y be the *FK* space of convergent sequences. The determining semi-norm for Y is

$$q(y) = \sup_n |y_n|$$

for $y = (y_1, y_2, \ldots)$. The conclusion is obtained by substitution in proposition 1.

As an immediate consequence of this corollary, we have

PROPOSITION 2. For every matrix $A = (a_{mn})$, the set $E(A)$ is an *FK* space with semi-norms

$$p_n(x) = |x_n|, \quad n = 1, 2, \ldots$$
$$q_m(x) = \sup_n \left| \sum_{k=1}^{n} a_{mk} x_k \right|, \quad m = 1, 2, \ldots .$$

Proof. By corollary 1 and the fact that the intersection of a countable set of *FK* spaces is an *FK* space with all the semi-norms of the factor spaces as the set of determining semi-norms.

Now, let A be an infinite matrix and consider the *FK* space $E(A)$. For every m, let

$$f_m(x) = \sum_{n=1}^{\infty} a_{mn} x_n.$$

Since

$$\left| \sum_{n=1}^{\infty} a_{mn} x_n \right| \le \sup_n \left| \sum_{k=1}^{n} a_{mk} x_k \right|,$$

the functionals f_m are continuous on $E(A)$. By proposition 1, we obtain

THEOREM 2. For every A, the set $C(A)$ is an FK space with semi-norms

$$\sup_n \left| \sum_{k=1}^{n} a_{mk} x_k \right|, \qquad m = 1, 2, \ldots,$$

$$|x_n|, \quad n = 1, 2, \ldots,$$

$$\sup_m \left| \sum_{n=1}^{\infty} a_{mn} x_n \right|.$$

The set $C(A)$ is called the **convergence field** of a matrix A. Thus, every convergence field is an FK space. We shall see that there are FK spaces, however, which are not convergence fields.

We shall consider, briefly, a special collection of regular matrix methods, the Cesaro methods. For every $\alpha > 0$, consider the matrix

$$a_{mn} = \begin{cases} \dfrac{\dbinom{m+n+\alpha-1}{m-n}}{\dbinom{m+\alpha}{m}}, & n \le m \\ 0 & n > m, \end{cases}$$

for $m, n = 0, 1, 2, \ldots$.

This method is designated the (C, α) method.

The following hold but will not be proved here (see Hardy [1] and Zeller [1]).

(a) For every $\alpha > 0$, the (C, α) method is regular.

(b) For every $\alpha, \beta > 0$, the (C, α) and (C, β) methods are consistent, i.e., if a sequence is in the convergence fields of both methods, its transformed sequences have the same limit.

(c) If $\alpha > \beta$, then (C, α) is strictly stronger than (C, β), i.e., the convergence field of (C, α) contains that of (C, β) as a proper subset.

Let E_α be the convergence field of (C, α). Then E_α is an FK space. It has the property:

(d) For every $x = (x_1, x_2, \ldots)$ in E_α, the sequence $x^n = (x_1, \ldots, x_n, 0, \ldots)$, $n = 1, 2, \ldots$, converges to x in E_α.

Let H consist of those sequences all of whose terms, except for a finite number, are zero.

PROPOSITION 3. If $\{A_n\}$ is a sequence of matrices such that, for every n, the convergence field E_n of A_n contains H and has property (d), and if

$E_{n+1} \subset E_n$, $E_{n+1} \neq E_n$, $n = 1, 2, \ldots$, then $E = \bigcap_{n=1}^{\infty} E_n$ is not the convergence field for any matrix, even though it is an *FK* space.

Proof. Let A be a matrix whose convergence field F contains E. We may then take sets of determining semi-norms for F and E such that the former is a subset of the latter. In particular,

$$q(x) = \sup_m \left| \sum_{n=1}^{\infty} a_{mn} x_n \right|$$

may be taken as one of the semi-norms for F. Since each E_i has a determining set of semi-norms $p_{i1}, p_{i2}, \ldots$, the totality p_{ij}, $i, j = 1, 2, \ldots$, is a determining set of semi-norms for E. It follows that

$$q(x) \leq M \sum_{k=1}^{n} p_{i_k j_k}(x)$$

for some M and n. Since the E_i are decreasing, for sufficiently large m,

$$q(x) \leq M \sum_{k=1}^{m} p_{i_n k}(x).$$

Thus A maps E, with the semi-norms $p_{i1}, p_{i2}, \ldots$, continuously into the space of convergent sequences. Since $E \supset H$, and E_i has the property (d), it follows that this mapping can be extended to E_i. Thus $F \supset E_i$, so that E is not the convergence field of A.

We may now state for the Cesaro methods

PROPOSITION 4. $\bigcup_{\alpha > 0} E_\alpha$ is not an *FK* space and so it is not a matrix convergence field. $\bigcap_{\alpha > 0} E_\alpha$ is an *FK* space but it is not a matrix convergence field.

5.10 Ordered Vector Spaces

Vector spaces of real functions have a natural order relation and there is an interplay between this order relation and topologies on such spaces.

It is just as easy to work in a more general setting. Let X be a vector space as well as an ordered set. By an **ordered set** X we understand that X has an order relation $\geq$ defined for some pairs in X such that

(i) $x \geq y$, $y \geq z$ implies $x \geq z$
(ii) $x \geq y$, $y \geq x$ implies $x = y$.

Then, X is said to be an **ordered vector space** if

(a) X is a vector space and an ordered set, and
(b) $x \geq y$ implies $x + z \geq y + z$ for every $z \in X$; $x \geq 0$ and $a \geq 0$ implies $ax \geq 0$.

An ordered vector space is said to be a **vector lattice** if for every $x, y \in X$, $\sup(x, y)$ and $\inf(x, y)$ exist.

We give some examples of ordered vector spaces.

(a) The vector space $C(X)$ of continuous functions, each having compact support, on a locally compact Hausdorff space X, with the order relation $x \geq y$ if $x(t) \geq y(t)$ for every $t \in X$. $C(X)$ is easily seen to be a vector lattice.

(b) The vector space D of differentiable real functions, defined on the reals with $x \geq y$ if $x(t) \geq y(t)$, for every real t, is an ordered vector space. It is not a vector lattice since the pair of functions t, $-t$ does not have a least upper bound in D.

(c) Let $(X, \mathcal{S}, \mu)$ be a measure space and $p \geq 1$. The corresponding L_p space is a vector lattice if, for equivalence classes x and y, $x \geq y$, if $x(t) \geq y(t)$ almost everywhere for every representative of x and of y.

Although some of the most important ordered vector spaces of functions are not vector lattices, we shall consider only the vector lattice case.

Let X be a vector lattice. For every $x \in X$, let $x^+ = \sup(x, 0)$ and $x^- = -\inf(x, 0)$. Since $y \geq z$ implies $x + y \geq x + z$, it follows easily that for every $x, y, z \in X$,

$$\sup(x + y, x + z) = x + \sup(y, z).$$

Then

$$\begin{aligned} x^+ - x &= \sup(x, 0) - x = \sup(x - x, -x) \\ &= \sup(0, -x) = -\inf(0, x) = x^-. \end{aligned}$$

Thus, every $x \in X$ has the decomposition

$$x = x^+ - x^-.$$

We define the **absolute value** of x by

$$|x| = x^+ + x^-.$$

We shall be interested in linear functionals on a vector lattice.

A linear functional x' on X is said to be **positive** if $x \geq 0$ implies $x'(x) \geq 0$.

A linear functional x' on X is said to be **bounded** if for every $x \in X$, x' is bounded on the set of y for which $|y| \leq |x|$.

We shall call the set X^ω of bounded linear functionals on X the **order dual** of X. The set X^ω is easily seen to be a vector space, a subspace of the algebraic dual X^* of X. If x' is positive, then

$$|y| \leq |x| \text{ implies } x'(|y|) \leq x'(|x|).$$

But

$$\begin{aligned} |x'(y)| &= |x'(y^+ - y^-)| = |x'(y^+) - x'(y^-)| \\ &\leq |x'(y^+)| + |x'(y^-)| = x'(y^+) + x'(y^-) \\ &= x'(|y|). \end{aligned}$$

Hence x' is bounded.

The main fact here is that every bounded linear functional is the difference between positive ones. This fact closely resembles the theorem on decomposition of signed measures.

We need

LEMMA 1. If $x_1, x_2, y_1, y_2 \geq 0$ and $x_1 + x_2 = y_1 + y_2$, there are $u, v, r, s \geq 0$ such that $x_1 = u + v$, $x_2 = r + s$, $y_1 = u + r$, $y_2 = v + s$.

Proof. Let $u = (x_1 - y_2)^+$, $v = x_1 - u$, $r = y_1 - u$, $s = u - (x_1 - y_2)$. It is easy to see that $u, v, r, s \geq 0$. The rest is trivial.

COROLLARY 1. If $x_1, x_2 \geq 0$ and $0 \leq y \leq x_1 + x_2$, there are $0 \leq y_1 \leq x_1$, $0 \leq y_2 \leq x_2$ such that $y = y_1 + y_2$.

Proof. Apply lemma 1 to $x_1, x_2, y, x_1 + x_2 - y$. There are $y_1, y_2, z_1, z_2 \geq 0$, with $x_1 = y_1 + z_1$, $x_2 = y_2 + z_2$, $y = y_1 + y_2$, $x_1 + x_2 - y = z_1 + z_2$.

We are now ready to prove the main theorem.

THEOREM 1. If X is a vector lattice, and f is a bounded linear functional on X, there are positive linear functionals g, h on X such that $f = g - h$.

Proof. Let $x \geq 0$. Define

$$g(x) = \sup [f(y) : 0 \leq y \leq x].$$

Now, let $x_1, x_2 \geq 0$. We show that

$$g(x_1) + g(x_2) = g(x_1 + x_2).$$

In the first place,

$$\begin{aligned} g(x_1) + g(x_2) &= \sup [f(y) : 0 \leq y \leq x_1] \\ &\quad + \sup [f(y) : 0 \leq y \leq x_2] \\ &= \sup [f(y_1 + y_2) : 0 \leq y_1 \leq x_1, 0 \leq y_2 \leq x_2] \\ &\leq g(x_1 + x_2). \end{aligned}$$

Conversely, let $0 \leq z \leq x_1 + x_2$. By the corollary 1, there are $0 \leq y_1 \leq x_1$, $0 \leq y_2 \leq x_2$ such that $y_1 + y_2 = z$. Then

$$f(z) = f(y_1) + f(y_2) \leq g(x_1) + g(x_2).$$

Since this holds for every $0 \leq z \leq x_1 + x_2$,

$$g(x_1 + x_2) \leq g(x_1) + g(x_2).$$

Now, for arbitrary $x \in X$, $x = x^+ - x^-$. Define

$$g(x) = g(x^+) - g(x^-).$$

Then,

$$g(x_1 + x_2) = g(x_1) + g(x_2)$$

for every $x_1, x_2 \in X$. That $g(ax) = ag(x)$ is evident.

Thus, g is a positive linear functional on X. Moreover, $g(x) \geq f(x)$ for every $x \geq 0$. So, $h = g - f$ is positive on X.

We remark that if we endow X^{ω} with the order relation $f \geq g$ if $f - g$ is positive, then X^{ω} is a vector lattice.

5.11 Banach Lattice

We now allow a vector space X to have both an order relation and a topology. To prevent any confusion we shall avoid using the term bounded to refer to continuous linear functionals in this discussion. Let X be a vector lattice. A semi-norm p in the vector space X is said to be **compatible** with the order relation in X if

$$|x| \geq |y| \text{ implies } p(x) \geq p(y).$$

A locally convex topology in X turns X into a **locally convex topological vector lattice** if it has a determining set of semi-norms compatible with the order relation in X. Now, a locally convex topological vector lattice has a topological dual X' and an order dual X^{ω}.

PROPOSITION 1. $X' \subset X^{\omega}$.

Proof. It is only necessary to show that if p is a semi-norm compatible with the order relation in X, then the semi-norm set of p is a subset of X^{ω}. Accordingly, let $f \in X^*$ be such that

$$|f(x)| \leq Mp(x)$$

for some M and all $x \in X$. Now, let $0 \leq y$. For $|z| \leq y$,

$$p(z) \leq p(y),$$

so that

$$|f(z)| \leq Mp(z) \leq Mp(y),$$

and $f \in X^{\omega}$.

On the other hand, we have

PROPOSITION 2. For every vector lattice X there is a locally convex topology on X such that X is a locally convex topological vector lattice for which $X' = X^{\omega}$.

Proof. Let $f \in X^{\omega}$ be positive and let p_f be defined by

$$p_f(x) = f(|x|)$$

for every $x \in X$. Then p_f is easily seen to be a semi-norm compatible with the order relation in X. But

$$|f(x)| = |f(x^+) - f(x^-)| \leq f(x^+) + f(x^-)$$
$$= f(|x|) = p_f(x),$$

so that $f \in X'$, where X' is the dual of X topologized by the semi-norms $[p_f : f \subset X^\omega]$. Since every positive $f \subset X^\omega$ is in X', it follows that $X^\omega \subset X'$.

Suppose now that X is a vector lattice and that there is a norm in X, which is compatible with the order relation, for which X is a Banach space. Then X is called a **Banach lattice.** We shall show that in a Banach lattice, $X' = X^\omega$. We first prove

LEMMA 1. If X is a Banach lattice, if $0 \leq x_1 \leq x_2 \leq \ldots$ and

$$\lim_{n \to \infty} \|x - x_n\| = 0,$$

then $x_n \leq x$, for every n.

Proof. Suppose $(x_n - x)^+ > 0$ for some n. For every $m > n$,

$$(x_m - x)^+ \geq (x_n - x)^+.$$

But,

$$|x_m - x| \geq (x_m - x)^+,$$

so that

$$\|x_m - x\| \geq \|(x_n - x)^+\|,$$

and the sequence $\{x_n\}$ does not converge to x.

PROPOSITION 3. If X is a Banach lattice, then $X' = X^w$.

Proof. Suppose, on the contrary, that f is a positive linear functional on X which is not continuous. Then, for every M, there is an $x \in X$ such that $\|x\| \leq 1$ and $|f(x)| \geq M$. Since $\| \, |x| \, \| = \|x\|$ and since

$$f(|x|) = f(x^+ + x^-)$$
$$\geq |f(x^+ - x^-)| = |f(x)|,$$

there is an $x > 0$ for which $\|x\| \leq 1$ and $f(x) \geq M$.

Now, let $\{x_n\}$ be a sequence of positive elements in X such that, for every n,

$$\|x_n\| \leq 1 \quad \text{and} \quad f(x_n) \geq 2^{2n}.$$

Let

$$y_n = \frac{1}{2} x_1 + \frac{1}{2^2} x_2 + \ldots + \frac{1}{2^n} x_n,$$

for every $n = 1, 2, \ldots$. Then $\{y_n\}$ is an increasing sequence of positive elements. But, for every n and k,

$$\|y_{n+k} - y_n\| = \left\| \frac{1}{2^{n+1}} x_{n+1} + \ldots + \frac{1}{2^{n+k}} x_{n+k} \right\|$$
$$\leq \frac{1}{2^{n+1}} + \ldots + \frac{1}{2^{n+k}} < \frac{1}{2^n}.$$

Thus, $\{y_n\}$ is a Cauchy sequence in X. Let $y = \lim_{n\to\infty} y_n$. By lemma 1, $y > y_n$, for every n. Hence,

$$f(y) > f(y_n) > \frac{1}{2^n} f(x_n) \geq 2^n, \text{ for every } n.$$

But this is impossible, so that f is continuous, and the proposition is proved.

We may now prove

THEOREM 1. If a vector lattice X can be normed as a Banach lattice, it can be done in only one way to within topological equivalence.

Proof. Since X^ω is a subspace of X^*, there is at most one norm on X, within topological equivalence, for which X is a Banach space and X^ω is the dual. But, for every compatible norm on X, for which X is a Banach space, X^ω is the dual. This proves the theorem.

We give some examples.

(a) Let S be a compact Hausdorff space, and let $C(S)$ be the vector lattice of continuous functions on S with the natural order relation $x \geq y$ if $x(t) \geq y(t)$ for every $t \in S$. Then the only norm, within topological equivalence, for which $|x(t)| \geq |y(t)|$, for every $t \in S$, implies $\|x\| \geq \|y\|$ and $C(S)$ is a Banach space, is the sup norm

$$\|x\| = \max\,[|x(t)| : t \in S].$$

(b) If $(S, \mathcal{S}, \mu)$ is a measure space, then the L_p norm is the only one, within topological equivalence, for the pth power integrable functions, $p \geq 1$, for which it is a Banach space in which $|x(t)| \geq |y(t)|$ almost everywhere implies $\|x\| \geq \|y\|$.

Vector lattices also play a role in the theory of integration. Let S be a compact Hausdorff space, and let $C(S)$ be the set of continuous real functions on S. Then, $C(S)$ is a Banach lattice with norm

$$\|x\| = \max\,[|x(t)| : t \in S]$$

and the natural order relation for functions. The **Radon measures** on S are, by definition, the continuous linear functionals on $C(S)$. But we have already seen that these are precisely the bounded linear functionals on $C(S)$. The

Radon measures on S may thus be defined either in terms of the topology or the order relation in $C(S)$.

More generally, if S is locally compact and $C(S)$ is the set of continuous functions with compact support on S, then a Radon measure on S is defined as a linear functional on $C(S)$, whose restriction to $C(K)$ is in the dual of the Banach space $C(K)$, for every compact $K \subset S$. There is a locally convex topology on $C(S)$ such that the Radon measures are the dual of $C(S)$ with this topology, but we shall not discuss it here. Moreover, this dual is again the same as the order dual of $C(S)$.

5.12 Köthe Spaces

We consider now a special class of locally convex topological vector lattices, the so-called Köthe spaces.

As a setting for this subject, we consider a totally σ finite measure space $(S, \mathcal{S}, \mu)$, and fix a sequence $\{S_n\}$ of sets in $\mathcal{S}$ such that $\mu(S_n) < \infty$, $n = 1, 2, \ldots$, and $S = \bigcup_{n=1}^{\infty} S_n$.

A real function f on S is said to be **locally summable** if the restriction of f to S_n is summable for every $n = 1, 2, \ldots$. (As usual, we do not discriminate between functions and equivalence classes of functions unless a special need arises.)

Let F be the set of all locally summable functions on S. All of the spaces of interest here will be subspaces of F.

Let $A \subset F$. We associate a subset $A^* \subset F$ with A, where $f \in A^*$ if and only if $f \in F$ and

$$\int_S |f(t)g(t)|\, dt < \infty$$

for every $g \in A$.

PROPOSITION 1. For every $A \subset F$, A^* is a vector lattice.

Proof. $\int_S |f(t)g(t)|\, dt < \infty$ implies $\int_S |af(t)g(t)|\, dt < \infty$ for every real a. If $\int_S |f_1(t)g(t)|\, dt < \infty$ and $\int_S |f_2(t)g(t)|\, dt < \infty$, then

$$\int_S |(f_1(t) + f_2(t))g(t)|\, dt < \infty, \quad \text{and} \quad \int_S |\max(f_1(t), f_2(t))g(t)|\, dt < \infty.$$

PROPOSITION 2. For every $A \subset F$, $A^{***} = A^*$, where $A^{**} = (A^*)^*$ and $A^{***} = (A^{**})^*$.

Proof. Let $f \in A$. Then $\int_S |f(t)g(t)|\, dt < \infty$ for every $g \in A^*$, so that $f \in A^{**}$. Thus $A \subset A^{**}$. In particular, $A^* \subset A^{***}$. Conversely, let $f \in A^{***}$. Then, for every $g \in A^{**}$, $\int_S |f(t)g(t)|\, dt < \infty$. Then, for every $g \in A$, $\int_S |f(t)g(t)|\, dt < \infty$. Hence $f \in A^*$. It follows that $A^* \supset A^{***}$.

A subset $K \subset F$ is called a **Köthe space** if $K = K^{**}$. For a Köthe space K, the space K^* is also a Köthe space and is called the **associate Köthe space** of K. If K^* is the associate of K, then, by proposition 2, K is the associate of K^*.

In particular, F itself is a Köthe space.

PROPOSITION 3. F^* is the vector lattice of bounded measurable functions on S each of which vanishes outside of some S_n.

Proof. Let $f \in F$ and let g be bounded, measurable, and vanishing outside of one of the S_n. There is an M such that $|g(t)| < M$, $t \in S$. Then

$$\int_S |f(t)g(t)|\, dt = \int_{S_n} |f(t)g(t)|\, dt \leq M \int_{S_n} |f(t)|\, dt < \infty.$$

Hence, $g \in F^*$.

Suppose g is unbounded. Then g is unbounded on some S_n (since we are dealing with equivalence classes, this means that every function equivalent to g is also unbounded on S_n). It is not hard to construct a summable function f on S_n such that $\int_S |g(t)f(t)|\, dt = \infty$. Then $g \notin F^*$.

Suppose the support of g is not in any S_n. Then, there is a sequence $\{S_{n_k}\}$, $a_k > 0$, $k = 1, 2, \ldots$, $T_k \subset S_{n_k} \sim S_{n_k - 1}$, $\mu(T_k) > 0$, and $|g(t)| > a_k$ on T_k. Let

$$f_k = \frac{1}{a_k \mu(T_k)} \chi_{T_k}.$$

Then $f = \sum_{k=1}^{\infty} f_k \in F$ and $\int_S |f(t)g(t)|\, dt = \infty$. Thus $g \notin F^*$.

If K and L are Köthe spaces and $K \supset L$, then $K^* \subset L^*$. Thus, F is the largest Köthe space and F^* is the smallest Köthe space.

If $p > 1$, then L_p and L_q, for $1/p + 1/q = 1$, are associated Köthe spaces. This holds since $\int_S |f(t)g(t)|\, dt < \infty$ for every $g \in L_p$ if and only if $f \in L_q$, and these particular spaces do not depend on the choice of $\{S_n\}$.

For every Köthe space K, locally convex topologies, which are compatible with the order relation in K, may be introduced in K in terms of the associate K^* of K. In this connection, we say that a set $A \subset K^*$ is admissible if, for every $f \in K$,

$$\sup \left[\int_S |f(t)g(t)|\, dt : g \in A\right] < \infty.$$

For every admissible $A \subset K^*$ and $f \in K$, let

$$p_A(f) = \sup \left[\int_S |f(t)g(t)|\, dt : g \in A\right].$$

PROPOSITION 4. For every admissible $A \subset K^*$, p_A is a semi-norm on K, compatible with the order relation in K.

The proof is a direct and simple computation.

Now, let $[A]$ be a collection of admissible sets in K^* which covers K^*, i.e., every $g \in K^*$ belongs to at least one $A \in [A]$. The set of semi-norms p_A, $A \in [A]$, then defines a locally convex topology in K.

In order to show this we must see that, for every $f \in K$, $f \neq 0$, there is an $A \in [A]$ such that $p_A(f) > 0$. Clearly, it suffices to show that for every $f \in K$ there is $g \in K^*$ such that $\int_S |f(t)g(t)|\, dt > 0$. Indeed, there is a $g \in F^*$ with this property. For, there is an S_n and $T \subset S_n$ such that $\mu(T) > 0$ and either $f > 0$ or $f < 0$ on T. Let $g = \chi_T$.

We thus have

PROPOSITION 5. For every collection $[A]$ of admissible sets in K^*, which covers K^*, the corresponding set of semi-norms p_A, $A \in [A]$, converts K into a locally convex topological vector lattice.

It is clear that the hypothesis of proposition 5 can be weakened and the conclusion still hold. For, in the proof, we only need that $[A]$ should cover F^*. Moreover, it should be apparent that it is enough that the scalar multiples of the $A \in [A]$ cover F^*.

We note that there is a finest topology in K obtained in this way from K^*. This is the one obtained by letting $[A]$ be the set of all admissible sets in K^*.

There is a special case of particular interest. This is the one where the finest topology is given by a single admissible set $A \subset K^*$. In this case, K is a normed space.

PROPOSITION 6. If K is a normed space, with respect to the finest topology defined in K by means of K^*, then K is a Banach space.

Proof. There is an $A \subset K^*$ such that

$$\|f\| = \sup \left[\int_S |f(t)g(t)|\, dt : g \in A\right].$$

Suppose $\{f_n\}$ is a Cauchy sequence in K. Since this is the finest Köthe topology in K, it follows that for every $g \in K^*$,

$$\lim_{n,m\to\infty} \int_S |f_n(t) - f_m(t)|\, |g(t)|\, dt = 0.$$

But $K^* \supset F^*$. So, by letting g be in turn the characteristic function of each S_m, it follows that $\{f_n\}$ is a Cauchy sequence in $L_1(S_m)$, for every m. By a diagonal process, $\{f_n\}$ has a subsequence, which we also designate by $\{f_n\}$, which converges almost everywhere to a function f. Now,

$$\int_S |f(t)g(t)|\, dt \leq \liminf_{n\to\infty} \int_S |f_n(t)g(t)|\, dt < \infty,$$

since $|fg|$ is the almost everywhere limit of $\{|f_n g|\}$.

Let $\epsilon > 0$. Let $g \in A$, and let n be such that

$$\int_S |f_m(t) - f_n(t)|\,|g(t)|\,dt < \epsilon$$

for every $m > n$. Then,

$$\int_S |f(t) - f_n(t)|\,|g(t)|\,dt \leq \liminf_{m \to \infty} \int_S |f_m(t) - f_n(t)|\,|g(t)|\,dt \leq \epsilon.$$

Since there is an N such that $n, m > N$ implies

$$\int_S |f_n(t) - f_m(t)|\,|g(t)|\,dt < \epsilon$$

for every $g \in A$, it follows that, for $n > N$,

$$\int_S |f(t) - f_n(t)|\,|g(t)|\,dt \leq \epsilon$$

for every $g \in A$. This shows that $\{f_n\}$ converges to f in our space.

In particular, if $K = L_p$, $p > 1$, then the unit ball $A \subset L_q = K^*$ is admissible and defines the L_p norm in K.

Let K be a Köthe space with the finest topology determined by K^*. With this topology, K has a topological dual K'. For every $g \in K^*$, the linear functional,

$$\varphi(f) = \int_S f(t)g(t)\,dt$$

is defined on K and is evidently continuous. Thus K^* has a natural imbedding in K'. In many cases, $K^* = K'$. In the particular case, $K = F^*$, K^* is generally a proper subset of K'.

We close this section with brief mention of two special cases.

(a) Let S be the interval $[0, 1]$ and let μ be Lebesgue measure. Then $F = L_1[0, 1]$ and F^* is the set of bounded measurable functions. In this case, the Köthe spaces, which are Banach spaces with the finest topologies as given by their associated spaces, may be defined as follows:

Let $A \subset L_1[0, 1]$ be a bounded set, in the L_1 norm, such that $1 \in A$ and, if $f \in A$ and $|g| \leq |f|$ then $g \in A$. Let

$$K = \left\{f : \sup \left[\int_0^1 |f(t)g(t)| : g \in A\right] < \infty\right\}.$$

It is not hard to see that K is a Köthe space,

$$K^* = \bigcup [cA : c \in R],$$

and that K, with the finest topology defined by K^*, is a Banach space with norm

$$\|f\| = \sup \left[\int_0^1 |f(t)g(t)|\,dt : g \in A\right].$$

We leave the details to the reader.

(b) Let S be the set of natural numbers, and let μ be the measure which is

1 for each number. The set F of locally summable functions consists of all sequences of real numbers, and F^* consists of those sequences which vanish except for a finite number of terms.

EXERCISES

1.1 Prove that a collection $\mathcal{B}$ of open sets is a base for a topological space $(X, \mathcal{G})$ if and only if for every $x \in X$ and $G \in \mathcal{G}$, such that $x \in G$, there is an $H \in \mathcal{B}$ such that $x \in H$ and $H \subset G$.

1.2 Let X be the set of pairs of real numbers, ordered by means of $(x, y) > (x', y')$ if either $x > x'$ or $x = x'$ and $y > y'$. Then X is a totally ordered set. Let the open intervals in X be a base for the topology in X. Show that the space obtained is metrizable.

1.3 Let X be the set of all real functions on R which vanish except at a well-ordered subset of R. For $f, g \in X$, let $f > g$ if $f(x) > g(x)$ at the first real number x where they are not equal. Then X is a totally ordered set. Show that X is not metrizable.

1.4 Let $X = R$. Let G be open in X if it is an open set in the usual topology, minus a countable set. Let $\mathcal{G} = [G]$. Show that $(X, \mathcal{G})$ is a Hausdorff topological space which is not metrizable.

1.5 If $S \subset R$ is measurable, then x is said to be a point of density of S if, for every $\epsilon > 0$, there is a $\delta > 0$ such that if I is an interval, $x \in I$, and $m(I) < \delta$, then

$$\frac{m(S \cap I)}{m(I)} > 1 - \epsilon.$$

Let G be open if G is measurable and every point of G is a point of density of G. Let $\mathcal{G} = [G]$. Show that $(S, \mathcal{G})$ is a Hausdorff topological space which is not metrizable.

1.6 Show that if $(X, \mathcal{G})$ has a countable base then every base has a countable subset which is also a base.

2.1 If X is a compact space and $\{f_n\}$ is a sequence of decreasing functions which converges to zero on X, show that the convergence is uniform.

2.2 A topological space is said to be Lindelöf if every covering of the space by a set of open sets has a countable subcover. Show that the space of exercise 1.5 is not a Lindelöf space.

2.3 For every $\alpha \in A$, let X_α be a 2 point space all of whose subsets are open. For what cardinal numbers for A is

$$\prod[X_\alpha : \alpha \in A]\cdot$$

metrizable? Prove.

2.4 Show that if X is a Hausdorff space, and $S \subset X$ is compact with the induced topology, then S is a closed subset of X. Conversely, if S is closed in a compact X, then it is compact with the induced topology.

2.5 If $S \subset X$, where X is a topological space, show that $x \in \bar{S}$ if and only if for every open set G such that $x \in G$, the set $G \cap S$ is non-empty.

2.6 Show that the closed unit ball B in $L_2[0, 1]$ is weakly compact in the sense that if $\{f_n\} \subset B$ is such that $\left\{\int_0^1 f_n g\right\}$ is Cauchy for every $g \in L_2$, then there is an $f \in B$ with $\int_0^1 fg = \lim_{n\to\infty} \int_0^1 f_n g$, for every $g \in L_2$.

3.1 Show that a topological vector space, satisfying (d), is a Hausdorff space.

3.2 State necessary and sufficient conditions for a collection of sets in a vector space X to be a base at 0 for a topology in X for which X is a topological vector space.

3.3 Show that if X is a topological vector space, for every $a \in R$, $a \neq 0$, the mapping

$$f_a : X \to X$$

defined by $f_a(x) = f(ax)$, $x \in X$, is a homeomorphism.

3.4 Define the product of a set of topological vector spaces and prove that it is a topological vector space.

3.5 For what cardinalities are products of metrizable topological vector spaces also metrizable?

4.1 If X is a vector space and X^* is its algebraic dual, show that $\dim X^* \geq \dim X$, equality holding if and only if $\dim X < \infty$.

4.2 If X is a finite dimensional vector space, show that the topology on X for which X is a topological vector space is unique.

5.1 Give details of the proof that the equivalence classes of measurable functions on [0, 1] are a metric space with metric

$$d(f, g) = \int_0^1 \frac{|f(t) - g(t)|}{1 + |f(t) - g(t)|}\, dt.$$

5.2 Give details of the proof that the above metric d and the metric δ defined in the text are equivalent.

5.3 Show that the space $L_p[0, 1]$ has a minimal total set for every $p \geq 1$. Hint: Try the Haar functions.

5.4 If $\{f_1, f_2, \ldots\}$ is a complete orthonormal set for $L_2[a, b]$, show that for every n there is a bounded g such that $\{gf_{n+1}, gf_{n+2}, \ldots\}$ is also total.

6.1 Show that the product of locally convex topological vector spaces is locally convex and define the semi-norms in the product in terms of those in the factors.

6.2 If X is a vector space and t_1 and t_2 are topologies, for each of which X is a locally convex topological vector space with the same topological dual, show that the closed convex sets in X are the same for the two topologies.

6.3 Let X be an infinite dimensional locally convex topological vector space whose topological dual is the entire algebraic dual of X. Show that X is not a Banach space.

6.4 Describe the dual of the product of topological vector spaces.

6.5 Let $\{X_n\}$ be an increasing sequence of Banach spaces such that, for every n, the identity mapping

$$i_n : X_n \to X_{n+1}$$

is continuous. Show that there is a finest topology in $X = \bigcup_{n=1}^{\infty} X_n$ such that the identity mappings

$$j_n : X_n \to X$$

are all continuous.

6.6 Define a system of semi-norms for the topology of the space in exercise 6.5.

6.7 Prove that the space defined in exercise 6.5 is not metrizable.

6.8 In a vector space X a **manifold** is a set $M = x + Y$, where Y is a subspace of X. M is a **supporting manifold** of a convex $S \subset X$ if $S \cap M \neq 0$, and for every $x \in S \cap M$, every open segment through x which is in S is also in M. A supporting manifold consisting of one point is called an **extreme point.** A supporting manifold is called a **supporting hyperplane** if, together with a point in X, it spans X. Show that a compact convex S in a locally convex topological vector space has a supporting hyperplane.

6.9 If X is a locally convex topological vector space, and $S \subset X$ is compact and convex, show that every supporting hyperplane of S contains at least one extreme point of S.

6.10 If X is a locally convex topological vector space, and $S \subset X$ is compact and convex, show that S is the smallest closed convex set containing its extreme points.

6.11 Give an example of a closed bounded convex set, in a locally convex topological vector space, which has no extreme points.

6.12 Let X be a locally convex topological vector space. A closed convex set $A \subset X$ is called a **barrel** if for every $x \in X$, $ax \in A$ for some $a \neq 0$ and if $x \in A$, $|a| \leq 1$ implies $ax \in A$. X is called a **barrel space** if every barrel in X contains an open set containing 0. Show that a vector space X admits at most one barreled topology for which a given subspace $Y \subset X^*$ is the dual of X.

6.13 Show that a locally convex topological vector space which is locally compact is finite dimensional.

7.1 Show that every Fréchet space is barreled.

7.2 Let X be a Fréchet space which is not a Banach space. Let X' be the dual of X. For every bounded set $A \subset X$, we define a semi-norm

$$p_A(x') = \sup\,[|x'(x)| : x \in A].$$

Show that, with this topology, X' is not a Fréchet space.

7.3 Prove that the space defined in exercise 6.5 is barreled.

7.4 Generalize the result of exercise 7.3.

7.5 A barreled space X is called a Montel space if every closed bounded set in X is compact. Show that the only Banach spaces which are Montel spaces are finite dimensional.

7.6 Show that the space of entire functions is a Montel space which is a Fréchet space.

7.7 Show that if X is a Montel space, then if semi-norms are defined in X' as in exercise 7.2, X' is also a Montel space.

7.8 Show that the results of exercises 7.6 and 7.7 assure the existence of Montel spaces which are not Fréchet spaces.

7.9 Let D be the vector space of real functions of a real variable each of which has compact support and is infinitely differentiable. Define a locally convex topology in D so that a sequence $\{f_n\}$ converges to zero in the topology if and only if $\{f_n\}$ has a common compact support, converges uniformly to zero and, for every $k = 1, 2, \ldots$, the sequence $\{f_n^{(k)}\}$ of kth derivatives converges uniformly to zero.

7.10 Show that the space D of exercise 7.9 is not metrizable but is barreled.

7.11 A continuous linear functional on D is called a distribution. Show that if f is locally summable (i.e., for every compact $K \subset R$, $f\chi_K$ is summable), then the functional $\varphi(g)$ is a distribution, where

$$\varphi(g) = \int fg.$$

7.12 For every distribution φ, the linear functional φ' defined by $\varphi'(f) = -\varphi(f')$ is called the derivative of φ. Justify the name derivative and show that φ' is a distribution.

7.13 A distribution φ is said to have compact support if there is an n such that for every $f \in D$, which vanishes on $[-n, n]$, $\varphi(f) = 0$. If φ has compact support, show there is an integer k and a distribution ψ, determined by a summable function as in exercise 7.11, such that φ is the kth derivative of ψ.

7.14 Let $\{\varphi_n\}$ be a sequence of distributions such that, for every $f \in D$, $\{\varphi_n(f)\}$ converges. Let $\varphi(f)$ be the limit functional. Show that φ is a distribition.

7.15 If μ is a regular measure on R, finite on compact sets (and such that for every bounded measurable E and $\epsilon > 0$ there are compact K and open G such that $\mu(G) - \epsilon < \mu(E) < \mu(K) + \epsilon$), show that

$$\varphi(f) = \int f\,d\mu$$

is a distribution. Characterize the functions whose derivatives are measures of this sort.

8.1 Give details of the proof that the space of all sequences is an FK space, with semi-norms $p_n(x) = |a_n(x)|$, $n = 1, 2, \ldots$, where $x = (a_1(x), a_2(x), \ldots)$.

8.2 Give details of the proof that if X and Y are Fréchet spaces and f is a continuous linear mapping of X into Y, then $f(X)$ is either all of Y or is of the first category in Y.

8.3 Prove that if X_n, $n = 1, 2, \ldots$, is a sequence of FK spaces, then $X = \bigcap_{n=1}^{\infty} X_n$ is an FK space.

9.1 Show that the convergence field of a regular matrix does not contain all bounded sequences.

9.2 Show that every sequence is in the convergence field of some regular matrix.

9.3 A matrix A is said to be row finite if every row contains only finitely many non-zero terms. If A is row finite, describe the semi-norms for $C(A)$.

9.4 Show that, for every $\alpha > 0$, the (C, α) method is regular.

9.5 Show that if $\alpha > \beta$, then (C, α) is strictly stronger than (C, β).

10.1 An ordered group G is a group and an ordered set such that $x \geq y$ implies $x + z \geq y + z$ for every $z \in G$. $x \in G$ is called **positive** if $x \geq 0$, the group identity. G is called **directed** if, for every $x, y \in G$, there is $z \in G$ such that $z \geq x$, $z \geq y$. Show that G is directed if and only if every $x \in G$ is the difference between positive elements.

10.2 An ordered group G is called **lattice ordered** if $x, y \in G$ implies $\sup (x, y)$ exists. Show how every directed ordered group may be imbedded in a lattice ordered group.

10.3 Show that in a lattice ordered group G, $x, y, z \in G$ implies

$$\sup [x, \inf (y, z)] = \inf [\sup (x, y), \sup (x, z)].$$

10.4 A lattice ordered group G is archimedean if $x > 0$ and $ny \leq x$, for all n, implies $y \leq 0$. Show that every archimedean lattice ordered group is abelian.

10.5 A lattice ordered group is said to be complete if every set with an upper bound has a least upper bound. Show that a lattice ordered group can be imbedded in a complete lattice ordered group if and only if it is archimedean.

10.6 In a lattice ordered group G, show that $x, y, z \in G$ implies

$$\sup (x + y, x + z) = x + \sup (y, z).$$

10.7 Show that the ordered group of differentiable real functions is not a lattice ordered group.

10.8 Show that the lattice ordered group of continuous functions on $[0, 1]$ is not complete. Describe its completion.

11.1 Describe a locally convex topology, on the continuous functions with compact support on a locally compact space, so that the dual is the set of Radon measures.

11.2 Show that this dual is the same as the order dual.

12.1 For the real line R, with Lebesgue measure, show that the set of locally summable functions depends on the choice of $\{S_n\}$.

12.2 If g is unbounded (not equivalent to any bounded function) and vanishes outside a set of finite measure, show that there is a summable f such that $\int fg \, d\mu = \infty$.

12.3 Show that the spaces $L_p, p \geq 1$, are independent of the decomposition $\{S_n\}$.

12.4 For every admissible $A \subset K^*$, show that the functional p_A, defined in the text, is a semi-norm compatible with the order relation in K.

12.5 Show that the spaces K of example (a) are Banach spaces.

12.6 If a space K of exercise 12.5 is such that $f \in K$ and $\lim_{n \to \infty} \mu(T_n) = 0$ implies $\{f\chi_{T_n}\}$ converges to 0 in K, show that K^* is the topological dual of K.

12.7 If a space K of example (*b*) is a Banach space, with the finest topology determined in K by K^*, and if $\{a_n\} \in K$ implies $\{a_1, \ldots, a_m, 0, \ldots\}$, $m = 1, 2, \ldots$, converges in K to $\{a_n\}$, show that K^* is the topological dual of K.

12.8 Let φ be right continuous for $t \geq 0$, positive, non-decreasing, with $\varphi(0) = 0$ and $\lim_{t \to \infty} \varphi(t) = \infty$, and let $\Phi(t) = \int_0^t \varphi(u)\, du$. Let ψ be defined by

$$\psi(t) = \sup\, [u : \varphi(u) \leq t],$$

and $\Psi(t) = \int_0^t \psi(u)\, du$. These are called **complementary functions.** Show that, for every $u, v \geq 0$,

$$uv \leq \Phi(u) + \Psi(v).$$

12.9 If there is an $M > 0$ and a t_0 such that $t \geq t_0$ implies $\Phi(2t) \leq M\Phi(t)$, show that the set of functions for which $\int \Phi(|f(t)|)\, dt < \infty$ is a Köthe space whose associate is the set of functions for which $\int \Psi(|f(t)|)\, dt < \infty$.

12.10 Show that in a space of exercise 12.9 a set is compact if and only if it is closed, bounded, and the integral means converge uniformly on the set (in the space).

CHAPTER 6

BANACH ALGEBRAS

6.1 Definition and Examples

Let X be a Banach space in which the standard vector space operations $(x, y) \to x + y$ and $(a, x) \to ax$ are defined and also a multiplication operation $(x, y) \to xy$.

The set X, together with the operations $(x, y) \to x + y$ and $(x, y) \to xy$, is assumed to be a ring and the norm is assumed to satisfy the Banach space conditions and

$$\|xy\| \leq \|x\| \, \|y\|.$$

The ring X need not have a multiplicative identity, but if it does have an identity e, then $\|e\| \geq 1$, since $\|x\| = \|ex\| \leq \|e\| \, \|x\|$.

It is easy to show that there is then an equivalent norm for which $\|e\| = 1$. (This means that the identity mapping of X with one norm onto X with the other norm is a homeomorphism.)

Such a system, with or without an identity, is called a **Banach algebra.**

We give some examples of Banach algebras.

(a) Let X be a compact Hausdorff space, and let $C(X)$ be the set of continuous, complex valued, functions on X. Then $C(X)$ is closed with respect to point-wise multiplication of functions, and multiplication of functions by scalars. With the usual norm

$$\|x\| = \max\,[|x(t)| : t \in X],$$

$C(X)$ is a Banach space. Moreover, a very easy computation shows that $\|xy\| \leq \|x\| \, \|y\|$ so that $C(X)$ is a Banach algebra. It is a commutative ring, and has an identity given by the function which is identically 1.

(b) Now, let X be a locally compact Hausdorff space, and let $C(X)$ be the set of complex valued continuous functions on X which vanish at infinity. This means that, for every $\epsilon > 0$, there is a compact set $K \subset X$ such that $|x(t)| < \epsilon$ for every $t \notin K$. With $\|x\| = \max\,[|x(t)| : t \in X]$, $C(X)$ is again a Banach algebra. It is commutative, but does not have an identity if X is not compact.

(c) Let X be the set of equivalence classes of summable complex valued functions on the real line R, i.e., $x \in X$ if x is measurable and $\int_{-\infty}^{\infty} |x(t)|\,dt < \infty$.

Then X is a vector space with the usual point-wise operations. However, the point-wise product of summable functions need not be summable. A natural definition of product is the convolution of functions. The convolution $x * y$ of summable functions x and y is summable, where the convolution is defined by

$$(x * y)(t) = \int_{-\infty}^{\infty} x(t-u)y(u)\,du.$$

Indeed,

$$\int_{-\infty}^{\infty} \left| \int_{-\infty}^{\infty} x(t-u)y(u)\,du \right| dt \le \int_{-\infty}^{\infty}\int_{-\infty}^{\infty} |x(t-u)y(u)|\,du\,dt$$

$$= \int_{-\infty}^{\infty} |x(u)|\,du \int_{-\infty}^{\infty} |y(u)|\,du.$$

Hence, not only is $x * y \in X$, but $\|x * y\| \le \|x\|\,\|y\|$, so that we have a Banach algebra. It is designated as $L_1(R)$.

$L_1(R)$ is commutative, but has no identity. However, there are sequences of functions in $L_1(R)$ which act as a substitute for the identity for certain purposes. As an example, let

$$e_n(t) = \begin{cases} n & \text{if } |t| < \dfrac{1}{2n} \\[2ex] 0 & \text{if } |t| \ge \dfrac{1}{2n}, \end{cases}$$

$n = 1, 2, \ldots.$

Then, for any $x \in L_1(R)$,

$$(x * e_n)(t) = n \int_{-1/2n}^{1/2n} x(t+u)\,du.$$

Then

$$\|x - x * e_n\| = \int_{-\infty}^{\infty} \left| x(t) - n \int_{-1/2n}^{1/2n} x(t+u)\,du \right| dt.$$

But we have seen in Chapter 3 that the limit of this integral is zero as n approaches infinity. It follows that for every $x \in L_1(R)$,

$$\lim_{n\to\infty} \|x - x * e_n\| = 0.$$

In general, an **approximate identity** for $L_1(R)$ is a sequence $\{e_n\}$ of non-negative functions which converges uniformly to zero on any set which does not have 0 as a limit point and $\lim_{n\to\infty} \int_A e_n = 1$, for every open set A containing 0.

(d) Let H be a complex Hilbert space, and let X be the set of bounded linear operators

$$T : H \to H.$$

We define the operations

$$T_1 + T_2 \text{ by } (T_1 + T_2)(x) = T_1(x) + T_2(x),\ aT \text{ by } (aT)(x) = aT(x),$$

and

$$T_1T_2 \text{ by } (T_1T_2)(x) = T_1(T_2(x)),$$

for every $x \in H$.

With $\|T\|$ defined in the usual way for bounded operators, X is easily seen to be a Banach algebra. X has an identity, but is not commutative.

(e) The subset Y, which consists of the completely continuous operators of H into H is easily seen to be a subalgebra of X.

(f) Example (d) holds for any Banach space.

6.2 Adjunction of Identity

If X is a Banach algebra without identity, it is an easy matter to extend X so that the extended Banach algebra has an identity. Let C be the set of complex numbers, and let $\tilde{X}$ be the set of all pairs (x, a), $x \in A$, $a \in C$. Define operations in $\tilde{X}$ as follows:

$$\begin{aligned}(x, a) + (y, b) &= (x + y, a + b),\\ c(x, a) &= (cx, ca),\\ (x, a)(y, b) &= (xy + ay + bx, ab).\end{aligned}$$

Then it is easy to check that $\tilde{X}$ is an algebra. Define a norm in $\tilde{X}$ by

$$\|(x, a)\| = \|x\| + |a|.$$

It is easy to check that this has the properties of a norm and that

$$\|(x, a)(y, b)\| \leq \|(x, a)\| \ \|(y, b)\|.$$

It is also easy to verify that a sequence $\{(x_n, a_n)\}$ in X is Cauchy if and only if $\{x_n\}$ is Cauchy in X and $\{a_n\}$ is Cauchy in C. If $x = \lim_{n\to\infty} x_n$ in X and $a = \lim_{n\to\infty} a_n$ in C, then $(x, a) = \lim_{n\to\infty} (x_n, a_n)$ in $\tilde{X}$. We thus know that $\tilde{X}$ is a Banach algebra. The mapping

$$\varphi : X \to \tilde{X},$$

for which $\varphi(x) = (x, 0)$, preserves the algebraic operations and the norm.

Since $(x, a)(0, 1) = (x, a)$, the element $(0, 1)$ is the identity in $\tilde{X}$.

Although this imbedding can always be made, it can usually be circumvented by other devices, as will be seen in the sequel.

6.3 Haar Measure

An analogue to $L_1(R)$ may be defined for every locally compact topological group G. For this purpose, we have to know what is meant by the Haar measure on G.

A **topological group** is a set G which is

(a) a group,
(b) a Hausdorff topological space, and
(c) the mappings $(x, y) \to xy$ of $G \times G$ onto G and $x \to x^{-1}$ of G onto G are continuous.

In particular, every topological vector space is a topological group. Moreover, just as in the special case of a topological vector space, a base for the open sets may be given as the left and right translations of a base for the open sets containing the identity e.

Recall that a topological space X is said to be compact if for every collection $U = [U_\alpha]$ of open sets such that $X = \bigcup U_\alpha$, there is a finite subcollection such that $X = \bigcup_{k=1}^{n} U_{\alpha_k}$. X is said to be **locally compact** if every $x \in X$ is contained in an open set whose closure is compact.

By a **left invariant measure** μ on a locally compact topological group G, we mean a measure on the σ algebra $\mathcal{S}$ generated by the compact sets, such that

(a) if E is compact, then $\mu(E) < \infty$; if E is open, then $\mu(E) > 0$,
(b) if E is measurable and $x \in G$, then xE is measurable and $\mu(xE) = \mu(E)$.

A right invariant measure is defined similarly. A left invariant measure exists and is unique, except for a constant factor. We shall indicate a proof of existence. The method is to compare the sets in $\mathcal{S}$ with a given compact set F with non-empty interior.

Let E be a compact set. For every open set U, containing e, we shall compare the "U size" of E with the "U size" of F. The "U size" of a set $A \subset G$ will be the smallest number of left translations of U whose union contains A. We shall designate the "U size" of A by

$$n(A, U).$$

We then let

$$\mu_U(E) = \frac{n(E, U)}{n(F, U)}.$$

Clearly, $\mu_U(E)$ is a finite positive number. Indeed, if we let $n(E, F)$ be the smallest number of left translations of F such that E is in their union, it is easy to see that $n(E, F)$ is finite. Moreover,

$$\text{(a)} \qquad 0 \le \mu_U(E) \le n(E, F)$$

follows from the easy to prove relation

$$n(E, F)n(F, U) \ge n(E, U).$$

The set function $\mu_U(E)$ has the further properties:

(b) $$\mu_U(xE) = \mu_U(E),$$

(c) $$\mu_U(E_1 \cup E_2) \leq \mu_U(E_1) + \mu_U(E_2),$$

and

$$\mu_U(E_1) \geq \mu_U(E_2) \quad \text{if} \quad E_1 \supset E_2.$$

Finally, suppose that $E_1U^{-1} \cap E_2U^{-1} = 0$. Then E_1 may be covered by left translations $x_1U, \ldots, x_nU$ of U and E_2 may be covered by left translations $y_1U, \ldots, y_mU$ of U such that

$$\left(\bigcup_{i=1}^{n} x_iU\right) \cap \left(\bigcup_{j=1}^{m} y_jU\right) = 0,$$

and we have

(d) $$\mu_U(E_1 \cup E_2) = \mu_U(E_1) + \mu_U(E_2).$$

The desired measure μ will be a "generalized limit of μ_U as U tends to zero." We proceed to indicate how this is defined.

Let X be the topological product of the closed intervals $[0, n(E, F)]$ as E varies over all compact sets. Then X is a compact space. For every open set U, containing e, the point ξ_U, whose coordinate at place E is $\mu_U(E)$, is in X. This is so because $0 \leq \mu_U(E) \leq n(E, F)$.

For every U, consider the set

$$S(U) = [\xi_V : V \text{ open}, e \in V, V \subset U].$$

The sets $\overline{S(U)}$ are closed, and it follows from the fact that the intersection of every finite set of open neighborhoods of e is an open neighborhood of e that the sets $\overline{S(U)}$ have the finite intersection property. Since X is compact, we have

$$\bigcup[\overline{S(U)} : U] \neq 0.$$

Let μ be any element of X contained in this intersection. Then μ is a set function defined on the compact subsets of G. It is a relatively easy matter to show that μ may be extended to $\mathcal{S}$ so that it is a left invariant measure, and we leave the details to the reader.

A discussion of uniqueness is not vital to our needs and will be omitted.

For the special case where G is abelian, every left invariant measure is also right invariant. We shall consider only abelian groups, and shall speak of invariant measure as **Haar measure**. Since the groups are abelian, we use the additive notation for the group operation.

Now, let $L_1(G)$ be the set of summable functions, on a locally compact abelian topological group G, with respect to a Haar measure on G. Then, just as for $L_1(R)$, $L_1(G)$ is a Banach algebra with multiplication given by

$$(f * g)(x) = \int f(x - y)g(y)\, dy$$

Also, just as for $L_1(R)$, it follows that

$$\|f * g\| \leq \|f\| \, \|g\|.$$

Moreover, it follows from the fact that G is abelian that $f * g = g * f$. We omit the proof, but give the proof of

PROPOSITION 1. $L_1(G)$ has an identity if and only if G is discrete (i.e., every point is an open set).

Proof. Suppose G is discrete. Then a Haar measure in G is one for which the identity e has measure 1. The function E, which is 1 at e and 0 everywhere else, is in $L_1(G)$. For every $f \in L_1(G)$, and every $x \in G$,

$$\begin{aligned}(E * f)(x) &= \int_G f(x - y)E(y)\, dy \\ &= \sum_{y \in G} f(x - y)E(y) = f(x).\end{aligned}$$

Hence E is the identity for $L_1(G)$.

Suppose, conversely, that $L_1(G)$ has an identity E. Let μ be a Haar measure for G. We show that there is a $k > 0$ such that, for every neighborhood U of e,

$$\mu(U) \geq k.$$

Suppose otherwise. Then, by absolute continuity of the integral, there is a neighborhood U of e such that

$$\int_U |E(x)|\, dx < \frac{1}{2}.$$

Let V be a neighborhood of e such that $V = -V$ and $V + V \subset U$. Then, for every $x \in G$,

$$|\chi_V(x)| = |(E * \chi_V)(x)| = \left| \int_G \chi_V(x - y)E(y)\, dy \right| \leq \int_U |E(x)|\, dx < \frac{1}{2},$$

where χ_V is the characteristic function of V. But this contradicts the fact that $\chi_V(x) = 1$ for every $x \in V$. This shows that there is a $k > 0$ such that every open set has measure greater than k.

Suppose there is an open set U in G which has finite measure but contains infinitely many points. Let n be a positive integer and let $x_1, x_2, \ldots, x_n$ be n distinct points in U. It can be shown that there are disjoint open sets $U_1, \ldots, U_n$, with $x_i \in U_i \subset U$, $i = 1, \ldots, n$. Now,

$$\mu(U) \geq \sum_{i=1}^{n} \mu(U_i) \geq nk.$$

Since $\mu(U) < \infty$, this is impossible. Hence every open set in G, of finite measure, must be finite. In other words, G is discrete.

6.4 Commutative Banach Algebras, Maximal Ideals

We shall now consider arbitrary commutative Banach algebras. We first suppose that the Banach algebra has an identity e.

PROPOSITION 1. If X is a commutative Banach algebra, with an identity e, then $\|e - x\| < 1$ implies x has an inverse.

Proof. Observe that, for every n,

$$\|e - x\|^n \geq \|(e - x)^n\|.$$

Then, the series

$$e + (e - x) + (e - x)^2 + \dots$$

is such that, for every n and p,

$$\|(e - x)^{n+1} + \dots + (e - x)^{n+p}\| \leq \|e - x\|^{n+1} + \dots + \|e - x\|^{n+p}.$$

This implies that the series converges to a point $y \in X$. But

$$xy = [e - (e - x)][e + (e - x) + (e - x)^2 + \dots]$$

and it is easy to see that the product on the right converges to e. Thus, y is the inverse of x.

COROLLARY. The set of points in X for which the inverse exists is open.

Proof. Suppose x has an inverse x^{-1}. Since the norm is continuous on X and the algebraic operations are continuous, there is a neighborhood U of e such that, for every $z \in xU$,

$$\|e - zx^{-1}\| < 1.$$

It follows, by proposition 1, that zx^{-1} has an inverse. But then z has an inverse.

It can be shown easily that the mapping $x \to x^{-1}$ on the open set for which the inverse exists is continuous, but we leave the proof to the reader.

We now consider X merely as an algebra. Certain subsets of X, called ideals, are of special importance. A subset $I \subset X$ is called an **ideal** if

(a) $$I \neq X,$$

(b) $$x, y \in I \text{ and } a, b \in C \text{ implies } ax + by \in I,$$

(c) $$x \in I,\ y \in X \text{ implies } xy \in I.$$

We consider the following examples of ideals.

(i) Let $C(S)$ be the algebra of continuous functions on a compact Hausdorff space S. Let $A \subset S$ be any non-empty subset of S. Let I be the set of $f \in C(S)$ which vanish on A. Then it is trivial to verify that I satisfies (a), (b), and (c) so that I is an ideal.

(ii) We consider the algebra $L_1(R)$. We shall assume some properties of the Fourier transform of a function $f \in L_1(R)$. The Fourier transform of a function $f \in L_1(R)$ is the function Tf defined by

$$Tf(t) = \frac{1}{2\pi} \int_{-\infty}^{\infty} f(x)e^{-itx}\, dx.$$

The properties we need are that for every $f, g \in L_1(R)$ and $a, b \in C$,

$$T(af + bg) = aTf + bTg$$

and

$$T(f * g) = Tf \cdot Tg,$$

where the product on the right is the point-wise product. We should note that this last fact is the reason for the naturalness of convolution as the multiplication operation in $L_1(G)$.

Let $A \subset R$ be any non-empty set. The set $I \subset L_1(R)$ of functions f for which Tf vanishes on A is then easily seen to be an ideal in $L_1(R)$.

(iii) Every algebra X has a trivial ideal consisting of the one element 0. This is called the null ideal. Suppose X has an identity e. Then, if X has no ideals besides the null ideal, it follows that X is a field. In order to prove this, let $x \in X$, $x \neq 0$. Consider the set

$$I = [xy : y \in X].$$

This set clearly has properties (b) and (c) of an ideal and contains elements different from 0; in particular, x itself. Since X contains no proper ideals (i.e., ideals different from the null ideal), it follows that $I = X$. Hence, there must be $y \in X$ with $xy = e$, so that x has an inverse and X is a field.

In an algebra with identity, e cannot be in any ideal. If I were an ideal and $e \in I$, then $x = ex \in I$ for every $x \in X$ so that $X = I$. But this is impossible.

A **maximal ideal** I is an ideal which is not a proper subset of any ideal in X.

PROPOSITION 2. If X is a commutative Banach algebra with identity, every ideal I in X is contained in a maximal ideal M in X.

Proof. Consider the ideals in X containing I to be ordered by inclusion. Let $\{I_\alpha\}$ be a maximal chain in this ordered set. We need only show that

$$M = \bigcup I_\alpha$$

is an ideal. For this, we first observe that since, for every α, $e \notin I_\alpha$, we have $e \notin M$. Hence, $M \neq X$.

Let $x, y \in M$ and $a, b \in C$. There are α, β such that $x \in I_\alpha$, $y \in I_\beta$. If $I_\beta \supset I_\alpha$, then $ax + by \in I_\beta \subset M$. Finally, let $x \in M$, $y \in X$. There is an α such that $x \in I_\alpha$. Then $xy \in I_\alpha \subset M$. Hence M is an ideal. If M were not a maximal ideal, then $\{I_\alpha\}$ would not be a maximal chain of ideals.

PROPOSITION 3. If X is a commutative Banach algebra with identity, then for every ideal $I \subset X$, the closure $\bar{I}$ of I is also an ideal.

Proof. We first note that I contains no element which has an inverse. For, if $x \in I$ and x^{-1} exists, then $xx^{-1} = e \in I$. Since the set of elements of X which have inverses contains e as an interior point, $e \notin \bar{I}$. Hence $\bar{I} \neq X$.

Let $x, y \in \bar{I}$ and $a, b \in C$. It follows readily from the continuity of the algebraic operations that $ax + by \in \bar{I}$. Moreover, it follows in similar fashion that if $x \in \bar{I}$ and $y \in X$, then $xy \in \bar{I}$. Hence $\bar{I}$ is an ideal.

PROPOSITION 4. x^{-1} exists if and only if x is in no ideal.

Proof. We have already observed that if x^{-1} exists, then x is in no ideal. For the converse, suppose x^{-1} does not exist. Let

$$I = [xy : y \in X].$$

Then $e \notin I$ so that $I \neq X$. It is easily verified that I has the other properties of an ideal.

As always in this section, let X be a commutative Banach algebra with identity. Let I be a closed ideal in X. We consider the quotient algebra X/I. We shall define a norm in X/I for which it turns out to be a Banach algebra.

The elements $\xi \in X/I$ are the subsets of X of the form $x + I$, $x \in X$. Then

$$\xi + \eta = [x + y : x \in \xi, y \in \eta],$$
$$a\xi = [ax : x \in \xi],$$
$$\xi\eta = [xy : x \in \xi, y \in \eta].$$

It is easy to see that these operations are clearly defined, and that X/I is an algebra with additive identity I and multiplicative identity E, where E is the element of X/I to which e belongs. We define a norm in X/I by

$$\|\xi\| = \inf [\|x\| : x \in \xi].$$

By the continuity of the group operation, every ξ is a closed subset of X. Suppose $\|\xi\| = 0$. Then there is a sequence $\{x_n\} \subset \xi$ with $\lim_{n\to\infty} \|x_n\| = 0$. Since ξ is closed, $0 \in \xi$, so that $\xi = I$. Thus $\|\xi\| = 0$ if and only if $\xi = I$.

For every $a \in C$ and $\xi \in X/I$,

$$\begin{aligned} \|a\xi\| &= \inf [\|ax\| : x \in \xi] \\ &= |a| \inf [\|x\| : x \in \xi] = |a| \cdot \|\xi\|. \end{aligned}$$

For every $\xi, \eta \in X/I$,

$$\begin{aligned}\|\xi + \eta\| &= \inf\,[\|x + y\| : x \in \xi, y \in \eta] \\ &\le \inf\,[\|x\| + \|y\| : x \in \xi, y \in \eta] \\ &= \inf\,[\|x\| : x \in \xi] + \inf\,[\|y\| : y \in \eta] \\ &= \|\xi\| + \|\eta\|.\end{aligned}$$

For every $\xi, \eta \in X/I$,

$$\begin{aligned}\|\xi\eta\| &= \inf\,[\|xy\| : x \in \xi, y \in \eta] \\ &\le \inf\,[\|x\|\,\|y\| : x \in \xi, y \in \eta] \\ &= \inf\,[\|x\| : x \in \xi] \cdot \inf\,[\|y\| : y \in \eta] \\ &= \|\xi\|\,\|\eta\|.\end{aligned}$$

In order to show that X/I is a Banach algebra, it remains only to prove completeness.

Let $\{\xi_n\}$ be a Cauchy sequence in X/I. Let $x_1 \in \xi_1$; there is $x_2 \in \xi_2$ such that $\|x_2 - x_1\| < 2\,\|\xi_2 - \xi_1\|$. There is $x_3 \in \xi_3$ such that $\|x_3 - x_2\| < 2\,\|\xi_3 - \xi_2\|$. In this way, for every n there is $x_n \in \xi_n$ such that $\|x_n - x_{n-1}\| < 2\,\|\xi_n - \xi_{n-1}\|$. The sequence $\{x_n\}$ is evidently a Cauchy sequence. Let $x = \lim_{n\to\infty} x_n$ and let ξ be the element of X/I to which x belongs. Since the mapping of X onto X/I which takes each x into the ξ to which it belongs is continuous, it follows that $\xi = \lim_{n\to\infty} \xi_n$. We have thus proved

PROPOSITION 5. If X is a commutative Banach algebra, and I is a closed ideal in X, then X/I is a Banach algebra with norm

$$\|\xi\| = \min\,[\|x\| : x \in \xi].$$

We need the fact from algebra that if X is a commutative ring with identity, and M is a maximal ideal in X, then X/M is a field. This is true because if there were $\xi \in X/M$, $\xi \neq M$, which had no inverse, then the multiples of ξ would be an ideal I in X/M. The set $\bigcup[\xi : \xi \in I]$ is then easily seen to be an ideal in X which has M as a proper subideal. This contradicts the assumption that M is a maximal ideal.

We are now ready to prove a theorem which is vital for the subject.

THEOREM 1. The only normed field X is the complex number field (within an isomorphism).

Proof. Let e be the multiplicative identity in X. We shall show that every $x \in X$ is of the form ae, $a \in C$. Suppose then that there were an $x \in X$ such that $x \neq ae$ for every $a \in C$. Since X is a field, it then follows that

$$(x - ae)^{-1}$$

exists for every $a \in C$.

By continuity of the inverse, it follows that

$$\frac{[x-(a+h)e]^{-1}-(x-ae)^{-1}}{h}=[x-(a+h)e]^{-1}(x-ae)^{-1}$$

has a limit, as h converges to 0 in C, and this limit is given by $(x-ae)^{-2}$. Thus the function $(x-ae)^{-1}$ of the complex variable a, with values in X, has a derivative everywhere. Moreover, as a tends to infinity,

$$\|(x-ae)^{-1}\|=|a^{-1}|\left\|\left(\frac{x}{a}-e\right)^{-1}\right\|$$

tends to zero.

Now, let f be any continuous linear functional on the Banach space X. By a direct computation, the complex function

$$f((x-ae)^{-1})$$

of the complex variable a has a derivative everywhere. Moreover, since as a tends to infinity, the norm of $(x-ae)^{-1}$ tends to zero, we have

$$\lim_{a\to\infty} f((x-ae)^{-1})=0.$$

Hence, $f((x-ae)^{-1})$ is a bounded entire function, therefore a constant, and the constant is obviously zero. But this holds for every $f\in X'$ so that, by the Hahn-Banach theorem, $(x-ae)^{-1}=0$. But this is impossible. Hence $x=ae$ for some $a\in C$, and X is isomorphic with C.

COROLLARY. If X is a commutative Banach algebra, with an identity, and M is a maximal ideal in X, then X/M is isomorphic with the complex numbers.

6.5 The Set $C[\mathcal{M}]$

Given a commutative Banach algebra X with identity, the last corollary allows us to associate, with every $x\in X$, a function on the set $\mathcal{M}$ of maximal ideals in X.

Let $x\in X$. For every $M\in\mathcal{M}$, let $\hat{x}(M)$ be the element of X/M to which x belongs. Then $\hat{x}(M)$ is a complex number. We thus have a mapping of X into the set of complex functions on $\mathcal{M}$. In particular, the image of 0 is the zero function, since $0\in M$ for every $M\in\mathcal{M}$, and the image of e is the function identically 1, since e is in the multiplicative identity of X/M for every $M\in\mathcal{M}$.

Let $x, y\in X$. Then, for every $M\in\mathcal{M}$,

$$\widehat{x+y}(M)=\hat{x}(M)+\hat{y}(M),$$

and

$$\widehat{xy}(M)=\hat{x}(M)\hat{y}(M),$$

and for $x \in X$, $a \in C$, and every $M \in \mathcal{M}$,

$$\widehat{ax}(M) = a\hat{x}(M).$$

This follows since the mapping of X onto X/M is a homomorphism for every $M \in \mathcal{M}$.

Hence, the mapping we have defined is a homomorphism of the algebra X into the function algebra of complex functions on $\mathcal{M}$ with the usual point-wise operations.

Now, for every $M \in \mathcal{M}$, and $x \in X$, let $x \in \xi \in X/M$. Then

$$|\hat{x}(M)| = \|\xi\| = \inf [\|y\| : y \in \xi] \leq \|x\|.$$

Thus $\hat{x}$ is bounded, and with the sup norm,

$$\|\hat{x}\| \leq \|x\|.$$

Thus the mapping which takes x into $\hat{x}$ is a norm decreasing homomorphism of the algebra X into the function algebra.

We shall show that there is a natural topology in $\mathcal{M}$ for which $\mathcal{M}$ is compact and the functions $\hat{x}$ are continuous.

However, we first give a simple and interesting application which does not use this fact.

Let X be the set of complex functions of a real variable of the type

$$x(t) = \sum_{n=-\infty}^{\infty} a_n e^{int}$$

where

$$\sum_{n=-\infty}^{\infty} |a_n| < \infty.$$

Then X is an algebra with respect to the usual point-wise operations for functions. If we define the norm by

$$\|x\| = \sum_{n=-\infty}^{\infty} |a_n|$$

then X is easily seen to be a commutative Banach algebra with identity.

We prove

THEOREM 1. If $x \in X$ is never zero, then its inverse exists.

Proof. Let M be a maximal ideal in X. For $x = e^{it}$, let $\hat{x}(M) = a$. Then

$$|a| \leq \|e^{it}\| = 1.$$

For $x = e^{-it}$, then $\hat{x}(M) = a^{-1}$. So,

$$|a^{-1}| \leq \|e^{-it}\| = 1.$$

It follows that $|a| = 1$. Let $a = e^{it_0}$. Consider next any finite series

$$x(t) = \sum_{n=-N}^{N} a_n e^{int}.$$

It is clear that

$$\hat{x}(M) = \sum_{n=-N}^{N} a_n e^{int_0}.$$

It follows readily, and we leave the easy details to the reader, that for every $x \in X$,

$$\hat{x}(M) = x(t_0).$$

Thus, M is the set of functions in X which vanish at t_0.

This means that if x does not vanish anywhere, then it is in no maximal ideal. But this implies that x^{-1} exists.

6.6 Gelfand Representation for Algebras with Identity

Again, let X be a commutative Banach algebra with identity, and let $\mathcal{M}$ be the set of maximal ideals in X. For every $x \in X$ there is a corresponding $\hat{x}$ which is a bounded function on $\mathcal{M}$. We now define a topology in $\mathcal{M}$. It is the coarsest topology for which the functions $\hat{x}$ are continuous. A subbase for the topology is given by the sets

$$x^{-1}(\sigma) \subset \mathcal{M}$$

for all open spheres σ in C and all $x \in X$.

We show that, with this topology, $\mathcal{M}$ is a compact Hausdorff space. Our method is to show that $\mathcal{M}$ may be imbedded as a weakly closed set in the unit sphere in X' with the weak topology. Since the latter space is a compact Hausdorff space, this will prove that $\mathcal{M}$ is compact.

Now, to every $M \in \mathcal{M}$ there corresponds an x'_M defined by

$$x'_M(x) = \hat{x}(M).$$

Then x'_M is a complex valued function on X. We show that $x'_M \in X'$.

x'_M is linear since

$$x'_M(ax + by) = \widehat{ax + by}(M) = a\hat{x}(M) + b\hat{y}(M) = ax'_M(x) + bx'_M(y).$$

x'_M is bounded since

$$|x'_M(x)| = |\hat{x}(M)| \leq \|x\|.$$

This shows that $x'_M \in X'$.

Moreover, the last inequality implies that $\|x'_M\| \leq 1$ for every $M \in \mathcal{M}$. We next show that

$$[x'_M : M \in \mathcal{M}]$$

is closed in X' with its weak topology. Accordingly, let x' be in the weak closure of this set. Since $x'_M(e) = 1$, for every $M \in \mathcal{M}$, $x'(e) = 1$. Hence

$x' \neq 0$. For every non-zero $x, y \in X$ and $\epsilon > 0$ there is an $M \in \mathcal{M}$ such that

$$|x'(x) - x'_M(x)| < \frac{\epsilon}{\|y\|}, \ |x'(y) - x'_M(y)| < \frac{\epsilon}{\|x\|}$$

and $|x'(xy) - x'_M(xy)| < \epsilon$.

Since $x'_M(xy) = x'_M(x)x'_M(y)$, we then have

$$\begin{aligned}\|x'(xy) - x'(x)x'(y)\| &\leq \|x'(xy) - x'_M(xy)\| \\ &+ \|x'_M(xy) - x'_M(x)x'_M(y)\| + \|x'_M(x)\| \, \|x'_M(y) - x'(y)\| \\ &+ \|x'(y)\| \, \|x'_M(x) - x'(x)\| \leq \epsilon + \|x\| \frac{\epsilon}{\|x\|} + \|y\| \frac{\epsilon}{\|y\|} = 3\epsilon.\end{aligned}$$

Since this holds for every $\epsilon > 0$, we have $x'(xy) = x'(x)x'(y)$. Thus x' is a bounded non-trivial homomorphism of the algebra X into the complex numbers. As such, its null space is a maximal ideal K in X. But then $x' = x'_K$. Thus

$$[x'_M : M \in \mathcal{M}]$$

is compact in the weak topology. But the open sets in $\mathcal{M}$ are precisely those induced by the weak topology in the unit sphere in X'. We thus have

THEOREM 1. If X is a commutative Banach algebra with identity, there is a norm decreasing homomorphism of X into the algebra $C(\mathcal{M})$ of continuous functions on the compact Hausdorff space of maximal ideals in X, topologized with the coarsest topology for which the images $\hat{x}$ of $x \in X$ are continuous.

6.7 Analytic Functions

In order to study further properties of $C(\mathcal{M})$, it is necessary to discuss analytic functions of a complex variable with values in a Banach algebra.

Let G be an open set in the complex plane. Let x be a mapping from G into a Banach algebra X. We say that x is analytic in G if, for every $c \in G$, there is a $\varphi \in X$ such that

$$\lim_{h \to 0} \left\| \varphi - \frac{x(c+h) - x(c)}{h} \right\| = 0.$$

We write $x'(c)$ for φ. It is called the **derivative** of x at c. x' is also a mapping from G into X.

PROPOSITION 1. If x is analytic on G and f is a continuous linear functional on X, then the function

$$c \to f(x(c))$$

is a complex valued analytic function on G.

Proof. Since

$$f\left(\frac{x(c+h) - x(c)}{h}\right) = \frac{f(x(c+h)) - f(x(c))}{h},$$

and f is continuous, the convergence in X of

$$\frac{x(c+h)-x(c)}{h} \quad \text{to} \quad \varphi = x'(c)$$

implies that

$$\lim_{h\to 0} \frac{f(x(c+h))-f(x(c))}{h} = f(x'(c)).$$

It is to proposition 1 and the Hahn-Banach theorem that much of the theory of an analytic function of a complex variable, with values in a Banach algebra, is due.

Let Γ be a curve of class C' in the complex plane, and let x be a continuous function on Γ with values in X. We define the integral of x on Γ.

Let Γ be given by a pair f, g of continuously differentiable functions on $[0, 1]$. Let

$$\pi = [0 = t_0 < t_1 < \ldots < t_n = 1]$$

be a partition of $[0, 1]$ and let $p_i = (f(\tau_i), g(\tau_i))$, $\tau_i \in [t_{i-1}, t_i]$, $i = 1, \ldots, n$, and $p = (p_1, \ldots, p_n)$. Let

$$\sigma(\pi, p) = \sum_{i=1}^{n} x(p_i)\sqrt{\{f'(\tau_i)\}^2 + \{g'(\tau_i)\}^2}(t_i - t_{i-1}).$$

We state without proof, the proof being just as in the standard case well known to the reader, that for every $\epsilon > 0$ there is a $\delta > 0$ such that if $\|\pi\| < \delta$, $\|\pi'\| < \delta$, then

$$\|\sigma(\pi, p) - \sigma(\pi', p')\| < \epsilon$$

for every choice of p and p'. It follows that for every sequence $\{\pi_n\}$, whose norms converge to zero, and every $\{p^{(n)}\}$, the sequence

$$\{\sigma(\pi_n, p^{(n)})\}$$

converges in X, and always to the same value. This value is

$$\int_\Gamma x(z)\, dz.$$

PROPOSITION 2. If x is analytic on an open set G, if Γ is a simple closed curve of class C' such that Γ and its interior are in G, then

$$\int_\Gamma x(z)\, dz = 0.$$

Proof. Let

$$y = \int_\Gamma x(z)\, dz.$$

Then for every continuous linear functional f on X, it follows from the definition of the integral as a limit of Riemann sums that

$$f(y) = \int_\Gamma f(x(z))\, dz.$$

Since $f \circ x$ is analytic, we have

$$\int_\Gamma f(x(z))\, dz = 0.$$

Hence $f(y) = 0$ for every $f \in X'$ so that, by the Hahn-Banach theorem, $y = 0$.

PROPOSITION 3. If x and Γ are as in proposition 2, then for every z in the interior of Γ, we have

$$x(z) = \frac{1}{2\pi i} \int_\Gamma \frac{x(\zeta)}{\zeta - z}\, d\zeta.$$

We omit the proof of this proposition and of the fact that, just as in the classical case, if x is analytic on an open set, then all its derivatives exist and the Cauchy formula

$$x^{(n)}(z) = \frac{n!}{2\pi i} \int_\Gamma \frac{x(\zeta)}{(\zeta - z)^{n+1}}\, d\zeta$$

is valid.

From this, we obtain

PROPOSITION 4. If x is analytic in the open disk $|z - z_0| < r$, then there is a Taylor series

$$x(z_0) + x'(z_0)(z - z_0) + \ldots + x^{(n)}(z_0)\frac{(z - z_0)^n}{n!} + \ldots$$

which converges absolutely to x in the disk.

Proof. Let $|z - z_0| < r' < r$. Then $\|x(\zeta)\|$ is continuous, hence bounded, on the circle $|\zeta - z_0| = r'$. Now,

$$\frac{x(\zeta)}{\zeta - z} = \frac{x(\zeta)}{\zeta - z_0} \cdot \frac{1}{1 - \dfrac{z - z_0}{\zeta - z_0}} = \sum_{n=0}^{\infty} \frac{x(\zeta)}{(\zeta - z_0)^{n+1}} (z - z_0)^n,$$

where the series converges uniformly on $|\zeta - z_0| = r'$. It follows that term by term integration is valid and, by the Cauchy integral formula, the proposition is proved.

6.8 Isomorphism Theorem for Algebras with Identity

Let $x \in X$. By the **spectrum** of x we mean the set of complex numbers a for which $(x - ae)^{-1}$ does not exist.

PROPOSITION 1. For every $x \in X$, the spectrum of x is equal to the set of values assumed by $\hat{x}(M)$, $M \in \mathcal{M}$.

Proof. Suppose $\hat{x}(M) = a$, $M \in \mathcal{M}$. Then $\hat{x}(M) = a\hat{e}(M)$, or

$$\widehat{(x - ae)}(M) = 0.$$

Thus $x - ae \in M$. This means that $(x - ae)^{-1}$ does not exist, so that a is in the spectrum of x.

Conversely, suppose $(x - ae)^{-1}$ does not exist. Then $x - ae \in M$ for some $M \in \mathcal{M}$, so that $\hat{x}(M) = a$.

PROPOSITION 2. For every $x \in X$,

$$\sup\,[\,|x(M)| : M \in \mathcal{M}] = \lim_{n\to\infty} \sqrt[n]{\|x^n\|}.$$

Proof. Let $\sup |\hat{x}(M)| = a$. Then, by proposition 1, $x - ze$ has an inverse whenever $|z| > a$. Thus

$$z(e - zx)^{-1} = \left(\frac{1}{z}e - x\right)^{-1}$$

exists whenever $|z| < 1/a$. It is analytic in this disk, as is seen by a direct computation of the derivative. Hence, it is given by the Taylor series

$$z(e - zx)^{-1} = z + z^2x + \ldots + z^{n+1}x^n + \ldots,$$

which converges absolutely for $|z| < 1/a$. Fix $0 < |z| < 1/a$. Then

$$\lim_{n\to\infty} \|z^nx^n\| = 0.$$

It follows that, for all n sufficiently large,

$$\|x^n\| \leq \frac{1}{|z|^n}.$$

Hence,

$$\limsup_{n\to\infty} \sqrt[n]{\|x^n\|} \leq \frac{1}{|z|}$$

whenever $|z| < 1/a$, so that

$$\limsup_{n\to\infty} \sqrt[n]{\|x^n\|} \leq a.$$

On the other hand, we know that

$$\|x\| \geq \sup\,[\hat{x}(M) : M \in \mathcal{M}],$$

for every $x \in X$. In particular,

$$\|x^n\| \geq \sup\,[\hat{x}^n(M) : M \in \mathcal{M}] = a^n.$$

Hence,

$$\sqrt[n]{\|x^n\|} \geq a,$$

and

$$\liminf_{n\to\infty} \sqrt[n]{\|x^n\|} \geq a.$$

We say that x is in the **radical** of X if

$$\lim_{n\to\infty} \sqrt[n]{\|x^n\|} = 0.$$

We then have

COROLLARY 1. x is in the radical of X if and only if x is contained in every maximal ideal.

Thus the radical of X is also the intersection of the maximal ideals in X.

An algebra is called **semi-simple** if the intersection of its maximal ideals consists of 0 alone.

COROLLARY 2. The canonical mapping of X into $C(\mathcal{M})$ is one-one if and only if X is semi-simple.

Suppose now that, for every $x \in X$, $\|x^2\| = \|x\|^2$. Since

$$\max\,[|\hat{x}(M)| : M \in \mathcal{M}] = \lim_{n\to\infty} \sqrt[n]{\|x^n\|} = \lim_{n\to\infty} \sqrt[2^n]{\|x^{2^n}\|} = \lim_{n\to\infty} \|x\| = \|x\|,$$

it follows that the canonical mapping of X into $C(\mathcal{M})$ is a norm preserving isomorphism. In particular, this condition implies semi-simplicity.

Suppose next that a mapping of X into X which takes each x into an x^* exists such that

$$(ax + by)^* = \bar{a}x^* + \bar{b}y^*,$$
$$(x^*)^* = x$$
$$(xy)^* = y^*x^*.$$

Such a mapping is called an **involution.**

For example, in $C(X)$, the mapping which takes each function into its complex conjugate is an involution. A similar remark holds for $L_1(G)$. In the Banach algebra of bounded operators on a Hilbert space, the mapping which takes each operator into its adjoint operator is an involution.

We prove

THEOREM 1. If X is a commutative Banach algebra, with an identity and an involution, such that for every $x \in X$,

$$\widehat{(x^*)} = (\bar{\hat{x}})$$

and

$$\|x^2\| = \|x\|^2,$$

then $\hat{X} = C(\mathcal{M})$, and the canonical mapping is an isometry.

Proof. We first note that if $M_1 \neq M_2$, there is an $x \in X$ such that $\hat{x}(M_1) \neq \hat{x}(M_2)$. For, let $x \in M_1$, $x \notin M_2$. Then $\hat{x}(M_1) = 0$, $\hat{x}(M_2) \neq 0$.

Let $x \in X$. Define

$$x_1 = \frac{1}{2}(x + x^*), \quad x_2 = \frac{1}{2i}(x - x^*).$$

Then,

$$x_1{}^* = \frac{1}{2}(x^* + x) = x_1$$

and

$$x_2{}^* = -\frac{1}{2i}(x^* - x) = x_2.$$

It follows that $\hat{x}_1$ and $\hat{x}_2$ are real, and

$$x = x_1 + ix_2.$$

Now, $\hat{x}(M_1) \neq \hat{x}(M_2)$ implies either

$$\hat{x}_1(M_1) \neq \hat{x}_1(M_2) \text{ or } \hat{x}_2(M_1) \neq \hat{x}_2(M_2).$$

Let R be the subset of $\hat{X}$ of real valued functions. Then R is an algebra over the reals which, as has been shown above, satisfies the Stone-Weierstrass condition. Hence, R is dense in the set of continuous real functions on $\mathcal{M}$. It follows that $\hat{X}$ is dense in $C(\mathcal{M})$.

But $\|x^2\| = \|x\|^2$ implies that the mapping $X \to \hat{X}$ is an isometry. Since X is complete, so is $\hat{X}$. Hence, $\hat{X} = C(\mathcal{M})$.

6.9 Algebras without Identity

We now remove the hypothesis that the commutative Banach algebra X has an identity.

First, let $\tilde{X}$ be the Banach algebra obtained by adjoining an identity to X. Let $\tilde{I}$ be any ideal in $\tilde{X}$ which is not a subset of X. Consider the set

$$I = X \cap \tilde{I}.$$

Then I is a proper subset of X. For, otherwise, $\tilde{I} \supset X$ and, since $\tilde{I} \neq \tilde{X}$, this implies $\tilde{I} = X$. But this contradicts the assumption that $\tilde{I}$ is not a subset of X. Next, $x, y \in I$ and $a, b \in C$ implies $ax + by \in I$ and $x \in I$, $y \in X$ implies $xy \in I$. Hence, I is an ideal in X.

Since $\tilde{I}$ is not a subset of X, it contains an element of the form

$$y = x - e, \; x \in X.$$

For every $z \in X$,

$$-z + xz = (x - e)z \in I,$$

since $(x - e)z \in \tilde{I}$ and $-z + xz \in X$. There is thus an $x \in X$ such that $xz - z \in I$, for every $z \in X$. Such an element is called an **identity modulo *I*.** If an ideal I is such that there is an identity modulo I in X, then I is called a **regular** ideal.

We have shown above that for every ideal $\tilde{I}$ in $\tilde{X}$, which is not a subset of X, $I = \tilde{I} \cap X$ is a regular ideal in X.

Conversely, let I be a regular ideal in X, and let x be an identity modulo I. Let

$$\tilde{I} = [z : z \in \tilde{X},\ xz \in I].$$

Now, $\tilde{I}$ is not a subset of X. For $x - e \notin X$. But

$$x(x - e) = x^2 - x \in I,$$

so that $x - e \in \tilde{I}$.

Next, $I = \tilde{I} \cap X$. For, $z \in \tilde{I} \cap X$ implies $z \in X$ and $xz \in I$. But

$$xz - z \in I,$$

so that $z \in I$. Conversely, $z \in I$ implies $z \in X$ and $xz \in I$ so that $z \in \tilde{I}$.

That $\tilde{I}$ is an ideal is an easy verification which we omit.

We show that if $\tilde{I}$ and $\tilde{J}$ are ideals in $\tilde{X}$, neither a subset of X, then $I = \tilde{I} \cap X$ equals $J = \tilde{J} \cap X$ implies $\tilde{I} = \tilde{J}$.

There are $x, y \in X$ such that $x - e \in \tilde{I}$, $y - e \in \tilde{J}$. Then

$$\begin{aligned} yx - y &= y(x - e) \in I \\ yx - x &= x(y - e) \in J = I, \end{aligned}$$

so that $y - x \in I$.

Suppose $ae + z \in \tilde{I}$. Then $ax + xz \in I$. But $-z + xz = z(-e + x) \in I$. Then $ax + z = ax + xz - z(-e + x) \in I$. Accordingly,

$$ae + z = a(e - y) + a(y - x) + ax + z \in \tilde{J}.$$

Hence $\tilde{I} \subset \tilde{J}$. In the same fashion, $\tilde{J} \subset \tilde{I}$, and we have proved

Proposition 1. The ideals in $\tilde{X}$ which are not subsets of X are in one-one correspondence with the regular ideals in X by the mapping

$$\tilde{I} \to I = \tilde{I} \cap X.$$

Corollary. The maximal ideals in $\tilde{X}$, different from X, are in one-one correspondence, by the above mapping, with the maximal regular ideals in X.

We note, moreover, that if X is a commutative Banach algebra without identity, then if M is a regular maximal ideal in X, the quotient algebra X/M is isomorphic to the complex field.

Indeed, suppose $\tilde{X}$ is the extension of X obtained by adjoining the identity, and let $\tilde{I}$ be any ideal in $\tilde{X}$ which is not a subset of X. Then every coset $\tilde{I} + (x + ae)$ of $\tilde{I}$ has non-empty intersection with X. For, if $a = 0$, then $x \in \tilde{I} + (x + ae)$. Suppose $a \neq 0$. Let $y + be \in \tilde{I}$, $b \neq 0$. Then,

$$-\frac{a}{b}y - ae \in \tilde{I} \text{ and } -\frac{a}{b}y + x \in X \cap [\tilde{I} + (x + ae)].$$

We next show that for every $x \in X$,

$$I + x = (\tilde{I} + x) \cap X.$$

For, $y \in I + x$ implies $y \in \tilde{I} + x$ and $y \in X$. Conversely, $y \in (\tilde{I} + x) \cap X$ implies $y = z + ae + x$ and $y = t \in X$. Then $a = 0$ and $y = z + x$. Then, since $z \in \tilde{I}$ and $z \in X$, we have $z \in I$ and $y \in I + x$.

This shows that $\tilde{X}/\tilde{I}$ is isomorphic with X/I. In particular, if M is a maximal regular ideal in X, then X/M is isomorphic with the complex field.

For every $x \in X$, and every maximal regular ideal M in X, let $\hat{x}(M)$ be the element of X/M to which x belongs. This defines a mapping from X into the set of complex functions on the set $\mathcal{M}$ of maximal regular ideals in X. But, from the isomorphism of X/M with $\tilde{X}/\tilde{M}$, the function $\hat{x}$ is also a function on the set $\tilde{\mathcal{M}} \sim \{X\}$, of maximal ideals in $\tilde{X}$. Since $x \in X$, and X is a maximal ideal in $\tilde{X}$, these are precisely the functions in $(\hat{\tilde{X}})$ which vanish at the maximal ideal X.

Thus $\hat{X}$ is an algebra of continuous functions on $\tilde{\mathcal{M}}$ which vanish at the point X. Identifying each maximal regular ideal M with the maximal ideal $\tilde{M}$ in $\tilde{X}$ which contains it, the space $\mathcal{M} = \tilde{\mathcal{M}} \sim \{X\}$, with the topology induced by the topology in $\tilde{\mathcal{M}}$, is locally compact. For every open set $G \subset \tilde{\mathcal{M}}$, with $X \in G$, the set $\tilde{\mathcal{M}} \sim G$ is a compact subset of $\tilde{\mathcal{M}}$. There is a G of this sort such that, for every $\tilde{M} \in G$, $|\hat{x}(\tilde{M})| < \epsilon$, given $\epsilon > 0$ and $\hat{x} \in \hat{X}$.

We say that a function f on a locally compact space vanishes at infinity if, for every $\epsilon > 0$, there is a compact set such that $|f(t)| < \epsilon$ on the complement of the set. By the above, the functions $\hat{x}$ on $\mathcal{M}$ are continuous and vanish at infinity.

The topology in $\mathcal{M}$ may be defined directly as the coarsest topology for which the functions $\hat{x}$ on $\mathcal{M}$ are continuous, but we do not discuss this approach here.

We have proved

THEOREM 1. *If X is a commutative Banach algebra without identity, there is a norm decreasing homomorphism of X into the algebra $C(\mathcal{M})$ of continuous functions, which vanish at infinity, on the locally compact Hausdorff space of maximal regular ideals in X, topologized by the coarsest topology for which the images $\hat{x}$ of $x \in X$ are continuous.*

(The proof that the mapping is norm decreasing is the same as for the case of an algebra with identity.)

PROPOSITION 2. *For every $x \in X$,*

$$\sup |\hat{x}(M)| = \lim_{n \to \infty} \sqrt[n]{\|x^n\|}.$$

Proof.

$$\sup |\hat{x}(\tilde{M})| = \lim_{n\to\infty} \sqrt[n]{\|x^n\|}\,.$$

But $\hat{x}(M) = \hat{x}(\tilde{M})$ and $\hat{x}(X) = 0$.

COROLLARY 1. If X is a Banach algebra without identity, then the radical of X is the intersection of its maximal regular ideals.

COROLLARY 2. The canonical mapping of X into the continuous functions on $\mathcal{M}$, which vanish at infinity, is one-one if and only if X is semi-simple.

We finally have

THEOREM 2. If X is a commutative Banach algebra with involution, but without identity, such that $\widehat{(x^*)} = \overline{(\hat{x})}$ and $\|x^2\| = \|x\|^2$, for every $x \in X$, then $\hat{X} = C(\mathcal{M})$, the algebra of continuous functions on $\mathcal{M}$, which vanish at infinity, and the canonical mapping is an isometry.

Proof. Just as for theorem 1, Sec. 6.8, except that the form of the Stone-Weierstrass theorem which is used says that, if the function 1 is not in the given algebra A, then A consists of all continuous functions on the compact space X which vanish at a given point.

6.10 Application to $L_1(G)$

We apply some of the above results to the case of the group algebra, $L_1(G)$, where G is a locally compact abelian group.

We first discuss approximate identities in $L_1(G)$. Let $\{U_\alpha\}$ be a directed set of neighborhoods of 0 in G which forms a base for the topology at 0. For each α, let F_α be a continuous function on G, such that

(a) $$F_\alpha(t) \geq 0 \text{ for all } t \in G,$$

(b) $$F_\alpha(t) = 0 \text{ for all } t \notin U_\alpha,$$

and

(c) $$\int_G F_\alpha(t)\,dt = 1.$$

Such a system of functions is called an **approximate identity** in $L_1(G)$. (For the case $L_1(R)$ we defined approximate identity somewhat more generally.) $L_1(G)$ has an identity only if G is discrete. The approximate identity assumes many of the roles of an identity.

PROPOSITION 1. For every $f \in L_1(G)$,

$$\lim_\alpha \|F_\alpha * f - f\| = 0.$$

Proof. We note that $(F_\alpha * f)(t) = \int_G f(t-u)F_\alpha(u)\,du$, and that $f(t) = \int_G f(t)F_\alpha(u)\,du$. Then

$$\begin{aligned}\|F_\alpha * f - f\| &= \int_G dt\left|\int_G [f(t+u) - f(t)]F_\alpha(u)\,du\right| \\ &\le \int_G dt \int_G |f(t+u) - f(t)|F_\alpha(u)\,du \\ &= \int_G \left\{\int_G |f(t+u) - f(t)|\,dt\right\}F_\alpha(u)\,du \\ &= \int_{U_\alpha} \left\{\int_G |f(t+u) - f(t)|\,dt\right\}F_\alpha(u)\,du.\end{aligned}$$

But, for every $\epsilon > 0$, there is an α_0 such that $\alpha > \alpha_0$ implies $\int_G |f(t+u) - f(t)|\,dt < \epsilon$ for every $u \in U_\alpha$. The result then follows.

For every $u \in G$, let f_u be the function given by $f_u(t) = f(t+u)$. We now state

COROLLARY. If $u \in G$ and $f \in L_1(G)$, then

$$\lim_\alpha \|F_{\alpha,u} * f - f_u\| = 0.$$

We shall need the characters of the group G. By a **character** χ of G we understand a continuous complex-valued function on G such that, for every $t \in G$, $|\chi(t)| = 1$, and for every t, $u \in G$, $\chi(t+u) = \chi(t)\chi(u)$. It is easily seen that the point-wise product of characters is a character and that the characters form a group.

It is an important fact that the characters in G are in one-one correspondence with the maximal ideals in $L_1(G)$, and we now discuss this relationship.

PROPOSITION 2. For every maximal regular ideal M in $L_1(G)$,

$$\lim_\alpha \hat{f}_{\alpha,t}(M)$$

exists, is a character in G, and is independent of the choice of approximate identity $\{f_\alpha\}$.

Proof. Let $f \in L_1(G) \sim M$. Then $\hat{f}(M) \neq 0$. Now,

$$\lim_\alpha \|f_t - f * f_{\alpha,t}\| = 0,$$

and

$$\lim_\alpha |\hat{f}_t(M) - \hat{f}(M)\hat{f}_{\alpha,t}(M)| = 0.$$

Hence,

$$\lim_\alpha \hat{f}_{\alpha,t}(M) = \frac{\hat{f}_t(M)}{\hat{f}(M)}.$$

Physical and Thermal Properties

r monolith materials are characterized using basically set of properties that is used for the description of low-through catalyst supports (Section 18.4). Certain f properties, however, play a more important role in noliths than in the flow-through substrates. Due to bility of the release of high quantities of heat during eneration, filter materials are required to demonstrate thermal properties in terms of resistance to both high ures and high temperature gradients. Insufficient tem- tolerance may result in melting of the material, while ent thermal shock resistance causes cracking.

trary to flow-through substrates in which support mate- parated from gases by a thick layer of washcoat, filter comes in much more direct contact with the exhaust terial pore characteristics (size, percent porosity, pore vity, open vs. closed pores, etc.) influence the physical on between gas and filter material and affect such prop- filtration efficiency and pressure drop. Furthermore, rability depends on the material resistance to chemical y exhaust gas components. In particular, materials need sistant to corrosion by the metal ash that may be part particulates.

al Expansion and Strength

coefficient of thermal expansion (CTE) is a measure much a given body expands (or contracts) on heating. ne filter monolith is exposed to thermal gradients, for , as a result of localized heat release during regen- the hotter parts expand relative to the colder regions g the material. This strain creates thermal stress in the , which increases with increasing thermal gradient (or lting temperature difference, ΔT), increasing CTE, and ng elastic modulus of the material (E). If the thermal xceeds the tensile strength of the material, the monolith ck. In practical applications, tensile stress in the substrate kept well below its strength value to provide sufficient for the material fatigue (i.e., a loss of strength when for a long time). Therefore, filter materials of low CTE E are preferred.

ile the tensile strength is important for thermal stress ty during filter operation, its compressive strength deter- he maximum allowable stress from the canning system, both packaging and operation. In general, the strength ters can be classified as follows:

e Strength

- ulus of rupture (MOR) in axial and tangential direction e part

ressive Strength

- xial crush strength along the axial (A), tangential (B), diagonal (C) directions
- 3-D isostatic strength, determined by pressurization of the whole substrate
- 2-D isostatic strength, determined by biaxial compressive crushing of the whole substrate

Thermal expansion and strength parameters were discussed in more detail in Section 18.4.

Thermal Shock Resistance

The most common measure of thermal shock resistance is the thermal shock parameter (TSP), which represents the ratio of mechanical strain tolerance to thermal strain imposed by temperature gradient. The higher the TSP, the better the thermal shock capability of the material. The TSP values presented throughout this section are calculated from the following formula:

$$TSP = MOR/(E \cdot CTE) \qquad (21.8)$$

where:

MOR = modulus of rupture, Pa
E = elastic modulus, Pa
CTE = coefficient of thermal expansion, 1/°C

TSP is a dimensionless factor; for the sake of simplicity, temperature difference (°C) has been omitted in Eq. (21.8).

Another method of quantifying the thermal shock resistance is the quenching or water immersion test [30]. Hot filter samples are immersed in water and evaluated for cracks. The quench test rating is determined as the temperature difference between the part and the water at which cracks appeared in the part (i.e., the higher the quench test temperature, the better the thermal shock resistance).

Melting Point and Maximum Operating Temperature

The melting point of the monolith material must be sufficiently high to withstand the normal use temperature, as well as temperature spikes occurring during controlled and uncontrolled regeneration. Uncontrolled regenerations can produce very high temperatures. In several applications the melting point of cordierite of 1450°C has been proven to be insufficient, thus driving the development of improved filter materials [31].

A quiet assumption was made in the older literature that the maximum operating temperature of a diesel filter was equal to its melting temperature (or a temperature where the material lost its properties, if a significant property loss occurred below the melting point). Newer studies show that the maximum practical use temperature of a diesel particulate filter may, in fact, be limited by the interaction of the material with ash, rather than by the melting point of the pure material [1]. These considerations are certainly valid with regard to long-term filter durability. Diesel PM ashes start sintering at about 900°C. If the temperature is further increased, the ash will adhere to the filter surface, thus making it impossible to periodically remove ashes from the filter, resulting in irreversible increase of pressure drop.

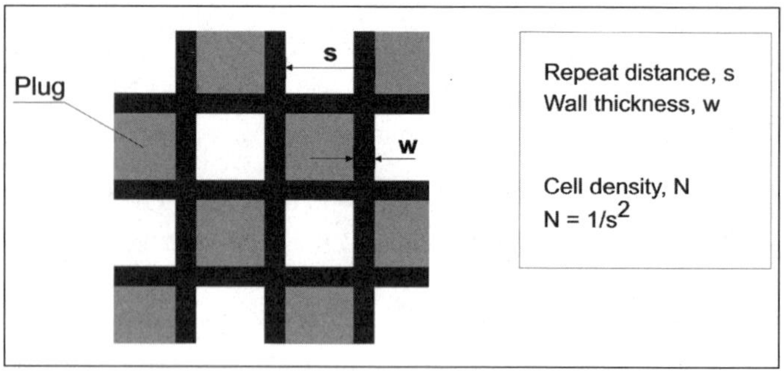

Figure 21.10 *Cell geometry.*

$$N = 1/s^2 \qquad (21.1)$$

Cell hydraulic diameter (d_h) is a measure of channel size in hydraulics, mass transfer, and heat transfer calculations. In the case of square cells, it can be simply expressed as the difference between repeat distance and wall thickness, as follows:

$$d_h = s - w \qquad (21.20)$$

Open frontal area (OFA) is defined as the part of the total substrate cross-sectional area that is available for the flow of gas (i.e., the cross-sectional area of the filter inlet channels). It is typically expressed relative to the total substrate cross section, as given by Eq. (21.3):

$$OFA = ½ [(s - w)/s]^2 = ½ (1 - w/s)^2 \qquad (21.3)$$

Specific filtration area (SFA) is the total channel surface area per unit of substrate volume. For square cells, it can be calculated from Eq. (21.4). It should be noted that the SFA increases with increasing cell density (N).

$$SFA = 2 (s - w)/s^2 =$$

Total filtration area (TFA) can b of SFA and the filter volume (V_f), as

$$TFA = SFA \cdot$$

$$V_f = ¼\pi D^2 L$$

where D is the filter diameter and L is ing on the required precision, L may b tive" channel length taken by the end

The filter cell configuration also integrity. Cell configurations of high have a higher mechanical integrity f wall porosity, the MIF is defined by E

$$MIF = w^2/[s (s -$$

Selected cellular properties calc monolith configurations (some comme opment) are listed in Table 21.4.

Configuration 200/12 from Table to provide the same OFA and, thus, in channels as the basic configuratio 200-cell substrates, however, use thic For example, Ibiden commercialized th a 0.014-in. wall. The Corning RC-20 more sturdy structure. In this case, the t introduced to provide certain thermal pr later), rather than just mechanical stren

TABLE 21.4
PROPERTIES OF MONOLITHS WITH DIFFERENT CELL CONFIGURATIONS

Cell Configuration	100/17	200/12	200/14	200/19
Commercial monolith	Corning EX-80	—	Ibiden	Corning RC-200/19
Cell density, N, cpsi	100	200	200	200
• 1/cm^2	15.5	31	31	31
Repeat distance, s, mm	2.54	1.80	1.80	1.80
• inch	0.100	0.071	0.071	0.071
Wall thickness, w, mm	0.43	0.30	0.36	0.48
• inch	0.017	0.012	0.014	0.019
Open frontal area, OFA	0.344	0.345	0.322	0.268
Specific filtration area, SFA, 1/cm	6.5	9.2	8.9	8.1
• in.^{2}/in.^{3}	16.6	23.4	22.6	20.6
MIF	0.035	0.034	0.048	0.098

At still higher temperatures, the ash attack may cause a loss of chemical durability and corrosion of the material.

Chemical Durability

The monolith material is exposed to widely ranging conditions, including high temperatures and both oxidizing and reducing conditions. In general, the filter life span depends on the chemical durability of its material(s).

As indicated earlier, special attention must be paid to one aspect of chemical durability: the resistance to metal ash. A number of metal oxides come from the fuel, engine lube oil, and fuel additives, as well as from engine wear deposit on the filter material. These oxides can have a twofold effect on the monolith material:

- *Eutectic melting:* The melting point of the material can be locally affected by the metal oxides. These deposits can act as a flux by forming eutectics that may have significantly lower melting point than the uncontaminated material [1].
- *Solid-state reactions:* Compounds produced in reactions between ashes and the monolith material can result in volume changes and local stresses that can compromise the integrity of the monolith. In the case of nonoxide materials that survive under oxidizing conditions thanks to a protective oxide layer at the surface, the reactions with impurities may result in damage to the oxide layer (or enhanced oxygen diffusion through the layer), exposing the underlying ceramics to oxidation—and ultimately failure.

Heat Capacity

The bulk volumetric heat capacity of the monolith is a product of the specific heat capacity of the ceramic filter material (material property) and the bulk density of the monolith (a function of material density, wall porosity, and cell structure). In the metric system, the unit of volumetric heat capacity is kJ/m^3 K. The specific heat of most ceramic materials increases with increasing temperatures.

The heat capacity determines the amount of thermal energy that can be absorbed by the filter monolith. For normal, controlled regeneration, a low heat capacity filter is desired. Such a filter enables rapid increases in the filter temperature and quick regeneration with minimal energy losses for heating of the monolith. On the other hand, a high heat capacity material is preferred for uncontrolled regenerations. In this case, the monolith should act as a heat sink to store the heat released during regeneration and still remain safely below its melting point (or below a temperature at which undesired ash reactions may occur). In conclusion, low heat capacity is preferable for filter regeneration. However, low heat capacity filters must have high melting temperatures and/or reliable regeneration control strategies to protect them from melting during uncontrolled regenerations.

Thermal Conductivity

The thermal conductivity of a material is a measure of how quickly it can conduct heat between regions of higher and lower temperatures. The metric unit of thermal conductivity is W/m K. In most materials, thermal conductivity decreases with increasing temperature.

High thermal conductivity allows for a more rapid heat transfer from the hot zones and creates more uniform temperature distribution throughout the filter, thus, resulting in better thermal shock resistance of filter materials. However, if the conductivity is too high, heat may be carried away from the regeneration zone and the regeneration may become extinguished. In high conductivity materials too much heat may also be transferred to the mat and can, possibly resulting in instability of the packaging system. Thus, for good regeneration, the monolith material would have both low heat capacity and low heat conductivity. To maintain good thermal shock resistance, the low conductivity material should have low CTE, low elastic modulus, and high strength.

Filtration Capacity

An important property of diesel filters is their filtration capacity or maximum soot loading, defined as the maximum amount of accumulated soot that will not result in filter damage during regeneration. Because the filter material has to be able to physically collect the soot, the filtration capacity is directly related to the TFA [Eq. (21.5)]. However, the maximum soot load is typically limited by the heat released during uncontrolled regeneration. Excessive heat release causes filter failure either due to high temperatures (melting, high-temperature ash reactions) or due to high-temperature gradients (cracking). Figure 21.11 shows peak temperatures measured in different materials during induced uncontrolled regeneration as a function of soot load [1].

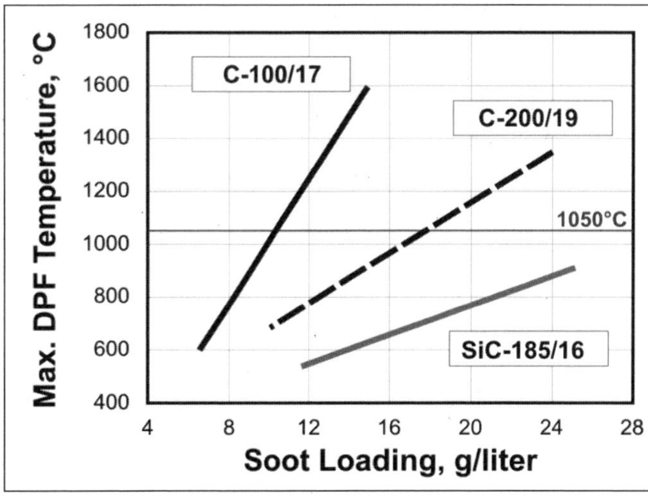

Figure 21.11 *Temperature limited maximum soot loading. (Based on Ref. 1.)*

The filtration capacity of the two cordierite formulations (C-100/17 and C-200/19 in Figure 21.11) is clearly limited by temperature, set in the graph at 1050°C, a maximum level to avoid adhesion of ash to the filter. As soot loading increases, the durability of cordierite in the presence of ash of about 1250°C is also exceeded in both samples. The SiC filter, on the other hand, never reaches dangerous temperature levels, thanks to its high thermal conductivity and heat capacity. This does not mean, however, that the SiC material can be loaded to the soot levels that appear to be indicated in the graph. As will be discussed later, in the case of SiC the filtration capacity is determined by the maximum allowable thermal gradient rather than by temperature.

Because the maximum soot load depends on the regeneration procedures and on a number of parameters during the experiment (regeneration gas velocity, oxygen concentration, presence of catalysts or additives, filter size, etc.), the filtration capacity is not a precisely defined term. Filtration capacities may provide a good comparison between various configurations from the same supplier, but should be used with caution for the purpose of filter system design or for comparisons between products from different manufactures.

Pore Characteristics

Pore characteristics, including volume percent porosity, size distribution, structure, and interconnectivity, determine the monolith's ability to filter particulates. Together with the cellular geometry, porosity characteristics also influence the monolith's hydraulic resistance to flow and the pressure drop. Some properties that are desired for high filtration efficiency (e.g., low porosity and small pore size) are opposite to those required for low pressure drop. Others, such as good pore interconnectivity and absence of closed, "dead-end" pores, are desired for both low pressure drop and high efficiency.

21.2.3 Commercial Materials

21.2.3.1 Overview

Cordierite and silicon carbide (SiC) are two monolith materials that have been commercialized. Cordierite monoliths, available since the 1980s, have been used in numerous applications, including light- and heavy-duty diesel engines, in passive and active filter configurations. A number of diesel filters utilizing cordierite monoliths are described in Chapter 22, which deals with various diesel filter systems . SiC filters were introduced in the 1990s, with the first large-scale commercial application—a light-duty filter system for a Peugeot car—launched in the year 2000.

A comparison of the materials is shown in Table 21.5. None of the materials represents a perfect combination of desired features. Cordierite is characterized by good thermal shock resistance and lower cost, but has an insufficient melting temperature. SiC has a higher operating temperature limit and favorable pore network structure, but low thermal shock resistance and a high cost.

TABLE 21.5
COMPARISON OF CORDIERITE AND SiC MATERIALS

Property	Cordierite[a]	SiC[b]	Comment
Melting temperature, °C	1450	2400[c]	Cordierite melting temperature is insufficient to protect the material during uncontrolled regenerations.
Thermal expansion, CTE, $\times 10^{-7}$/°C	7	45	Cordierite has superior thermal shock resistance (favorable, lower CTE and E compensate for lower strength producing higher TSP).
Elastic modulus, axial E, GPa	4.7	33.3	
Strength, axial MOR, MPa	2.6	18.6	
Thermal shock parameter, TSP	790	124	
Permeability, $\times 10^{-12}$ m^2	0.50	1.24	SiC has uniform network of open pores, resulting in higher permeability and lower ΔP.
Bulk volumetric heat capacity, kJ/m^3 K	500	950	Higher heat capacity and conductivity in SiC are believed to be unfavorable for sustaining regeneration, but provide greater margin of safety in uncontrolled regeneration.
Thermal conductivity, W/m K	<2	20	
Relative cost	Low	High	SiC requires more expensive raw materials and is processed at higher temperatures.

Source: Numerical data for both materials is after Cutler and Merkel [31].
a. Corning EX-80, 100/17, monolith.
b. Ibiden 200/14, cemented sections.
c. Sublimation.

Ongoing attempts have been made to develop new materials that have a more favorable combination of properties. Such materials would provide more flexibility in designing regeneration strategies and, in general, diesel filter systems. One of the more interesting candidates was a zirconium phosphate material of a generic formula $XZr_4P_8O_{24}$, where X denotes an alkaline earth atom. This family of materials, named NZP, combined excellent thermal shock resistance (TSP > 1200) with high melting temperature (1900°C) and a number of other favorable properties [31]. Unfortunately, the material has not been commercialized. It has been subjected to an extensive ash reactivity testing program [1], which, apparently, has not provided enough confidence as to the suitability of NZP as a commercial filter material.

21.2.3.2 Cordierite Monoliths

Cordierite is the most common material used for the wall-flow diesel filter monoliths. Cordierite is a ceramic composition of the formula $2MgO\text{-}2Al_2O_3\text{-}5SiO_2$ which was originally developed for automotive flow-through catalyst substrates. The wall-flow trap technology is an extension of the automotive substrate technology and similar materials are used for both products (see also Section 18.4). More attention is devoted to pore structure control in diesel filter technology. Due to its favorable thermal properties and mechanical strength, cordierite is well suited for diesel filter applications. It is characterized by a very low thermal expansion coefficient and can thus tolerate high thermal stress and thermal cycling, as well as resistance to cracking from material fatigue. Cordierite can be easily coated with a platinum or other noble metal catalyst, either through direct impregnation of the monolith or by using an intermediate washcoat layer.

The melting temperature of cordierite is about 1450°C. On contact with ash at temperatures greater than 1250°C, cordierite was reported to corrode with the formation of pinholes in the channel walls and pitting [1]. The maximum temperature in the filter should be controlled at an even lower level of about 1050°C in order to avoid permanent adhering of sintered ash to the walls [1]. However, considering the amount of existing field experience, in most applications the resistance of cordierite to ash attack should be deemed satisfactory.

The melting point of 1450°C has been considered insufficient under certain uncontrolled regeneration conditions, where monolith melting has been recognized as a problem by cordierite filter manufacturers [31]. Another drawback of cordierite is its imperfect pore structure. The pore characteristics are typically controlled by using pore precursors in the manufacturing process. Even though the material is engineered to result in a highly interconnected pore network, it is difficult to eliminate closed pores and obtain perfect pore distribution.

Filter monoliths are manufactured in a two-stage process. First, monoliths with open channels are extruded and fired in an oven. Combustible pore precursor substances (polymers, graphite, etc.) that can be added to the raw material are oxidized during the firing process, thus creating the pore network in the monolith walls. Then the channel ends are plugged with ceramics (cordierite) and the filters are fired again. Plugged parts are checked for leaks by filling with a fine granular material (e.g., black glass beads) or else through a diesel exhaust test. If leaking channels are detected, they are plugged with ceramic cement.

The largest diameter that can be extruded is approximately 300 mm. Bigger units are assembled of several square cross-sectional blocks that are glued together using a ceramic cement. After gluing, plugging channel ends, and firing, the assemblies are machined to a cylindrical shape and an outside skin is applied to the sides of the cylinder using air-setting cordierite cement. A finished cordierite wall-flow monolith is shown in Figure 21.12. The alternative plugged cell patterns for different cell configurations are clearly visible in the enlarged section.

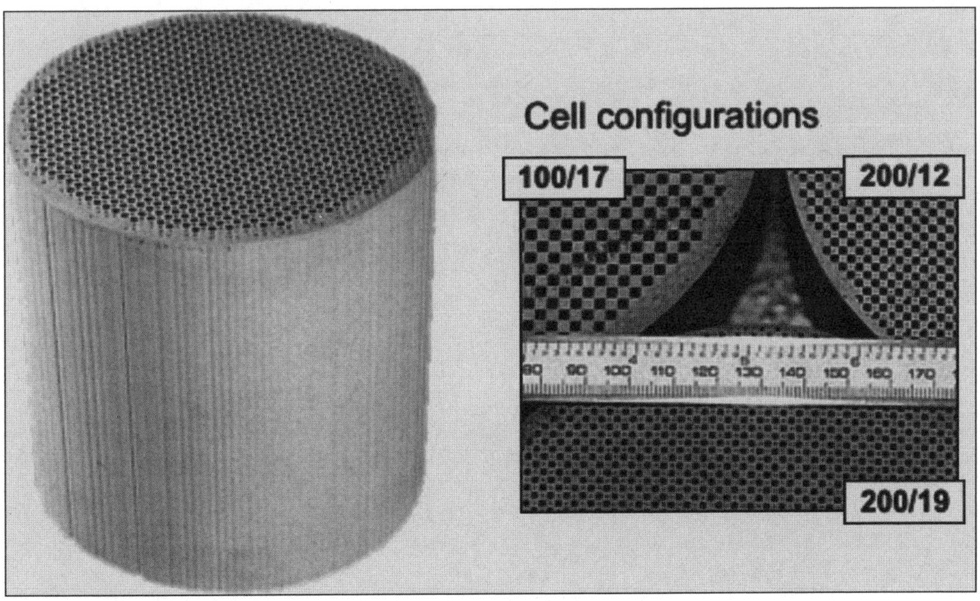

Figure 21.12 *Cordierite wall-flow monolith.*

A commercial line of cordierite filter monoliths has been developed by Corning. The first Corning diesel filter formulations were EX-47, EX-54, and EX-66. All of them had the same cell configuration (100 cpsi, 0.017 in. = 0.43-mm wall), but different pore sizes, resulting in different filtration efficiencies. Acid leaching, which was used in the manufacturing process of these early formulations to help reduce the coefficient of thermal expansion, had an adverse effect on mechanical durability, fatigue, and cost [29]. EX-80, a new material that was developed with the goal of improving long-term thermal durability, is the current commercial cordierite formulation by Corning [32, 33]. The EX-80 maintained the old cell geometry of 100/17. The material and cell structure of wall-flow monoliths are commonly coded as (formulation)/(cell density)/(wall thickness). In the case of the basic EX-80 formulation, the material designation is EX-80/100/17.

Faced with instances of filter melting in some applications, Corning developed a modified cell geometry material named DuraTrap RC ("Robust Cordierite") [34]. An important feature of the new material is increased heat capacity, which is realized through thick channel walls; the cell geometry is 200 cpsi and it has a 0.019-in. wall thickness. It was demonstrated that, thanks to the heat absorption in the monolith body, high heat capacity parts experience lower temperature peaks during uncontrolled regenerations [35]. Presumably, the high heat capacity parts are to some degree less effective in supporting normal regeneration. At this time the RC-200/19 monoliths are not commercially available.

The physical properties of different cordierite filter materials are listed in Table 21.6. Intrinsic material properties for cordierite in standard monolith sizes available from Corning are listed in Table 21.7.

21.2.3.3 Silicon Carbide Monoliths

Silicon carbide monoliths are manufactured through extrusion from SiC powder. The pore network is formed during the extrusion process, as a result of the granular geometry of the

TABLE 21.6
PHYSICAL PROPERTIES OF CORDIERITE DIESEL FILTER MATERIALS

	Composition					
	EX-47	EX-54	EX-66	EX-80	RC-200/19	
Reference	[29, 36][a]				[34]	
Configuration	100/17				200/19	
Porosity, %	50	50	50	48	49	45
Mean pore size, μm	13.4	24.4	34.1	13.4	13.0[b]	13.0[b]
Bulk density,[c] g/cm^3	0.39	0.45	0.54	0.42	0.46	0.70
Axial CTE (25–800°C), $10^{-7} K^{-1}$	8.8	8.7	10.5	3.3	4.3	6.0
Axial MOR, MPa	2.52	2.21	2.88	2.41	3.16	4.67
• psi	365	320	418	350	458	677
E-MOD, GPa	5.59	5.72	7.31	5.17	5.59	9.10
• 10^6 psi	0.81	0.83	1.06	0.75	0.81	1.32
Initial efficiency, %	84	70	55	87	—	—
Final efficiency, %	90	90	90	91	—	—
Pressure drop @ 40 g soot, in. Hg	2.9	3.7	3.7	2.9	—	—
A-axis strength, MPa	10.6	9.3	11.3	9.9	—	—
• psi	1539	1350	1642	1434		
3D isostatic strength, psi	585	460	660	500	—	—
2D isostatic strength, psi	500	390	560	425	—	—
TSP	516	459	383	1414	1315	855
Total raw penetration, cm^3/g	0.418	0.421	0.410	0.35–0.44	—	—
Bulk volumetric heat capacity,[d] $kJ/m^3 K$	—	—	—	—	540	820

Key: CTE, coefficient of thermal expansion; MOR, modulus of rupture; TSP = MOR/(CTE × E – MOD) – thermal shock parameter.

a. Properties determined from 10.5-in. diameter × 12-in. parts.

b. Median diameter.

c. Intrinsic density of cordierite = 2.51 g/cm^3.

d. Intrinsic specific heat of cordierite @ 500°C = 1.11 kJ/kg K.

TABLE 21.7
DIESEL FILTER SIZES
(CORNING, EX-80/100/17)

Size, Diameter × Length		Filter Volume, dm^3	Filtration Surface Area, m^2	Maximum Soot Load,[a] g
in.	mm			
One-Piece Extrusions				
5.66 × 6	143.8 × 152.4	2.5	1.66	15
7.5 × 8	190.5 × 203.8	5.8	3.83	35
9.0 × 12	228.6 × 304.8	12.5	8.27	75
10.5 × 12	266.7 × 304.8	17.0	11.26	100
11.25 × 12	285.8 × 304.8	19.6	12.93	120
11.25 × 14	285.8 × 355.6	22.8	15.08	140
12 × 15	304.8 × 381.0	27.8	18.39	170
Assemblies				
15 × 15	381.0 × 381.0	43.4	28.73	250
20 × 15	508.0 × 381.0	77.2	51.08	460

a. Approximately 6 g/L.

raw material, although additives may be also used to modify pore characteristics [30]. Extruded monoliths are fired at temperatures above 2500°C in an inert atmosphere, followed by processing at about 1000°C at oxidizing conditions, to form an approximately 50-nm-thick SiO_2 layer that protects the nonoxide material from corrosion [37]. SiC material (Figure 21.13) features favorable, uniform pore structure, characterized by narrow distribution of pore sizes, a high number of pores per unit surface area, and the absence of closed, "blind" pores.

SiC filters appear to be compatible with catalysts, although experience in this area is still limited. Due to the smooth surface of SiC grains, a refractory oxide layer ("washcoat") is probably necessary before a noble metal catalyst can be applied. Through promoting crystallization of cristobalite and tridymite phases, as opposed to amorphous silica, the SiO_2 surface layer may be optimized to provide some surface roughness and satisfactory washcoat adhesion. NoTox experimented with zirconia-alumina coatings on such surface-optimized SiC [37]. Taoka *et al.* [38] coated SiC of pore sizes ranging from 9 to 40 μm with alumina washcoats with loadings up to 25 g/dm^3.

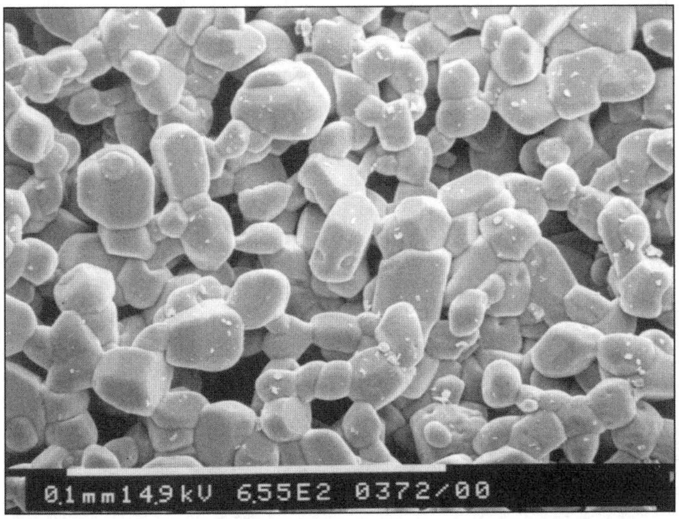

Figure 21.13 *SiC filter material: electron microscope photograph of channel surface.*

One of the most important advantages of SiC material is its high-temperature limit. SiC does not melt; rather it sublimates at a temperature, depending on the literature source, between 1800 and 2400°C. Again, as was the case with cordierite, the maximum use temperature of SiC is limited by ash reactions, which were reported to occur at around 1400°C [30]. According to Merkel [1], by 1300°C ash forms a glassy surface layer on SiC that seals the porosity of the material. However, considering the very high thermal conductivity of SiC and its high heat capacity, temperatures needed for the formation of such glaze are not easily reached in SiC filters, even during uncontrolled regeneration.

In general, SiC has better chemical resistance than cordierite, but, as a nonoxide material, it has lower resistance to oxidation [30].

Physical properties of a number of SiC materials are compared in Table 21.8. A 300/12 formulation being developed by Ibiden has the same pore characteristics as the 200/14 material. Due to possible differences in test methods and sample preparation, not all parameters are directly comparable.

TABLE 21.8
PROPERTIES OF SiC FILTER MATERIALS

Material and Configuration		
Supplier	Ibiden	NoTox[a]
Reference	[39]	[40]
Configuration	200/14	172/31
Mean pore size, µm	8.7	15
Porosity, %	42	45
Permeability, $\times 10^{-12}$ m^2	1.24	1.30
Mechanical Properties		
Axial MOR, MPa	>40	25
A-axis crush strength, MPa	>6	60
B-axis crush strength, MPa	>5	—
C-axis crush strength, MPa	>1.5	—
Isostatic strength, MPa	5	—
E modulus, GPa	48	50
Thermal Properties		
Sublimation temperature	2200	>1800°C
Thermal expansion, $\times 10^{-7}$/°C	40	39[b]
Thermal shock parameter (TSP)	>208	128
Thermal conductivity @ 25°C, W/m K	73	11
Thermal conductivity @ 1000°C, W/m K	19	7[c]
Specific heat @ 25°C, kJ/kg K	—	0.75
Specific heat @ 800°C, kJ/kg K	—	1.25
Maximum safe soot load, g/dm^3	—	6

a. Based on NoTox L6 substrate (θ•190 mm × 205 mm, four segments); there are slight differences in cell geometry and other properties between different NoTox parts.

b. @ 650°C – axial; 42 × 10^{-7} 1/°C @25°C – radial.

c. @ 630°C.

The fundamental disadvantage of SiC materials is their high CTE and high degree of brittleness (i.e., high E modulus), which result in poor thermal shock tolerance, as apparent in the relatively low TSP values (compare with Table 21.5). Attempts have been made to improve these properties through advanced material technologies. One such material—a composite Si-SiC—utilizes silicon metal as the bonding agent between SiC grains [30]. Compared with the conventional, recrystallized SiC, the composite material showed substantially reduced thermal conductivity (by 71%) and E-modulus (by 80%). Even though the fracture strength was also reduced (by 77%), the TSP in the new material increased from 250 to 300. Measured by the water quenching test, the Si-SiC showed thermal shock resistance of 1200°C, compared to 800°C in the standard SiC.

The most common method of coping with the thermal shock issue is to limit the diameter of SiC monoliths. Larger filters (including those utilizing the above-mentioned composite SiC material) are designed of several longitudinal segments, with spaces between them filled using a bonding and/or sealing material. Bonding materials based on such materials as silica-alumina fibers, silica sol, carboxymethyl cellulose, and SiC powder have been developed, which have both good adhesion and elastic properties and, thus, can absorb the thermal strain from SiC segments they are holding together [41]. Figure 21.14 shows a filter monolith assembled from square cross-sectional, 33- × 33-mm segments using this technique. Dimensional parameters of a commercial line of SiC filters utilizing this type of segmentation [42] are listed in Table 21.9.

One could expect that, due to the high temperature limit of SiC, its filtration capacity would be much higher than that of cordierite. Such a conclusion, however, would not be valid. In the case of SiC, the soot load is limited by the temperature gradient caused by uncontrolled regeneration. As a consequence of the low thermal shock resistance, cracking of SiC filters occurs when the temperature gradient exceeds about 100°C/cm

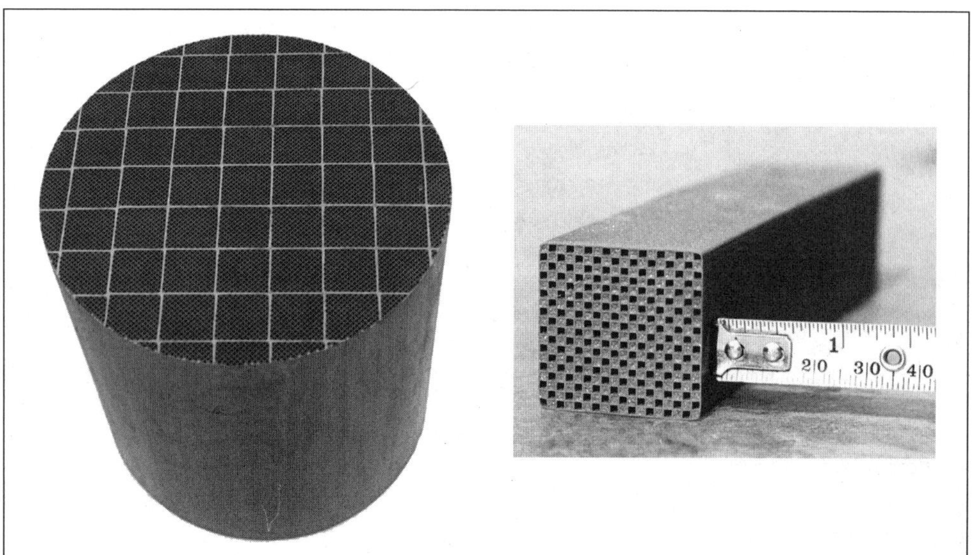

Figure 21.14 SiC wall-flow monolith made of cemented sections (Model G6 in Table 21.9).

TABLE 21.9
SiC FILTER SIZES
(IBIDEN, 200/14)

Model	Diameter mm	Diameter inch	Length mm	Length inch	Assembly Type	Number of Segments	Cross-Sectional Area, cm^2	Volume, dm^3
A1	118.4	4.66	150.5	6	4×4	14	110.1	1.66
A2			177.8	7				1.96
A3			203.2	8				2.24
A4			228.6	9				2.52
A5			254.0	10				2.80
A6			300.0	12				3.30
B1	143.8	5.66	150.5	6	4×4	14	162.4	2.44
B2			177.8	7				2.89
B3			203.2	8				3.30
B4			228.6	9				3.71
B5			254.0	10				4.12
B6			300.0	12				4.87
C1	165.0	6.5	150.5	6	5×5	23	213.8	3.22
C2			177.8	7				3.80
C3			203.2	8				4.34
C4			228.6	9				4.89
C5			254.0	10				5.43
C6			300.0	12				6.41
D1	190.5	7.5	150.5	6	6×6	32	285.0	4.29
D2			177.8	7				5.07
D3			203.2	8				5.79
D4			228.6	9				6.52
D5			254.0	10				7.24
D6			300.0	12				8.55
E1	228.6	9	150.5	6	7×7	45	410.4	6.18
E2			177.8	7				7.30
E3			203.2	8				8.34
E4			228.6	9				9.38
E5			254.0	10				10.42
E6			300.0	12				12.31
F1	266.7	10.5	150.5	6	8×8	60	558.6	8.41
F2			177.8	7				9.93
F3			203.2	8				11.35
F4			228.6	9				12.77
F5			254.0	10				14.19
F6			300.0	12				16.76
G1	285.8	11.25	150.5	6	9×9	69	641.5	9.65
G2			177.8	7				11.41
G3			203.2	8				13.04
G4			228.6	9				14.67
G5			254.0	10				16.29
G6			300.0	12				19.26

(Ibiden 200/14). The maximum safe soot load varies typically between 7 and 10 g/L, depending on various operational and geometric parameters [39].

21.2.4 Pressure Drop

The high filtration efficiencies of wall-flow filters are obtained at the expense of a relatively high pressure drop, which increases with the filter soot load. Pressure drop buildup in a wall-flow monolith under steady-state engine conditions [43] (i.e., steady-state gas flow and soot input) is illustrated in Figure 21.15. Initially, the filter is clean and its pressure drop is at the ΔP_{clean} level. As the particulates start depositing within the pores in monolith walls (depth filtration), the pressure drop starts increasing with time in a nonlinear manner. This phase is called the initial loading phase, during which pore properties like permeability and filter porosity continuously change due to the increasing soot deposit inside the pore network.

After the filtration capacity of the pores becomes saturated, soot starts depositing as a layer inside the inlet monolith channels (cake filtration phase). A linear increase in pressure drop with time (and with soot load) is observed during this period. The permeability and porosity of the monolith walls and the soot layer can be assumed constant during this stabilized loading period. The only property that changes is the thickness of the soot layer. Some authors also distinguish an intermediate short transition phase, from the moment the particulates start depositing on the channel surface until the soot layer is fully established [44, 45].

The pressure drop of the clean filter can be divided into three components:

- Pressure drop due to sudden contraction and expansion at the inlet and outlet from the filter, $\Delta P_{in/out}$
- Pressure drop due to channel wall friction, $\Delta P_{channel}$
- Pressure drop due to wall permeability, ΔP_{wall}

$$\Delta P_{clean} = \Delta P_{in/out} + \Delta P_{channel} + \Delta P_{wall} \qquad (21.9)$$

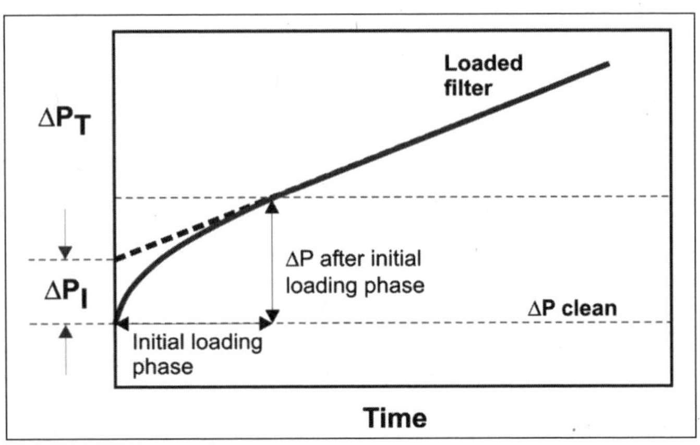

Figure 21.15 *Pressure drop regimes.*

Pressure drop across the monolith wall is most commonly modeled using the Darcy's law equation for viscous effects:

$$\Delta P_{wall} = (\mu/k_w) \cdot v_w \cdot w \qquad (21.10)$$

where:

μ = dynamic viscosity of exhaust gas, Pa·s
k_w = permeability of filter material, m^2
v_w = wall velocity of gas (exhaust flow per unit of filtration area), m/s
w = wall thickness, m.

The wall permeability, k_w, is usually specified by the filter manufacturer. It can be also estimated from other pore properties using the Kozeny-Carman correlation [39]:

$$k_w = (1/5.6)\, \varepsilon^{5.5}\, d_p^{\,2} \qquad (21.11)$$

where ε is the porosity of the filter wall and is dimensionless, and d_p is the pore diameter in meters.

The values for $\Delta P_{channel}$ and $\Delta P_{in/out}$ are calculated using the same methods used for pressure drop calculations in flow-through catalyst substrates (see Section 18.3.4). Pressure drop in the wall, ΔP_{wall}, is typically the most significant contribution to the total pressure drop. However, in substrates with long channels of small hydraulic diameter, the channel contribution ($\Delta P_{channel}$) may be also significant. The pressure drop effects of the inlet contraction and outlet expansion are small.

Pressure drop modeling in clean filter substrates is quite a straightforward task. Relatively simple models that have been developed show excellent agreement with experiment [46]. Most of the filter pressure drop in real applications, however, is created by the soot deposit. The importance of the soot contribution can be illustrated theoretically by comparing the permeability for cordierite (0.5×10^{-12} m^2) with that of diesel soot (2.2×10^{-15} – 12.8×10^{-15} m^2 [43]). As the soot layer builds up in the channels, its pressure drop will quickly exceed that of the wall, as one can calculate from Eq. (21.10) by substituting wall parameters with those of the soot layer. In practical applications, the pressure drop of the clean wall-flow filter is typically in the range of 1 to 2 kPa, while a loaded filter pressure drop of, say, 10 kPa can be considered low to moderate.

The total pressure drop of the particulate loaded filter, ΔP, can be divided into four components:

- *Pressure drop due to sudden contraction and expansion at the inlet and outlet from the filter, $\Delta P_{in/out}$:* Similar to the same component in the clean filter, except that the effective channel size (hydraulic diameter) is now smaller due to the soot layer, resulting in more gas contraction.
- *Pressure drop due to channel wall friction, $\Delta P_{channel}$:* Also increases relative to the clean filter scenario, due to the decrease in the channel hydraulic diameter. With thick soot layers, $\Delta P_{channel}$ can become a very significant contributor to the total pressure drop.

- *Pressure drop due to permeability of particulate layer, $\Delta P_{particulate}$:* Likely the highest contributor to the total pressure drop. Can be calculated from Eq. (21.10) by replacing wall parameters with soot layer parameters.
- *Pressure drop due to wall permeability, ΔP_{wall}:* Now also higher than in the clean filter, because the wall pores are partly filled with soot. The increase in ΔP_{wall} that can be attributed to the initial soot loading phase in the pores is represented by ΔP_I in Figure 21.15.

The total pressure drop can be expressed as follows:

$$P = \Delta P_{in/out} + \Delta P_{channel} + \Delta P_{particulate} + \Delta P_{wall} \qquad (21.12)$$

Mathematical modeling of the pressure drop in soot-loaded diesel filters becomes a complex and difficult task. Important properties of soot, such as the permeability and packing density, depend on the application, engine operating conditions, and other parameters. Efforts are ongoing to simulate pressure drop in wall-flow filters and increasingly more sophisticated models are being developed [43, 47]. Note also that predicting filter pressure drop in real-life operation requires not only a working pressure drop model, but also the knowledge of the particulate mass in the filter at any given moment. This soot load is constantly changing depending on the real application, duty cycle, and regeneration method. Predicting the actual soot loading would require a theoretical model of the regeneration process itself.

21.2.5 Canning

Diesel wall-flow filter monoliths are packaged into stainless steel housings. The recommended housing materials are ferritic stainless steels AISA 400 series, such as 409, which are characterized by low thermal expansion coefficients. The packaging methods are similar to those used for catalytic converters, with a tourniquet being the recommended technique. Particulate filters are typically larger than catalytic converters for the same application. Due to these large sizes and additional forces resulting from the higher pressure drop, supporting rings are usually provided inside the housing, which hold the monolith and prevent its axial movement.

Monolithic filters can be installed on a vehicle either horizontally or vertically. The horizontal installation provides better mechanical durability of the filter system, especially for larger and heavier units. Vertically installed monoliths have a tendency to slide down inside the can and have most of their weight supported by the bottom ring. Vehicle vibration and the exhaust gas pressure may cause the ring to gradually work its way into the monolith by rubbing the ceramic material.

Heavy packaging mats should be used to securely hold the relatively heavy weight of the filters and to evenly distribute the canning pressure. Modern filter materials have sufficient strength to withstand canning pressure. The mounting gap densities of the mat are usually only slightly lower than those used for catalytic converters. For example, gap densities of about 1 g/cm^3 are typically possible with the Interam 6200 g/m^2 mat. When designing the packaging system, the mat pressures should be estimated by calculation and compared with the monolith isostatic strength specified by the filter manufacturer. If intumescent mats are used, filter units should be fired after canning to allow for mat expansion. This requirement is related to the low diesel exhaust gas temperatures, which, in most cases, are not sufficient to expand the mat during vehicle operation.

21.3 Ceramic Fibers and Cartridges

21.3.1 Fiber Materials

Filter cartridges for filtering of diesel particulates can be assembled using high-temperature continuous ceramic fibers. A number of such fibers, typically polycrystalline metal oxide fibers made from alumina and silica, have been developed for textiles used in high-temperature operating environments. Typical applications include insulation for thermocouple wires, belts for high-temperature furnaces, filter bags for hot gas filtration, components of space shuttle tiles, and flame curtains. Traditional fiber materials include leached silica, fused silica, asbestos, or glass. In general, these fibers are characterized by high strength, high rigidity, low coefficients of thermal conductivity and expansion, and large active surface areas.

Nextel, a brand of synthetic fiber developed by 3M, is an example of a high-temperature fiber that was also extensively tested for diesel filter applications. Typical properties of Nextel fibers [48], compared with a commercial SiO_2 fiber [7], are listed in Table 21.10. The data indicate that Nextel fibers exhibit favorable strength and flexibility.

The Nextel fibers have a virtually round cross section and their surface is extremely smooth and uniform. In filter applications, the adhesion of filtered particles to the fibers can be greatly enhanced through special treatment to increase the surface roughness and the active surface, such as leaching of the fibers. A special texturization process has been developed for the manufacturing of the Nextel 312 fibers used for diesel filtration [6]. In the process, the raw fiber bundles are separated and the individual fibers are spread in a controlled manner. Spreading and separation produces "expanded" material with a number of 10- to 12-μm continuous fibers. The degree of expansion can be controlled, which allows the material to be manufactured with consistency.

Fiber diameter is one of the most important parameters that determine the filter performance. Generally, small-diameter fibers have better filtration efficiencies. Coarse fibers tend to show low efficiency, critical overloading phenomena, and blow-off [7]. However, very small fiber diameters may present a health hazard, as well as increased costs and problems with handling. A typical diameter for diesel filter fibers is in the range of 10 μm.

Fibers of 10 μm in diameter are generally considered not to be in the respirable range and to present a minimal health hazard. This opinion, based on current theory and knowledge, was also confirmed by the Nextel fiber manufacturer through experiments with animals [48]. The Nextel fibers are very uniform in diameter. In contrast, an asbestos sample presents

TABLE 21.10
PHYSICAL PROPERTIES OF NEXTEL 312 AND SILICA FIBERS

Material	Nextel 312	Silica
Form	Polycrystalline	Amorphous
Composition	Al_2O_3 - 62%; SiO_2 - 24%; B_2O_3 - 14 wt. %	SiO_2
Fiber length	Continuous	Continuous
Fiber diameter	10–12 μm (0.394–0.472 mil)	7–13 μm
Fiber density	2.7 g/cm^3 (0.097 lb/in.^{3})	3.0 g/cm^3
Surface area	<1 m^2/g	0.1–0.3 m^2/g
Elongation	1.2 %	
Filament tensile strength	1.72 GPa (250 × 10^3 psi)	0.80 GPa
Filament tensile modulus of elasticity	138 GPa (20 × 10^6 psi)	66 GPa
Continuous use temperature[a]	1204°C (2200°F)	1100°C
Short-term use temperature[a]	1371°C (2500°F)	
Melting point	1800°C (3272°F)	
Specific heat	1046.7 J/(kg K) (0.25 Btu/[lb °F)]	
Lineal shrinkage 1093°C (2000°F)	1.25%	
Thermal expansion coefficient 25–500°C (77–932°F)	3.0 × 10^{-6} 1/°C (1.7 × 10^{-8} 1/°F)	

a. in diesel filter application, fiber degradation occurs above 900°C due to ash melting and bonding of fibers together.

an inhomogeneous collection of fibrils with diameters 100 to 1000 times smaller than those of Nextel fibers.

21.3.2 Filtration Mechanism

Fibers in a fiber filter cartridge are relatively loosely spaced [8]. Average distance between fibers is typically many times larger than the fiber diameter, resulting in a rather "empty" appearance of the filtration volume. For this reason, fibers make true deep-bed filters with depth filtration as the dominant type of filtration.

Particles are deposited directly on the fibers, forming various size clusters. Contrary to monolithic filter substrates, a continuous soot layer (filtration cake) never develops under normal operating conditions in properly designed fiber filters. This depth filtration mechanism is responsible for most of the performance differences relative to monolithic filters, as discussed later.

21.3.3 Filter Cartridges

Fibers fill the working volume of filter cartridges, which can be welded and used as the building blocks for filter systems. The cartridge design typically involves perforated metal tubes with a layer of ceramic yarn around them. It is important that the exhaust gas does not bypass the fibers or develop channels of lower fiber density across the yarn layer or at the ends of the cartridge. Such bypassing or channeling would result in a dramatic deterioration of filtration efficiency. A possible technique of assembling the filter cartridge involves winding of the fibers around the tube. The diamond winding pattern visible in the example cartridge in Figure 21.16 was developed to ensure a uniform and leak-free flow of exhaust gases through the fibers [48].

Another design approach for durable and leak-free fiber beds utilizes knitting of ceramic fibers into three-dimensional structures. Special technologies were designed that allowed the fibers to be knit without stretching them. Fibers in the knitted structure could be held together to provide uniform flow distribution while maintaining elasticity and avoiding fiber stress.

Several techniques, such as flat multiple layering or cylindric rolling, are possible to fabricate filter cartridges from knitted fibers. The best results in terms of minimizing bypass leakage and pressure drop were reported using the following method [7]. The knitting was circular; that is, a tube was knitted from a single fiber thread. This tube was then wound into itself to form a torus-shaped structure. The toroids, basic units of the filter, were placed in a perforated can (between two concentric, perforated tubes) to form the filter cartridge.

In a filter system, several cartridges are packaged into a steel housing, which typically replaces the vehicle's muffler. An example filter system with fiber cartridges is discussed in Section 22.5 on electrically regenerated filters.

Figure 21.16 (a) Wound fiber cartridge and (b) diamond winding pattern of same cartridge.

21.3.4 Advantages and Disadvantages of Fiber Filters

Performance, design parameters, and other features of fiber filters are usually compared to those of cordierite wall-flow monoliths, which became the benchmark in diesel filter materials. The advantages of fiber filters can be summarized as follows:

- Resistant to damage by thermal or mechanical stress.
- Less risk of clogging with soot and/or ash.
- Tolerant to additive ash accumulation when used with fuel additive regeneration. The ash, uniformly deposited throughout the entire media, would not cause high pressure drop or clogging.
- Good noise attenuation [8].
- Good efficiency for particle number reduction. Under some conditions, fiber filters were shown to outperform monolithic filters [9].

On the negative side, fiber filters show a number of disadvantages, as follows:

- Incompatibility with catalysts. Catalyst coating of fiber filters was attempted [49], but no satisfactory results have been reported. Catalyzing fibers may require more sophisticated methods than washcoating, such as chemical vapor deposition (CVD) [50]. Even if CVD coating is successful, washcoated catalyst formulations would not be portable onto fiber filters, so new catalysts would have to be developed.
- Vulnerability to damage by ash. Fiber filters are easily damaged by sintered ashes that coat and bond fibers together [2]. Such coated and aggregated fibers are subject to vibrational failure, resulting in voids in the fiber bed and gas leakage. Fibers may also undergo chemical attack by some additives [51].
- Large size. Fiber filter systems are typically larger than monolithic filters.
- Lower filtration efficiency. Fiber filters show somewhat lower gravimetric filtration efficiency than monolithic filters. They may be also susceptible to blow-off of accumulated soot. This may result in visible black smoke during accelerations and further reduced efficiency on transient test cycles.
- Secondary fiber emissions. In case of fiber breakage, fiber fragments are emitted as a secondary emission.
- High cost of cartridges. There is no indication that fiber filters can be cost competitive with cordierite monoliths.

Fiber filter systems have been commercialized on a very limited scale. Aftermarket fiber filters are available for some industrial engine and urban bus applications, primarily in Europe. The systems are regenerated by fuel additives, fuel burners, or electrically.

21.3.5 Example Design: Nextel Fiber Cartridges

21.3.5.1 Types of Cartridges

Design details and performance of fiber filters can be illustrated using the example of Nextel fiber cartridges developed by 3M Company [6]. There are three lines of Nextel filter cartridges: (1) basic cartridges, (2) electric cartridges, and (3) a concentric tube pack. These cartridges are the building blocks available for the designer of diesel filter system. The electric cartridges were designed specifically with electric filter regeneration in mind. The nonelectric cartridges can be combined with different methods of regeneration. Recommended methods include fuel burners and fuel additives.

21.3.5.2 Basic Cartridge

The basic filter cartridge, illustrated in Figure 21.17, consists of a perforated support tube with ceramic fibers wound around it. A continuous fiber thread is wound onto the support tube in a diamond-shaped pattern. This pattern is wound in many layers to create the depth of fiber that is used in the filtration process.

The support tube is made of 1.22-mm (18-gauge), perforated, 50% open, Type 304 stainless steel. A 30- to 50-mm section at each end of the tube is not perforated, which allows

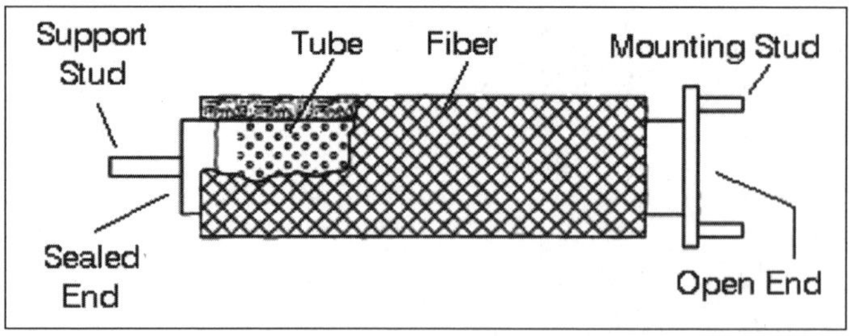

Figure 21.17 Wound fiber cartridge. (Based on Ref. 48.)

a gas seal to be created during the fiber winding process. One end of the tube is closed to force the exhaust gas through the wound fiber layer. The direction of the exhaust gas flow might be either outside-in or inside-out.

Basic cartridges can be manufactured in a wide range of dimensions:

- Cartridge diameter: 25 to 300 mm
- Cartridge length: 75 to 1500 mm
- Variable load time, filtration efficiency and pressure drop.

Dimensions of selected standard cartridges are listed in Table 21.11 [48].

Soot holding capacity and the filtration efficiency of a basic cartridge is shown in Figure 21.18. The data have been collected by the cartridge manufacturer in a laboratory test on a Cummins 6BTA5.9 engine under steady-state conditions.

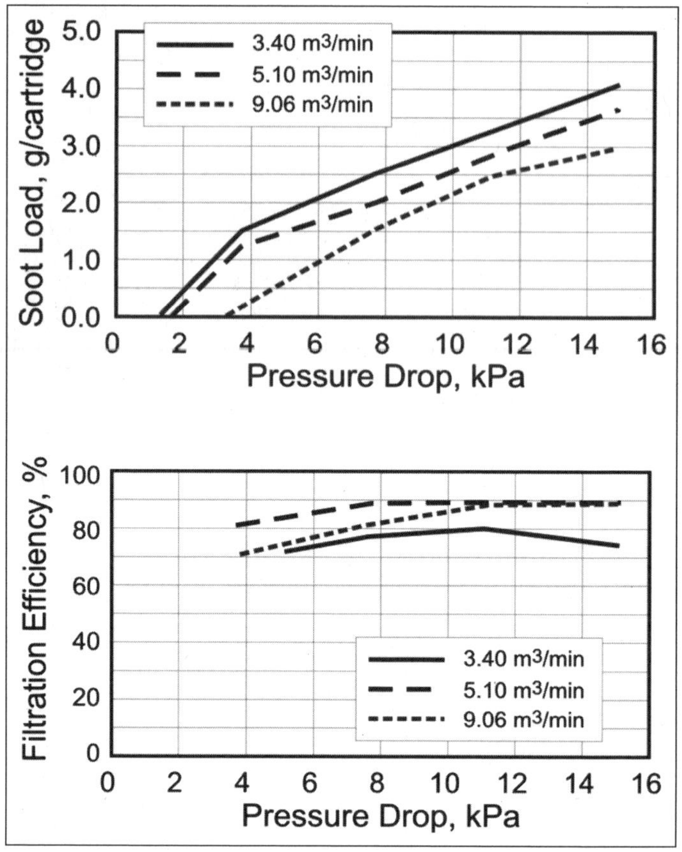

Figure 21.18 Soot holding and filtration efficiency of Nextel fiber cartridge. Flow rates at 280°C. Cartridges have a 32-mm tube diameter, a 61-mm diameter over fiber, and are 445 mm long (3M #XW3C-054).

21.3.5.3 Electric Cartridge

The basic cartridge design was modified by incorporating an electric heater to facilitate filter regeneration [52]. Early designs of the cartridge utilized a resistance wire incorporated into the wound fiber. This design was causing local overheating and breakage of fiber. It has been replaced with an improved configuration where the fiber support pipe itself serves as the heating element. The larger surface area of the heater, compared to the surface area of the wire, results in smaller energy densities and lower surface temperatures.

Figure 21.19 shows an electric filter cartridge. Ceramic fiber is wound around an electrically resistive heater support structure. The heater support is made of 0.46-mm-thick (26-gauge) Inconel 600 nickel-based alloy. This Inconel sheet is punched in a "V-slot" pattern of about 50% open area. The design of the pattern may vary depending on the required electrical resistance. The last 40- to 70-mm sections of each end of the material are not punched to allow for a gas seal to be created during the winding process. The punched Inconel sheet is rolled into a tubular form and welded. Mounting hardware is attached to the heater support tube to fit specific mounting and electrical connection requirements. The heater support tube is wound with filter fiber using the same techniques as the basic filter cartridge discussed earlier.

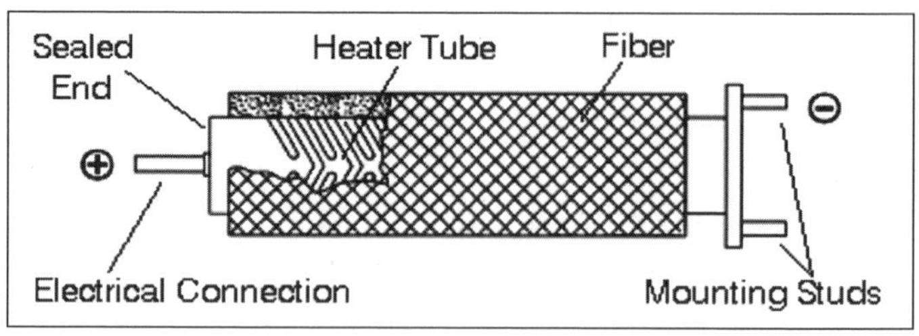

Figure 21.19 Filter cartridge with internal heater. (Based on Ref. 48.)

TABLE 21.11
STANDARD NEXTEL FIBER FILTER CARTRIDGES (BASIC DESIGN)

3M Part Number	Flow Range, m³/min	Tube Length, mm	Tube Outside Diameter, mm	Diameter Over Fiber, mm
XN-454	7.1–15.6	965.2	76.2	96.5
XW3C-053	2.8–5.6	500	40	68
XW3C-054	2.0–4.0	445	32	61

Only one exhaust flow direction, from the inside to the outside, is possible through the electrically regenerable cartridge. In this arrangement, the flow carries the heat from the heater into the filter media. For best process control and low power consumption, the regeneration should be done using a small, controlled amount of gas. During regeneration, the main exhaust gas stream bypasses the filter and a controlled amount of air or exhaust gas is supplied to the cartridge. The recommended regeneration flow rate is about 2.5% of the nominal exhaust gas flow capacity of the filter.

The electric cartridges offer much less design flexibility than the basic design. The following are possible ranges for the design parameters:

- Diameter: 38 to 76 mm
- Length: 200 to 300 mm
- Voltage: 12 or 24 V
- Power requirement: 400 to 1000 W
- Variable load time, filtration efficiency, and pressure drop.

Standard electric cartridges developed by 3M (part number XW3H-058) have tube lengths of 254 mm, tube outside diameters of 51.7 mm, and a diameter over the fiber layer of 70 mm. They are intended for exhaust gas flow between 0.6 and 1.4 m^3/min.

21.3.5.4 Concentric Tube Pack

The concentric tube pack (CTP) is made up of several concentric filter cartridges. Each filter cartridge is slightly larger than the previous one. They are separated by annular rings that block the space between the concentric support tubes at alternate ends. This configuration forces exhaust gases to flow through the filter media, trapping particulate in the fiber layer. This design has been developed to maximize the filtration surface area for a given volume. Figure 21.20 shows a concentric filter cartridge assembly with five concentric filter cartridges installed.

The canister is made from 1.6-mm (16-gauge) stainless steel. These filter cartridge assemblies can be made with outside diameters up to 321 mm (12.62 in.) and lengths up to 711 mm (28 in.). Standard packs consist of either three or five concentric cartridges and cover exhaust flow ranges from 7.1 to 50.9 m^3/min (actual).

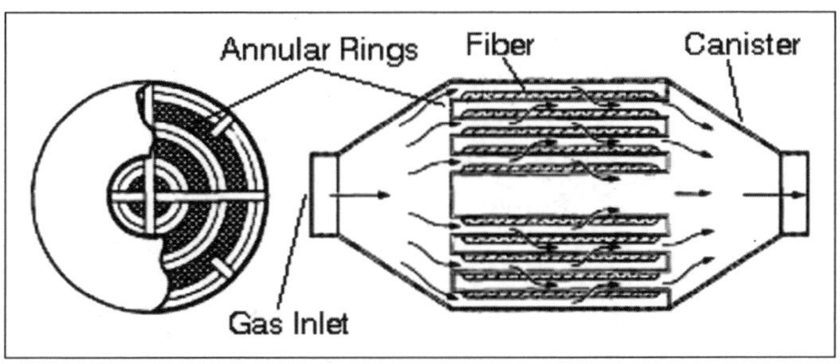

Figure 21.20 *Nextel Concentric Tube Pack. (Based on Ref. 48.)*

References

1. Merkel, G.A., W.A. Cutler, and C.J. Warren, "Thermal Durability of Wall-Flow Ceramic Diesel Particulate Filters," SAE Paper No. 2001-01-0190, SAE International, Warrendale, PA, 2001.
2. Child, D., and J. Cioffi, "An Electronically Controlled Diesel Particulate Filter System Employing Electrically Regenerated Ceramic Fiber Wound Cartridges," SAE Paper No. 942265, SAE International, Warrendale, PA, 1994.
3. Hickman, D.L., S. Ebener, and U. Zink, "Reduktion der Partikelemissionen bei Diesel PKW und Nutzfahrzeugen," in *22nd Int. Wiener Motorsymposium,* Fortschritt-Berichte VDI (VDI Verlag, Duesseldorf), Reihe 12, Nr. 455, Band 2, pp. 267–285, 2001.
4. Murtagh, M.J. "Diesel Particulate Filters (DPF): A Short Course," Diesel Particulate and NOx Emissions Course (University of Leeds), Ann Arbor, MI, October 2002.
5. Outland, R.J., "Ceramic Filters for Diesel Exhaust Particulates," U.S. Patent No. 4,276,071 (General Motors), 1981.
6. Bloom, R., "The Development of Fiber Wound Diesel Particulate Filter Cartridges," SAE Paper No. 950152, SAE International, Warrendale, PA, 1995.
7. Mayer, A., and A. Buck, "Knitted Ceramic Fibers—A New Concept for Particulate Traps," SAE Paper No. 920146, SAE International, Warrendale, PA, 1992.
8. Kahlert, B. "First Use of Additive-Regenerated Particulate Filters in German Underground Mines," in *Proceedings from the Mining Diesel Emissions Conference (MDEC),* CANMET, Toronto, November 1999.
9. Mayer, A., *et al.,* "Particle Size Distribution Downstream Traps of Different Design," SAE Paper No. 950373, SAE International, Warrendale, PA, 1995.
10. Chung, H.Y., K.B. Dema, B. Keh, and M.A. Barris, "Ceramic Filter Construction and Method," U.S. Patent No. 5,250,094 (Donaldson), 1993.
11. Sakaguchi, T., *et al.,* "Development of High Durability Diesel Particulate Filter by Using SiC Fiber," SAE Paper No. 1999-01-0463, SAE International, Warrendale, PA, 1999.
12. Miller, R., W.C. Haberkamp, K.M. Bedeau, Z.G. Liu, R.C. Shirk, and T. Wood, "Design, Development and Performance of a Composite Diesel Particulate Filter," SAE Paper No. 2002-01-0323, SAE International, Warrendale, PA, 2002.
13. Haberkamp, W., and T.M. Yonushonis, "High Temperature Composite Ceramic Filter," U.S. Patent No. 6,444,006 (Fleetguard), 2002.
14. Mann, G., "Diesel Exhaust Particulate Traps," U.S. Patent No. 4,264,346 (General Motors), 1981.
15. Mizrah, T., *et al.,* "Open Pore Ceramic Foam as Diesel Particulate Filter," SAE Paper No. 890172, SAE International, Warrendale, PA, 1989.
16. Yoshida, K., S. Makino, S. Sumiya, and G. Muramatsu, "Simultaneous Reduction of NOx and Particulate Emissions

from Diesel Engine Exhaust," SAE Paper No. 892046, SAE International, Warrendale, PA, 1989.

17. MacMahon, M.A., *et al.,* "Alumina Coated Metal Wool as a Particulate Filter for Diesel Powered Vehicles," SAE Paper No. 820183, SAE International, Warrendale, PA, 1982.
18. Fang, C.P., and D.B. Kittelson, "The Influence of a Fibrous Diesel Particulate Trap on the Size Distribution of Emitted Particles," SAE Paper No. 840362, SAE International, Warrendale, PA, 1984.
19. Meinig, U., *et al.,* "Tests on a New Diesel Particulate System with Traps of Metal Fleece," SAE Paper No. 982597, SAE International, Warrendale, PA, 1998.
20. DePetrillo, F., *et al.,* "Rypos Trap Active Diesel Particulate Filter System: Field Demonstration," paper presented at *8th Diesel Emissions Reduction Conference* (DEER), San Diego, CA, August 2002, retrieved from www.orau.gov/deer2002/Session11/DePetrillo1.pdf.
21. "Sintered Metal Filter for Commercial Vehicle Exhaust Aftertreatment," Purem GmbH, Menden, Germany, February 2002.
22. Zelenka, P., *et al.,* "Towards Securing the Particulate Trap Regeneration: A System Combining a Sintered Metal Filter and Cerium Fuel Additive," SAE Paper No. 982598, SAE International, Warrendale, PA, 1998.
23. Yoro, K., *et al.,* "Diesel Particulate Filter Made of Porous Metal," SAE Paper No. 980187, SAE International, Warrendale, PA, 1998.
24. Ban, S., Y. Nagai, K. Kobashi, and H. Yanagihara, "Particulate Trap for Diesel Engine," U.S. Patent No. 5,908,480 (Sumitomo Electric Ind.), 1999.
25. Majewski, W.A., and E.P. Martin, "Diesel Particulate Filter with a Disposable Pleated Media Paper Element," SAE Paper No. 930370, SAE International, Warrendale, PA, 1993.
26. Majewski, W.A., *et al.,* "On-Vehicle Exhaust Gas Cooling in a Diesel Emissions Control System," SAE Paper No. 921676, SAE International, Warrendale, PA, 1992.
27. Wright, M.E., A.D. Klein, and E.S. Stesniak, "A Diesel Exhaust Filter System for Industrial Diesel Forklifts," SAE Paper No. 911852, SAE International, Warrendale, PA, 1991.
28. Ambs, J.L., *et al.,* "Disposable and Reusable Diesel Exhaust Filters," Information Circular IC9324, U.S. Bureau of Mines, 1992, p. 67.
29. Gulati, S.T., "Ceramic Catalyst Supports and Filters for Diesel Exhaust Aftertreatment," in *Structural Catalysts and Reactors,* A. Cybulski (Ed.), Marcel Dekker, New York, 1998, pp. 501–541.
30. Miwa, S., F. Abe, T. Hamanaka, T. Yamada, and Y. Miyari, "Diesel Particulate Filters Made of Newly Developed SiC," SAE Paper No. 2001-01-0192, SAE International, Warrendale, PA, 2001.
31. Cutler, W.A., and G.A. Merkel, "A New High Temperature Ceramic Material for Diesel Particulate Filter Applications," SAE Paper No. 2000-01-2844, SAE International, Warrendale, PA, 2000.
32. Murtagh, M.J., *et al.,* "Development of a Diesel Particulate Filter Composition and Its Effect on Thermal Durability and Filtration Performance," SAE Paper No. 940235, SAE International, Warrendale, PA, 1994.
33. "Corning Diesel Products," Corning Incorporated, New York, 1997.
34. Merkel, G.A., D.M. Beall, D.L. Hickman, and M.J. Vernacotola, "Effects of Microstructure and Cell Geometry on Performance of Cordierite Diesel Particulate Filters," SAE Paper No. 2001-01-0193, SAE International, Warrendale, PA, 2001.
35. Hickman, D.L., "Diesel Particulate Filter Regeneration: Thermal Management through Filter Design," SAE Paper No. 2000-01-2847, SAE International, Warrendale, PA, 2000.
36. McLean, R., "EX-80 New Diesel Filter Composition," Corning Incorporated, New York, June 1993.
37. "NoTox Silicon Carbide Diesel Particulate Wall Flow Filters," in *Mini-Manual of Design Considerations with the Use of Catalytic Active Coatings,* NoTox Corporation, September 1997.
38. Taoka, N., *et al.,* "Effect of SiC-DPF with High Cell Density for Pressure Loss and Regeneration," SAE Paper No. 2001-01-0191, SAE International, Warrendale, PA, 2001.
39. Ohno, K., *et al.,* "Characterization of SiC-DPF for Passenger Car," SAE Paper No. 2000-01-0185, SAE International, Warrendale, PA, 2000.
40. "NoTox Silicon Carbide Diesel Particulate Wall Flow Filter," 4th ed., Technical Product Information, NoTox Corporation, August 17, 1995.
41. Naruse, K., *et al.,* "Ceramic Structural Body," U.S. Patent No. 5,914,187 (Ibiden/Nissan), 1999.
42. "SiC-DPF Standard Product List," Ibiden, December 25, 2000.
43. Suresh, A., A. Khan, and J.H. Johnson, "An Experimental and Modeling Study of Cordierite Traps—Pressure Drop and Permeability of Clean and Particulate Loaded Traps," SAE Paper No. 2000-01-0476, SAE International, Warrendale, PA, 2000.
44. Tan, J.C., C.N. Opris, K.J. Baumgard, and J.H. Johnson, "A Study on the Regeneration Process in Diesel Particulate Traps Using a Copper Fuel Additive," SAE Paper No. 960136, SAE International, Warrendale, PA, 1996.
45. Versaevel, P., *et al.,* "Some Empirical Observations on Diesel Particulate Filter Modeling and Comparison Between Simulations and Experiments," SAE Paper No. 2000-01-0477, SAE International, Warrendale, PA, 2000.
46. Masoudi, M., A. Heibel, and P.M. Then, "Predicting Pressure Drop of Wall-Flow Diesel Particulate Filters—Theory and Experiment," SAE Paper No. 2000-01-0184, SAE International, Warrendale, PA, 2000.
47. Masoudi, M., *et al.,* "Validation of a Model and Development of a Simulator for Predicting the Pressure Drop of Diesel Particulate Filters," SAE Paper No. 2001-01-0911, SAE International, Warrendale, PA, 2001.

48. "3M Nextel Diesel Filter Cartridges for Particulate Emission Control," Designer Guide, 3M Company, February 1995.
49. Mayer, A., *et al.,* "Passive Regeneration of Catalyst Coated Knitted Fiber Diesel Particulate Traps," SAE Paper No. 960138, SAE International, Warrendale, PA, 1996.
50. Mayer, A., "Microfiber Knits for Catalytic Converters," SAE Paper No. 970479, SAE International, Warrendale, PA, 1997.
51. Bloom, R.L., *et al.,* "Fiber Wound Diesel Particulate Filter Durability Experience with Metal Based Additives," SAE Paper No. 970180, SAE International, Warrendale, PA, 1997.
52. Shirk, R., R. Bloom, Y. Kitahara, and M. Shinzawa, "Fiber Wound Electrically Regenerable Diesel Particulate Filter Cartridge for Small Diesel Engines," SAE Paper No. 950103, SAE International, Warrendale, PA, 1995.

Chapter 22

Diesel Filter Systems

22.1 Overview

22.1.1 Design Considerations

Diesel filter systems are designed by combining different filter materials with selected regeneration methods. The major system design considerations include emission performance, pressure drop, durability, maintenance, and cost. Filter system design targets can be listed as follows:

- High PM collection efficiency: gravimetric particulate collection efficiency, sulfate formation, penetration of small particles (<0.1 μm), particle number emissions
- Reductions of gaseous emissions (CO, HC, etc.)
- No generation of secondary emissions in the filter: increase of gas emissions (CO, HC, NO_2, etc.), new gas emissions (dioxins, etc.), additive ash emission, fiber loss (with fibrous media)
- Low pressure drop (average, maximum)
- High durability, total life span
- Low maintenance requirements, cleaning/service frequency
- Low cost: total investment cost, maintenance cost.

22.1.2 Classification of Systems

Figure 22.1 shows a classification scheme for diesel filter systems based on the principle of regeneration. An acceptable diesel filter system has to provide means of regeneration, preferably fully automatic regeneration. All regenerating diesel filter systems of practical importance utilize thermal regeneration, during which the particulates are oxidized to produce gaseous products. The temperature of diesel exhaust gas is, however, too low to sustain autoregeneration of the trap. That problem may be solved by either (1) decreasing the required soot combustion temperature to a level that is reached during regular engine operation or (2) increasing the temperature of the trapped soot to a point where it starts oxidizing. The first approach is used in *passive* filter systems, the second in *active* filter systems.

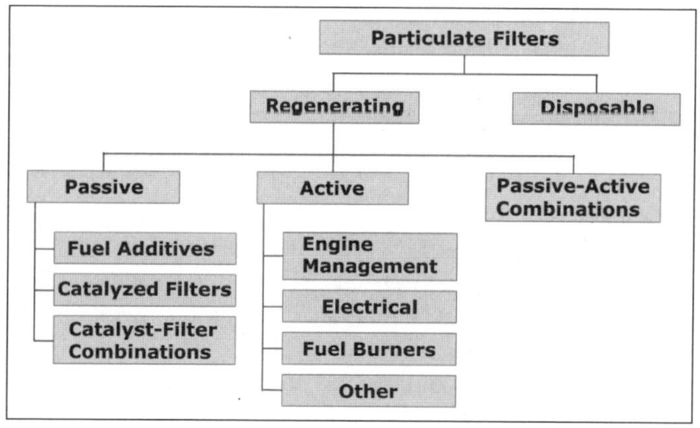

Figure 22.1 *Classification of filter systems by regeneration method.*

In passive systems the soot oxidation temperature is lowered to a level that allows for autoregeneration during regular vehicle operation. This can be achieved by introducing an oxidation catalyst to the system. The catalyst can be placed directly onto the filter media surface or added to the diesel fuel as a fuel additive. Catalyzed filters can utilize platinum-based or base metal catalysts. Active ingredients of available fuel additives include iron, cerium, copper, platinum, or mixtures of metals. A different principle has been utilized in the CRT filter. The catalyst, placed upstream of the filter, is used to generate nitrogen dioxide, which then oxidizes the collected soot.

The second approach is to actively trigger regeneration by raising the temperature in the trap. Most of the active filter systems use either electric heaters or fuel burners . These filters can be further divided into full- and partial-flow systems. In full-flow systems the entire exhaust gas stream is heated to the required temperature. In partial-flow systems the main flow is diverted from the filter, and only a small portion of exhaust gas is heated for regeneration. A modification of the partial-flow system is hot air regeneration. Here, the exhaust gases bypass the soot-loaded filter, which is regenerated by a small stream of air preheated to the necessary temperature. Partial-flow and hot air regenerated systems are more complicated mechanically (valves, air blowers, etc.) but require less heating energy.

A number of electrically regenerated systems have been proposed. In these systems the filter temperature is increased by an electric heater. The heater can be placed in either an (1) upstream or (2) in-bed position. Another classification, based on the source of heater power, divides these filters into (1) on-board and (2) off-board regeneration systems. The off-board systems have to be connected to an external power source or special device for regeneration. This method is applicable for vehicles operating in confined spaces (diesel forklift fleets, etc.), which have access to external power sources.

In newer technology engines, exhaust gas temperature can be also increased through engine management methods, such as by common-rail, post-injections of fuel. This particular method is usually complemented by the use of a "warm-up" catalyst, which further increases temperature through the oxidation of hydrocarbons resulting from the post-injection. Other elements of engine management strategies may include altering fuel injection timing and rate, throttling of intake air, and using EGR.

The active trap systems are much more complex than passive filters. They require sophisticated hardware, including an electronic control unit to trigger and control the regeneration process. Their operation may also involve high fuel economy penalties associated with the energy required to achieve temperatures needed for regeneration. The passive filters, due to their simplicity, appear to be a more attractive approach. However, the regeneration of passive systems is always dependent on the vehicle duty cycle. If the engine is operated at low load (e.g., idling or slow city driving) for a prolonged period of time, the filter will become overloaded with soot and problems will occur. In many applications, the essentially passive filter regeneration can be supplemented with a varying degree of active support, to prevent such problems. In this passive-active approach, fuel economy penalty may be also minimized, due to the lower temperatures required for regeneration. An example of such system would be a catalyzed filter, which is also equipped with an electric heater. During periods of high-temperature operation, the system would passively self-regenerate. During periods of low engine loads and temperatures, the electric regeneration would be triggered, but the required temperature would be, say, around 400°C instead of the 600°C needed with an uncatalyzed filter. Regeneration at a lower temperature also results in lowering thermal stress and increasing life span of the filter media.

22.1.3 Regeneration Strategy, Durability and Reliability

Regeneration is certainly the most critical outstanding issue in the design of diesel particulate filters. The multitude of regeneration systems that have been developed proves that the technology is still far from being perfect. Unfortunately, diesel filters have been troubled by problems that negatively affect their reliability and durability. Most of the diesel trap issues are related to the filter regeneration method and to its compatibility with the filter material.

Most filter regeneration problems are caused by (1) poor, slow regeneration, or (2) rapid, "uncontrolled" regeneration. In an ideal situation, particulates that enter the filter are oxidized in a continuous or almost continuous manner. The filter maintains an approximately constant, moderate soot load, which produces an acceptable pressure loss. Continuous regeneration does not produce high temperature peaks due to the exothermic combustion of soot. Thus, there is little thermal stress on the filter material.

Poor regeneration leads to increasing soot load in the filter and, hence, increasing pressure drop. Eventually, it may lead to complete clogging of the unit. This is a common problem with many passive systems, which self-regenerate during periods of sufficiently high exhaust gas temperature. If the engine operation pattern changes in a way that results in lower loads and exhaust temperatures, the filter regeneration may become inadequate and, ultimately, the unit may become clogged with soot.

Uncontrolled regeneration occurs when a high load of soot becomes "ignited." Such regeneration, also called "runaway" regeneration, is the opposite of the ideal, continuous regeneration mode. The "ignited" soot load burns rapidly, releasing high quantities of heat. The rate of heat release during uncontrolled regeneration exceeds the rate at which released heat is being removed from the filter. This causes increased filter temperatures and, eventually, damage to the filter material. Typical modes of failure are (1) melting, which is caused by high temperatures, and (2) cracking, which is caused by a high temperature gradient. An example of such failure is shown in Figure 22.2. The filter experienced a number of meltdowns, as indicated by the black channels visible at its outlet face. It is seen in the cutout that the black channels correspond to localized melted areas inside the monolith.

Because melted/cracked filters usually do not cause adverse engine effects, they may remain undetected by the vehicle operator. It is not always clear if a failure occurred in a single uncontrolled regeneration incident, or was instead the result of

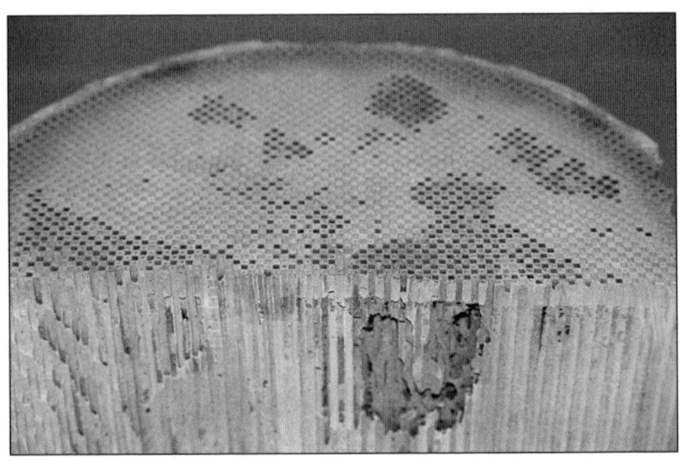

Figure 22.2 Melted ceramic diesel filter.

a gradual deterioration over a number of regeneration events. In theory, a filter with a few melted channels may still maintain decent filtration efficiency. A single uncontrolled regeneration event, however, can most certainly cause a catastrophic failure of the filter. An interesting recording of filter temperatures and pressure drop [1] is shown in Figure 22.3. An uncontrolled regeneration occurred during the recording, causing a catastrophic failure as indicated by the rapid change in the filter pressure drop. The exhaust temperature at the filter outlet reached 959°C; much higher temperatures were likely to exist in "hot spots" inside the filter.

Uncontrolled regeneration can be a problem with both passive and active filter systems. In passive systems, uncontrolled regenerations may occur when the filter becomes overloaded with soot and then exposed to a high-temperature spike that "ignites" the accumulated soot load. In fact, passive filters in low-temperature applications are as prone to melting as they are to clogging. In active systems, such as those utilizing electric heaters or fuel burners to trigger regeneration, the maximum soot load is controlled and, theoretically, should never be exceeded. Precise control of the soot load and regeneration process, however, is a real challenge for the system designer. In real operation, many active filter systems were reported to suffer damage caused by regeneration of soot-overloaded units.

It is a good design practice to instrument filter systems with a pressure drop monitor and/or alarm for monitoring the filter performance and early detection of problems.

22.1.4 Future Trends

22.1.4.1 Passive vs. Active Systems

Passive, catalytic filters are generally favored for most diesel engine applications due to their simplicity and relatively low cost. In light-duty applications, however, the exhaust gas temperature is generally not sufficient to ensure proper filter regeneration. Existing light-duty systems utilize the "passive-active" approach and incorporate active regeneration strategies, primarily through engine management [2]. Future systems are likely to follow the same trend.

An important question remains: whether passive filters can be further developed to enable their wide-scale application in heavy-duty engines. While it is likely that better, more thermally durable filter materials and configurations will be developed, there appears to be little chance that the minimum regeneration temperature limit can be significantly lowered. This minimum regeneration temperature is currently about 275 to 300°C (provided filters are installed on new, low particular emission engines and ultra-low-sulfur fuel is used).

In retrofit applications, exhaust gas temperatures can be measured and analyzed to ensure that filters have a reasonable chance for successful operation. This type of approach was reported by the New York City transit authority, which was able to retrofit buses powered by DDC Series 50 four-stroke engines, but found unacceptably high failure rates of the same passive filter system when installed on 6V92 DDEC two-stroke engines [3].

Particulate filters in new engine application, however, must provide reliable operation under any operating conditions. Some heavy-duty engine applications experience exhaust temperatures that are too low to support continuous filter regeneration. An example exhaust temperature distribution [4] from a distributor vehicle in city traffic is shown in Figure 22.4. A passive filter, if installed on this engine, would likely regenerate when operated over the ETC test cycle, but would most certainly create problems during the real-field operation.

Future particulate filter systems for new engines will have to account for such low engine load duty cycles, as well as for operation in cold climates. In general cases, totally passive filters carry excessive risk of the filter clogging and eventually immobilizing the vehicle (or filter failure). Therefore, filter systems for heavy-duty vehicles are also likely to require some

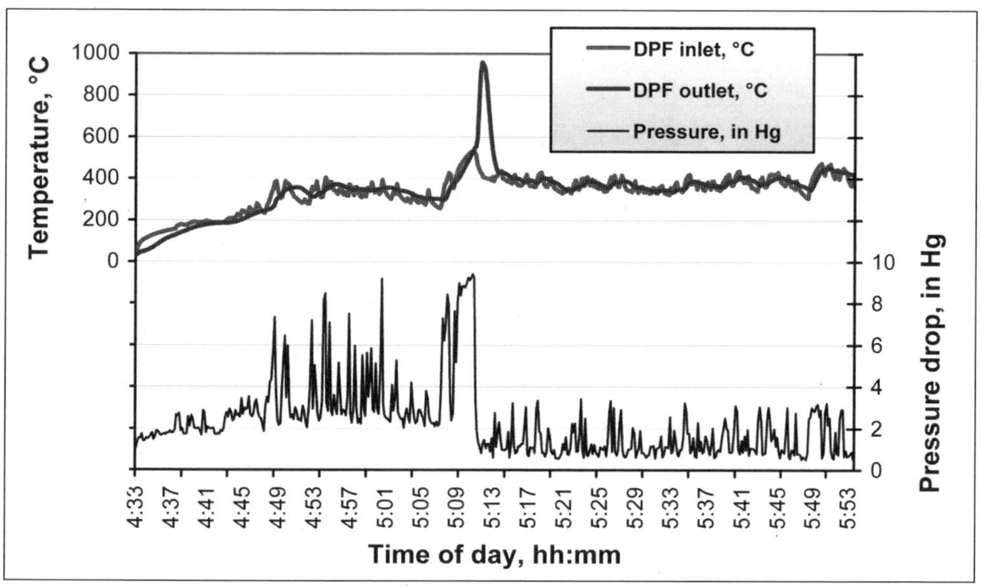

***Figure 22.3** Catastrophic failure of diesel particulate filter. Passive, catalyzed DPF on urban, hybrid diesel-electric bus.*

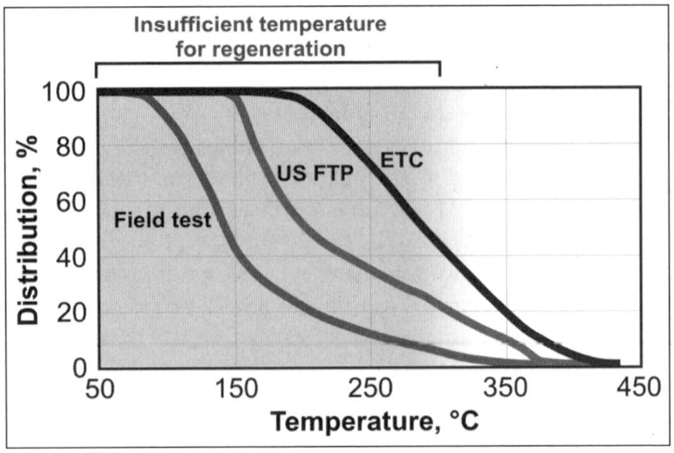

Figure 22.4 *Exhaust temperature distribution in a commercial heavy-duty diesel vehicle.*

form of active regeneration support, most likely through engine management.

22.1.4.2 Diesel Filter Challenges

Considering current state-of-the-art technologies in particulate filter and related technologies, successful integration of filters with the diesel engine is facing the following major challenges:

- *Controlled regeneration/duty cycle effects:* Controlled regeneration is absolutely essential for the durability of filter systems. Particulate filter regeneration critically depends on the vehicle duty cycle, especially in passive systems; the filter system must be designed to work reliably under all possible duty cycles. Depending on the type of filter, emission performance may be also influenced by the duty cycle.
- *Fuel economy/NV driveability:* The use of filters is always associated with a certain fuel economy penalty due to increased pressure drop and, in the case of active regeneration strategies, to the regeneration energy requirement. Filters and/or certain regeneration strategies (e.g., through engine management) may also cause noise-vibration-harshness (NVH) and driveability problems.
- *Ash management:* The ash portion of diesel particulates cannot be regenerated and will accumulate in the filter. Low ash oils are highly desirable for filter-equipped diesel engines.

22.2 Catalyzed Diesel Filters

22.2.1 Introduction

Application of a catalyst has been attempted with virtually all diesel filter media in order to develop passive trap systems. The function of the catalyst in the catalyzed diesel particulate filter (CDPF) is to lower the soot combustion temperature to facilitate regeneration of the filter by oxidation of diesel particulate matter (DPM) under exhaust temperatures experienced during regular operation of the engine/vehicle, typically in the 300 to 400°C range. In the absence of the catalyst, DPM can be oxidized at appreciable rates at temperatures in excess of 500°C, which are rarely seen in diesel engines during real-life operation. Reported catalyst applications include cordierite and silicon carbide wall-flow monoliths, wire mesh, ceramic foams, and ceramic fiber media. The most common type of CDPF is the catalyzed ceramic wall-flow monolith.

Catalyzed ceramic traps were developed in the early 1980s. Their first applications included diesel-powered cars and, later, underground mining machinery. Catalyzed filters were commercially introduced for Mercedes cars sold in California in 1985. Mercedes models 300SD and 300D with turbocharged engines were equipped with 5.66-in. diameter × 6-in. filters (Figure 22.5) fitted between the engine and the turbocharger [5].

The use of diesel traps on cars was later abandoned, due to such issues as insufficient durability, increased pressure drop, and filter clogging. Today, even though not all of these problems have been solved, catalyzed ceramic traps remain one of the most important diesel filter technologies. CDPFs are being increasingly used in a number of heavy-duty applications, such as urban buses and municipal diesel trucks. For a number of years, limited quantities of catalyzed filters have also been used in underground mining (North America, Australia) and in certain stationary engine applications.

Catalyzed ceramic filters are commercially available for a number of highway, off-road, and stationary engine applications as both OEM and aftermarket (retrofit) products. The list of suppliers includes Engelhard, OMG (formerly Degussa, dmc^2), as well as several smaller emission control manufacturers who specialize primarily in the off-road markets.

22.2.2 Filter Design

A schematic of a catalyzed ceramic diesel filter is shown Figure 22.6. The main component of the filter is a ceramic (cordierite, SiC) wall-flow monolith. The porous walls of

Figure 22.5 *CDPF for Mercedes Benz 300D. (Courtesy of Per Stobbe)*

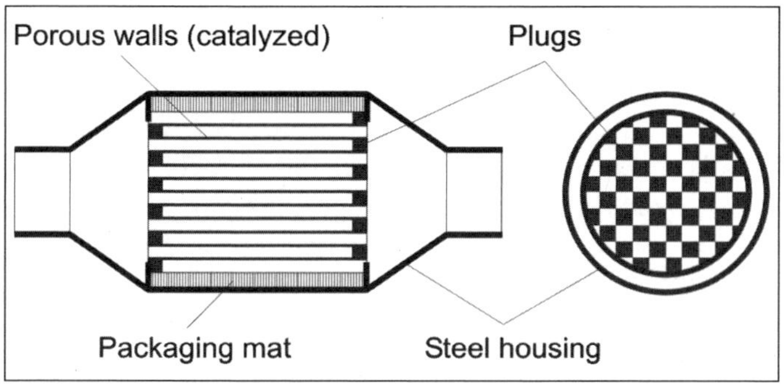

Figure 22.6 *Catalyzed diesel particulate filter.*

TABLE 22.1
TYPICAL CDPF SIZING RANGES IN HEAVY-DUTY ENGINES[a]

	Engine PM Emission	
Parameter	**Low (≤0.05 g/bhp-hr)**	**High (≥0.30 g/bhp-hr)**
Space velocity, 1/h	80,000	25,000
Filter volume/engine displacement ratio	1.5	4
Filter volume/engine rated power, cm^3/hp	40	150

a. Based on 100/17 wall-flow monolith geometry. Smaller filter sizes may be sufficient if monoliths having a higher specific filtration surface area are used.

the monolith are coated with an active catalyst. As the diesel exhaust aerosol permeates through the walls, the soot particles are deposited within the wall pore network, as well as over the inlet channel surface. The catalyst, through a number of possible mechanisms discussed later, facilitates DPM oxidation by the oxygen present in exhaust gas.

The ceramic monolith is packaged into a steel housing. Typical wall-flow monolith packaging techniques are used. To minimize the total exhaust system back pressure, soot filters are usually installed without mufflers. Wall-flow monoliths have their own noise attenuation properties and are in most cases a sufficient replacement for the muffler. To minimize installation costs, the filter housing may be designed as a direct-fit replacement for the vehicle muffler. Such CDPF–muffler design is common in retrofit urban bus applications.

Because filter regeneration critically depends on high temperatures, the unit should be installed as close to the engine exhaust manifold as possible. To further prevent heat losses, thermal insulation of the filter and the inlet exhaust piping is often provided. Typically, insulation can prevent about a 10°C or more temperature decrease per meter of exhaust pipe length due to heat losses. In OEM applications, where the filter system can be fully integrated with the engine, the pre-turbo position of the particulate filter would provide exposure to the highest possible exhaust temperatures.

CDPF sizing varies greatly, depending on the engine technology, duty cycle, fuel quality, and other factors. The most important parameter affecting the filter sizing is the engine-out PM emission, which influences both the regeneration process itself and the soot holding capacity requirement of the filter. Old, high PM-emission engines typically require much larger filter sizes than cleaner engines. Approximate filter sizing rules depending on engine technology are listed in Table 22.1. The numbers listed under the high engine PM emission category roughly correspond to filter sizing used for older, mechanical engines in underground mine applications. The sizing rules given for low PM emission engines describe the smallest filter volumes as used in certain U.S. 1994-compliant or better highway engines.

22.2.3 Catalyst Systems

22.2.3.1 Overview

Various catalyst systems used for diesel filters utilize noble metals, base metals, and mixtures of noble and base metals. Platinum is the most active and the most commonly used noble metal, but palladium, rhodium, or ruthenium catalysts, usually in mixtures, are also possible. The list of common non-platinum-group metals used in diesel filters includes vanadium, magnesium, calcium, strontium, barium, copper, and silver.

The most convenient method of applying catalyst to the ceramic monolith is impregnation with water-based solutions of catalyst precursors, followed by drying and calcining. The final catalyst (usually metal or metal oxide) is formed from the precursor at elevated temperatures during the calcining process. The use of water solutions provides uniform distribution of the catalyst throughout the porous filter walls. Some diesel filter catalysts are also applied from water suspensions of insoluble oxides or salts. Several formulations containing vanadium [6, 7] are examples of catalysts applied from suspensions. Whenever suspensions are used, the particle size distribution of the suspended material, as well as other process parameters, must be carefully controlled to prevent clogging of the filter pores with the catalyst. This is especially important for newer ceramic monolith materials, which are characterized by much smaller pore diameters (about 10 μm) than those used in the past.

Platinum-group metals, including platinum and palladium, are compatible with cordierite and can be applied to ceramic monoliths without a transition metal oxide carrier or washcoat layer. Although various catalyst carriers, such as alumina, silica and Zirconia, have been suggested [8], most of today's diesel filters do not include a high surface area catalyst carrier. Application of the transition oxide catalyst carriers or, even more so, a washcoat layer to filter monoliths of small pore diameters is very problematic. The resulting "carrier-less" diesel filter catalysts, however, have very low physical surface area, resulting in poor catalyst dispersion. Theoretically, such catalysts should be troubled by a number of durability problems, including sintering or reactions between the catalytic components and the ceramic substrate. The published data are still insufficient to determine if such phenomena indeed happen in CDPFs.

Even less experience exists with catalytic coating of SiC filter substrates. The surface of silicon carbide grains that form the wall-flow monolith has a very smooth, almost glossy appearance that appears to be incompatible with catalysts. Several authors have suggested applying some kind of intermediate oxide layer composed of alumina or zirconia-alumina before applying platinum-group catalysts on SiC parts [9, 10]. For some oxides, it is possible to apply such an intermediate layer through impregnation using water-based precursors and, thus, avoiding the inconvenient washcoat (i.e., suspension-based) technologies.

22.2.3.2 Noble Metal Catalysts

The most common noble metal catalyst used in diesel filters contains platinum with promoters, such as alkaline earths. Typical platinum loadings in filters used for off-road engines through the 1990s were between 35 and 50 g/ft^3. These filters, installed on relatively high polluting engines, required minimum temperatures of nearly 400°C for regeneration. Later, when catalyzed filters were applied to much cleaner urban bus and other highway vehicle engines, it was found that they were able to regenerate at much lower temperatures. However, higher platinum loadings were needed to support the low-temperature regeneration. Filters used in clean engine, low-temperature applications have typically platinum loadings of 50 to 75 g/ft^3.

An example platinum catalyst developed by Engelhard is composed of 5 to 150 g/ft^3 Pt/Rh at 5:1 ratio and 30 to 1500 g/ft^3 of MgO [11]. The catalyst is impregnated onto cordierite monoliths from water-based solutions. A filter coated with the catalyst requires exhaust temperatures of 375 to 400°C to regenerate. The function of rhodium in the preceding formulation is to suppress the catalytic oxidation of SO_2 and, thus, the sulfate make in the catalyst. The final effect depends on the Pt/Rh ratio. Table 22.2 lists some SO_2 oxidation data generated in a laboratory flow-through reactor. However, even though effective in controlling sulfates, the addition of Rh to Pt catalysts results in a significant increase in the filter regeneration temperature [12].

TABLE 22.2
CATALYTIC OXIDATION OF SO_2[a]

Pt:Rh Ratio, wt. %	SO_2 Conversion, %
100% Pt	41
10 Pt : 1 Rh	33
5 Pt : 1 Rh	13
3 Pt : 1 Rh	6
1 Pt : 1 Rh	7

a. Catalyst loading, 1.77 g/dm^3 (50 g/ft^3); catalyst volume, 0.9 dm^3; gas flow, 1.7 N m^3/h; 300 ppm SO_2; temperature, 400°C.

Theoretically, the sulfate make in active noble metal filter catalysts can be controlled using a number of methods that were developed for selective diesel oxidation catalysts, for example, through the use of low Pt loadings or sulfate-suppressing additives. However, as was the case for the preceding Pt/Rh catalyst example, one always has to accept a penalty in the catalytic activity when designing a more selective catalyst. In particulate filters, such a penalty manifests itself as a deterioration in the filter regeneration performance. In other words, sulfate suppressed CDPFs require higher exhaust temperatures for problem-free operation. Because in most CDPF applications no regeneration penalty can be accepted, the preferred method of sulfate control in catalyzed filters is through the use of ultra-low-sulfur fuel.

22.2.3.3 Base Metal Catalysts

Vanadium is one of the most popular base metal catalysts used for diesel particulate filters. The CDPF commercialized in Mercedes cars in California in the mid-1980s, mentioned earlier (see Figure 22.5), was coated with a vanadium pentoxide (V_2O_5) catalyst by Degussa. This catalyst provided filter regeneration at temperatures of 380 to 400°C.

One of the potential problems with vanadium-based catalysts is the volatility of V_2O_5 at higher temperatures, such as those that could be encountered during filter regeneration. The evaporation of V_2O_5 may lead to gradual catalyst losses and shorten the filter life span, not to mention vanadium emission problems. Catalysts were developed that utilized vanadium compounds other than V_2O_5, for example, silver or copper vanadates. An example copper vanadate-based metal catalyst was developed by Heraeus [6]. The catalyst was prepared by doping and calcining copper vanadate $Cu_3V_2O_8$ with potassium carbonate in the molar ratio Cu:V:K 3:2:0.13. Diesel filters were then coated with the catalyst from a water suspension. The catalyst loading was between 10 and 80 g/m^2 of the filtration surface area. The soot ignition temperature measured by the differential thermal analysis (DTA) amounted to 365°C with this catalyst and to 500°C without catalyst.

22.2.4 Emission Performance

22.2.4.1 DPM Filtration Efficiency

As with any type of diesel filter, the filtration efficiency for solid fractions of diesel particulates, including inorganic carbon and metal ashes, is determined by the properties of the filter substrate. Most catalyzed particulate filters use wall-flow monoliths of proven, high efficiency for capturing solid particulates. The filtration efficiency increases with increasing soot load in the filter. Somewhat lower efficiencies are seen when soot is collected within the pore network in the walls during the initial filtration period. After a layer of soot stabilizes in the monolith channels, the efficiency for solid particulates increases to typically between 95% and nearly 100%.

The large differences in reported total filtration efficiency of catalyzed filters, which usually varies between some 60% and 95%, can be explained by the filter activity toward nonsolid fractions of diesel particulates—sulfate particulates and the organic fraction of DPM, known as SOF. As discussed before, most of these compounds are passing through the filter as vapors, especially at higher exhaust temperatures. Being a part of the gas phase, these compounds cannot be trapped and retained in the filter the way solid particulates are. Rather, their transformation in the CDPF depends entirely on chemical reactions that may occur on contact with the filter catalyst.

At high exhaust temperatures, CDPFs may generate sulfates by catalytic oxidation of the exhaust SO_2 to SO_3, the precursor of sulfuric acid. At high temperatures, the gaseous SO_3 can penetrate the porous walls and freely leave the filter. Later, it combines with water to form sulfate particulates when cooled down in the DPM sampling system. The "sulfate make" may be a big problem with catalyzed filters. When fuels of higher sulfur content are used on clean diesel engines, catalyzed traps may even increase the total particulate matter (PM) output; that is, they may exhibit an overall negative filtration efficiency. This will happen when the quantity of generated sulfates is higher than the amount of particulates trapped in the filer.

Among all of the diesel filter catalysts, platinum has the strongest tendency to form sulfates. Most sulfates are typically formed over platinum catalysts at relatively high exhaust temperatures of about 350 to 450°C. Active, high Pt loading catalysts can effectively control total DPM emissions only in conjunction with ultra-low-sulfur fuels, which contain insufficient amount of sulfur to cause problems, even if most of it would be oxidized to sulfates. The exact allowed level of sulfur, which depends on the catalyst activity, can be estimated using the formulas given in Section 19.2.3. Active catalysts can convert as much as 50% and more of the exhaust SO_2 to sulfates. Fuels of 300 to 500 ppm wt. S, commonly referred to as "low-sulfur fuels," can still cause considerable sulfate emission problems.

DPM conversion efficiency in a noble metal catalyzed filter at different levels of sulfur in the fuel is shown in Figure 22.7 [13]. Particulate emissions were measured over three engine test procedures, as indicated in the graph, all of which were characterized by high temperatures that favored the sulfate make. DPM reduction efficiencies between 93% and 95% were measured with 3 ppm S fuel. When 30 ppm S fuel was used, the efficiency dropped to between 72% and 80%. The CDPF efficiency decreased to zero at about 150 ppm sulfur content (with some differences between the test cycles). At that point, the amount of sulfate generated in the CDPF became approximately equal to the amount of solid particulates captured in the filter. The efficiency of the CDPF with the U.S. ultra-low-sulfur fuel of 15 ppm S interpolated over the ESC test would be 88%.

We should emphasize that the sulfate considerations depend on the regulatory definition of diesel particulate matter. It is common in occupational health regulations to exclude sulfates from the DPM definition and measure DPM as either total carbon (i.e., inorganic carbon + SOF; example: U.S. non–coal mining [14]) or elemental carbon (example: German occupational regulations [15]). If sulfates are neglected, catalyzed filters will show excellent filtration efficiencies, even with extremely high levels of sulfur in the fuel. Care must be taken, however, that active catalysts do not cause air quality problems with other pollutants, such as SO_3 or NO_2.

The organic fraction (SOF) is also vaporized in the filter during high-temperature operation and cannot be physically trapped. The SOF performance of the CDPF will depend on the activity and the temperature of the catalyst. An active catalyst, such as Pt, will oxidize the SOF vapors. However, if a low-activity catalyst is used, such as some base metal catalysts, the SOF vapors may pass through the filter. As a result, the CDPF may show decreased DPM filtration efficiency.

CDPFs are also very effective in reducing particle (and nanoparticle) number emissions, provided the nanoparticles are solid. If liquid nanoparticles, such as those composed of

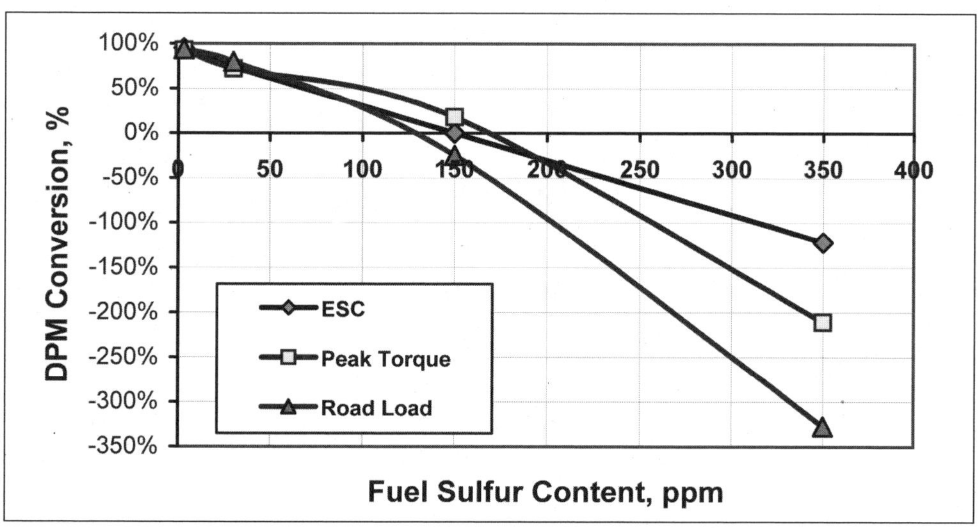

Figure 22.7 Conversion of DPM in Pt-based CDPF at different fuel sulfur levels. CDPF: 10.5-in. diam. × 12-in., 100-cpsi, 17-mil wall; engine: CAT 3126, six cylinders, 7.2 L, 205 kW (275 hp) @ 2200 rpm. (Data from Ref. 13.)

sulfuric acid or hydrocarbons, are created through condensation downstream of the filter, they obviously cannot be controlled by the CDPF. This issue, which applies to all filter types, was discussed in Chapter 20. The formation of nanoparticles after the CDPF increases with the fuel sulfur content and with temperature. In a study with fuel of 371 ppm S, a CDPF was found to decrease particle numbers by about 80% in the low-temperature mode, and increase particulate numbers by some 140% (due to an estimated more than 300% increase in sulfates) in the high-temperature mode [16].

22.2.4.2 Gaseous Emissions Performance

Catalyzed particulate filters also facilitate a numbers of oxidation reactions in the gas phase, which are similar to those occurring in the diesel oxidation catalyst. Gaseous emission performance of a CDPF varies depending on the type and activity of the catalyst. In general, platinum-based catalysts, especially those with high Pt loadings, will be very active in the oxidation of hydrocarbons and carbon monoxide. On the other hand, some base metal catalysts used in diesel filters are quite inactive with respect to the gas-phase components. They do not cause significant changes in the concentrations of CO and HC.

Platinum-based CDPFs typically achieve about 80% to 90% conversion of CO. The example data in Figure 22.8 showing 95% CO conversion were generated on the ESC test cycle [13], which is characterized by high average load factors and high exhaust temperatures that favor catalytic reactions. Note that the carbon monoxide performance also depends on the regeneration cycle. If a filter overloaded with particulates rapidly regenerates, there might be temporary shortage of oxygen leading to partial combustion of soot. Under such conditions secondary CO emissions can be generated in the filter. We are not familiar with any literature data showing how these temporary conditions may affect the total CO emission reduction over the vehicle duty cycle.

Platinum CDPFs typically show a 60% to 70% conversion of hydrocarbons due to catalytic oxidation, as shown in Figure 22.8 [13] and confirmed by other studies [16].

A slight reduction in NO_x emissions, which is typically measured in CDPFs (between 4.9% and –1.5% in Figure 22.8), is likely caused by internal exhaust gas recirculation due to increased pressure levels in the exhaust manifold, rather than by catalytic reactions. However, just like other oxidation catalysts, CDPFs also change the proportion between NO_x constituents (NO_2/NO) by increasing the NO_2 emissions through the oxidation of NO. Although there is no change in the total NO_x emissions, that phenomenon is perceived as a disadvantage, due to the high toxicity of nitrogen dioxide (occupational health!) and possible effects on ozone formation. Example data from chassis dynamometer tests of CDPF-equipped school buses [17] are shown in Figure 22.9. The particulate filter increased NO_2 levels from below 5% to about 30% of the total NO_x. This increase is generally comparable to that measured in the CRT filter, which produced NO_2 levels ranging from 26% to 34% when tested using the same test cycle and fuels (but different vehicles) [17]. It can be speculated that filters with Pt catalysts at high loadings cause a significant increase in NO_2, while some base metal formulations may be relatively inactive. The results in Figure 22.9 were obtained using platinum-containing CDPFs of undisclosed Pt loadings.

Experiments with fuel sulfur levels changing from 3 to 350 ppm concluded that fuel sulfur does not significantly affect the CO, HC, and total NO_x performance of a Pt-based CDPF [13].

Catalytic conversion of hydrocarbons also results in conversions of particulate SOF and other hydrocarbon derivatives, such as aldehydes or particle- and vapor-phase PAH. The conversion of aldehydes is typically higher than that of HCs and may exceed 90%. Conversions of PAHs after a CDPF were measured anywhere from less than 40% to as much as 97%, depending on the PAH type, association (particle or vapor phase), and test mode [16]. The filter used in the study had a relatively low Pt loading of 5 g/ft^3; CDPFs of higher Pt loadings would likely produce higher PAH conversions. Many PAHs were measured at levels below the minimum detectable limit of the measuring equipment, making it difficult to precisely quantify the effect of the CDPF. The same study found an increase of 1-nitropyrene

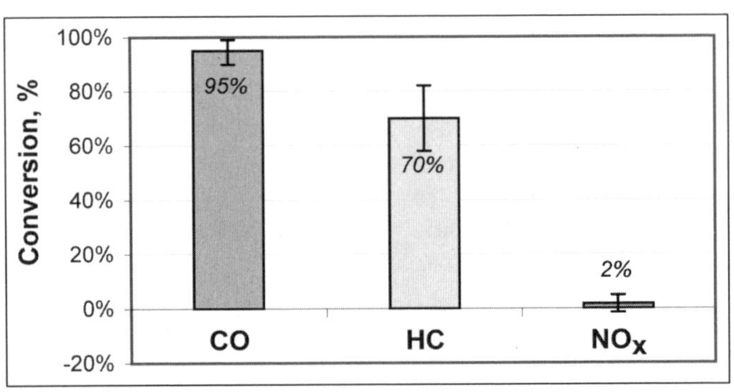

Figure 22.8 Conversion of gases in Pt-based CDPF. ESC (OICA) test cycle; CDPF and engine as in Figure 22.7; data shown represent averages from several measurements with fuel sulfur changing from 3 to 350 ppm.

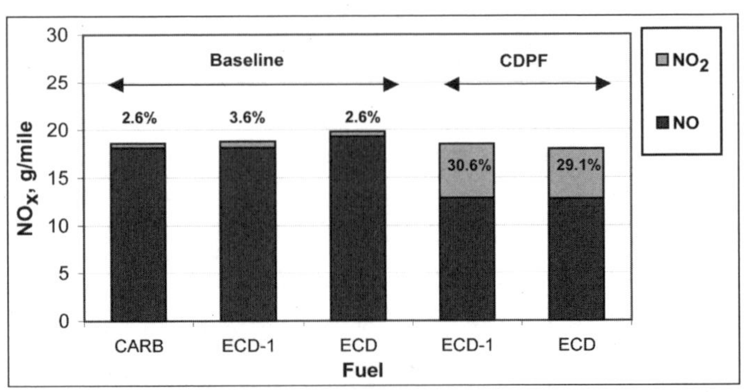

Figure 22.9 Impact of CDPF on NO_2 emissions. CSHVR dynamometer driving cycle; average results from two school buses; CARB and ARCO (BP) EC diesel fuels.

after the CDPF at one engine mode. This nitro-PAH species could have been produced in catalytic reactions in the particulate filter, but could also be an artifact generated by reactions taking place in the PM sampling filter.

The impact of the CDPF on the mutagenic activity results, as determined by the Ames test, was inconclusive [16]. However, the samples in general showed very little mutagenic activity.

22.2.5 *Regeneration and Pressure Drop*

Catalyzed filters in their pure, passive form rely on the exhaust gas temperature for thermal regeneration. The temperatures required for soot oxidation, normally in the 500 to 600°C range, are lowered through the use of the catalyst to levels that are seen in real-life operation of the engine, typically in the 320 to 420°C range. The exact mechanisms of the catalytic regeneration are not entirely understood. Heterogeneous catalytic processes in the gas phase occur on contact of gas molecules with the solid catalyst, that is, the reactions take place on the surface of the solid phase. It is doubtful, however, that solid particulates, several orders of magnitude larger and heavier than gas molecules, have sufficient mobility to physically contact the catalyst. Furthermore, it would be only those particles that are deposited within the catalyzed pore network inside the walls or directly on the wall surface that would have such a chance. Obviously, particles that form layers inside the channel walls can never come in contact with the catalyst. A number of mechanisms are likely involved in the regeneration of catalyzed filters, as follows:

- *Catalytic oxidation by oxygen:* Particles are oxidized by oxygen atoms adsorbed on catalytic sites. This mechanism is limited to particles that physically contact catalytic sites. Oxidation of particle layers in the inlet channels would require that soot deposits be moved inside the filter by the impact of gas flow, as particles already in contact with the catalyst are oxidized to gaseous products.
- *Oxidation by nitrogen dioxide:* NO_2, which is generated through catalytic oxidation of NO in the filter, can oxidize diesel soot at much lower temperatures than oxygen [12]. In a CDPF, this mechanism would apply primarily to particles positioned very close to or downstream of catalyst sites, generally within the pore network. Oxidation of the soot layer in the inlet channels, upstream of the catalyst would have to rely on a diffusional transfer of NO_2 countercurrently to the gas flow—not a very effective reactor design. This NO_2 effect, which is utilized to its full potential in the CRT filter, is blocked by sulfur and requires the use of ultra-low-sulfur fuel.
- *Thermal oxidation by oxygen:* Catalytic oxidation of hydrocarbons and other gas species, as well as the carbon particulates themselves, can create local zones of increased temperature due to the exothermal heat of reaction. The increased temperature may be sufficient (>500 to 600°C) to support noncatalytic, thermal oxidation of soot deposits. This oxidation mode is certainly important during uncontrolled regeneration, when temperatures increase well above 1000°C; its significance during continuous regeneration is not known.

As is the case with catalytic processes, exhaust gas temperature is the most important factor influencing filter regeneration. The rate of soot oxidation increases with the filter temperature. If the temperature is too low, soot accumulates in the filter, causing excessive flow restriction, high exhaust gas pressure losses, and, eventually, clogging of the unit. Units overloaded with soot are also prone to uncontrolled regenerations, which may be caused by a sudden "ignition" of the soot load accumulated in the CDPF, leading to a mechanical failure of the filter substrate (melting, cracking) due to excessive temperature or temperature stress.

In general, the minimum exhaust temperature required to sustain adequate self-regeneration of CDPFs varies between about 320 and 420°C. In practice, the temperature requirement it typically defined as the filter inlet temperature that must be reached over a certain minimum percentage of the engine operating time. Using an arbitrary example, one could require that, for a given CDPF application, a temperature of 350°C be reached or exceeded over at least 30% of the duty cycle. Commercial CDPFs offered for a number of nonroad applications over the 1990s typically required exhaust temperatures of 380 to 420°C over some 25% of the time. Lower exhaust temperatures are sufficient for filters used on newer, cleaner engines, especially in conjunction with ultra-low-sulfur fuel. One commercial retrofit systems, the DPX filter by Engelhard, requires an average exhaust gas temperature of at least 225°C for use with a maximum 15 ppm S fuel on post-1994 highway engines in California [18]. Even though an average temperature specification cannot be directly compared with the previously given numbers, this application does represent a lower temperature requirement. Indeed, DPX filters having high Pt loadings were demonstrated to operate on hybrid diesel-electric urban buses on duty cycles where exhaust temperatures hardly exceeded 300°C (winter months in New York City) [1]. Remember that the hybrid bus application is characterized by especially low engine-out PM emission due to the absence of rapid engine accelerations and other transients in the hybrid powertrain. Balance temperatures of the Engelhard filter, as measured in a laboratory as a function of the fuel sulfur content using a nonroad engine of higher PM emissions [13], are shown in Figure 22.10. The balance temperature was determined using the DECSE five-mode balance point test.

Laboratory-determined balance temperatures, even though not necessarily a good approximation of the temperature requirements in the field, are very valuable in studying the relative changes of the balance point in response to changing system variables. Figure 22.10 illustrates such changes in response to changing fuel sulfur content and engine speed (exhaust gas flow). There are more variables influencing the CDPF requirements in regards to the minimum operation temperatures. The most important factors can be summarized as follows:

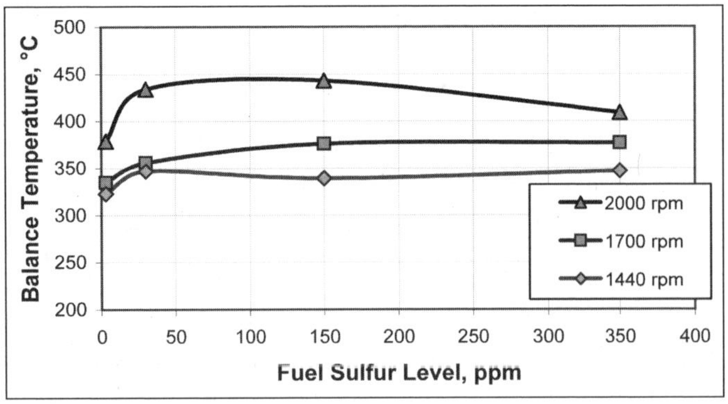

Figure 22.10 CDPF balance temperature at different fuel sulfur levels. See Figure 22.7 for CDPF and engine information.

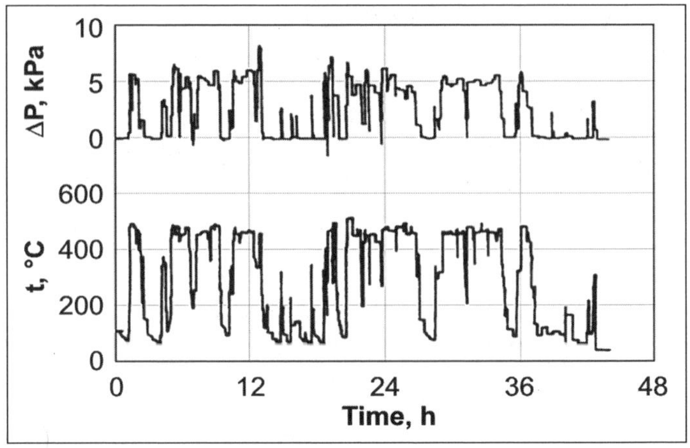

Figure 22.11 Pressure drop and temperature in a high-temperature CDPF application. (Reprinted from Ref. 19.)

- *Engine-out PM emissions:* CDPFs can regenerate at lower temperatures on cleaner engines. This finding is consistent with the theory of diesel filter regeneration.
- *Sulfur content in the fuel:* Decreasing the sulfur level from 30 to 3 ppm provided about 30°C improvement in the balance temperature (Figure 22.10). However, the balance temperature remained relatively constant when sulfur levels were further increased beyond 30 ppm.
- *Exhaust gas space velocity:* Increasing the engine speed and, as a consequence, the exhaust gas flow were found to increase the balance temperature (Figure 22.10). This finding would suggest that larger filters can regenerate at lower temperatures. However, increasing the filter size beyond a certain point is likely to deteriorate the regeneration due to the "extinguishing" effect of the filter heat capacity on transient temperature spikes.

In a suitable, high-temperature application, the regeneration of catalytic soot filters is practically continuous and loading and regeneration phases are not well pronounced. Rather, the unit is operating close to balance conditions and the amount of DPM oxidized in the trap is approximately equal to the amount entering the trap. This kind of application is illustrated in Figure 22.11 [19]. Exhaust gas pressure drop and temperature in the catalytic filter were recorded during more than 40 h of operation on an underground mining vehicle. The engine, a CAT 3306 PCTA rated 175 kW at 2200 rpm, was equipped with a ceramic monolith filter with a 15-in. diameter × 15 in., catalyzed with a base metal catalyst. Temperatures during periods of high engine load were constantly above 450°C. The corresponding pressure drop was fluctuating around 5 kPa (20 in. H_2O) with no apparent tendency to increase or decrease.

Applications with lower exhaust temperatures may experience a different regeneration and pressure drop pattern. Soot may accumulate during low-temperature periods and then rapidly regenerate when the temperature increases, resulting in periodic regenerations. This type of CDPF operation is illustrated in Figure 22.12, which shows maximum engine back-

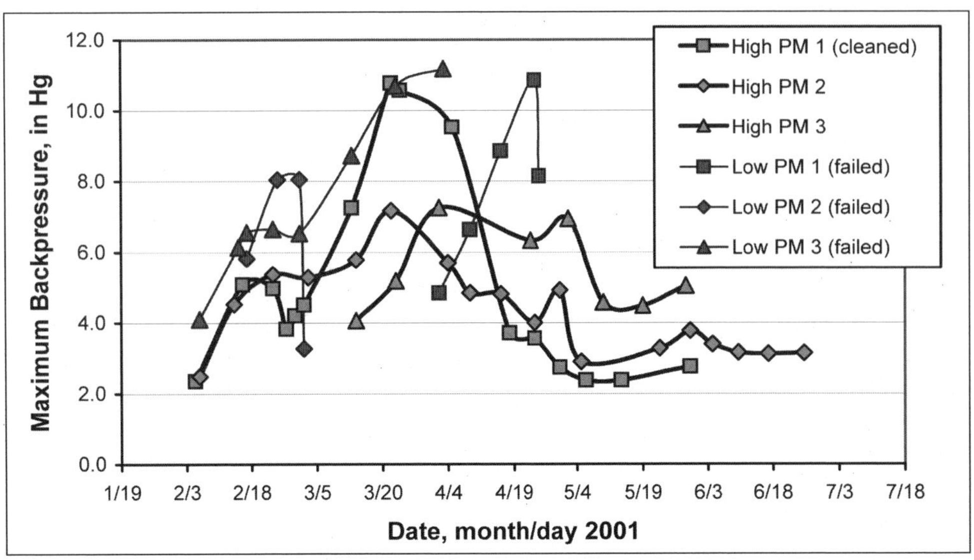

Figure 22.12 Pressure drop in a low-temperature CDPF application.

pressure data collected from six CDPF-equipped hybrid diesel electric urban buses during the New York City study [1]. The fluctuation of CDPF pressure drop provides a clear indication that regeneration occurred periodically in all filter units.

Such a periodic regeneration pattern carries the danger of an uncontrolled regeneration, which occurs when excess soot becomes ignited during momentary high load operation. If engine speed drops to idle after the soot ignites, the exhaust flow becomes too low to carry away the heat released in the reaction. As a result, internal filter temperatures and/or temperature gradients become high enough to damage the filter substrate (the typical failure mode is melting for cordierite and cracking for SiC filters). Among the six filters shown in Figure 22.12, three failed due to uncontrolled regeneration. All of the failed units were filters of low precious metal (platinum) loading, as indicated in the graph. However, even one of the high-Pt loading units, which all survived the 6-month field test, had to be removed from the vehicle for cleaning to lower its excessive pressure drop.

Exhaust gas temperatures were very low during the New York City hybrid study. In the winter months (February and March), the exhaust temperatures on most buses exceeded 300°C over only about 10% of the engine running time. During the following warmer months, the 10% temperature oscillated around 325°C. Filter regeneration was observed only once in several bus runs during cold weather conditions.

High pressure drop is one of the inherent features of all passive filters, including catalyzed diesel filters. Pressure drop on properly regenerating units in high-temperature applications typically exceeds 5 kPa (20 in. H_2O). Low-temperature applications experience higher exhaust gas pressure losses. Maximum engine back-pressure levels exceeding 6 in. Hg (20 kPa = 80 in. H_2O) and more are not uncommon in Figure 22.12. Pressure loss levels of similar magnitude are reported in a number of other field test reports. The filter pressure drop is causing a certain fuel economy penalty associated with this technology, which is usually estimated to be between 2% and 4%. A fuel economy penalty up to 2% was associated with particulate filters by the DECSE program [13], which carried out its measurements on a test bench, with filters properly regenerating on relatively hot test cycles. An average fuel economy penalty measured for the CDPF device on the ESC (OICA) cycle amounted to 0.5%.

22.2.6 *Durability*

Long-term durability is the most important issue with catalyzed particulate filters, especially in the fully passive configuration. Because filter regeneration depends on the duty cycle, it is very difficult, if not impossible, to ensure failure-free operation under all possible operating conditions and duty cycles. Even on hot engines, a single period of extended idling can lead to overloading of the filter with soot, followed by an uncontrolled regeneration and catastrophic failure.

Proper engine maintenance is also of critical importance in ensuring long filter life. In particular, filter catalysts are typically sensitive to poisoning by lubricating oil and its additives. CDPF-equipped engines must be well maintained to avoid excessive oil consumption. Engine failures resulting in oil leaks to the exhaust system may result in immediate deactivation of the catalyzed filter.

In OEM applications CDPFs will most likely require some kind of active regeneration support in cases where excessive soot load (pressure drop) is detected in the filter. Such measures may involve, for example, increasing the exhaust gas temperature through special engine management strategies. In retrofit applications, a long filter life span can be ensured through (1) carefully verifying (recording) of the exhaust temperature during real operation of the vehicle prior to retrofitting, (2) providing a high level of engine maintenance (oil consumption!) after the retrofit, and (3) monitoring the exhaust back-pressure on the retrofitted vehicle for early detection of problems. Filter pressure drop monitors, which can be connected to dashboard-installed warning lights and/or the engine control module, are available with most commercial CDPF systems. The use of such monitors with CDPFs is actually required in some retrofit programs [18]. Excessively high pressure drop is an indication that the filter accumulated too much soot and may be facing an uncontrolled regeneration or plugging. In either case, the unit must be taken off the vehicle and cleaned. Low pressure drop levels, on the other hand, may be an indication that the filter has already lost its mechanical integrity and is leaking unfiltered exhaust gas. It must be realized that proper implementation of such a CDPF monitoring program may result in a noticeable increase in the amount and cost of the engine and vehicle maintenance.

Experience from underground mining indicates that the lifetime of a CDPF installed on a suitable, high-temperature application varies between 5,000 and 10,000 engine hours. However, the average filter life is shorter due to premature filter failures in unsuitable applications (cold engine duty cycle, engine maintenance problems, etc.). According to U.S. Bureau of Mines estimates from the early 1990s, the average CDPF durability in underground mines was about 2500 h.

Filters used for OEM applications have to meet regulatory emission durability requirements. The first CDPF-equipped engine that was emission certified in the United States—the MY2001 Navistar 530 for school bus applications—has been classified as a medium-heavy-duty engine with a 185,000 miles required emission durability. In the certification process, the manufacturer demonstrated this durability through accelerated 1000-h engine dynamometer aging of the CDPF. Particulate filter manufacturers who verified CDPF systems for the California diesel retrofit program, announced in 2000, had to demonstrate emission durability of 150,000 miles and provide product warranty for 100,000 miles. At this time data are insufficient to estimate the average CDPF life and failure rates in real-field operation.

Another source of increased maintenance in CDPF-equipped engines is related to ash emissions. As with other types of filter systems, CDPFs retain most of the metal oxide ash emissions that are generated (primarily) by lube oil additives. Because ashes are incombustible, their accumulation causes a steady increase in the filter pressure drop. This problem may be partially alleviated by periodic cleaning of the filter using compressed air, steam, water, or other means. (Cleaning operations should conform to the manufacturers' recommendations to avoid any detrimental effect to the catalyst.) Early experience from retrofit installations in the United Kingdom in the late 1990s indicates that de-ashing of CDPFs may be needed as often as every 30,000 km [20]. On the other hand, some field studies of CDPF retrofits in the United States found that maintenance/de-ashing intervals may be as long as 240,000 km (150,000 miles), exceeding the CDPF manufacturer recommendation of 60,000 miles [17]. These huge differences are likely caused by different engine technologies and lubricating oil quality, but more experience is needed before any conclusions can be drawn. Filter cleaning intervals can be certainly maximized through the use of low ash lube oils.

22.3 CRT Filter

22.3.1 Principle of Operation

22.3.1.1 Reactions and Configuration

The CRT (Continuously Regenerating Technology; originally introduced as the Continuously Regenerating Trap) is a trade name for a catalytic, two-stage, passive particulate filter system capable of regenerating at temperatures of below 300°C on a suitable application and with the use of ultra-low-sulfur diesel fuel.

The principle of CRT regeneration is based on the finding, discovered and patented by Johnson Matthey, that diesel particulate matter is more easily oxidized by nitrogen dioxide (NO_2) than by oxygen [12, 21]. Carbon in the form of soot is oxidized by oxygen with noticeable reaction rates at temperatures greater than 550°C. With NO_2, the process occurs at temperatures as low as 250°C, as illustrated by the laboratory reactor data shown in Figure 22.13 [22].

A schematic of the CRT configuration is shown in Figure 22.14. The CRT is composed of two devices: an oxidation catalyst (upstream) and a ceramic wall-flow diesel filter (downstream). The NO_2 necessary for filter regeneration is generated in the oxidation catalyst from nitric oxide (NO) present in diesel exhaust, according to the following reaction:

$$NO + \frac{1}{2}O_2 \leftrightarrow NO_2 \tag{22.1}$$

Diesel particulate matter trapped in the filter is continuously oxidized by NO_2, as follows:

$$NO_2 + C \rightarrow NO + CO \tag{22.2}$$

$$NO_2 + C \rightarrow \frac{1}{2}N_2 + CO_2 \tag{22.3}$$

where C (carbon) represents the combustible portion of DPM.

Theoretically, in the process of PM oxidation NO_2 may be reduced to both NO and N_2 as shown in Eqs. (22.2) and (22.3), respectively. In practice, most of the NO_2 reacts with carbon to re-create the nitric oxide [Eq. (22.2)]. The oxidation of soot is further enhanced by the exhaust water vapor, presumably due to the formation of HNO_3 from NO_2 and H_2O, which also reacts with carbon [21]. From the existing literature reports it is not clear if nitrous oxide (N_2O) is also present among the reaction products.

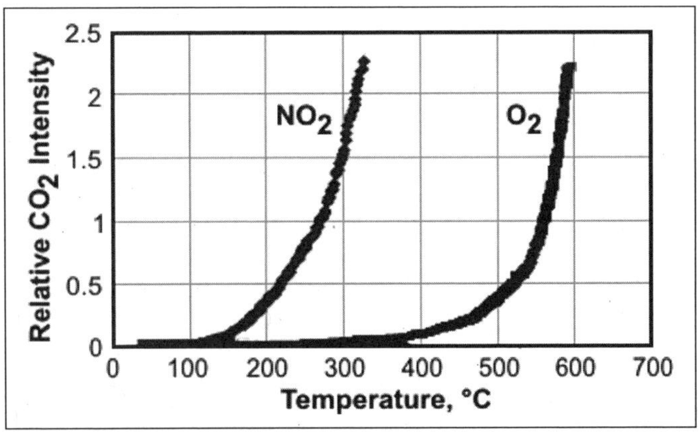

Figure 22.13 Oxidation of PM by oxygen and nitrogen dioxide. Laboratory reactor preloaded with controlled amount of soot from a diesel engine; temperature of sample gradually increased at a steady flow of oxidizing gas; outlet CO_2 measured by a mass spectrometer.

22.3.1.2 Catalytic Oxidation of NO in CRT Filter

Engine-out NO_x emissions from diesel engines are typically composed of about 95% NO and 5% NO_2. Depending on temperature, the catalyst part of the CRT can increase the NO_2 fraction to about 50% or more of total NO_x. The oxidation of NO to NO_2, as expressed by Eq. (22.1), is an equilibrium process controlled by the reaction kinetics at lower temperatures, and by thermodynamic constraints at high temperatures [23], as illustrated in Figure 22.15. The four equilibrium lines shown in the graph were computed for different oxygen concentrations. When operating at low temperatures, in the kinetically controlled regime, the conversion rate increases with the catalyst activity and temperature. Thus, NO conversion can be increased by using a more active catalyst or by increasing the catalyst size. At higher temperatures the NO/NO_2 shift is constrained thermodynamically. Increasing exhaust gas temperature in that region causes a decrease in NO conversion. The highest NO conversion occurs at medium temperatures of about 250 to 350°C, where,

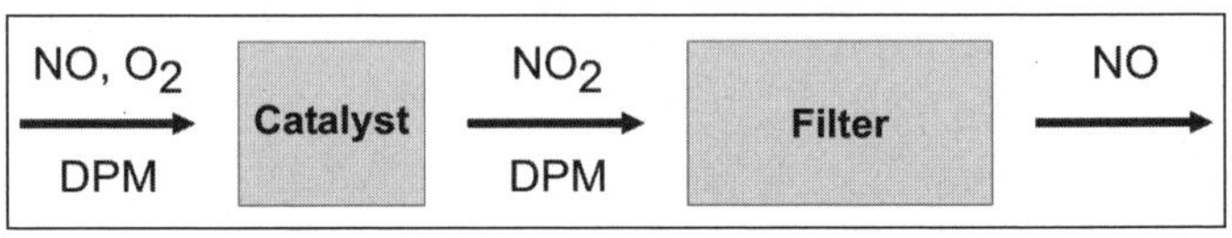

Figure 22.14 CRT filter: configuration and principle of operation.

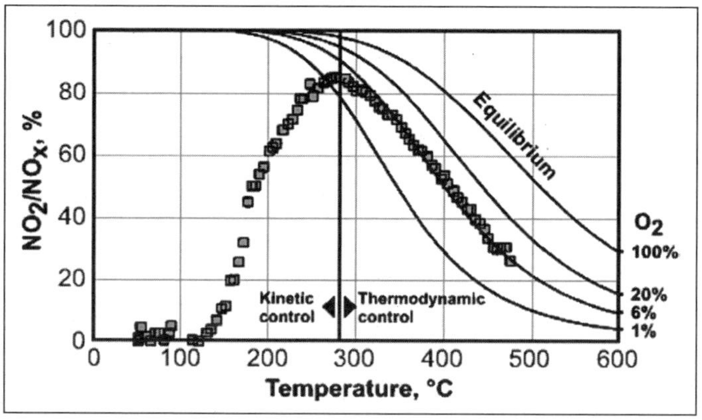

Figure 22.15 *NO/NO$_2$ shift in oxidation catalyst. Catalyst: Pt, 50 g/ft^3; gas: 270 ppm NO, 6% O$_2$, 10% H$_2$O in N$_2$.*

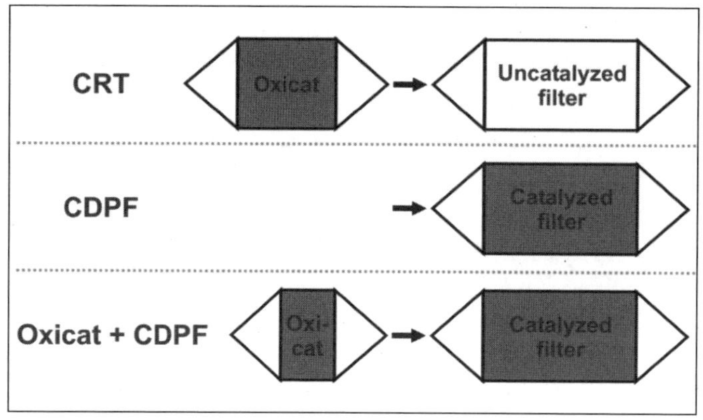

Figure 22.16 *Configurations of catalytic filters.*

with the use of a sufficiently large and active catalyst, the NO_2 fraction can be increased to 70% to 80% of total NO_x.

Once the NO_2 enters the filter part of the CRT, it reacts with soot. The rate of this reaction (or the regeneration rate of the CRT) increases with temperature and with the NO_2 concentration. Thus, a certain excess in NO_2 is always needed to maintain appreciable reaction rates. The NO_2 levels at the CRT outlet are believed to be typically above 20% (but sometimes as high as 60%) of the total NO_x, which represents an increased level relative to the engine-out emission. Experiments show that the total NO_x emission is not changed over the CRT, indicating that most of the NO_2 reverses to NO, according to Eq. (22.2). The reaction to N_2 given by Eq. (22.3) apparently plays little role in CRT regeneration.

The regeneration of the system can be enhanced, within certain limits, by increasing the size of the catalyst and the filter and/or by increasing the noble metal loading in the catalyst. Nevertheless, successful passive operation of the CRT requires that the exhaust gas reach a sufficient temperature and meet certain conditions, as discussed later. As is the case with regeneration of all passive filter systems, CRT operation depends on the vehicle's duty cycle. If the application is unsuitable or the duty cycle is too cold, the filter may be plugged with soot or experience uncontrolled regenerations.

22.3.1.3 Comparison with Alternative Configurations

Passive catalytic filter systems can be configured as the CRT, the catalyzed diesel particulate filter (CDPF), or as a combination of both. These three configurations are schematically shown in Figure 22.16. The "classic" CRT involves a combination of an oxidation catalyst and an uncatalyzed filter. In the CDPF, the catalyst is coated directly onto the wall-flow monolith. The combined system includes an oxidation catalyst, followed by a CDPF.

As discussed in the preceding section, the CRT depends exclusively and completely on the NO_2 effect for regeneration. CDPF regeneration, on the other hand, occurs due to a number of mechanisms relying on both oxygen and NO_2 as the oxidants, as discussed in Section 22.2. In particular, the NO_2 regeneration mechanism would also occur, to some degree, in CDPFs coated with active noble metal catalysts, such as platinum. The platinum catalyst will inadvertently generate some NO_2 with a beneficial effect on the filter regeneration. The CRT, however, features a superior reactor configuration that maximizes the NO_2 effect. In the CRT, the NO_2 is generated upstream of the filter and, thus, fully available for the oxidation of soot. In the catalyzed filter, NO_2 is generated within the monolith walls and is not available for the oxidation of the upstream portion of soot that may be collected in the inlet monolith channels. As a result, when CRT filters are operated on suitable, sufficiently high NO_x engines, they show a regeneration temperature advantage over catalyzed filters.

On the other hand, an NO_2 molecule in the CRT reactor can be used only once for the oxidation of soot. Once it reacts with carbon to form NO, it cannot be oxidized again to nitrogen dioxide and used for oxidation of further carbon particles. This is a drawback of the CRT configuration, which can be overcome by combining both devices. In such a combination, denoted as "Oxicat + CDPF" in Figure 22.16, the gas entering the filter is enriched in NO_2, just as it is in the CRT. However, as NO molecules re-created during regeneration pass through the catalyzed filter pores, they can be oxidized back to NO_2 over the active Pt catalyst sites. This second-generation NO_2 can oxidize further carbon particles, thus increasing the regeneration performance in comparison to the classic CRT system. Regeneration rates in this configuration may be also further increased due to catalytic oxidation of carbon by oxygen.

Indeed, laboratory tests of the combined oxicat + CDPF configuration confirmed improved regeneration performance, both in terms of lowered balance temperatures and increased regeneration rates [22]. Balance temperatures, as determined on an engine dynamometer test with a 3 ppm S fuel, are listed in Table 22.3. The CRT shows a 15°C advantage over the CDPF. The combined oxicat + CDPF configuration yields an additional 10°C improvement in the balance temperature.

TABLE 22.3
BALANCE TEMPERATURE COMPARISON

Configuration	Balance Temperature, °C
CRT (oxicat + uncatalyzed DPF)	265
CDPF	280
Oxicat + CDPF	250

Source: *Data from Ref. 22.*

Increased regeneration rates in the oxicat + CDPF configuration, measured as the rate of pressure drop reduction over time in soot-preloaded filters, is shown in Figure 22.17. All three filters were preloaded to the same soot loading of about 6 g/L (the CDPF pressure drop is slightly less than in the other configurations due to the absence of the upstream catalyst substrate). Clearly, the combined configuration shows a significant advantage over both the CDPF and the CRT. The Pt loadings in the filters were not disclosed, but the catalyst volume in the oxicat + CDPF configuration was half of that in the CRT, while the Pt loading per unit of catalyst volume was constant, resulting in a decrease in the total amount of platinum in the combined system, relative to the CRT.

The preceding data indicate that the oxicat + CDPF system may very well be the best configuration for a passive, catalytic diesel particulate filter, which could combine the advantages of both the CRT and the CDPF at no cost increase in comparison to a CRT. Johnson Matthey has been working on the optimization of the balance of catalytic activity between the precatalyst and the catalyzed filter. The system is expected to be commercialized under the trade name CCRT™ and offered in parallel with the CRT configuration.

22.3.2 *Application Limits*

Exhaust gas temperature, NO_x/PM ratio, and fuel sulfur content are the major factors that influence the CRT emission performance and regeneration. The following are the CRT application limits:

- *Exhaust gas temperature:* 275°C (minimum)
- *Sulfur content in the fuel:* 50 ppm wt. (maximum)
- *Exhaust NO_x/PM ratio:* between 8:1 and 25:1 (by weight, minimum).

These numbers, based on recommendations from CRT manufacturer Johnson Matthey, should be considered simplified guidelines. They do not account for such important aspects of real-life operation as variability in engine-out PM emissions or fluctuations of parameters under transient conditions. There is also a certain trade-off between various parameters, such as the exhaust gas temperature requirement may be lowered by lowering the fuel sulfur content, increasing the NO_x/PM ratio, or operating the CRT on a cleaner (lower PM emission) engine.

To sustain fully passive operation (regeneration), the CRT filter requires a certain minimum level of exhaust gas temperature. The CRT filter system verified by the California Air Resources Board (CARB) for retrofitting post-1994 highway engines requires exhaust temperatures exceeding 270°C over at least 40% of the operating time (fuel sulfur ≤ 15 ppm) [24]. Example CRT balance temperatures [13], determined on a CAT 3126 industrial engine using the DECSE method, for a range of fuel sulfur levels and engine speeds are shown in Figure 22.18. As seen in the graph, the filter could regenerate at low temperatures of around 300°C, provided ultra-low-sulfur fuel was used. The engine used in this study had a relatively high PM baseline; it is possible that somewhat lower balance temperatures would be measured on a cleaner diesel engine.

As long as ultra-low-sulfur fuel is used, the CRT has a balance temperature advantage over the catalyzed filter. A relatively high average balance temperature advantage of 60°C was found by the DECSE study [13]. As shown earlier in Table 22.3, a much smaller advantage of 15 °C was reported by others [22]. Because balance temperatures depend on a number of engine-specific parameters (e.g., NO_x/PM ratio), any generalization is difficult.

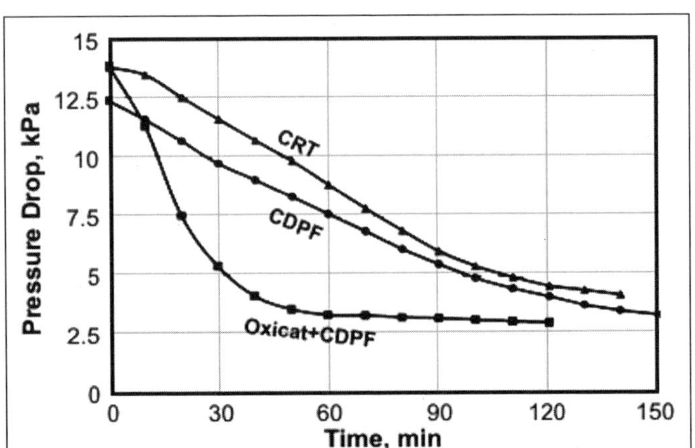

Figure 22.17 *Comparison of regeneration rates in various catalytic filter configurations.*

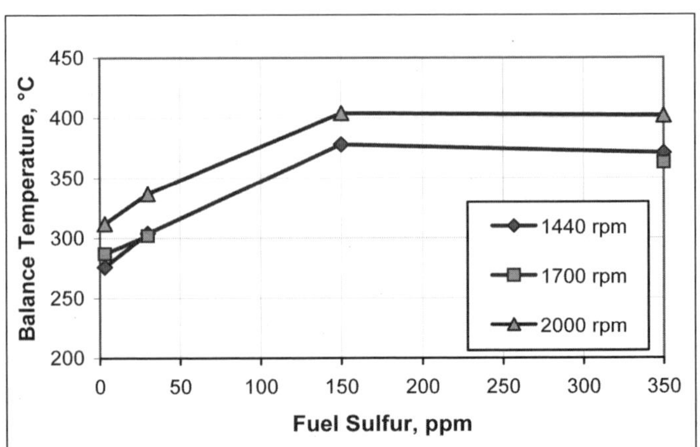

Figure 22.18 *CRT balance temperature as a function of fuel sulfur level.*

At higher sulfur levels, CRT regeneration deteriorates. It is believed that sulfur dioxide (SO_2) adsorbs in strong competition to NO on the catalytically active sites in the catalyst element of the CRT. This competitive adsorption is blocking the active sites and lowers the NO/NO_2 shift in the catalyst [Eq. (22.1)]. Because insufficient amounts of NO_2 are generated, the trap requires higher exhaust temperatures to regenerate. This deactivation by sulfur has a temporary character. Filters deactivated by operation on high-sulfur fuels regain their ability to regenerate when operated for a period of time on ultra-low-sulfur fuels [13]. Most of the experience with CRT systems has been accumulated using fuels having a maximum sulfur content of 10 ppm, such as the Swedish Environmental Class 1 diesel. Tests with higher sulfur levels indicate that fuels of up to 50 ppm S may be still satisfactory.

Another reason for using ultra-low-sulfur fuels with the CRT system is catalytic oxidation of SO_2. To maximize the NO/NO_2 conversion, a very active oxidation catalyst formulation is used in the CRT. That catalyst is also effective in oxidizing SO_2 to SO_3 and, thus, in producing sulfate particulates. The nucleation of sulfate particulates occurs after gases leave the filter, leading to an observed decrease in the PM filtration efficiency. This effect depends on the fuel sulfur content and the exhaust temperature, as shown in Figure 22.19. Experiments with 500 ppm S fuel showed that at temperatures above 400°C the filter was increasing the total particulate emission to double the engine PM baseline [25].

An important limitation of the CRT regeneration is its dependence on the NO_x/PM ratio in the engine-out emissions. The CRT configuration in its "classic" form uses an uncatalyzed filter and, thus, depends exclusively on NO_2 for regeneration. Higher NO_x/PM ratios result in higher NO_2 concentrations and better filter regeneration.

The theoretical minimum NO_x/PM ratio can be calculated based on the stoichiometry of reactions given by Eqs. (22.1) and (22.2). If all NO was converted to NO_2, which then would entirely revert to NO during regeneration, the required NO_x/PM ratio would be 3.83:1 (NO_x as NO_2 equivalent, PM as pure carbon). In practice, however, neither all of the engine-out NO can be converted to NO_2 (compare Figure 22.15 and the following discussion), nor can the entire NO_2 that was generated be used in the regeneration process. In several earlier publications, the minimum NO_x/PM ratio required by the CRT was typically listed at 8:1 [26]. Minimum NO_x/PM ratios as high as 20:1 to 25:1 have been referenced as "sufficient in most applications" in newer literature [27].

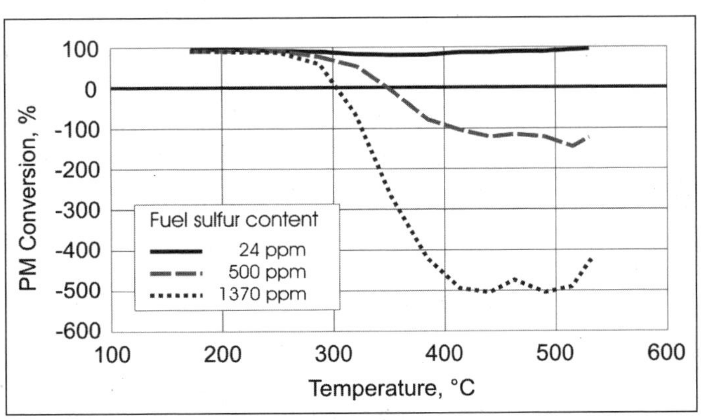

Figure 22.19 *Effect of fuel sulfur content on PM conversion in the CRT.*

Again, there is some trade-off between the different parameters that influence CRT performance. For instance, installations with high NO_x/PM ratios may be able to regenerate at lower exhaust temperatures. On the other hand, low NO_x emission engines may not be the ideal candidates for CRT application.

22.3.3 Filter System Design

The system consists of two functional components packaged into one housing. The first component is an active oxidation catalyst, the second component is a ceramic wall-flow diesel filter. For the oxidation catalyst usually a 400-cpsi ceramic substrate is used. Because the best activity for NO oxidation is achieved in the Pt/Al_2O_3 system, the catalyst is usually a high loading of platinum on alumina washcoat. The Pt loading in commercial products is not disclosed, but catalysts of 80 g/ft^3 Pt were considered during CRT system development [21]. The particulate filter in most applications has been a commercial, uncatalyzed cordierite wall-flow honeycomb of 100 cpsi with a 17-mil wall thickness. Other filter materials, such as SiC monoliths or sintered metal, can be also used in the CRT configuration.

In systems used for retrofitting of heavy-duty Euro 0 and Euro 2 truck and bus engines, the volume of the catalyst approximately matched the engine displacement, with the filter having twice the volume of the catalyst. For example, systems with catalysts having an 8.5-L volume (267-mm diameter × 152 mm = 10.5-in. diameter × 6 in.) and filters having a 17-L volume (267-mm diameter × 305 mm = 10.5-in. diameter × 12 in.) were used for retrofitting of engines having between 7 and 11 L displacement [28]. The same size of CRT unit was used for testing on the CAT 3126, 7.2-L, 205 kW @ 2200 rpm engine in the DECSE program [13].

A schematic of the CRT is shown in Figure 22.20. The CRT housing includes four stainless steel modules: inlet module, catalyst, filter, and the outlet module. Modules are clamped together by three V-clamps. A bypass that opens in case of filter clogging may be provided in the inlet module.

The standard design uses the same diameter of oxidation catalyst and particulate filter. The system usually replaces the production silencer of the vehicle. The size of the system is similar to that of the silencer although the weight is higher. The modular design, shown in Figure 22.20, allows for simple replacements in case of a defect and minimizes the total number of different catalyst and filter models. The filter module can be axially turned to blow out the inorganic ash buildup and, thus, increase the filter life. An alternative design is a direct-fit replacement for the original muffler. In that case, the system is very easy to retrofit, but the direction of flow through the filter cannot be changed.

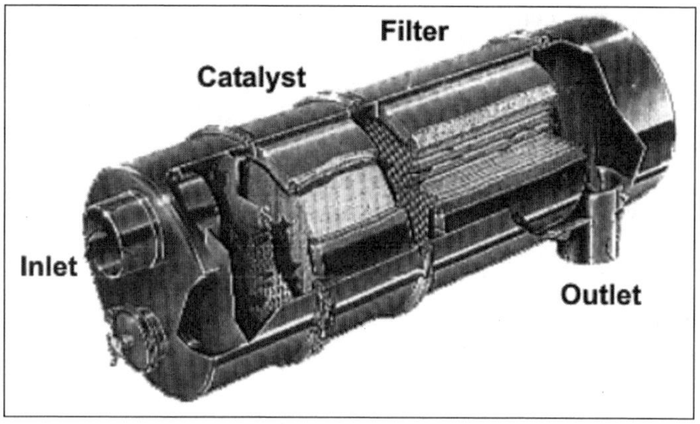

Figure 22.20 Cutaway view of CRT. (Courtesy of Johnson Matthey.)

22.3.4 Performance

22.3.4.1 Gases

The CRT, being a combination of catalyst and filter, influences the emissions of gaseous pollutants as well as the PM. CO and HC emissions, based on test results from several heavy-duty Euro 1 and Euro 2 engines, are reduced by about 85% to 95% when tested on the ECE R49, ESC, or ETC test cycles. Due to lower exhaust temperatures, the CO and HC performance is somewhat lower over the U.S. FTP Transient cycle, where conversions of about 70% to 90% were measured [28].

Aldehydes, including formaldehyde and acetaldehyde, can be reduced by 50% to 90% and more. In a study on urban buses with 1999 DDC Series 50 engines, emission reductions greater than 90% were reported for CO and HC, along with a more than 99% reduction in carbonyls (aldehydes, ketones, etc.), an up to 80% PAH reduction, and a more than 90% reduction in nitro-PAHs, all measured on CBD and NYB chassis dynamometer test cycles [29].

Because most of the nitrogen dioxide consumed in regeneration reacts to NO rather than to nitrogen, there is no significant change in the total NO_x emissions over the CRT. The CRT system, like all systems incorporating active oxidation catalysts, changes the proportions between the nitrogen oxides, producing increased NO_2 emissions. NO_2 emissions may be disadvantageous in some situations, especially in occupational health environments such as underground mining. The NO_2 emission may also cause problems in other areas with poor air movement, for example, in city streets surrounded by high-rise buildings. NO_2/NO_x tailpipe ratios between 20% and 60% were measured with the CRT on different test cycles, as compared to about 5% in untreated diesel exhaust [30]. NO and NO_2 emissions measured from a CRT-equipped urban bus [31] are illustrated in Figure 22.21.

22.3.4.2 Particulate Matter

The overall PM mass reduction efficiency depends on the conversion of the three fractions of diesel particulates: (1) solid carbon and ash, (2) SOF (heavy hydrocarbons), and (3) sulfate particulates. Because the CRT utilizes well-proven wall-flow monoliths, with more than 95% efficiency for solids, and an active oxidation catalyst that is fairly efficient in oxidizing SOF, the overall PM performance depends on sulfate particulates. The filtration efficiency decreases with increasing sulfur content in the fuel and with increasing exhaust gas temperature during the test cycle. Some results measured over the ESC (OICA) test cycle and at the peak torque condition [13] are shown in Table 22.4. The ESC cycle, due to its high average engine load and high exhaust temperature, may be considered disadvantageous for the CRT performance.

A number of studies using 10 to 30 ppm S fuel reported total PM reductions in the range of 90%. These reductions make it possible to meet Euro 4 and Euro 5 PM emission standards using a CRT-equipped Euro 2 engine or to meet the U.S. 2007 limits on a U.S. 1994 engine. On today's light-duty diesel engines of about 0.2 g/mile PM emissions, the CRT can reduce

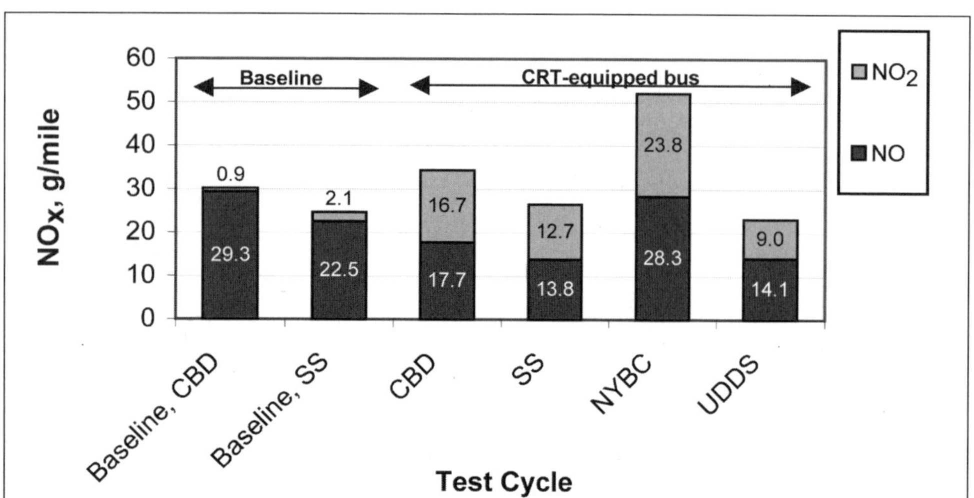

Figure 22.21 CRT effect on diesel bus NO_x emissions for a 1998 DDC Series 50 engine.

TABLE 22.4
PM REDUCTION EFFICIENCY IN CRT FILTER (CAT 3126 ENGINE)

Fuel Sulfur Level, ppm	PM Reduction, %	
	ESC Test	Peak Torque
3	95	91
30	72	73
150	−3	19
350	−155	−139

PM to below 0.02 g/mile, which satisfies the requirements of both the Euro 4 and U.S. Tier 2 regulations [32].

The CRT trap was also reported to decrease the nanoparticle number emissions, as illustrated in Figure 22.22. The data were generated on a six-cylinder, DI, turbocharged, and intercooled Euro II diesel engine, rated 169 kW @ 2400 rpm [30]. Some filter systems increase the nanoparticle numbers, due to hydrocarbon and sulfate nucleation after the trap. It can be speculated that little hydrocarbon material is available for such nucleation in the CRT, due to its high hydrocarbon and SOF removal efficiency.

22.3.5 Experience and Commercial Status

In countries that introduced ultra-low-sulfur diesel fuels, the CRT system is commercially available as an aftermarket kit from either Johnson Matthey or its licensed suppliers. The systems have been installed on buses and trucks mainly in Sweden but also in the United Kingdom, Germany, and other EU countries. Nearly all vehicles operate on Swedish Class 1 (10 ppm S) or equivalent fuel. Some systems were also tested on 50 ppm S fuel, and a few on sulfur-free biodiesel fuel. After 15 ppm S diesel fuel became commercially available in California in 2000, the CRT system also became available in the United States. About 19,000 vehicles have been equipped with CRT filters since the introduction of the system in 1994 through the year 2001 (18,000 in EU; 1000 in the United States) [27].

CRT retrofits for heavy-duty engines regenerate as fully passive systems. As such, they require careful selection of the engine and duty cycle to ensure sufficient exhaust gas temperatures, as well as NO_x/PM ratios, for problem-free operation. In Europe, the manufacturer recommended that CRT systems be installed on turbocharged engines no older than model year 1985.

A durability study was conducted to determine the emission performance of in-use CRT systems [28]. Seven CRT systems that accumulated between 106,000 and 600,000 km in bus, truck, and train operation were evaluated. PM performance measurements, as determined over a number of engine test cycles, yielded conversion efficiencies anywhere between about 80% and 95%. The exhaust gas pressure drop over the in-use CRT filters was not reported in the study.

Field experience indicates a gradual increase in exhaust back-pressure levels due to the filter becoming loaded with ash [26], as shown in Figure 22.23. Periodic reversing of the filter module is recommended as a means of cleaning the trap of these ash deposits. When operated with typical EU heavy-duty diesel oils of approximately 1.8% residual ash, the CRT needs to be reversed typically once per year [28].

The first CRT filter study in North America was conducted by the New York City transit authority, which retrofitted 50 urban buses (25 with 1999 DDC Series 50 four-stroke engines and 25 with 1993 6V92 DDEC two-stroke engines) and operated them over an 8-month evaluation period using 30 ppm S diesel fuel. The filters regenerated properly on the newer Series 50 engines, but experienced several instances of filter plugging on the older, two-stroke 6V92 engines [3]. In-service exhaust temperature recordings from the Series 50 buses indicated that the exhaust temperatures were above 275°C for 60% of the operating time. The filters maintained stable pressure drop levels with maximum readings of about 14 kPa (4 in. Hg) [29, 33].

Published test data from heavy-duty retrofit applications indicate a fuel penalty of 1% to 3%, due to the increased system back-pressure. This is consistent with the 2% fuel economy penalty, as measured in the laboratory by the DECSE program [13].

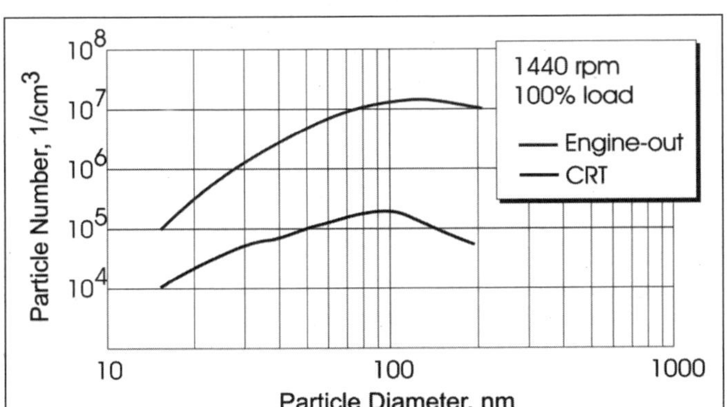

Figure 22.22 *Particle size distribution with CRT trap.*

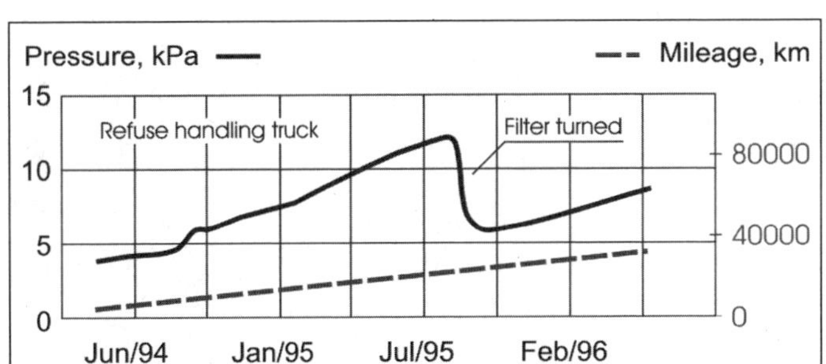

Figure 22.23 *Exhaust back-pressure with CRT trap.*

Filter systems based on the two-stage CRT configuration have been also developed for new diesel passenger car applications. However, contrary to heavy-duty engine retrofits, which can operate as totally passive systems (at least in some applications), light-duty engines require some form of active regeneration support. Exhaust temperatures were evaluated on both the ECE+EUDC and the U.S. FTP-75 cycles, as well as during real city driving, to determine the feasibility of CRT use. In all cases the temperatures were too low to sustain continuous filter regeneration [32].

A prototype filter system for the Volkswagen D1 V10 5.0l engine (10-cylinder TDI passenger car engine, rated 230 kW at 4000 rpm [34]) is shown in Figure 22.24. The system includes two two-stage catalyst-filter units (one per exhaust bank) in the underfloor position. It relies on an active strategy, involving engine management and a close-coupled "warm-up" catalyst, to periodically increase filter temperature. At the time Volkswagen unveiled the engine, no decision had been made regarding whether this filter system would be commercialized.

22.3.6 SCRT System

A combination of the CRT filter with a urea-SCR (selective catalytic reduction) catalyst, termed SCRT, has been proposed to address the need for simultaneous reduction of PM and NO_x emissions [35]. The CRT was positioned upstream of the SCR catalyst to enhance SCR performance by increasing the NO_2 content of the exhaust gas. (This effect was discussed in Section 19.3.)

It has been shown that today's engines equipped with the SCRT system could meet the Euro 4/5 PM and NO_x limits. The list of issues for future system optimization includes N_2O and ammonia slip, high exhaust gas pressure drop (up to 20 kPa), secondary PM emissions (presumably ammonium nitrate/sulfate), and system durability.

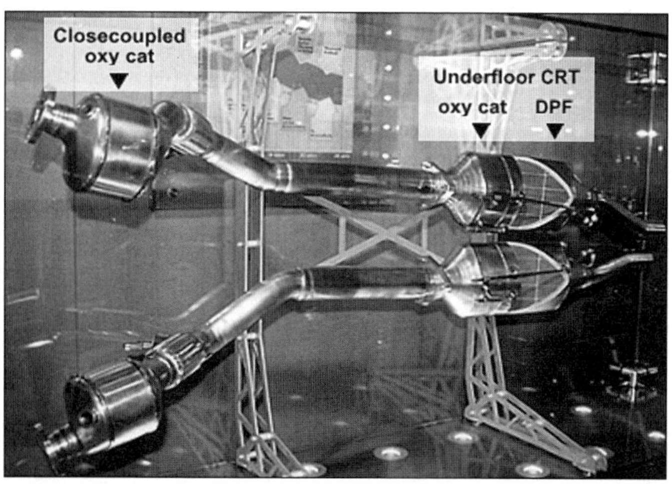

Figure 22.24 Prototype of light-duty CRT system.

22.4 Filters with Fuel Additives

22.4.1 Fuel Additives

Metal-based fuel additives were first studied as smoke suppressants and cetane improvers [36]. Several metals, including Ba, Ca, Fe, and Mn, have been found effective in lowering the amount of soot formed during combustion in both diesel and SI engines. This effect was explained by a combination of the following three mechanisms:

- Ions or radicals of the alkali or alkaline earth metals used as additives remove particle precursors.
- The ions or radicals inhibit the nucleation of particle precursors.
- Many transition metals catalyze the oxidation of particulates at a later stage of the soot formation process.

The most effective smoke suppressants, Ba and Ca, are believed to work primarily through the first two mechanisms.

Currently, the main interest in metal-based fuel additives is related to diesel particulate filters. Numerous metal additives have been investigated as soot oxidation catalysts that would facilitate regeneration in passive diesel filter systems. Fuel additives used for that purpose are also called fuel-soluble catalysts. As the additive is combusted in the engine cylinder, its metal component leaves the combustion chamber in the form of the corresponding metal oxide or other inorganic compound (e.g., sulfate). These compounds can form particles of their own or can be incorporated into diesel particulates. After being collected in the particulate filter, the catalytic metal is distributed throughout the diesel particulate phase and can effectively catalyze the oxidation of carbon particles.

The mechanism of trap regeneration is similar to that of a catalyzed filter, except the use of additives provides better contact between the catalyst and the carbon particles. This is probably why additive regenerated traps regenerate at somewhat lower temperatures than the catalyzed traps. However, even though fuel additive trap systems may have the lowest regeneration temperatures, they are not likely to provide the ultimate solution of the particulate problem, especially in light-duty applications. As illustrated in the following examples of filter systems, there are new "passive-active" approaches that support the additive-induced regeneration with some engine management measures in order to periodically increase the exhaust gas temperature.

Although many papers have been published on diesel filter additives, there is little or no information on such issues as the actual interactions between the additive and diesel particulates, catalytic mechanisms, additive particle size, and its distribution within the diesel particle phase.

Fuel additives are fuel-soluble organometallic compounds. Because the additive compound is destroyed by oxidation during the combustion process, its exact chemical formulation is, from the catalytic activity point of view, of limited importance. The additive formulation is, however, very important for its stability and miscibility properties. Organometallic compounds used for diesel trap additives that were reported in the literature include fatty acid soaps, acetyl acetonates, alcoholates, beta-diketonates, sulfonates, carbonates, carboxylates, dicyclopentadienes, and naphthenates [37, 38].

Tens of compounds of various transition, alkaline, and noble metals have been tested as fuel additives to facilitate diesel trap regeneration. The list of metals tested includes sodium, lithium, calcium, barium, manganese, iron, nickel, cerium, copper, lead, silver, vanadium, and platinum. For the purpose of diesel trap regeneration, additives are usually added to the fuel at concentrations from a few to about 100 ppm wt. of the respective metal. The choice of additive level depends on the activity of the particular additive, exhaust gas temperature, and the quantity and composition of particulate emissions. Generally, higher additive concentrations allow the trap to regenerate at lower temperatures. The disadvantages of high additive levels are (1) increased filter pressured drop due to the accumulation of ash and (2) higher operating costs.

Additive concentrations much higher than 100 ppm have been reported in several laboratory studies. Such high concentrations may have only academic importance and are not practical. Even the concentration of 100 ppm can result in significant pressure drop buildup in the filter due to accumulation of the additive ash and appears to be too high for practical purposes.

Fuel additives for diesel trap regeneration that have successfully passed laboratory evaluations and are either commercial products or are undergoing field evaluations are listed in Table 22.5.

Keep in mind that commercial use of additives in different countries may require formal approval or certification. Thorough health and biological studies are sometimes required by regulatory agencies to grant such approvals. In the United States, additives require a U.S. Environmental Protection Agency (EPA) registration to be legally sold for on-road use. (With the exception of underground mining, no additive approval is required in the United States for off-road use.) Not all products from Table 22.5 have received such a registration.

The following is a list of potential drawbacks and unresolved issues that have to be addressed by designers of fuel additive regenerated diesel filter systems:

TABLE 22.5
FUEL ADDITIVES FOR DIESEL FILTER REGENERATION

Active Metal/Compound	Brand Name	Supplier
Iron (Fe/ferrocene)	Satacen	Octel (Pluto)
Iron-strontium (Fe/Sr)	Octimax 4800	Octel
Cerium (Ce)	Eolys	Rhodia
Platinum (Pt), platinum-cerium (Pt/Ce)	Platinum Plus	Clean Diesel Technologies
Copper (Cu)	—	Lubrizol[a]

a. Lubrizol copper additive is no longer commercially available. Lubrizol has exclusively licensed the Clean Diesel Technologies "Platinum Plus" additive for sale in Europe.

- The necessity of introducing the catalyst to fuel. In most fuel distribution systems, doping the fuel with additive is not practical. On-board automated dosing devices are being developed to introduce the additive directly to the vehicle's tank after the vehicle is fueled. However, such devices increase the system's complexity and, in effect, turn passive traps into more complicated and expensive active systems.
- Ash deposits on the filter. Most of the additive is trapped on the filter in the form of an inorganic oxide and/or salt. With time, it accumulates to considerable quantities. The ash deposits contribute to increased pressure drop.
- Ash emissions. Additive ash that is not trapped, due to limited filtration efficiency or filter failure, is emitted to the atmosphere and may have adverse health or environmental effects. A formal approval or certification of additives is required by law in some countries.
- Impact on engine-out emissions. Additives can improve or deteriorate the engine-out gaseous and particulate emissions. Various emission characteristics may be affected, such as PM size distribution. Additives may also create new secondary emissions.
- Impact on the engine or its components. Some additives are known to cause fuel injector fouling. Generally, additives have to be tested for engine wear problems. For engine warranty reasons, the use of additives should be always approved by the engine manufacturer.
- Fuel stability. Blending the additive with fuel may result in deposit formation, an increase in sedimentation from the fuel itself, and/or increased deposit formation when water is added to the doped fuel.

The importance of proper additive testing for its possible engine impact is best illustrated by an incident of additive-induced engine problems that was reported in Germany, where several 6-ton forklift trucks were equipped with a passive, additive-regenerated, commercially available diesel particulate filter system [39]. The systems were installed on Deutz BF6M1012 engines, rated 85 kW at 2200 rpm. After about 2000 engine hours, reduced power, increased smoke, and other engine problems were reported. It was found that the additive caused severe injector fouling by oil-derived coke and magnetite compounds. As illustrated in the microscope photographs of the contaminated injectors shown in Figure 22.25, several injector holes are completely plugged by a thick layer of deposits. Actual additive levels in the fuel, as measured in the field, were found to be between 8 and 50 ppm.

22.4.2 Catalytic Metals

22.4.2.1 Iron

Ferrocene

Iron, in the form of ferrocene (dicyclopentadienyl iron, $(C_5H_5)_2Fe$), has been commercially available as a combustion improver to increase fuel economy, primarily for bunker fuel in heating applications. It is also suitable as a diesel trap additive, as demonstrated by several studies [40, 41]. A

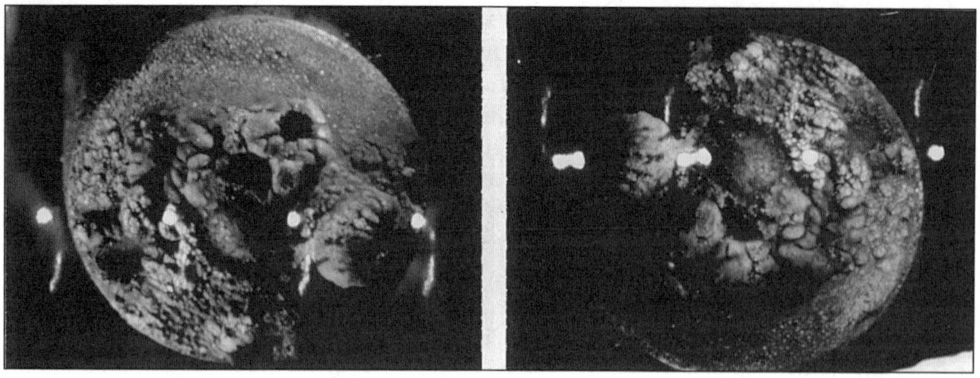

Figure 22.25 Contaminated fuel injector tips. (Courtesy of Deutz.)

ferrocene-regenerated trap system for industrial forklift trucks has been commercially offered in Germany by GfA.

The molecular structure of ferrocene is shown in Figure 22.26. The molecule includes a central iron atom which is bonded with two symmetrical cyclopentane rings. Several ferrocene derivatives that have been manufactured as fuel economy improvers are listed in Table 22.6.

For the purpose of diesel trap regeneration, ferrocene levels of about 18 ppm wt. as Fe (60 ppm wt. ferrocene) have been recommended by additive suppliers. Fe levels of 36 ppm wt. (120 ppm ferrocene) were used in the VERT program for retrofitting of construction diesel engines with passive, additive-regenerated systems [40].

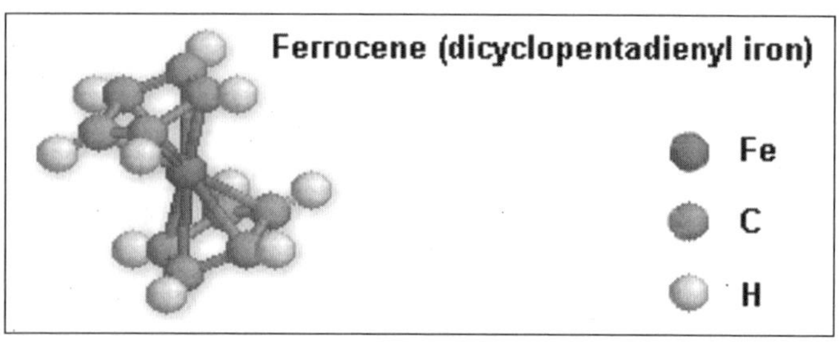

Figure 22.26 Molecular structure of ferrocene.

TABLE 22.6
FERROCENE DERIVATIVES USED AS FUEL ADDITIVE

Dicyclopentadienyl iron	Di(methylcyclopentadienyl) iron
Di(ethylcyclopentadienyl) iron	Methylferrocene
Ethylferrocene	*n*-Butylferrocene
Dihexylferrocene	Phenylferrocene
m-Tolylferrocene	Didecylferrocene
Dicyclohexylferrocene	Dicyclopentylferrocene

Fe/Sr

A new iron-based, nonferrocene additive has been introduced by Octel. The additive has been specifically designed as a diesel trap additive and includes iron and strontium as active components. According to the manufacturer, the required additive dosage is between 10 and 20 ppm wt. of combined metal (Fe+Sr). This additive was tested in the laboratory and in a number of road trials [42, 43].

Octel was also developing a strontium-sodium additive for trap regeneration. The Sr/Na additive exhibited very good activity as a soot oxidation catalyst. However, it was shown to chemically attack and impair the mechanical durability of several diesel filter media, including cordierite and fiber fibers [44].

22.4.2.2 Cerium

A cerium-based diesel filter additive has been developed by Rhodia (former Rhone-Poulenc). It has been commercialized under the brand name Eolys. The additive has been tested with diesel filters in urban bus, car, and other applications, primarily in Europe. The advantages of cerium claimed by Rhodia include its biological inertness and nontoxic character. An improvement in engine-out emissions is also attributed to cerium [45].

The Ce additive was proven in several studies to be effective in diesel trap regeneration and compatible with filer media [40, 41]. Additive dosage levels used in projects where heavy-duty diesel engines were retrofitted with passive, cerium-regenerated systems were in the range of 100 to 120 ppm wt. Ce, which tends to be higher than additive levels reported for iron/ferrocene.

A study on the compatibility of metal oxide ashes with diesel filter materials found that cerium ash had a less deteriorating effect on cordierite wall-flow substrates than iron-based ashes [46].

22.4.2.3 Platinum

Platinum additive was introduced under the brand name Platinum Plus by Clean Diesel Technologies (CDT). According to the manufacturer, the Pt additive is suitable for both decreasing the engine-out emissions and regenerating diesel traps.

A study on the activity of Pt fuel additives was conducted at the University of Delft [47]. It was found that a synergistic effect exists between Pt and Ce, when used as diesel trap additives. Furthermore, the lowest balance temperatures were seen on traps that were precatalyzed with Pt in conjunction with a small dose of Pt/Ce fuel additive. Minimum balance temperatures for different trap/additive systems, as determined on a laboratory Lister Petter LPW2, two-cylinder, DI, NA, water-cooled engine, are listed in Table 22.7. The minimum balance temperature was defined as the minimum temperature at which a stable pressure drop level could be achieved during steady-state engine operation. The corresponding pressure drop was 15 to 20 kPa (60 to 80 in. H_2O). All experiments used 500 ppm S fuel.

A series of laboratory reactor experiments was used to illustrate that NO chemistry played an important role in the overall process [47]. However, no clear explanation exists as to the mechanism of the synergistic effect between Pt and Ce or to the superior performance of traps with supported platinum catalyst. A definite advantage of the Pt/Ce additive would be its low dosage level, apparently as little as 5 ppm, minimizing the fouling effect of the additive ash on traps.

Keep in mind that performance data on Pt additives were generated under laboratory conditions. The advantages of this promising product, including its high activity at low dosage level, have to be confirmed by evaluations under field conditions.

22.4.2.4 Copper

A copper-based additive for trap regeneration was developed and extensively tested by Lubrizol [48]. The copper additive was also found to be the most effective of the additives tested by the VERT program [40]. Additive levels of about 60 ppm wt. Cu were used in the VERT study.

The copper additive, however, caused some negative effects on engine-out emissions. The biggest concern was the generation of secondary emissions of dioxins and furanes from engines equipped with Cu-regenerated traps [49]. Other additives, including Fe and Ce, were not producing dioxins or furanes. These health concerns have made the copper additive less attractive. The copper additive was also reported to cause severe fouling of fuel injectors. Gold-plated injectors were developed in an attempt to solve this problem [48].

22.4.3 Peugeot Filter System

PSA Peugeot Citroen developed a diesel particulate filter system for passenger cars that entered serial production in May 2000. Peugeot first launched the filter as a standard component on its Model 607 automobile, which was powered by the DW12 TED4, 2.2-L, 98-kW, common-rail diesel engine. In the following years the system was introduced on several other car models with Peugeot engines. By the end of 2002, the PSA Group had sold nearly 400,000 filter-equipped vehicles [50].

The filter system utilizes a fuel additive to lower the soot combustion temperature to support regeneration. Because the PSA system also employs an active engine strategy to periodically increase the exhaust gas temperature, it hardly meets the definition of a passive filter. Rather, it is a sophisticated active particulate filter, which can be considered a textbook example of a system approach to diesel emission control where the emission target is achieved through tight integration of a new technology, low-emission diesel engine with advanced exhaust gas aftertreatment.

The Peugeot system utilizes a silicon carbide wall-flow filter substrate (DPF in Figure 22.27) and an oxidation catalyst upstream of the filter (OxyCat) [2, 51]. A cerium-based fuel additive (Eolys by Rhodia) is used to lower the combustion temperature of the soot, allowing the filter to regenerate at temperatures below 450°C. This temperature, however, is usually not reached during regular operation of the car. Therefore, the system includes an active means of increasing the exhaust temperature in order to trigger regeneration.

The regeneration of the filter is controlled by the engine electronic control unit (ECU). Once the ECU determines that the filter needs regeneration, it increases the exhaust gas temperature through a combination of two effects:

1. *Late fuel injection:* Three injections of fuel are performed in the regeneration mode. The preinjection initiates the heat release in order to control noise. The second injection produces the engine power (BMEP). The third injection

TABLE 22.7
BALANCE TEMPERATURE FOR DIFFERENT TRAP/ADDITIVE SYSTEMS

Additive	Concentration, ppm wt.	Filter	Temperature, K	Temperature, °C
None	—	EX-80	810–830	537–557
None	—	Pt EX-80	690–700	417–427
Cerium	100	EX-80	705	432
Platinum/cerium	0.5–5	Pt EX-80	600	327
Platinum/copper	0.5–5	Pt EX-80	620	347
Platinum/iron	0.5–22	Pt EX-80	630	357

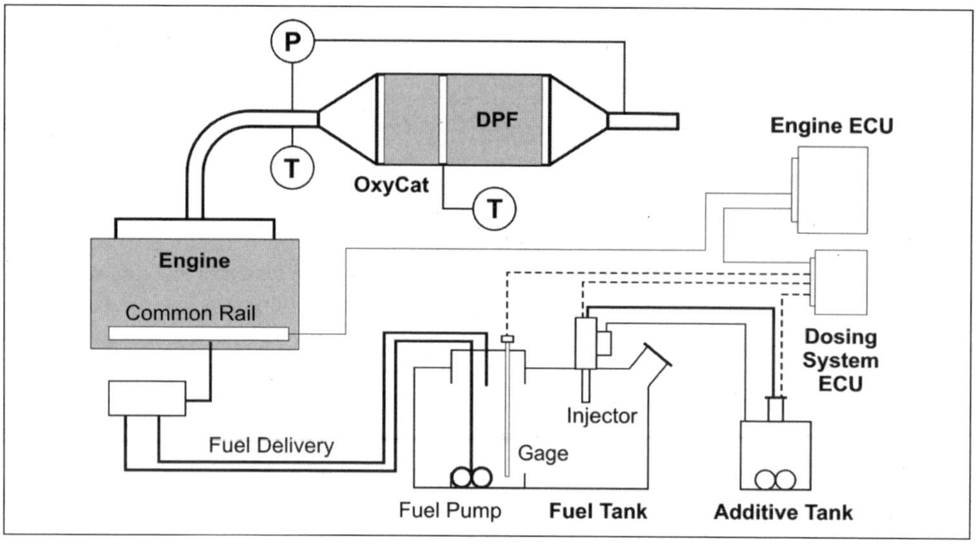

Figure 22.27 Schematic of Peugeot filter system. T, temperature sensor; P, pressure sensor.

extends into the expansion phase and causes postcombustion in the cylinder, increasing the temperature by 200 to 250°C. Development of the common-rail fuel injection strategy also involved special turbocharger calibration, to control enthalpy increases at the turbine and the intake air temperature.

2. *Catalytic combustion:* Complementary postcombustion, generated by the oxidation catalyst upstream of the filter, deals with the unburned hydrocarbons resulting from the post-injection. The temperature can be increased by more than 100°C. The catalytic combustion mechanism was necessary under low engine speed and load, where exhaust temperatures of 450°C could not be reached without a BMEP increase.

The combination of both effects increases the exhaust gas temperature to about 450°C. Filter regeneration occurs periodically every 400 to 500 km, depending on the driving pattern of the particular vehicle.

Start of regeneration is determined by a software algorithm based on an estimation of the mass of carbon material in the filter. This is computed based on the filter pressure drop and the composition of the particulates in the filter. The major fractions of the material trapped in the filter are (1) carbonaceous particulates, (2) ash from the additive, and (3) ashes from lube oil. The oil consumption was reduced by elaborate liner machining to limit particulate formation. A typical oil consumption value for the engine was 0.1 L/1000 km and the ash content of the oil was 8 to 10 g/L, making the oil ash buildup relatively slow. The additive residue in the filter depends on the treat rate and the fuel consumption. The quantity of each type of ash is calculated at every moment by the ECU to estimate the corresponding pressure drop. The final step is to estimate the mass of carbonaceous soot in the filter. It was found that the pressure drop depends on the shape of particulate aggregates, which in turn depends on engine operating condition. Thus, the soot mass is determined from the filter pressure drop (corrected for the ash contribution) and a map of engine operating conditions stored in the ECU.

The additive is stored on the vehicle and automatically dosed. When the vehicle fuel tank is topped up, the fuel additive system—controlled by a dedicated ECU—injects the required quantity of the additive. For a full tank of 60 L the system will inject 37.5 mL of the solution containing 1.9 g of ceria. This ratio corresponds to a Ce concentration in fuel of 25 ppm wt. The additive reservoir's 5-L capacity gives a range of 80,000 km. After the 80,000 km the additive must be replenished and the filter cleaned from the additive ashes by flushing with pressurized water.

The filter system shows very good emission performance. Tested over the ECE+EUDC test, filter-equipped cars produce PM emissions equivalent to the PM levels in the dilution air. For instance, PM levels from the Peugeot 605 HDi + DPF were reported at 0.004 g/km, way below the Euro 4 limit of 0.025 g/km. Reference tests with dilution air yielded PM levels between 0.002 to 0.004 g/km [2]. Steady-state PM emissions at high-temperature modes produce sulfate particulates in the catalyst. For example, PM emission of 0.095 g/km was measured at a 120 km/h cruise condition using 500 ppm S fuel. Particle number emissions were reduced by three to four orders of magnitude over the entire particle size range. Excellent PM emissions reductions, both mass and numbers, were also confirmed by independent tests [52].

A combined fuel economy penalty of up to 5%, depending on the vehicle driving pattern and frequency of regeneration, is associated with the increased pressure drop in the filter and with the post-injections of fuel. A 2% reduction of the engine power output is also seen at full air flow, due to the filter pressure drop.

22.4.4 Fiber Filters with Fuel Additives

22.4.4.1 Additive Comparison

A study on the compatibility of the Nextel fibers with fuel additives (or fuel-soluble catalysts) for regeneration was undertaken by 3M Company [41]. The study included laboratory filter regeneration tests, engine bench durability tests, and field tests with some of the additives. The work was targeted toward heavy-duty bus and truck engines. Three additives based on the following active metals were evaluated: (1) copper (Cu), (2) iron (Fe), and (3) cerium (Ce).

Filter regeneration using these three additives was compared on a Cummins NTC-350 engine under identical test conditions (Figure 22.28). The media dimensions of the evaluated wound fiber cartridge were 140 mm in diameter × 1270 mm in length. The engine was operated on an automated speed/load cycle to allow filter loading (light load) and also to force regeneration on each cycle (high speed and load). The curve profiles in Figure 22.28 reflect these loading and forced regeneration segments of the test cycle. Additive dose rates listed in the graph are all on metal basis.

The results shown in Figure 22.28 indicate that the copper-based additive promotes the highest regeneration rate, resulting in the lowest average pressure drop in the filter. Iron promoted regeneration at the slowest rate while cerium was in between Cu and Fe.

Additional laboratory testing revealed an important difference between the regeneration behavior induced by copper and that with Ce or Fe. When tested at constant exhaust gas temperature, the copper additive had a tendency to promote periodic soot buildup and regeneration cycles. Soot load in the filter was increasing up to a certain level, and then ignited and rapidly combusted, releasing heat and resulting in high-temperature peaks. The cerium and iron additives, on the other hand, promoted continuous filter regeneration. During the continuous regeneration, the filter operates under balance conditions in which the amount of soot entering the filter equals that which is being oxidized in the filter. There is no rapid combustion and no high-temperature spikes in the filter media.

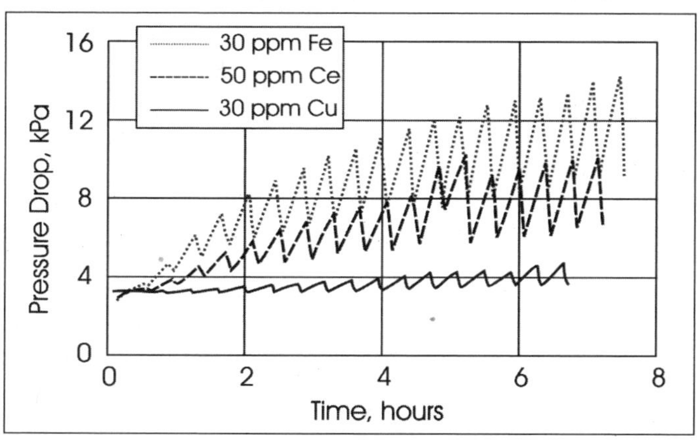

Figure 22.28 Regeneration behavior with Fe, Ce, and Cu additives.

22.4.4.2 Copper-Based Additive

Further laboratory tests conducted with the copper-based additive revealed that the Nextel filters exhibited a high copper ash collection efficiency of 95% to 99% but no fiber deterioration was observed in the lab. During the subsequent field testing, however, it became apparent that filter systems were failing after about 30,000 km of use. Visible smoke was observed after the filter and investigation revealed a drop in filtration efficiency from 90% to about 50%.

On disassembly of the filters, it was discovered that copper ash produced a sticky coating on the filter media that obstructed filter passages and bonded fibers together. Analysis of fibers confirmed that high temperatures during regeneration resulted in glazing of the copper and oil ash and, hence, bonding of the fibers. The matrix of fiber and glassy ash was brittle, leading to fiber damage by the thermal, mechanical, and vibration stresses experienced during operation. It was concluded that ceramic fiber wound filters are not compatible with copper-based fuel additives.

22.4.4.3 Iron-Based Additive

Laboratory evaluations of the iron additive were conducted on a 1994 EPA- and CARB-certified, 5.9-L, DI, turbocharged, air-charge-cooled diesel engine, rated 119 kW at 2500 rpm. A concentric tube pack filter was used including three wound fiber cartridges nested one inside another. The engine test cycle was composed of repeated 1-h segments. Each segment included 20 iterations of a simulated urban bus driving cycle (54% duty cycle based on rated speed fuel rate) and 2 steady-state sections (7 min at torque peak and 8 min at rated speed conditions). The dose rate of iron additive during the test was about 4 ppm. The average exhaust gas temperature increased over the 1000-h test from about 380 to 440°C as a result of increasing exhaust pressure. The average filter pressure drop and the total mass (ash + soot) accumulated in the filter are plotted versus time in Figure 22.29.

The following observations were made during the 1000-h test with iron fuel additive:

- After some initial period, recorded soot filtration efficiencies were higher than 90%. Slightly higher efficiencies were observed during steady-state conditions than during transients.
- The filter regenerated properly. There was no damage or deterioration to the fibers. There was no evidence of material sticking to the fibers in photomicrographs taken after tests.
- Ash analysis indicated that a phase change (glazing) occurs at 900°C. Because the ash in the filter was not glazed, regeneration temperatures must have been below that level.
- Most of the iron ash was retained in the filter (87.6% collection efficiency of Fe_2O_3 was measured).
- Oil and iron ash mass balance calculations confirmed that the increase of filter mass and its pressure drop (beyond some

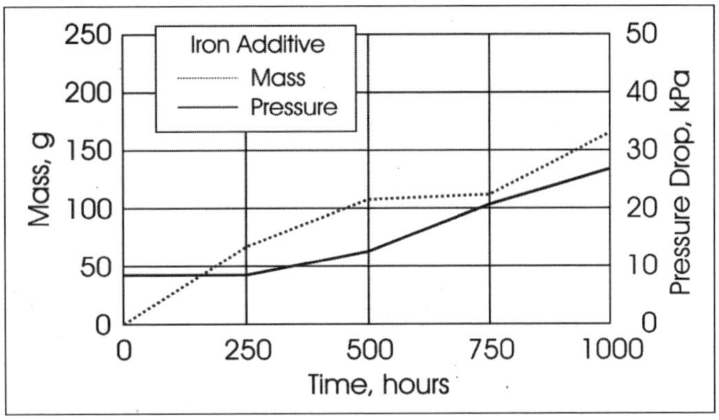

Figure 22.29 *Total collected mass and pressure drop (Fe test).*

initial period of less than 250 h) was due to the oil and iron ash accumulation (Figure 22.29). Analysis revealed that 43% of the total noncombustible deposit found in the filter was derived from the iron additive.

- A 14% increase in engine-out particulate emission due to increased exhaust pressure was recorded over the 1000-h test.
- A 0.85% increase of fuel consumption was recorded during the test.

Field testing of Nextel filters with iron additives commenced on 15 city buses in Bucharest, Romania. All of the buses were powered by RABA-MAN D2156 H6U engines. Some of the buses were equipped with a concentric tube pack filter, whereas others used assemblies of eight nonconcentric Nextel cartridges. The additive dose was 30 ppm Fe (100 ppm ferrocene). During 17 months of testing, the fleet buses accumulated between 48,000 and 120,000 km (average 80,000 km per bus). One of the systems failed mechanically in that time period. Other systems maintained their performance as indicated by smoke number measurements. The average system performance is listed in Table 22.8.

22.4.4.4 Cerium-Based Additive

A cerium-based additive was evaluated under laboratory conditions similar to those used for the iron-based additive testing. The additive dose rate was higher than that in the Fe test, and amounted to 12 ppm. Exhaust temperature increased during the test from 380 to 470°C. Pressure drop and mass curves taken during a 700-h test are shown in Figure 22.30. A faster pressure buildup was observed than that in the test with iron.

The following observations were made during the 700-h test with a cerium-based fuel additive:

- After the initial period, soot filtration efficiencies were higher than 90%.
- The filter regenerated properly and no damage or fiber deterioration occurred. Fiber photomicrographs showed layers of ash material flaking from the fiber.
- Ce-oil ash analysis indicated that the glazing process occurs at 900°C, similarly to the Fe ash. No glazed ash was found in filters, so the regeneration temperatures must have been below that level.
- Most of the cerium ash was retained in the filter (90.5% collection efficiency of CeO_2 was measured).
- The increase of filter mass and its pressure drop (beyond some initial period of less than 250 h) was due to the accumulation of ash from oil and cerium additive (Figure 22.29). Analysis revealed that 55% of the total noncombustible deposit found in the filter after tests was derived from the cerium additive.
- A decrease in engine-out particulate emission of 21% was recorded over the 700-h test, despite the increased pressure in the exhaust system.
- A 4.7% increase in fuel consumption was recorded during the test.

TABLE 22.8
AVERAGE VEHICLE PERFORMANCE

Smoke number, no filter	5.1
Smoke number with filter	1.2
Pressure drop across filter	6 kPa
Filter inlet temperature	300°C
Filter outlet temperature	260°C
Noise level, muffler, but no filter	87.5 dB(A)
Noise level, filter, but no muffler	88.0 dB(A)
Increase in fuel consumption	6%

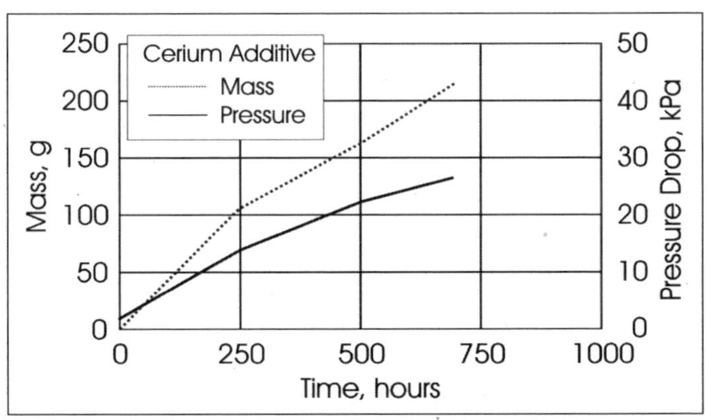

Figure 22.30 *Total collected mass and pressure drop (Ce test).*

22.5 Electrically Regenerated Filters

22.5.1 Electric Regeneration

22.5.1.1 Methods of Energy Deposition

Thermal regeneration of diesel particulate filters requires energy input to increase the temperature within the reaction zone between soot—solid carbon particles—accumulated in the filter and oxygen (and/or other oxidants) present in the exhaust gas. In general, the energy can be supplied to the exhaust gas, to the filter substrate, directly to the accumulated soot, or through a combination of these methods. These different methods may show different energy efficiency, due to a different degree of energy utilization in the reaction zone, as opposed to heat loss. Electricity offers perhaps the most flexibility among other forms of energy, and can be applied in the following ways:

- *Substrate heating:* This method requires that the filter substrate be a conductor of electricity, serving as a resistive heater itself. Filter concepts that have been investigated include sintered metal fiber cartridges [53, 54] or silicon carbide segments [55]. If materials of low heat capacity can be used, this method may show good energy efficiency. Electrically heated filter substrates remain in the product development stage.
- *Soot heating:* Energy can be deposited directly to the soot through the use of microwaves. Development of this type of system, which involves conversion of electric energy into electromagnetic waves, is described in more detail in Section 22.7 dealing with microwave regeneration.
- *Exhaust gas/air heating:* The most common electric regeneration method involves heating of the exhaust gas or a stream of regeneration air by a resistive heater. This approach has been demonstrated in numerous active filter projects and commercialized in selected applications, mainly for off-road industrial vehicles.

The use of an electric heater to increase the temperature of either the exhaust gas or a stream of air used for regeneration is a relatively simple method of triggering the regeneration of a soot-laden diesel filter. It can be used with a wide range of filter substrates. In systems utilizing wall-flow monolith filters, the heater is usually placed upstream of the filter substrate. Depending on the filter media, electric heaters may be also incorporated as an integral part of the filter; in those cases the deposited energy may be split in varying proportions between the flowing gas and the filter media.

22.5.1.2 Classification of Systems

Based on system configuration and the source of power, electrically regenerated filters can be divided into three groups:

- On-board regeneration systems
- Shore power regeneration systems
- Off-board regeneration systems.

In on-board regeneration systems, the power needed for regeneration is drawn from the vehicle's electrical system. This is the only practical filter system concept applicable to highway vehicles such as trucks and buses. The prime concern in on-board regeneration systems is energy consumption. The energy required by the heater puts an extra load on the vehicle electrical system and also presents an additional source of fuel penalty. The amount of heat needed to increase the exhaust gas temperature by 1°C per 1 hp of the engine rated power can be estimated as follows:

Assumptions:

Exhaust gas flow: m = 5 kg/hp·h (at rated engine speed)
Exhaust gas specific heat: C_p = 1.028 kJ/kg·K (air @ 400°C)

Heat required:

$Q = mC_p/3.6 = 1.43$ W/hp·K

In the most simple concept of the electric particulate filter, the so called "full-flow" system, the entire stream of exhaust gas is heated, so the filter inlet temperature reaches about 550 to 700°C as may be required to oxidize the soot. Electrical energy required to heat up the entire exhaust gas stream from a 100-hp engine by 100°C would be 100 × 100 × 1.43 = 14.3 kW or 19.2 hp. Assuming that the efficiency of converting mechanical energy at the engine shaft into electricity is 50%, 38.2 hp of the engine power output—or nearly 40%—would be consumed by the filter system. In most real-life situations, a 100°C temperature increase would be insufficient to regenerate the filter. It is easy to calculate that increasing the exhaust gas temperature by 250°C would require that approximately 100% of the engine mechanical power output be converted to electricity and consumed by the heater. In other words, the particulate filter system during regeneration may consume no less but the full rated power output of the engine. Clearly, full-flow electric regeneration systems are not feasible. Even if the regeneration was performed at a lower engine speed, the energy requirement in the full-flow system could not be lowered to realistic levels.

Most of the energy consumed by the full-flow filter is lost when hot gas is discharged at a high flow rate from the tailpipe. To minimize energy consumption, most on-board regeneration systems utilize either a partial-flow layout or hot air regeneration. In the partial-flow system, only a portion of the exhaust gas stream is used for regeneration, while most of the gases bypass the filter. In the hot air system, a dedicated blower supplies the regeneration air. The exhaust gas has to bypass the filter for the regeneration period. If exhaust gas filtration is required over 100% of the vehicle operating time, partial-flow and hot air systems require dual filter units, so the flow can be switched between the filters for the filtration and regeneration modes.

Shore power systems require that the filter system be connected to an external power source, such as a wall power outlet in a garage or maintenance shop, for regeneration. Thus, the vehicle is immobilized for the duration of the regeneration cycle. Shore power regeneration systems are applicable to vehicle fleets with easy access to a source of external power, for example, a fleet of construction vehicles operated within a construction site. This type of system would be impossible to use, or at least very inconvenient, on most highway vehicles.

Several variants of the shore power system are possible. In most cases, the filter is regenerated by hot air, but using the vehicle's exhaust gas while idling is also possible. If hot air is used, the blower and heater in the "pure" shore power configuration are part of the on-board filter system. This configuration (shown later in Figure 22.34) provides the most portability, because the only external connection that is needed is the electrical power. In some systems certain components are moved to an off-board regeneration unit. For instance, the heater may be still a part of the on-board filter, but the blower may be located off-board. In such a case, both power cord and air hose connections need to be made for the duration of regeneration. Moving components to external regeneration units allows for space savings on the vehicle itself. Cost savings may be also possible if more vehicles can share the same regeneration unit. However, the use of such systems is even less flexible than most other methods and in most cases limited to short-range vehicle fleets within one facility, for example, a fleet of diesel forklifts operated in a single warehouse.

In off-board systems the particulate filter is removed from the vehicle and placed in an off-board unit for regeneration. If spare filters are available, a soot-laden filter can be quickly replaced with a regenerated one, allowing for less equipment downtime than is the case for the shore power system, at a price of higher maintenance intensity.

Shore power and off-board systems are maintenance intensive and their proper operation depends on the timely intervention of the vehicle operator, who must initiate regeneration. These are fundamental shortcomings of these types of filters, which attracted rightful criticism from their users, as well as engine and vehicle manufacturers [56]. To gain general acceptance, particulate filter systems must be fully automated, integrated with the vehicle powertrain, and "invisible" to the vehicle operator.

22.5.1.3 Design Considerations

Certain design considerations apply to electrical heating of exhaust gases or air in all of the preceding filter configurations. Depending on their energy density (W/cm^2), the surface of electric heaters may reach considerably high temperatures. This causes increased heat transfer rates by radiation, as opposed to convection, calling for special precautions to control heat losses and other phenomena. The following are a few selected considerations; example filter systems, which are discussed later, may provide additional insight into these matters:

- Heat losses present an additional source of fuel economy penalty or, in off-board systems, additional energy cost and a possible safety issue. Hot surfaces require efficient heat insulation and may also require heat shields. To control heat radiation in the direction opposite to the exhaust gas flow, certain systems employ a heat reflector upstream of the electric heater. Such a reflector could be a disk of a large pore size ceramic foam, facilitating convective transfer of the absorbed heat to the gas.
- Temperature measurement provides important feedback to the system controller, which oversees the operation of the heater. If the tip of the temperature sensor can "see" the hot surface of the heater, it will absorb some of the radiated heat. In effect, the temperature of the sensor will be higher than that of the surrounding gas, possibly resulting in very substantial errors. This effect is extensively discussed in texts on heat transfer [57].
- Flow distribution over the heater and the filter inlet face, as well as the geometric configuration, must provide uniform temperature distribution at the filter inlet to achieve uniform regeneration. Some systems have used flow straighteners upstream of the heater.

22.5.2 *Off-Board Regenerated Filters*

An off-board regenerated system as shown in Figure 22.31 consists of (1) a removable filter unit, (2) a fixture installed at the vehicle's tailpipe that allows for relatively easy removal and installation of the filter, (3) an off-board regeneration unit, and (4) a pressure drop measuring kit (not shown), typically including a dashboard-installed indicator light or gauge. The pressure signal line connects to the coiled tubing visible on the fixture (2).

When the filter reaches its maximum soot holding capacity, as determined by the measurement of its pressure drop, the vehicle operator is notified by a dashboard-mounted indicator

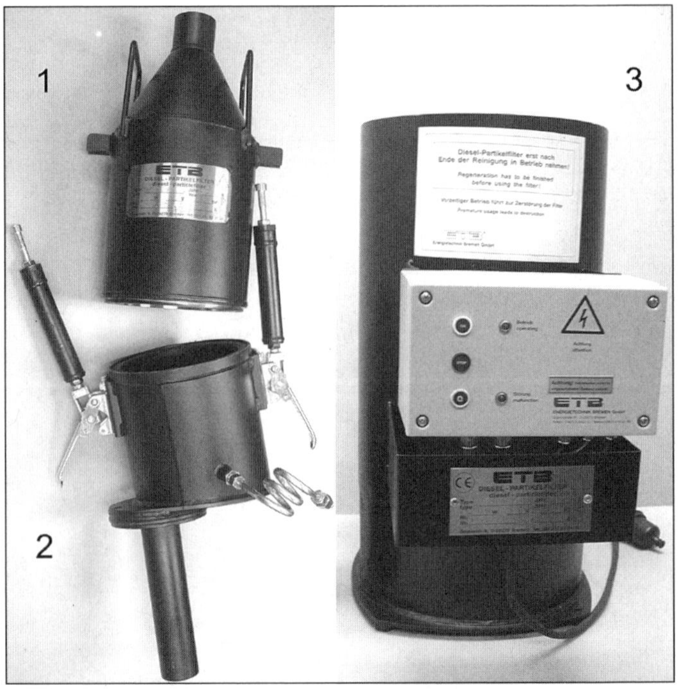

Figure 22.31 *Off-board regenerated filter system. 1, removable particulate filter; 2, vehicle fixture; 3, electric regeneration unit.*

light that filter regeneration is required. The frequency of regeneration depends on the filter size, engine PM emission, vehicle duty cycle, and so forth. Most systems are designed to allow for at least one shift of operation between regenerations. The system shown in Figure 22.31, manufactured by ETB, utilized a 7.5-in. diam. × 8-in. EX-80/100/17 substrate and was designed for diesel forklifts with up to approximately 50-hp engines. Its regeneration unit, fitted with a 3-kW heater, allowed for regeneration within approximately 1 h.

A schematic of a generic off-board regeneration unit is shown in Figure 22.32. The walls of the unit in contact with hot gas are heavily insulated. Once the soot-laden filter is placed in the top chamber, the regeneration sequence is started manually by pushing a button. The regeneration cycle—run by an electronic controller—is usually fairly simple, involving a ramp to a predetermined filter inlet temperature of typically 600 to 700°C, which temperature is then maintained over a period of time, typically on the order of 1 h. After the cycle is completed, the hot filter must be allowed to cool down to ensure safe handling.

In applications where equipment downtime must be minimized, spare filters allow for a relatively quick exchange of soot-laden units with regenerated ones. Still, the off-board filter systems require a considerable amount of labor, often every day, to replace and regenerate filters. Larger units, which are available in 15-in. (381-mm) or even 20-in. (508-mm) filter substrate diameters, must be replaced by two workers. The durability of off-board regenerated filters critically depends on the vehicle operator, who must regenerate the unit on time. If filters are overloaded with particulates, damage to the filter media may occur due to excessive heat release during regeneration. Failures of off-board filters have been reported under such circumstances [58].

Multiple small size filters can be also installed on larger equipment. Figure 22.33 shows three units installed on a concrete mixing truck—two units serving the propulsion engine and one unit serving the auxiliary concrete mixer engine. Using the

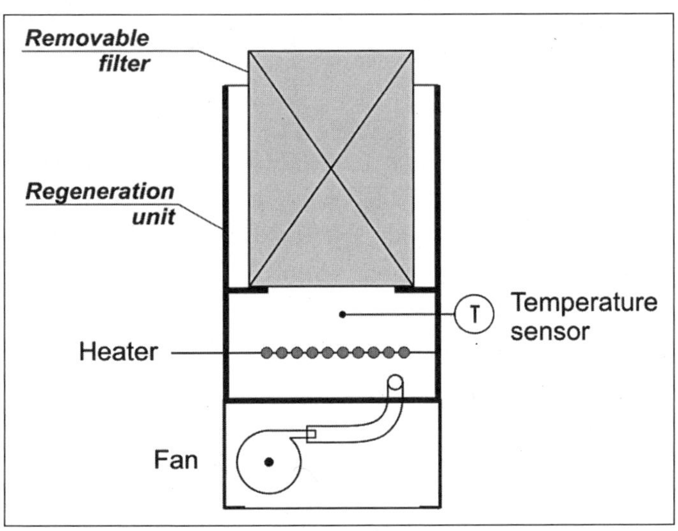

Figure 22.32 *Schematic of off-board regeneration unit.*

Figure 22.33 *Three particulate filter units on a concrete truck at a Swiss tunnel construction site, August 2002.*

same size units allows the number of filter sizes used within a fleet to be minimized. However, considering the drawbacks of the off-board regenerated filter concept, the most compelling reason for such installations is probably the lack of reliable and cost-effective automated filter systems.

22.5.3 Shore Power Regenerated Filters

22.5.3.1 Overview

Contrary to the off-board concept, shore power regenerated filters remain on the vehicle during regeneration. The filter system mounted on the vehicle may include all functional components such as air blower and heater, but the system must be connected to an external power source for regeneration. In comparison to the off-board systems, shore power filters require less maintenance (no filter removal is needed), but the vehicle is immobilized for the regeneration time resulting in more equipment downtime. An example of a shore power system is shown in Figure 22.34. The shore power cable and connector can be seen in the picture.

The regeneration cycles and control strategies in most shore power and off-board filters are fairly similar. Also in both cases the problem-free operation depends on the vehicle operator, who must promptly initiate the regeneration when signaled by the dashboard indicator.

As discussed, a variety of configurations exist, where certain components (e.g., the air blower) are moved from the vehicle to an off-board regeneration unit. Such systems could be in fact classified somewhere in between the pure shore power and the off-board system configurations. While the filter itself

Figure 22.34 Shore power regenerated filter on construction machinery.

remains on-board, the system must be connected not only to shore power, but also to a regeneration unit, which is the source of air for regeneration.

To illustrate a number of issues pertinent to the design of shore power regenerated filters, we will discuss in detail a filter system developed by Engelhard and marketed in Europe under the commercial name STX Diesel Soot Filter System [59].

22.5.3.2 STX Diesel Soot Filter System

The STX filter system utilizes ceramic fiber cartridges with shore power electrical regeneration [60, 61]. The system, designed for industrial forklift trucks, was available in several sizes covering engines up to 6 L in displacement. The layout of the STX filter system is shown in Figure 22.35.

The filter system includes three components: (1) the fiber cartridge filter assembly, (2) the regeneration power unit (RPU), and (3) the electronic control unit (ECU). The filter assembly consists of up to six individual, electrically regenerated filter cartridges utilizing ceramic Nextel 312 fibers (see Section 21.3). The central support of an individual filter cartridge is formed by a metal heating element. The filtration medium is made up of fibers wound around that element. During the regeneration cycle, the heating element supplies heat to the regeneration air, as well as to the adjacent layers of filter media, initiating and sustaining the oxidation of soot. A cut-out model of the STX filter assembly is shown in Figure 22.36. The exhaust gas enters inside the fiber cartridge tubes (heating elements) at the right-hand side. The exhaust gas outlet, electrical, and thermocouple connections are visible at left.

The RPU is a wall-mounted unit providing electrical power and a constant supply of combustion air to the filter during regeneration. The output power is limited to a potential of 24 V because of safety considerations and compatibility with the vehicle electric system. During regeneration, the unit

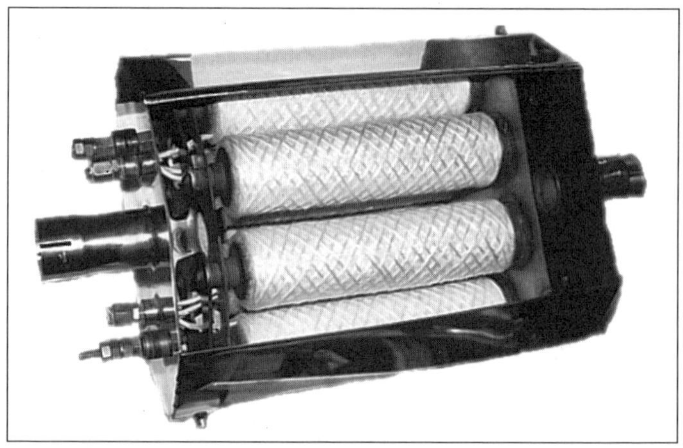

Figure 22.36 Cutout view of the STX soot filter assembly.

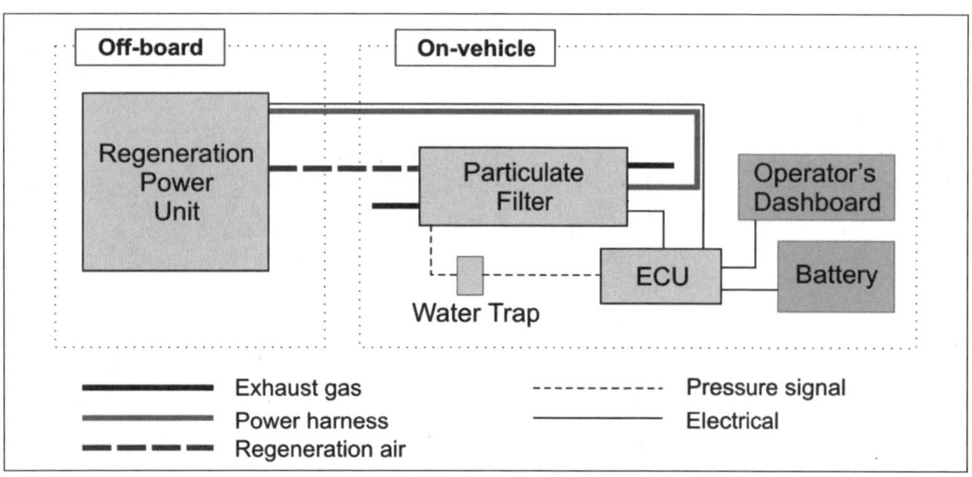

Figure 22.35 STX filter system layout.

is controlled by the ECU, which is mounted on the vehicle. It performs all monitoring and control functions of the system during vehicle operation and filter regeneration.

A major problem encountered during the development of the system was cartridge failure due to fiber damage. A reduction in filtration performance was observed after 300 to 1000 h of operation. Cartridges developed visible voids in the filter medium, allowing for the bypass of unfiltered exhaust. The reason for the failure, as revealed by laboratory analysis, was molten ash from the engine oil, which coated the fibers during regeneration. The ash-coated fibers became brittle and prone to vibration failure in the pulsating gas stream of the engine exhaust system. The melting temperature of the oil ash was 950°C. It was concluded that to prevent premature degradation of the ceramic fibers, their temperature must be kept below this level at all times.

Maintaining filter integrity was achieved by keeping the cartridge regeneration temperature below 950°C. Soot combustion starts at temperatures of 550 to 600°C. The available regeneration temperature range, therefore, is between 600 and 950°C. It was found that to maintain a minimum temperature of 600°C at the outer surface of the fiber layer, a temperature of 900°C is required at the interface between the heater element and the fibers. To control the fiber temperature, each filter cartridge has been equipped with a thermocouple. Thermocouple signals are passed to the ECU as a feedback signal used for heater power control to maintain safe fiber temperature.

During operation of the system, the filter pressure drop gradually increases due to the soot collected in the fiber cartridges. The exhaust pressure is continuously monitored by the ECU. After the pressure reaches a predetermined level, the ECU signals to the operator that regeneration of the filter is necessary. To perform regeneration, the operator drives the truck to the RPU, makes power and air connections, and starts the filter cleaning process. A small flow of air is provided by the RPU to the filter. The air is heated in the filter cartridges and oxidizes the collected soot to form carbon dioxide and water vapor. The two key functions of the ECU are to monitor the temperatures of the filter cartridges and to control the application of heating energy to the cartridge assembly during regeneration. Cartridge temperatures and the corresponding power profile during regeneration are illustrated in Figure 22.37.

The system was tested in the field on several makes of forklifts, with particular systems accumulating from 200 to more than 2500 engine hours. The following conclusions were made after field testing:

- Achieved filtration efficiencies were better than 80%, typically about 90%.
- The regeneration temperature was always maintained below 950°C. There was no runaway regeneration, even with heavily loaded cartridges.
- Vehicle operation between regenerations was from 6 to 8 h.
- Time necessary to regenerate filters did not exceed 14 min.

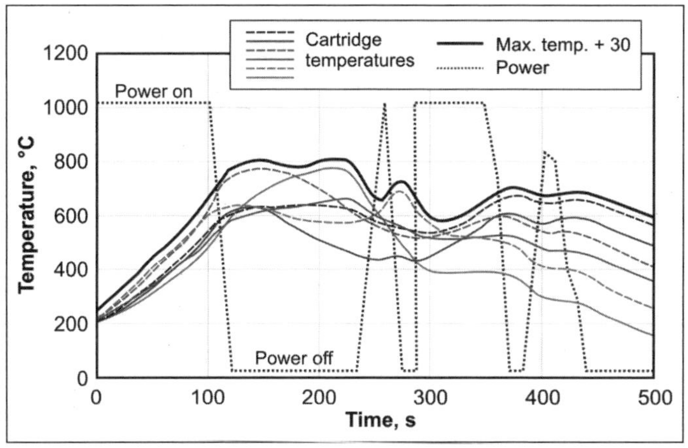

Figure 22.37 *Temperatures and power during filter regeneration.*

22.5.4 Filters with On-Board Regeneration

22.5.4.1 Overview

On-board regenerated systems draw electric power needed for regeneration from the vehicle's electrical system. They must be fully automated and their operation invisible to the vehicle driver or operator. As shown earlier, electric heating of the entire exhaust gas stream (full-flow regeneration) is not feasible due to insufficient on-board power. For this reason, on-board filter systems must use a partial-flow configuration, which in most cases involves two filter units—one for operation, one for regeneration—and the resulting switching of gas flow by a system of valves. These complexities, together with the required sophisticated electronic control, result in high system costs, especially at low production volumes.

The shortage of an on-board power supply remains the critical issue for electric filters, even in the partial-flow or hot air regeneration systems. Drawing additional energy from the vehicle electrical system is likely to require larger alternators and switching to a higher than 24-V voltage system. In large vehicles it is also possible to use an auxiliary generator of electricity powered by a small diesel engine, which could be operated during filter regeneration. This approach may be compatible with those highway truck idling reduction initiatives that promote the use of auxiliary generators for such purposes as cabin heating or air conditioning during parking periods [62].

Although no fully automatic on-board electric systems are commercially available today, the technical feasibility of such a system was demonstrated by Donaldson in the 1990s. The Donaldson system, one of the most advanced particulate filter systems with electric regeneration, was developed for urban buses in the United States. Two bus engines equipped with the system, DDC 6V92TA and Cummins L10, were EPA and CARB certified for 1992 and 1993 to PM levels between 0.018 and 0.041 g/bhp-hr.

Donaldson diesel particulate filters were installed as a retrofit and operated on more than 2000 buses and trucks in 33 different cities, with the New York City Transit Authority being the biggest test site [63, 64]. Total mileage of more than 20,000,000 miles was accumulated during the tests. The system was also being adapted for Class 8 trucks [65]. Despite the amount of effort put into the system's development and despite the increasing system reliability, Donaldson finally abandoned the product. In 1994 the U.S. EPA relaxed the expected bus PM standard of 0.05 g to 0.07 g/hp·h. The 0.07 g standard could be achieved using other control strategies that were more cost effective than the Donaldson filter system, which is discussed in more detail next.

22.5.4.2 Donaldson Filter System

System Layout

The Donaldson system utilized a dual filter layout [66], as shown in Figure 22.38. The system filtered the exhaust gas and periodically disposed of the collected PM by controlled oxidation. The two filters facilitated uninterrupted gas filtering during regenerations by switching the exhaust flow between filters. The system was electronically controlled and entirely automatic. The main components of the system were the ceramic wall-flow filter elements used to filter the exhaust gas. The other components, which constituted the regeneration and control systems, included the electrical heater system, a blower system, sensors, valves, and a microprocessor-based ECU. For the sake of simplicity, the schematic shows sensors on only one of the filter units. The ECU is not shown in the schematic. Filter-equipped buses did not have mufflers.

The filters are cordierite wall-flow monoliths. Cordierite monoliths used in the systems were Corning formulations EX-54 and EX-47 and NGK formulations DHC-221 and C415E. The filters fill with diesel particulates until the optimum loading is reached, at which time the unit is cleaned by means of hot air regeneration. The regeneration components include sensors to monitor particulate loading and hardware to perform controlled regenerations. The particulate loading is determined from measurements of the intake air flow, filter pressure drop, and the filter inlet temperature. When the particulate loading reaches a predetermined level, the blower system and the heater perform the regeneration. The diverter valves and associated solenoids carry out the task of switching the flow between two filters.

The electronic control unit monitors the operation of the system and operates the regeneration components when a regeneration is required. The ECU initially diverts flow to the more heavily loaded filter and proceeds monitoring that filter. When the ECU determines that the filter requires regeneration, it switches the exhaust gas flow to the other filter and activates all needed components to perform the regeneration. The ECU also warns the driver of system malfunctions and provides some diagnostic functions.

Another variation on the system utilized only one filter element. In that variant, the exhaust gas bypassed the filter and was released untreated for the duration of the regeneration cycle. The one-filter system was more cost effective but somewhat compromised the average filtration efficiency.

Emission Performance and Regeneration

The emission performance of the filter system, tested on a Detroit Diesel engine, is given in Table 22.9. A particulate matter efficiency of 85% was measured, which allowed the bus engine

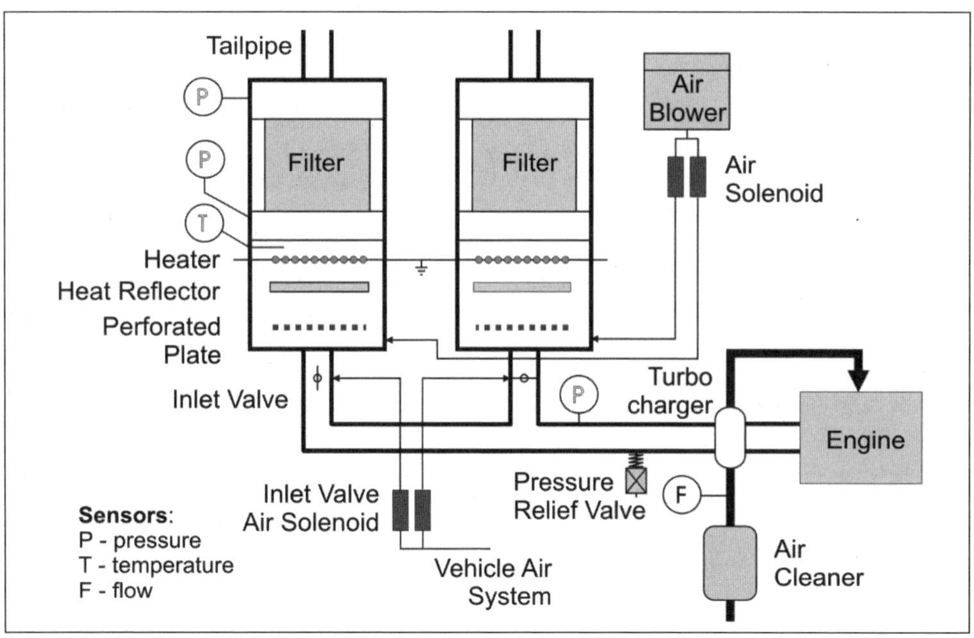

Figure 22.38 Donaldson system schematic.

TABLE 22.9
EMISSION PERFORMANCE ON A DDC 6V92TA DIESEL ENGINE, FEDERAL HD URBAN BUS CYCLE, 0.08% SULFUR #2 FUEL

	Engine Baseline, g/bhp-hr	Filter, g/bhp-hr	Efficiency, %
NO_x	5.0	4.3	14
HC	0.7	0.6	14
CO	2.5	3.0	–20
PM	0.350	0.053	85
• SOF	—	—	63
• Non-SOF	—	—	92

Source: Data from Ref. 66.

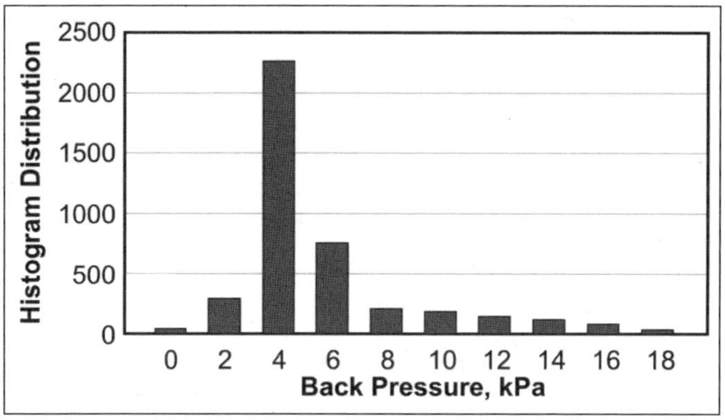

Figure 22.39 *Back-pressure distribution on DPF-equipped buses.*

to be certified at the 0.05 to 0.07 g/bhp-hr particulate level in 1993. There was practically no change in NO_x and HC, and some generation of CO, due to filter regenerations. Emission tests on Detroit Diesel and Cummins engines showed no deterioration in emission performance with time. The target emission durability of the filter was 8 years/290,000 miles. An average regeneration interval for the filter was 4.2 h. The duration of regeneration was 6.5 min. The heater power consumption during that period was 150 A at 24 V.

Back-Pressure and Fuel Penalty

Figure 22.39 shows a histogram of engine back-pressure measured during the New York field tests. The average back-pressure was 4.1 kPa (16.4 in. H_2O). The numbers represent the pressure drop of filter system and exhaust piping.

The fuel economy penalty of the filter system was caused by the increased filter pressure drop (above that of a muffler), as well as by the electrical power drawn from the alternator and batteries. Table 22.10 lists theoretical fuel penalties modeled by Donaldson for different PM emission levels. According to the experience of the New York City Transit Authority, a 3% fuel penalty was seen on 0.25 g/bhp-hr PM engines. The modeled fuel economy penalty for that PM level engine was 1.2%.

System Reliability

Poor reliability of the system was a serious concern, especially in the initial stage of the program. Reliability tracking was conducted on bus fleets in New York and Dayton, Ohio. Information was collected weekly from filter-equipped buses in the field. The overall reliability of the system increased during the program from 54% to 66% (Table 22.11).

22.6 Filters with Fuel Burners

22.6.1 Introduction

A fuel burner can be used to increase the temperature of the exhaust gas stream to the level of approximately 650°C that would facilitate regeneration of the trap by oxidation (burning) of the accumulated soot. The use of fuel burners is one of the most important methods for the regeneration of active diesel filter systems. Although in general cases the burner can be powered by

TABLE 22.10
FILTER SYSTEM PERFORMANCE—FUEL PENALTY

PM, Engine Baseline, g/bhp-hr	Regeneration Frequency, hr	PM Filter, g/bhp-hr	Fuel Penalty, %		
			Electric	Pressure	Total
1.00	1.4	0.150	2.5	0.4	2.9
0.80	1.6	0.120	2.1	0.4	2.5
0.60	2.0	0.090	1.7	0.4	2.1
0.40	2.9	0.060	1.2	0.4	1.6
0.25	4.4	0.038	0.8	0.4	1.2
0.15	7.2	0.023	0.5	0.4	0.9

Source: Data from Ref. 66.

TABLE 22.11
SYSTEM AND COMPONENT RELIABILITY

Component	Reliability		
	Initial	Final	Target
Entire system	0.5421	0.6609	0.9000
Filter assembly	0.7481	0.8309	0.9513
Diverter valve assembly	0.9345	0.9960	0.9884
Piping	0.9950	0.9960	0.9991
Sensors	0.9282	0.9335	0.9873
Electronic controller	0.8627	0.9624	0.9749
Control harness	0.9932	0.9739	0.9988
Heater power	0.9981	0.9778	0.9997
Combustion air	0.9969	0.9765	0.9995
Sensor harness	0.9916	0.9960	0.9986
Power harness	0.9932	0.9908	0.9988
Brackets/miscellaneous	—	0.9730	0.9999

Source: Data from Ref. 22.

any fuel, for example propane, diesel fuel is the most obvious choice, due to its availability on the vehicle.

Fuel burners have the advantage of utilizing chemically bound energy. Not limited by energy availability considerations, burners usually heat up the entire exhaust gas stream. Contrary to the partial-flow electric filters, exhaust gases do not bypass the filter element during regeneration. However, while some systems are designed to perform the regeneration at any engine operating conditions, others require that the engine operate at low idle speed for the time of regeneration. According to that difference, filter systems utilizing burners can be divided into two categories: (1) single-point burner systems and (2) full-flow burner systems.

Single-point filter systems accumulate soot in the filter element until a predetermined soot load is reached. That moment is usually determined by a measurement of the exhaust back-pressure. When the filter control system detects that the maximum soot load has been reached, it activates a warning light that indicates the "need for regeneration." The vehicle operator is then supposed to initiate the regeneration and to run the engine at low idle for 15 to 20 min of the regeneration cycle. The burner operates at a constant power that has been calculated to increase the exhaust gas temperature to a desired level at the low idle engine speed.

Single-point systems are offered in limited numbers for special applications, such as for vehicles operating in enclosed spaces (e.g., fleets of diesel forklifts). They are not convenient for on-road use due to the need for operator's intervention and the required idling period during regeneration. Single point burner systems are supplied, among others, by Eberspächer and HUG.

Full-flow burner systems feature fully automated regeneration at the actual exhaust gas flow during the regular duty cycle of the vehicle. The particulate collection phase proceeds, quite similar to the single point system, until a predetermined soot load is reached. At that point, however, the filter controller triggers regeneration without any intervention, or even knowledge, of the vehicle operator. The filter control unit starts the burner and controls its operation to maintain the required filter temperature. After the filter is cleaned, the controller stops the burner and the filter enters the next loading cycle.

Full-flow fuel burner filters require complex electronic control units to maintain thermally balanced regeneration that minimizes thermal stress to the filter monolith. Here the term *thermal balance* means that the thermal energy produced by the burner corresponds to the soot load in the filter and to the actual exhaust gas flow conditions. The soot load generates quantities of additional heat during regeneration, which may result in significant increase of the filter temperature and in overheating of the unit. Therefore, the filter controller has to watch the actual soot load in the filter, as well as the actual engine speed, to effectively control the exhaust gas temperature. That control is realized by varying the amount of thermal energy produced by the burner. In addition, the regeneration is performed by cycling through a sequence of temperature profiles, rather than maintaining the filter face at a constant elevated temperature.

The burner itself is another critical component of the system. Developers of filter systems reported a number of burner-related problems. A successful design always requires a significant amount of experience and feedback from field trials.

A certain fuel economy penalty is associated with the burner operation. On the other hand, because the regeneration can be triggered at any preset soot loading, the burner filters operate at a lower engine back-pressure than most passive filter systems. Therefore, contrary to the passive systems, the burner systems do not experience the fuel penalties associated with high engine back-pressures.

Another concern sometimes raised with burner filter systems is the additional emission generated by the burner itself. In view of the relatively short duration of regeneration, these concerns are probably not justified, as indicated by the performance of example systems discussed later (see Figure 22.43).

22.6.2 Full-flow Filter Systems

Advanced filter systems with burner regeneration are being developed by several manufacturers [67–69]. These systems feature automatic, full-flow, in-line regeneration supported by an integrated diesel fuel burner. An example schematic of a fuel burner filter system for heavy-duty vehicles developed by Zeuna is shown in Figure 22.40 [70].

The exhaust gases pass through the ceramic monolith filter and are cleaned. The interval between regenerations depends on the PM emissions level from the engine and the size of the filter substrate (PM load is maintained below 5 to 6 g/dm^3). An ECU monitors the filter condition through sensors and processes the information together with engine data. When certain parameters are reached (e.g., pressure drop, collecting time), regeneration starts automatically by activating components that are contained in an aggregate box. Diesel fuel sucked from the vehicle's fuel

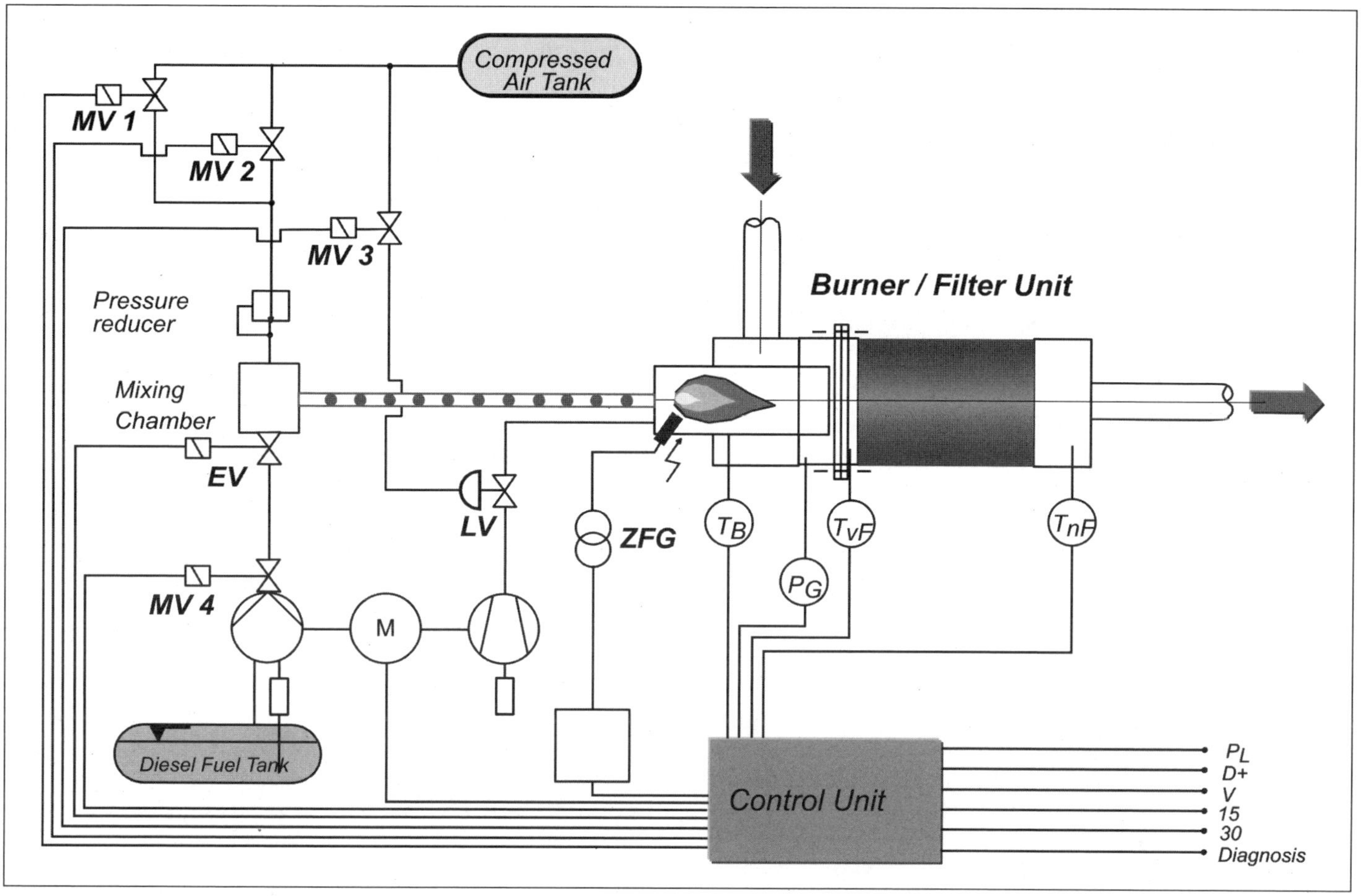

Figure 22.40 *Full-flow burner filter system developed by Zeuna. (Reprinted from Ref. 70.)*

tank is mixed with the atomized air from the compressed air tank, injected into the combustion chamber via an injector, and ignited by ignition electrodes. To ensure excess air operation, primary combustion air is supplied by an electrically driven compressor. This way a soot-free flame is facilitated irrespective of the engine operating conditions and, thus, oxygen concentration. The exhaust gas is heated to the predetermined temperature—normally 650 to 700°C—in a ramp-like fashion. The continuous temperature ramp is achieved by pulse-width modulation of the injection valve, which maintains the preset regeneration temperature within narrow limits regardless of the exhaust gas temperature. The burner output required to raise the temperature to 650°C varied, depending on operating conditions, between about 25 and 70 kW on an 11-L DI/TCI engine with cooled EGR.

A critical component in full-flow filter systems is the integrated burner/particulate filter unit. A schematic of such a unit developed by Deutz is shown in Figure 22.41. This system, known as the Diesel Particulate Filter System DPFS II, was commercially introduced in 1994 for retrofitting of city buses, municipal vehicles, and industrial engines, as well as stationary and railroad applications [69]. Several hundred filters have been built, with some individual units accumulated more than 6000 engine hours on industrial engines or more than 160,000 km on urban buses [71]. Two filter sizes in "mono" and "double" versions cover an engine power range from 70 to 440(500) kW.

The filter unit integrates the ceramic wall-flow monolith (versions with ceramic fiber filters were also developed), the mixing chamber, and the burner. The burner is based on the principle of fuel atomization by air blast. Diesel fuel enters the burner through an electronically controlled fuel injector that cycles at 100 Hz. A variable ratio of opening and closing times of the cycling valve is used to control the supplied fuel quantity and, thus, the burner power output. The burner is ignited by a spark ignition unit consisting of a high-voltage generator and two electrodes. The burner is capable of handling variable exhaust flow rates, which change with the engine speed and load.

Burner-related problems troubled the earlier version of the system, DPFS I. To prevent intrusion of soot particles into the burner and contamination of the air blast atomizer, cleaning of the burner components was required every 500 h of operation. In practice, however, cleaning of the burner was not always performed as scheduled. A malfunctioning burner could not

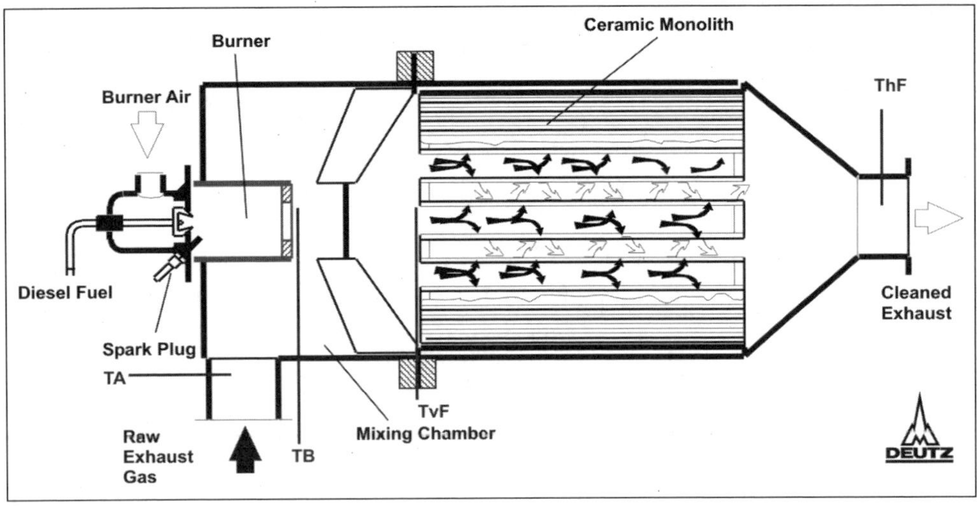

Figure 22.41 Schematic of the DPFS filter. (Courtesy of Deutz.)

perform regenerations when required, resulting in overloading of the trap. Eventually, when the required burner maintenance was performed and the burner became operational, runaway regeneration of the overloaded filter often took place, damaging the ceramic monolith. The burner is maintenance free in the new system, due to the introduction of a burner sweep air support during the loading phase. The spark plugs, however, still have to be changed after a specified number of regenerations.

Filter regeneration is controlled by an ECU that monitors four temperatures via four thermocouples including exhaust inlet (Figure 22.41, TA), burner discharge (TB), filter face (TvF), and temperature downstream of the filter (ThF), as well as the engine speed. The burner power output is closed-loop controlled to maintain the preprogrammed temperature profile of the gas mixture at the filter face (TvF). The loading phase of the DPFS system, depending on engine type, engine maintenance condition, duty cycle, and so on, lasts anywhere between 6 and 10 h. During that time the filter collects the exhaust gas particulates. After the nominal soot load is reached, the ECU triggers and controls the regeneration process, which takes approximately 12 min.

To determine the regeneration start, the ECU needs to know the filter loading factor. The control unit continuously determines the loading condition of the filter by comparing the actual engine speed and exhaust back-pressure readings against the maps stored in memory. More precisely, the ECU is calculating the amount of soot based on the monitored engine duty cycle and the memorized engine emission map. Then the expected back-pressure is calculated from the soot load. That calculated back-pressure is compared with the actual, measured back-pressure corrected for the actual engine speed condition.

The main functional improvements introduced in the DPFS II were in the area of burner power output control. A closed-loop digital controller reads the temperature of the gas mixture entering the filter face (TvF). The required burner power output is determined by comparing the actual temperature to a preprogrammed profile. That temperature profile has been developed to achieve a complete, thermally balanced regeneration and to minimize the exposure of the filter material to high-temperature excursions. An example regeneration temperature profile is shown in Figure 22.42.

The emission performance of the DPFS system is illustrated in Figure 22.43. The emissions were measured from a Euro 0 engine rated at 188 kW @ 2300 rpm. The test cycle was the German BMU trap test (ECE R-49 modified for diesel filter testing). Note that (1) the emissions are plotted on a logarithmic scale and (2) the load phase data represent total emissions from 20 load cycles, whereas the regeneration emission bars are just one-cycle emissions. That representation of data was chosen to illustrate the relative emission contributions, including those from the engine and burner during regeneration, over an entire duty cycle composed of 20+1 identical engine runs.

Gaseous emissions including NO_x, CO, and HC are not affected by the filter. The total particulate matter collection efficiency is about 90%. Collection efficiency for the solid fraction of PM (elemental carbon) is about 99%. Reported data indicate that the filter also reduces nanoparticle emissions, as well as a portion of the PAH (chrysene, benzo(a)pyrene and benzo(g,h,i)perylene).

The experience gathered in bus fleet tests by the filter manufacturer indicates that the system induces a fuel penalty of 1% to 2% due to the burner regeneration.

22.7 Microwave-Regenerated Filters

22.7.1 Microwave Heating

22.7.1.1 Process Principle

Microwaves, also known as radio-frequency (rf) waves, are electromagnetic waves with frequencies in the range of 500 MHz to 100 GHz. Microwaves have been used primarily in the field

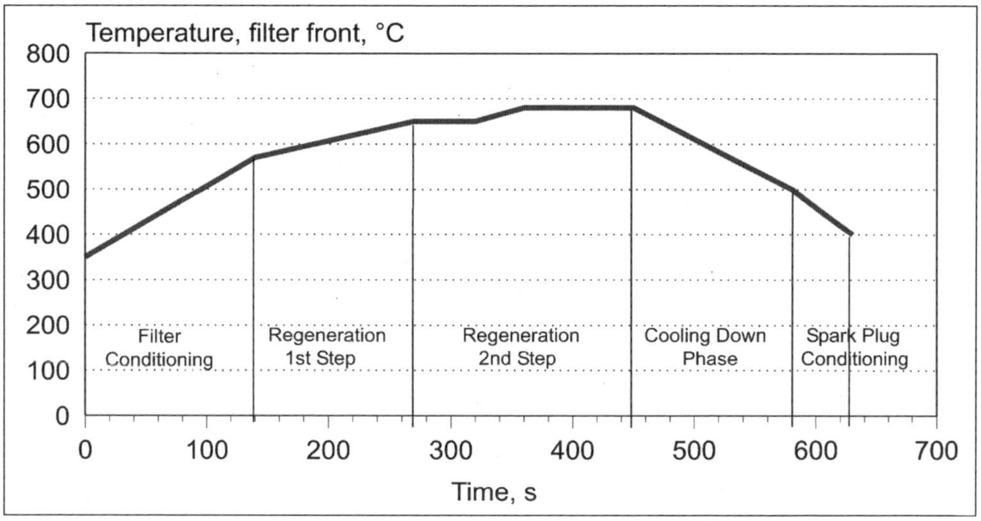

Figure 22.42 *Regeneration temperature profile.*

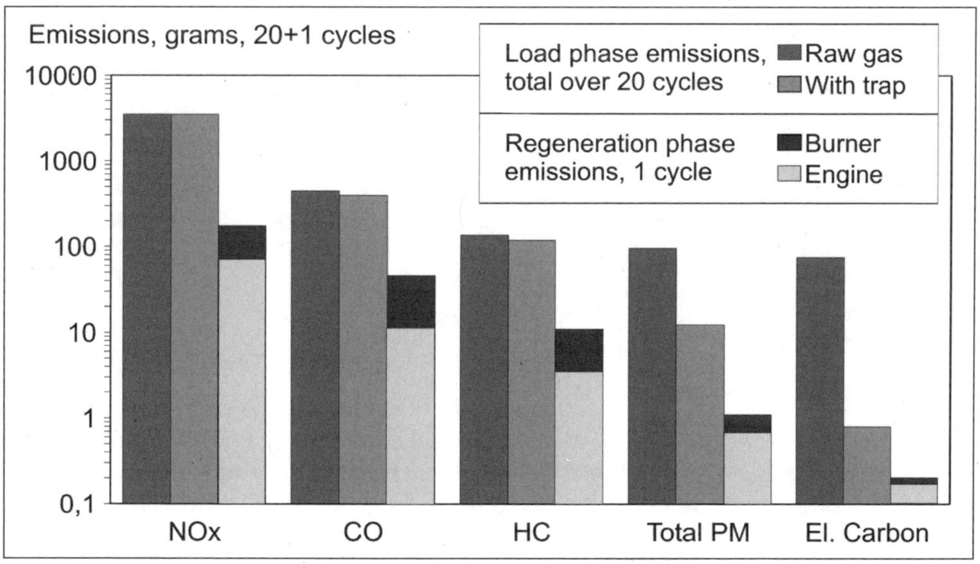

Figure 22.43 *DPFS emission performance.*

of telecommunications. After the invention of the magnetron, an efficient high-frequency microwave generator, microwaves have been increasingly used for heating applications with the first microwave ovens introduced to the market in late 1960s.

Microwave heating is based on absorption of rf energy by dielectric materials. Molecules of polar components, called susceptors, vibrate when placed in electromagnetic fields. This vibration causes an increase in the molecule's kinetic energy, which is dissipated as heat.

The dielectric properties of a material are characterized by its complex permittivity ε [72]:

$$\varepsilon = \varepsilon' - i\,\varepsilon'' \tag{22.4}$$

where the real part (ε′) is the *dielectric constant* and the imaginary part (ε″) is the *dielectric loss factor*. The dielectric constant is an indication of the amount of energy that can be stored in a material in the form of electric field, whereas the dielectric loss factor is a direct measure of how much energy a material can dissipate in the form of heat.

When a material is heated by microwave irradiation, the amount of rf power converted to heat per unit volume of a susceptor is a function of the electric field frequency, its intensity, and the dielectric loss factor, as given by the following equation:

$$P = 2\pi f \varepsilon_0 \varepsilon'' E^2 \tag{22.5}$$

where:

P = power, W/m^3
f = field frequency, Hz
ε_0 = absolute permittivity, 8.854×10^{-12} F/m
ε'' = dielectric loss factor, dimensionless
E = electric field intensity, V/m.

Materials that exhibit magnetic properties, for example, ferrite, can be heated by both the electric (E) and magnetic (H) components of an electromagnetic field. The rf power associated with the H-field component can be expressed by a relationship similar to that of Eq. (22.5), in which the dielectric loss factor (ε'') and the electric field intensity (E) are replaced by magnetic loss factor (μ'') and magnetic field intensity (H), respectively.

Microwave heating differs significantly from conventional heating. During conventional heating of a solid material, the heat is first transferred to its surface, typically by a combination of convection and radiation mechanisms. The inside of the material is then heated through conduction from the surface. In microwave heating, however, the energy deposition is concentrated in the material itself, resulting in heating from inside out.

Another characteristic feature of microwave heating is its selectivity. Some materials are strong absorbers of rf energy, while others are not. Materials of high dielectric constants and high dielectric loss factors are most effectively heated by microwaves. Examples of such materials include water, carbon black, as well as diesel soot. Most ceramic materials, due to their low dielectric constant and loss factor, are practically transparent to rf energy. This class of materials includes cordierite, a ceramic used in some diesel particulate filter substrates, which is virtually transparent to microwaves. Metals, on the other hand, are nearly perfect reflectors of microwave energy. Therefore, metal ducts can be used as waveguides for rf energy.

Specific data on complex permittivities of materials is somewhat difficult to find in the published literature. Available data are often incomplete. It is also inconsistent, presumably due to the use of different measurement techniques and parameters by different authors. The dielectric properties of some selected materials, including diesel soot, are listed in Table 22.12. These data, despite some discrepancies, confirm that diesel particulates are an rf absorber, while cordierite is not. (This characteristic can be also used for the measurement of soot loading in ceramic diesel particulate filters by microwaves [73].)

TABLE 22.12
DIELECTRIC PROPERTIES OF SELECTED MATERIALS

Material	Dielectric Constant, ε'	Dielectric Loss Factor, ε''	Reference
Diesel soot 1	10.695	3.561	[74]
Diesel soot 2	9.578	2.854	
Diesel soot 3	8.6	7.4	[75]
Cordierite 1	2.873	0.138	[74]
Cordierite 2	1.0	0.00006	[75]
Gamma-Al_2O_3	3.006	0.170	[74]
SiO_2	3.066	0.215	
TiO_2	7.020	0.430	
ZrO_2	4.214	0.186	

The deposition of microwave energy within a heated medium is dependent on the medium geometry, sometimes called an rf cavity (in the case of a diesel filter, the cavity would be defined by the metallic filter container). If rf energy launched down a metal waveguide strikes a metal wall, the energy will be reflected off the wall, setting up a pattern of E- and H-field standing waves. The E-field distribution in a microwave oven can be pictured as a two-dimensional array of E-field peaks of high electrical field intensity and valleys where the field intensity is low. Materials with dielectric loss properties will be heated only when placed in an E-field standing wave peak and not if located in a valley. Consumer microwave oven manufacturers attempt to compensate for this uneven heating by moving the food through the electrical field on a rotating platform and/or by utilizing rotating metal vanes to periodically redistribute the field.

22.7.1.2 Energy Efficiency

The efficiency of a consumer microwave oven for conversion of electrical energy to heat in the rf load (food) is about 50%. Because consumer ovens have to accommodate a wide range of load sizes and ranges, their load and rf power source parameters are difficult to match. Higher efficiencies of up to 60% to 70% may be possible in a specialized, single-use system [76].

22.7.1.3 Commercial Components and Costs

Electromagnetic field frequencies used for microwave heating have been established through historical evolution and recommendations by international committees that review the use of the electromagnetic spectrum. The frequencies that are normally employed in the industry include 915, 2450, and 3000 MHz, with the 2450-MHz frequency band being most common in consumer ovens.

Due to extensive development and mass production, component costs are relatively low. For example, the manufacturing costs for a typical 650-W magnetron have been reported to be only about \$15 in quantity with the associated power supply unit at about \$20 to \$40 [76]. A 650-W magnetron has an average service life of about 5000 h.

22.7.2 Diesel Filter Experience

22.7.2.1 Microwave Filter Systems

There are two approaches to the microwave regeneration of diesel filters: (1) heating of the soot accumulated in the filter by microwaves and (2) heating of soot and the microwave-absorbing filter material. The latter approach requires that either (1) a microwave susceptor actually be incorporated into the filter substrate or coated onto the substrate surface or (2) the filter substrate itself be made of rf absorbing material.

Diesel Engines," Paper presented at 7th Diesel Emissions Reduction Workshop (DEER), Portsmouth, VA, August 2001, U.S. Department of Energy, Washington, DC, retrieved from www.osti.gov/hvt/deer2001/chatterjee.pdf.

28. Allansson, R., *et al.,* "European Experience of High Mileage Durability of Continuously Regenerating Diesel Particulate Filter Technology," SAE Paper No. 2000-01-0480, SAE International, Warrendale, PA, 2000.
29. Lanni, T., *et al.,* "Performance and Durability Evaluation of Continuously Regenerating Particulate Filters on Diesel Powered Urban Buses at NY City Transit," SAE Paper No. 2001-01-0511, SAE International, Warrendale, PA, 2001.
30. Hawker, P., *et al.,* "Effect of a Continuously Regenerating Diesel Particulate Filter on Non-Regulated Emissions and Particle Size Distribution," SAE Paper No. 980189, SAE International, Warrendale, PA, 1998.
31. Ayala, A., N. Kado, and R. Okamoto, "ARB Study of Emissions from Late-model Diesel and CNG Heavy-duty Transit Buses," California Air Resources Board, Sacramento, CA, 2001.
32. Zelenka, P., M. Egert, and W. Cartellieri, "Ways to Meet Future Emission Standards with Diesel Engine Powered Sport Utility Vehicles (SUV)," SAE Paper No. 2000-01-0181, SAE International, Warrendale, PA, 2000.
33. Chatterjee, S., "Performance and Durability Evaluation of Continuously Regenerating Particulate Filters on Diesel Powered Urban Buses at NY City Transit—Part II," SAE Paper No. 2002-01-0430, SAE International, Warrendale, PA, 2002.
34. Westphal, C., *et al.,* "Der V10 TDI von Volkswagen," 22. Internationales Wiener Motorsymposium, Fortschritt-Berichte VDI (VDI Verlag, Duesseldorf), Reihe 12, Nr. 455, 2001, pp. 102–120.
35. Chandler G.R., *et al.,* "An Integrated SCR and Continuously Regenerating Trap System to Meet Future NOx and PM Legislation," SAE Paper No. 2000-01-0188, SAE International, Warrendale, PA, 2000.
36. Howard, J.B., and W.J. Kausch Jr., "Soot Control by Fuel Additives," *Prog. Energy Combust. Sci.* **6,** 1980, p. 263.
37. Peter-Hoblyn, J.D., *et al.,* "Methods for Reducing Harmful Emissions from a Diesel Engine," U.S. Patent No. 6,003,303 (Clean Diesel Technologies), 1999.
38. Neeft, J.P.A., M. Makkee, and J.A. Moulijn, "Diesel Particulate Emission Control," *Fuel Proc. Techn.* **47,** 1996, pp. 1–69.
39. Sauerteig J.E., "Low Emission Technologies for Underground Mining Engines," Paper presented at SAE Clean Diesel Engine Technology TOPTEC Conference, St. Pete Beach, FL, September 1999.
40. Mayer, A., "VERT—Curtailing Emissions of Diesel Engines in Tunnel Sites," TTM Report W11/12/97, Switzerland, November 1997.
41. Bloom, R.L., *et al.,* "Fiber Wound Diesel Particulate Filter Durability Experience with Metal Based Additives," SAE Paper No. 970180, SAE International, Warrendale, PA, 1997.
42. Richards, P., B. Terry, M.W. Vincent, and S.L. Cook, "Assessment of the Performance of Diesel Particulate Filter Systems with Fuel Additives for Enhanced Regeneration Characteristics," SAE Paper No. 1999-01-0112, SAE International, Warrendale, PA, 1999.
43. Vincent, M.W., *et al.,* "The Long Distance Road Trial of a Combined Diesel Particulate Filter and Fuel Additive," SAE Paper No. 2000-01-2849, SAE International, Warrendale, PA, 2000.
44. Vincent, M.W., P. Richards, and S.L. Cook, "Particulates Reduction in Diesel Engines through the Combination of a Particulate Filter and Fuel Additive," SAE Paper No. 982654, SAE International, Warrendale, PA, 1998.
45. Pattas K., *et al.,* "Cordierite Filter Durability with Cerium Fuel Additive: 100000 km of Revenue Service in Athens," SAE Paper No. 920363, SAE International, Warrendale, PA, 1992.
46. Merkel, G.A., W.A. Cutler, and C.J. Warren, "Thermal Durability of Wall-Flow Ceramic Diesel Particulate Filters," SAE Paper No. 2001-01-0190, SAE International, Warrendale, PA, 2001.
47. Jelles, S.J., M. Makkee, J.A. Moulijn, *et al.,* "Diesel Particulate Control: Application of an Activated Particulate Trap in Combination with fuel Additives at an Ultra Low Dose Rate," SAE Paper No. 1999-01-0113, SAE International, Warrendale, PA, 1999.
48. McKinnon, D.L., D.A. Pavlich, T. Tadrous, and D. Shephard, "Results of North American Field Trials Using Diesel Filters with a Copper Additive for Regeneration," SAE Paper No. 940455, SAE International, Warrendale, PA, 1994.
49. Heeb, N.V., "Influence of Particulate Trap Systems on the Composition of Diesel Engine Exhaust Gas Emissions," EMPA Report Nr. 167985, Duebendorf, Switzerland, January 11, 1998.
50. "PSA Peugeot Citroen Exceeds 2002 Sales Target," Press Release, PSA Peugeot Titroen, January 7, 2003.
51. Quigley, M., and T. Seguelong, "Series Application of a Diesel Particulate Filter with a Ceria-Based Fuel-Borne Catalyst: Preliminary Conclusions after One Year of Service," SAE Paper No. 2002-01-0436, SAE International, Warrendale, PA, 2002.
52. Ahlvik, P., "Environmental and Health Impact from Modern Diesel Cars," Ecotraffic, Report for the Swedish National Road Administration, May 2002, retrieved from www.vv.se/publ_blank/bokhylla/miljo/2002_62/2002_62.pdf.
53. Meinig, U., *et al.,* "Tests on a New Diesel Particulate System with Traps of Metal Fleece," SAE Paper No. 982597, SAE International, Warrendale, PA, 1998.
54. DePetrillo, F., *et al.,* "Rypos Trap Active Diesel Particulate Filter System: Field Demonstration," Paper presented at 8th Diesel Emissions Reduction Conference (DEER), San Diego, CA, August 2002, U.S. Department of Energy, Washington, DC, retrieved from www.orau.gov/deer2002/Session11/DePetrillo1.pdf.

55. "Porotherm," Product brochure, Heimbach GmbH, Dueren, Germany, May 2000, retrieved from www.heimbach.com/PDF-Produkte/englisch/porotherm%20-%20GB.pdf.
56. "Investigations into the Feasibility of PM Filters for Nonroad Mobile Machinery," Joint EMA and EUROMOT Report, August 31, 2002, retrieved from www.euromot.org/download/positions/NRMM_PM_Filters_310802.pdf.
57. Bennet, C.O., and J.E. Myers, *Momentum, Heat, and Mass Transfer,* McGraw-Hill, New York, 1962.
58. Stachulak, J., "Diesel Particulate Filter Study at INCO's Stobie Mine," Paper presented at Mining Diesel Emissions Conference (MDEC), CANMET, Toronto, ON, October 28–31, 2002.
59. "Emission Control for Materials Handling," Product brochure, Engelhard Corporation, Iselin, NJ, 1994.
60. Child D., and J. Cioffi, "An Electronically Controlled Diesel Particulate Filter System Employing Electrically Regenerated Ceramic Fiber Wound Cartridges," SAE Paper No. 942265, SAE International, Warrendale, PA, 1994.
61. Chadderton, J., *et al.,* "Particulate Filter, and System and Method for Cleaning Same," U.S. Patent No. 5,551,971, 1996.
62. Lewis, S., "Emissions from Idling Trucks for Extended Time Periods," Paper presented at 8th Diesel Emissions Reduction Conference (DEER), San Diego, CA, August 2002, U.S. Department of Energy, Washington, DC, retrieved from www.orau.gov/deer2002/Session7/lewis.pdf.
63. Norwich, W., *et al.,* "Particulate Trap Technology Demonstration at New York City Transit Authority," SAE Paper No. 920138, SAE International, Warrendale, PA, 1992.
64. Ha, K., *et al.,* "Particulate Trap Technology Demonstration at New York City Transit Authority," SAE Paper No. 910331, SAE International, Warrendale, PA, 1991.
65. Ha, K., *et al.,* "Demonstration of Durable Retrofit Diesel Particulate Trap Systems on an Urban Bus and Class 8 Truck," SAE Paper No. 910332, SAE International, Warrendale, PA, 1991.
66. Kojetin, P., *et al.,* "Production Experience of a Ceramic Wall Flow Electric Regeneration Diesel Particulate Trap," SAE Paper No. 930129, SAE International, Warrendale, PA, 1993.
67. Zelenka, P., S. Schmidt, and G. Elfinger, "An Active Regeneration Aid as a Key Element for Safe Particulate Trap Use," SAE Paper No. 2001-01-3199, SAE International, Warrendale, PA, 2001.
68. Okazoe, H., *et al.,* "Development of a Full-Flow Burner Regeneration Type Diesel Particulate Filter Using SiC Honeycomb," SAE Paper No. 960130, SAE International, Warrendale, PA, 1996.
69. Houben, H., *et al.,* "The Optimized Deutz Service Diesel Particulate Filter System DPFS II," SAE Paper No. 942264, SAE International, Warrendale, PA, 1994.
70. Zelenka, P., C. Telford, D. Pye, and N. Birkby, "Development of a Full-Flow Burner DPF System for Heavy Duty Diesel Engines," SAE Paper No. 2002-01-2787, SAE International, Warrendale, PA, 2002.
71. Breuer, K., "Deutz Particulate Filter Systems," Paper presented at Canadian Mining Diesel Conference (CANMET), Toronto, October 1998.
72. Meredith, R., *Engineers Handbook of Industrial Microwave Heating,* Institution of Electrical Engineers, London, 1998.
73. Walton F.B., *et al.,* "On-Line Measurement of Diesel Particulate Loading in Ceramic Filters," SAE Paper No. 920564, SAE International, Warrendale, PA, 1992.
74. Ma J., M. Fang, P. Li, *et al.,* "Microwave-Assisted Catalytic Combustion of Diesel Soot," *Appl. Catal. A* **159,** 1997, pp. 211–228.
75. Babu, V.S., L. Farinash, and M.S. Seehra, "Carbon in Diesel Particulate Matter: Structure, Microwave Absorption, and Oxidation," *J. Mater Res.* **10,** 5, 1995, pp. 1075–1078.
76. Walton, F.B., P.J. Hayward, and D.J. Wren, "Controlled Energy Deposition in Diesel Particulate Filters during Regeneration by Means of Microwave Irradiation," SAE Paper No. 900327, SAE International, Warrendale, PA, 1990.
77. Gautam, M., S. Popuri, B. Rankin, and M. Seehra, "Development of A Microwave Assisted Regeneration System for A Ceramic Diesel Particulate System," SAE Paper No. 1999-01-3565, SAE International, Warrendale, PA, 1999.
78. Ning, Z., and Y. He, "Experimental Study on Microwave Regeneration Characteristics of Diesel Particulate After-Treatment Systems," SAE Paper No. 1999-01-1470, SAE International, Warrendale, PA, 1999.

Chapter 23

Plasma Exhaust Treatment

23.1 Introduction

Nonthermal plasma (NTP) discharges in exhaust gas have been studied as a potential way of reducing NO_x and particulate matter (PM) emissions in diesel exhaust as well as NO_x and cold start hydrocarbons (HCs) in lean gasoline exhaust. Vehicle exhaust gases, both diesel and gasoline, undergo chemical changes when exposed to plasma. Logically, oxidation processes dominate in the presence of oxygen. These reactions include oxidation of hydrocarbons, carbon monoxide, and, to a degree, diesel particulate matter (DPM). Nitric oxide (NO) can be oxidized by plasma to NO_2. The oxidation properties of plasma have been utilized in the treatment of flue gases from power plants [1]. In the power plant flue gas treatment, the purpose of the plasma is to oxidize NO to NO_2 and subsequently to nitric acid. The desired products, in the form of ammonium salts, are then obtained by reacting the formed acid with ammonia. Industrial plasma systems have also been demonstrated for VOC removal.

Obviously, this method of NO_x removal is not applicable to trucks or cars. The objective in the plasma treatment of exhaust gases from internal combustion engines is the reduction, as opposed to oxidation, of NO_x. Contrary to some earlier literature reports, there is now a wide consensus that plasma alone, due to its oxidizing character, is not a viable NO_x control method. However, combinations of plasma with catalysts, referred to as *plasma-assisted catalysts* or simply *plasma catalysts,* have been suggested for NO_x reduction. The plasma is believed to show potential to improve catalyst selectivity and removal efficiency. Current state-of-the art plasma catalysts have efficiencies comparable to those of active DeNOx systems, removing about 50% of NO_x at a fuel economy penalty of less than 5% [2]. While such performance would not be sufficient to reduce NO_x in heavy-duty engines from the U.S. 2004 level of about 2 g/bhp-hr to the 2010 level of 0.20 g/bhp-hr, it could be sufficient to reach NO_x levels of about 1 g/bhp-hr, as will be needed during the 2007–2009 phase-in period of the NO_x standard.

In the case of diesel exhaust, removal of PM emissions would also be a valuable benefit of plasma systems. Plasma systems have been shown to be capable of reducing DPM by means of low-temperature oxidation. It is not currently clear whether the NO_x and PM control functions by plasma can be combined in one device, that is, if NTP reactors can be designed for the simultaneous control of NO_x and PM.

An increasing number of research reports have been published in the literature that have aided our understanding of plasma chemistry in engine exhaust gases. However, this technology still has a novel character and published results need to be evaluated with caution. Because many studies are conducted as small-scale laboratory experiments, as opposed to full-flow engine experiments, erratic interpretation of data is frequently suspected. It is very easy to overlook a formation of unidentified chemical compounds in the plasma or to confuse adsorption and storage of material in the test equipment with its steady-state removal. For vehicle plasma applications, it is very important to make a distinction between NO removal by chemical oxidation and NO removal by chemical reduction. The desired overall process is chemical reduction to benign products, such as nitrogen and oxygen. In the plasma processing literature many authors use the term *NO reduction* even when the NO removal is accomplished by oxidation to NO_2 and nitric acid. It is not sufficient for a plasma experimental work to record a decrease of the NO or NO_x concentration. Possible reaction products may include many other nitrogen species that may be not acceptable. Several of these by-products may be also difficult to detect in the laboratory setup. Besides nitrous and nitric acids, these products may include nitrates, nitrites, or organonitrites that can be deposited on reactor walls, on particulates, or on the pellet material if a packed-bed reactor is used. Even if chemical reduction of NO dominates, the products may include nitrous oxide (N_2O), which, although not a regulated emission, is not an acceptable product. Commercialization of NTP technology for emission control from mobile sources requires significant advancements and much more development work. The plasma may or may not become a viable choice for lean NO_X or PM removal systems.

23.2 Nonthermal Plasma

23.2.1 *What Is Plasma?*

Plasma is a gas consisting of positive and negative charges that has an overall tendency to remain electrically neutral over large length scales. It is composed of free electrons, ions, radicals, atoms, and molecules in various states of excitation. Plasma is divided into "thermal" or "hot" plasma and "nonthermal" or "cold" plasma. In the thermal variety the kinetic energy (temperature) of charged particles and the kinetic energy (temperature) of the background gas are similar. In this type of plasma the electrons' energy is transferred through collisions to other components of the gas stream. Because of the high temperatures, a considerable amount of heat must be added to the system to maintain thermal plasma. An example of thermal plasma is the electrical arc discharge.

In the nonthermal plasma electrons have a kinetic energy higher than the energy corresponding to the random motion of the background gas molecules. The intent of using a nonthermal plasma is to selectively transfer the input electrical energy to the electrons, which would generate free radicals through collisions and promote the desired chemical changes in the exhaust gas. These reactions can be accomplished at a fraction of the energy required in the thermal plasma system. An example of NTP is the gas filling a fluorescent tube. Its temperature is only about 40°C, but the temperature of free electrons in the system exceeds 10,000°C.

Nonthermal plasmas are favored by low pressures. At ambient pressure high local electrical fields are needed to maintain NTP. However, there is also a maximum field beyond which NTP cannot be maintained and will develop a thermal plasma discharge. The highest electric field that can be applied while still maintaining a nonthermal plasma is known as the *electrical breakdown threshold.* Under atmospheric pressure conditions in an N_2 or air discharge, the electrical breakdown threshold corresponds to an electron mean energy of around 4 eV. Nonthermal plasma reactors typically operate in the regime where the average kinetic energy of the electrons is in the 3- to 6-eV range.

Typically, space electrical charges build up in the NTP, which cancels the effect of the external electrical field. Therefore, to maintain a continuous NTP discharge, the reactors are powered by either alternating current (AC) of pulse generators. Ultrashort pulse supplies (less than 50 ns) were reported to increase the breakdown threshold and, as a consequence, the energy transfer to the gas [3]. However, current technology to produce ultrashort pulses is inefficient. Stationary systems use readily available 60-Hz power, which, due to large and heavy high voltage transformers, is not practical for mobile applications. The ozonator industry uses 1.2-kHz transformers, which are readily available and have been adapted in some engine studies [4].

23.2.2 *Plasma Generators*

Nonthermal plasmas can be produced in a number of ways including a variety of electrical corona discharges, radio-frequency discharges, microwave discharges, or electron beams. The following NTP technologies are considered for emission reductions in engine exhaust streams [3]:

- Corona discharge
- Surface plasma discharges
- Dielectric barrier discharges
- Dielectric packed bed reactor
- Electron-beam reactor.

23.2.2.1 Corona Discharge

The corona discharge (Figure 23.1) is the simplest type of plasma generator. A feature of the corona discharge, which differentiates it from the other discharges, is that no dielectric is involved. Instead, an electron avalanche is initiated from a sharp metallic surface where the radius of curvature is small. The electric field has to be pulsed to prevent the plasma from going into the thermal mode and forming an arc. The electric fields in corona reactors are in the order of 50 kV/cm.

23.2.2.2 Surface Plasma Discharges

In the surface plasma discharge configuration, shown in Figure 23.2, the electric fields are parallel to the surface. One of the electrodes is attached to one side of a dielectric. The other electrode is placed at the other side but does not cover the dielectric completely. The plasma is created adjacent to this dielectric surface, which is in contact with the gas. In operation, the surface plasma covers the entire dielectric surface during the pulse. A feature of this discharge is that after a few nanoseconds charge starts to build up at the dielectric surface, which has the effect of reducing the electric fields outside the dielectric, eventually extinguishing the discharge.

23.2.2.3 Dielectric Barrier Discharge

Dielectric barrier discharges (DBDs) are distinct from the surface discharges in that one of the electrodes is detached from

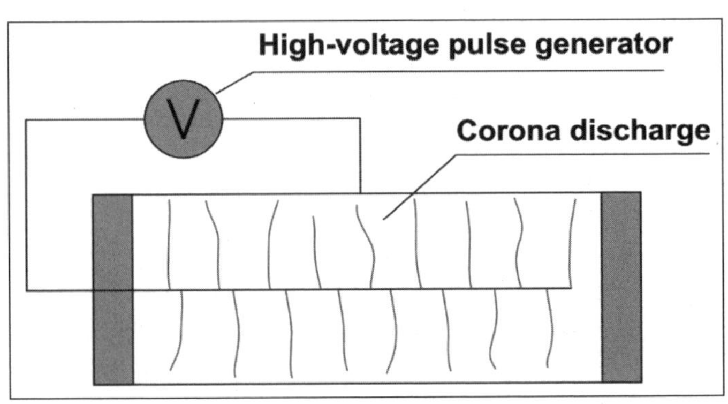

Figure 23.1 *Corona discharge.*

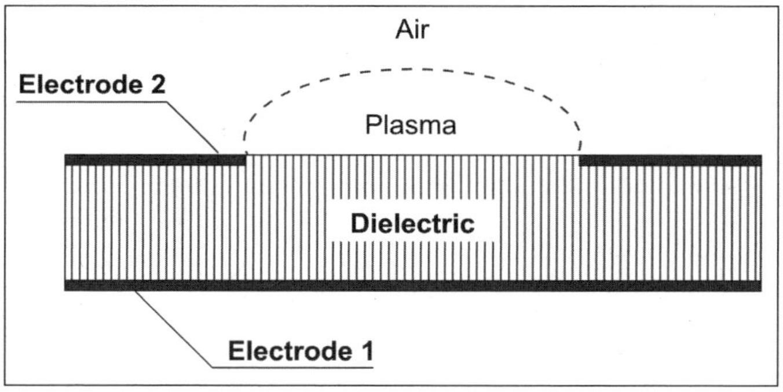

Figure 23.2 *Plasma surface discharge.*

the dielectric (Figure 23.3). Like the corona discharges, small-scale electron streamers are formed. When the electric field is perpendicular to the dielectric, streamers form with a density of about $100/cm^3$. After several nanoseconds of operation, space charge builds up on the dielectric surface, locally cancelling the externally applied electric field and eventually extinguishing the discharge. In the DBD mode of operation, the threshold electric field is relatively low and low energy electrons are formed.

DBD reactors can be also configured as concentric cylinders (similar to the corona reactor, but with surfaces coated with a dielectric layer) or as arrays of alternating positive and negative tube-shaped, dielectric coated electrodes. The parallel plate architecture, however, features the most simple and cost-effective design. In a comparative study, a parallel plate reactor have a 1-mm gap distance and 0.5-mm dielectric barriers provided the best balance between the performance and the exhaust gas pressure drop [5].

A big advantage of the parallel plate geometry is the possibility to stack plates together to build larger reactors with higher power deposition to the gas. The alternating positive and negative electrodes in this design must have a dielectric layer on each side. A prototype of such a stacked reactor has been built by Delphi [4]. Alumina plates, as used in the electronic industry, were selected as the dielectric barrier. The conductive electrode was "inked" onto the alumina using an Ag/Pd ink and covered with another alumina plate. A number of such plates were stacked on top of each other using 1-mm-thick spacers between plates along two opposite edges of the structure. The plate stack was glued together using high-temperature glass encapsulant, wrapped with ceramic wool and stuffed into a stainless steel can. A full-flow system for a 2-L Opel diesel car was powered by a 3000-VA transformer with a 6-kV output at 1.2 kHz.

23.2.2.4 Dielectric Packed-Bed Reactor

In this configuration the dielectric in the form of small pellets is placed between electrodes. The pellets can be spherical, pebble shaped, or other shape chunks. As a consequence of Gauss's law and the difference in dielectric constant between the pellets and the surrounding gas, local electric field enhancements are observed outside the dielectric. A cylindrical configuration for the packed-bed reactor is shown in Figure 23.4. Another configuration where the pellet bed is placed between two flat electrodes is also possible.

The advantages of the packed-bed reactor include its simplicity and the possibility of using catalyst pellets, which makes this type of reactor a possible choice for plasma catalyst systems. The disadvantages of the packed bed are high pressure drop and attrition of the pellets.

23.2.2.5 Electron-Beam Reactor

An electron beam is formed in a separate generator, such as a cathode tube, and injected into the exhaust gas stream. The energy of electrons can be much higher in the electron-beam reactor than in the other reactors. The electron energy can be also better controlled, resulting in improved gas-phase reaction efficiency. Disadvantages of the electron-beam reactor are the need for special hardware (the electron generator) and poor efficiency in transferring the electrons into the gas.

23.2.3 Energy Efficiency

One of the critical issues in the use of plasma is the consumption of electrical energy. The overall efficiency of an NTP reactor can be divided into the following two components: (1) electrical conversion efficiency and (2) chemical processing efficiency. The electrical conversion efficiency refers to the

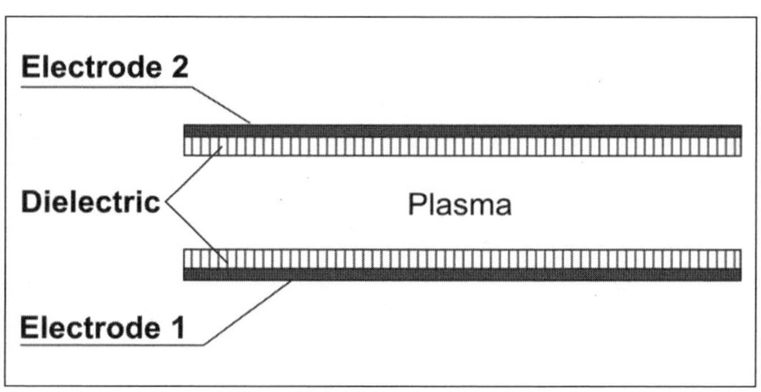

Figure 23.3 *Dielectric barrier discharge.*

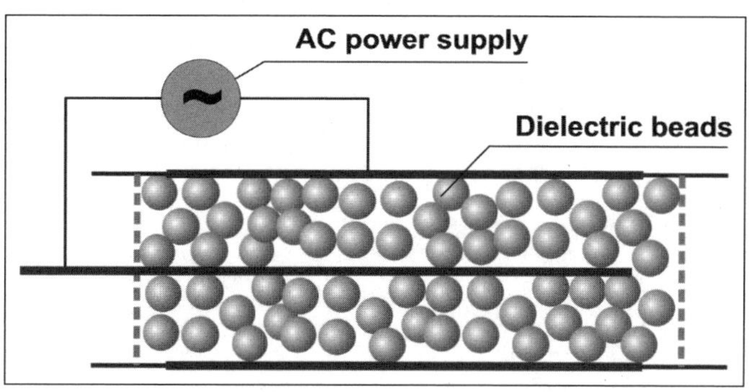

Figure 23.4 *Packed-bed reactor.*

efficiency for converting the input electrical power into the power deposited by electrons into the plasma. The chemical processing efficiency refers to the amount of pollutant removed or decomposed for a given amount of energy deposited into the plasma. The latter can be expressed in terms of the specific energy consumption in such units as electron volts (eV) per molecule of NO_x or grams of NO_x per kilowatt-hour. Obviously, the chemical processing efficiency is of fundamental importance and must satisfy user requirements before high electrical efficiency reactors are constructed.

Energy efficiencies reported by different studies vary greatly, depending on the reactor design, its operating parameters, targeted chemical processes, and, in the case of plasma catalyst systems, on the catalyst formulation and configuration. A number of ideas are being pursued to improve the efficiencies of NTP reactors:

- *Optimization of the electrode structure of the plasma reactor* to increase the electrical conversion efficiency by minimizing energy losses. In the case of the parallel plate DBD device, the main electrical losses are related to the heating of the dielectric barrier and to surface streamers.
- *Optimization of the voltage wave form applied to the reactor* to control various operating parameters, such as the electron energy.
- *Optimization of plasma catalysts* to take advantage of heterogeneous reactions.

It is customary in plasma research to express the amount of energy consumed by plasma devices relative to the amount of treated gas. This energy, often referred to as the energy deposition or energy density, is expressed in units of joules per standard (STP: 101.325 kPa, 0°C) liter of gas and commonly abbreviated as J/L. Energy deposition needed to facilitate various plasma processes will be given in the following discussion of particular chemical reactions.

For those who specialize in engine and emission control technologies, the unit of J/L sounds fairly abstract. To better understand the energy requirement of plasma devices, it should be related to more common terms, such as the engine power output and the fuel economy penalty. It can be assumed that the diesel engine produces about 1.4 L (STP) per second of exhaust gas per 1 kW of its output power. In such a case, the energy deposition of 10 J/L is equivalent to 1.4% of the engine rated power. To generate that amount of electricity, the engine will experience a certain fuel economy penalty. To estimate the fuel cost of electrical energy, one has to assume the conversion efficiency of the vehicle's alternator and the efficiency of the power supply from the typical 12 V to the high plasma voltage. Because the combined efficiency of these devices is about 70% [6], the cost of the electrical energy at 10 J/L is equivalent to an approximately 2% fuel economy penalty. Plasma processes that use less than, say, 20 J/L of electrical energy may be therefore considered realistic. Processes that use up to some 30 to 40 J/L may be realistic, provided progress in NTP reactor efficiency is made. Many NTP devices that have been studied use much higher energy depositions, sometimes in excess of 100 J/L. The importance of these processes appears to be purely academic.

23.3 Chemical Reactions

23.3.1 Plasma Chemistry

A number of products have been detected in plasma-treated diesel exhaust gases, including a mixture of nitrogen oxides and acids, CO, aldehydes, and ozone (O_3). If the treated gas contains propene, often introduced to laboratory systems as a representative HC, formic acid, formaldehyde, and acetaldehyde have been identified among the products. The underlying chemistry is not well understood. The following set of reactions, adopted mostly from Thomas *et al.* [7], represents a probable hypothesis rather than proven mechanisms. Numerous other reactions are also possible, but are believed to be less likely than the reaction pathways presented next. The reactions apply to soot-free exhaust gas. The presence of DPM has been shown to influence the plasma chemistry, but the chemical mechanisms remain even more obscure. Some discussion is given in the section on the plasma treatment of diesel PM.

The chemical processes are initiated via electron induced dissociation of oxygen:

$$O_2 + e \rightarrow O + O + e \qquad (23.1)$$

The electron energy range in NTP reactors, typically at 3 to 6 eV, corresponds to a maximum production of oxygen radicals in air or exhaust gases [Reaction (23.1)]. This is why NTP reactors are effective ozone generators:

$$O_2 + O \rightarrow O_3 \qquad (23.2)$$

Oxygen atoms are also very effective in oxidizing NO to NO_2. In the absence of hydrocarbons, the efficiency of NO_2 production [Reaction (23.3)] is reduced by the reverse reaction of Reaction (23.4), and also by the conversion to acids of Reactions (23.5) and (23.6):

$$NO + O \rightarrow NO_2 \qquad (23.3)$$

$$NO_2 + O \rightarrow NO + O_2 \qquad (23.4)$$

$$H_2O + e \rightarrow OH + H + e \qquad (23.5)$$

$$NO_2 + OH \rightarrow HNO_3 \qquad (23.6)$$

Hydrocarbons are attributed an important role in plasma. In the presence of hydrocarbons, the efficiency of NO_2 production is greatly increased. A possible explanation is that HCs act as a sink for O atoms and OH radicals. The resulting products react with O_2, to form peroxy radicals (HO_2), which further convert NO to NO_2. These processes are illustrated by Reactions (23.7) through (23.12), with propene as the example hydrocarbon species:

$$C_3H_6 + O \rightarrow C_2H_5 + HCO \tag{23.7}$$

$$C_3H_6 + O \rightarrow CH_2CO + CH_3 + H \tag{23.8}$$

$$C_3H_6 + O \rightarrow CH_3CHCO + H + H \tag{23.9}$$

$$HCO + O_2 \rightarrow CO + HO_2 \tag{23.10}$$

$$H + O_2 \rightarrow HO_2 \tag{23.11}$$

$$NO + HO_2 \rightarrow NO_2 + OH \tag{23.12}$$

The hydrocarbons may also react with the OH radicals. This reaction provides a second method of promoting the NO/NO_2 shift, which, under the temperatures in engine exhaust gases, may become predominant, as shown here:

$$C_3H_6 + OH \rightarrow C_3H_6OH \tag{23.13}$$

$$C_3H_6OH + O_2 \rightarrow C_3H_6OHOO \tag{23.14}$$

$$C_3H_6OHOO + NO \rightarrow C_3H_6OHO + NO_2 \tag{23.15}$$

The produced alkoxy radical [Reaction (23.15)] may decompose to acetaldehyde and CH_2OH [Reaction (23.16)], which, in turn, may react with oxygen to formaldehyde:

$$C_3H_6OHO \rightarrow CH_3CHO + CH_2OH \tag{23.16}$$

$$CH_2OH + O_2 \rightarrow HCHO + HO_2 \tag{23.17}$$

Finally, the acetaldehyde may react with OH radicals to formic acid, which may be broken down to such products as CO, CO_2, and H_2O:

$$CH_3CHO + OH \rightarrow HCOOH + CH_3 \tag{23.18}$$

$$HCOOH + OH \rightarrow CO_2 + H_2O + H \tag{23.19}$$

$$HCOOH + OH \rightarrow CO + OH + H_2O \tag{23.20}$$

$$HCOOH + O \rightarrow CO + OH + OH \tag{23.21}$$

Even though these reactions are hypothetical, an agreement exists in the literature on the importance of hydrocarbons in plasma chemistry. It is believed that hydrocarbons play the following important functions in NTP [8]:

- The HCs lower the electrical energy requirement for oxidation of NO to NO_2.
- The HCs minimize the formation of acid products.
- The HCs suppress the oxidation of SO_2 to SO_3.

Because very few exhaust gas components are unaffected by the NTP, plasma devices were being developed to target a number of pollutants. Some of these approaches turned out to be either impossible or to require excessively high energy densities; others still show promise and are under continuing development, as summarized in Table 23.1.

23.3.2 NO Reduction by Plasma

Theoretically, nitric oxide can be chemically reduced through a number of plasma reactions, such as (1) direct NO dissociation followed by the formation of N_2 and O_2 molecules from radicals, or (2) dissociation of nitrogen into atoms, which in turn can react with NO to N_2 and oxygen. Earlier plasma research attempted to maximize this "reduction channel" of plasma over the "oxidation channel" described by Reactions (23.1) through (23.21). Unfortunately, this task was never achieved and NO reduction by plasma could not be convincingly demonstrated by experiments [9]. A consensus has formed in the current literature that the process is not feasible. This notion is also supported by the following theoretical argument, which estimates the NO_x abatement potential that can be offered by plasma treatment based on energy considerations [10].

TABLE 23.1
POSSIBILITIES OF POLLUTANTS CONTROL BY PLASMA

Process	Development Status
NO reduction	Kn_{ox} reduction by plasma alone is not believed to be a feasible process. Despite years of research, no convincing experimental demonstrations have been shown. Theoretical modeling indicates that NO reduction by plasma would require prohibitively high energy deposition.
NO oxidation	Oxidation of NO to NO_2 remains the most promising plasma process, which can be realized using realistic energy depositions. The plasma-generated NO_2 facilitates Kn_{ox} reductions in plasma-catalyst systems.
HC oxidation	Technically possible, but requires high energy rates. Not feasible for diesel, but investigated for cold start HC removal from gasoline engines.
PM control	NTP can oxidize carbonaceous PM if long residence times are provided, for example, through particulate "trapping" in the plasma device.

Three minimum energy limits can be considered with regard to NO reduction by plasma:

- NO dissociation energy limit
- Nitrogen radical production limit
- Effect of oxygen.

The dissociation energy limit is related to the minimum amount of energy needed to dissociate an NO molecule. In an ideal situation, the kinetic energy of the electrons is dissipated entirely in the dissociation of NO molecules into N and O radicals, which would then form exclusively N_2 and O_2 molecules. The energy required to dissociate an NO molecule is 6.5 eV. It can be calculated that this corresponds to the dissociation of 40 ppm of NO per J/L of input energy density. In other words, 10 J/L energy deposition, estimated at about 2% fuel economy penalty, would allow dissociation of 400 ppm of NO.

In reality, the concentration of NO is very low and direct dissociation of NO is not probable. The kinetic energy of the electrons is deposited primarily into the major exhaust gas components, N_2 and O_2. The electrons could also lose considerable energy through other reactions, such as vibrational excitation of N_2, that do not promote dissociation of NO.

A radical production energy limit can be estimated by assuming the more realistic reaction path through the formation of N and O radicals. The following reactions represent the dissociation of nitrogen and oxygen by plasma:

$$e + N_2 \rightarrow e + N(^4S) + N(^4S) \qquad (23.22)$$

$$e + N_2 \rightarrow e + N(^4S) + N(^2D) \qquad (23.23)$$

$$e + O_2 \rightarrow e + O(^3P) + O\,(^3P) \qquad (23.24)$$

$$e + O_2 \rightarrow e + O(^3P) + O\,(^1D) \qquad (23.25)$$

where $N(^4S)$ and $N(^2D)$ are ground-state and metastable excited-state nitrogen atoms, respectively, and $O(^3P)$ and $O(^1D)$ are ground-state and excited-state oxygen atoms, respectively.

$N(^4S)$ is the only plasma species that can chemically reduce NO. Let us suppose that it is the only radical produced in the system; that is, Reactions (23.23) through (23.25) do not occur. Furthermore, let's assume that all nitrogen atoms can be used entirely for the reduction of NO according to the following reaction:

$$N + NO \rightarrow N_2 + O \qquad (23.26)$$

In this case the energy required to reduce NO is simply determined as the energy required to produce N from the electron-impact dissociation of N_2. The number of N_2 dissociations per unit of input energy increases with the electron mean energy. Because of the electrical breakdown threshold, the electron mean energy in NTP reactors is around 4 eV. Lets assume that the electrical breakdown limit is increased to a very high value of 10 eV. It can be calculated that for a 10-eV electron mean energy about 1.25 N_2 molecules dissociate for each 100 eV of input energy. This translates to the consumption of 40 eV of electrical energy for each N atom produced. This is the minimum energy required under the most ideal plasma conditions to produce 1 atom of nitrogen and, according to Reaction (23.26), to reduce 1 molecule of NO. It can be further calculated that this corresponds to the reduction of 6.5 ppm of NO per J/L of input electrical energy density. This time a 10 J/L input energy deposition (2% fuel economy penalty) results in the reduction of only 65 ppm of NO. Reducing NO levels of 650 ppm, as might be seen in heavy-duty diesel engines, would require a prohibitive input energy deposition of 100 J/L.

The effect of oxygen, although difficult to estimate in numbers, will further increase the preceding minimum energy requirement for NO reduction. This is caused for at least three reasons:

- The dissociation energy of O_2 is smaller than that of N_2 resulting in a significant fraction of the input electrical power being dissipated in the dissociation of O_2 to produce oxidative radicals [Reactions (23.24) and (23.25)]. The generated ground-state oxygen atoms will oxidize NO to NO_2.
- The metastable nitrogen atoms, $N(^2D)$, react preferentially with oxygen to produce NO. If the electron mean energy is increased to favor dissociation of N_2 rather than O_2, even more excited-state nitrogen atoms are generated.
- The metastable oxygen atoms, $O(^1D)$, react with H_2O to produce OH radicals, which in turn react with NO and NO_2 to convert them to nitrous acid and nitric acid, respectively.

In light of the high minimum energy limit for NO reduction, it is not believed that plasma treatment alone, in the absence of heterogeneous reactions, can become an effective means of exhaust NO_x abatement. Rather, plasma has to be combined with other technologies, such as emission control catalysts.

23.3.3 NO Oxidation

Plasma was found to be effective in selective oxidation of NO to NO_2. Although this reaction does not bring any emission benefit in itself, the generated NO_2 is a very reactive species, which can be utilized for selective NO_x reduction in plasma catalyst systems or for PM oxidation in plasma devices.

The presence of hydrocarbons is critical to the NO/NO_2 shift in plasma, especially at higher temperatures, such as those encountered in exhaust gases. The optimum HC_1/NO_x ratio is believed to be about 6 [11]. At this ratio, the energy density to achieve maximum oxidation efficiency at high NO concentration (600 ppm) is below 30 J/L. At lower NO concentrations (100 ppm), the maximum oxidation efficiency can be achieved at an energy deposition of only 6 J/L. An example relationship between the energy density and the NO/NO_2 shift for two initial NO concentrations is shown in Figure 23.5. The lines denote modeling predictions, and the points represent experimental measurements.

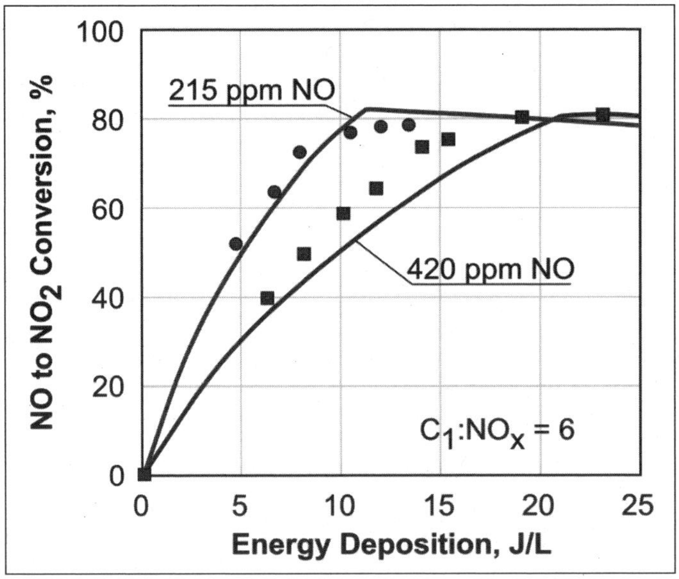

Figure 23.5 *Oxidation of NO to NO_2 by plasma (10% O_2, 10% CO_2, 5% H_2O, balance N_2; propene additive; t = 200°C).*

The HC_1/NO_x ratio of 6 is much higher than that in the native diesel exhaust, which is always below 1. Therefore, the use of plasma-induced NO/NO_2 shifts is likely to require some enrichment of the exhaust gas with hydrocarbons (e.g., using diesel fuel) at an additional fuel economy penalty.

23.3.4 HC Oxidation

As shown earlier, the chemistry of HC oxidation and the chemistry of NO oxidation in plasma are strongly coupled. However, much higher energy deposition is required to completely oxidize hydrocarbons to CO_2 and H_2O. At energy densities required for the NO oxidation, the post-reaction mixture contains such products of partial HC oxidation as aldehydes, ketones, organic acids, and CO.

For example, it is possible to convert 100 ppm of NO to NO_2 with less than 10 J/L, but complete oxidation of the resulting 50 ppm of formaldehyde requires more than 150 J/L [11]. Therefore, in view of the more effective technologies such as the diesel oxidation catalyst, plasma cannot be considered an attractive option for the abatement of diesel HC. However, plasma can be effective at lower temperatures than those required by catalyst. From this reason, plasma is still an option for cold start HC oxidation in gasoline engines [12].

23.4 Plasma Catalysts

23.4.1 DeNOx Catalyst

Nonthermal plasma can be used to enhance the performance of DeNOx (lean NO_x) catalysts promoting selective catalytic reduction of NO_x by hydrocarbons. The plasma catalyst system has two possible configurations:

- Single-stage reactor, usually in the form of the packed-bed NTP reactor, where the catalyst is present in the plasma reactor bed
- Two-stage reactor, with a monolithic catalyst positioned downstream of the NTP reactor. This configuration, utilizing various versions of the parallel plate DBD plasma reactor, appears to be preferred in more recent research.

One of the first studies of that type of plasma catalyst was published by Tonkyn *et al.* [13]. The experimental work was conducted in a small laboratory packed-bed plasma reactor that was filled with different dielectric packings. Simulated exhaust gas containing 7.5% O_2, 4% CO_2, 2% H_2O, 0.2% CO, 750 ppm C_3H_6, and 270 ppm NO was used. The space velocity in the reactor was approximately 7500 1/h. Investigated packing materials included Zirconia, barium titanate, and various types of zeolites. Experiments were conducted at the bed temperature of 100, 140, and 180°C. Zeolite material was investigated in a fresh and sintered state.

Concentrations of NO and NO_x were measured at the reactor inlet and outlet. The decrease of NO was always higher than the decrease of NO_x, indicating that NO was partially oxidized to NO_2. The results, presented as normalized concentration profiles for NO and NO_x as a function of the input energy density, are shown in Figure 23.6. The solid lines represent material sintered at 800°C with little surface area left. The dashed lines represent fresh zeolite. The "fraction reduced" lines indicate the ratio of NO_x concentration change to the NO concentration change. A similar decrease of NO is observed in both cases, whereas the NO_x performance is dramatically altered in the sintering process. The fresh zeolite reduces about 50% of the total NO_x at high energy densities. The sintered material is capable of reducing only about 20% of NO_x.

Additional experiments showed that the catalyst performance was also improved by increasing oxygen and propene concentrations. Added oxygen, as well as added propene, improved the energy efficiency; that is, NO_x conversions occurred at lower energy densities. Increasing oxygen or propene concentrations, however, did not have any effect on

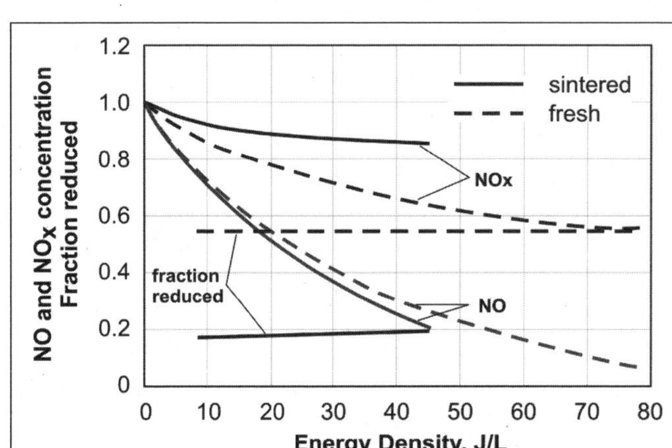

Figure 23.6 *NO_x performance of fresh and sintered zeolite.*

the fractions of NO that were oxidized and reduced. This performance is similar to the behavior of zeolite-based lean NO_x catalysts that have been reported to improve performance with increasing concentrations of oxygen and hydrocarbons.

A number of later studies, including experiments in the two-stage NTP-catalyst configuration, generally confirmed the preceding findings. It can be concluded that NO in the plasma catalyst undergoes a two-step reaction process, as follows [14]:

1. *Oxidation of NO to NO_2 by plasma in the presence of hydrocarbons:* The products include NO_2, as well as partially oxidized HC, which may be also enhancing the catalyst performance.
2. *Selective catalytic reduction of NO_2 to N_2 using hydrocarbons:* It is suggested that NO_2 is the main species that reacts with HC on the catalyst sites, in a process similar to that in the DeNOx catalyst. However, even though the importance of the NO_2 intermediate appears to be confirmed in the literature, the exact mechanisms remain unclear and may be more complicated than a simple selective reduction by hydrocarbons [2, 9]. Catalyst formulation has significant influence on the performance in this second step.

Many catalyst systems have been tested in conjunction with plasma, from noble metal catalysts, through various base metal formulations, to a simple γ-alumina catalyst. Best performance has been reported with some proprietary, nonprecious metal zeolite formulations [2]. Interestingly, the copper-substituted ZSM5 catalyst (Cu-ZSM5), so effective as a DeNOx catalyst, was found to reduce the NO_2 back to NO, with no useful overall NO_x conversion.

The experience with plasma catalyst systems can be summarized as follows [2]:

- Plasma catalysts can achieve about a 50% to 60% NO_x reduction at an overall fuel economy penalty (electrical + secondary HC injection) of about 5%. In this respect, plasma catalyst and active DeNOx catalysts offer similar performance [6], as illustrated in Figure 23.7.
- Zeolite catalysts used in plasma systems convert about 80% of NO_x to N_2, and 20% to N_2O.
- Plasma catalysts oxidize only about 5% of exhaust SO_2 to SO_3, resulting in no sulfate particulate formation.
- Hydrocarbons, aldehydes, and CO generated by plasma are not affected over the catalyst. An oxidation catalyst is required downstream of the plasma catalyst.
- To achieve significant NO_x conversions, plasma catalysts have to be operated with an electrical energy deposition of at least 15 J/L, an HC:NO_x ratio of at least 5, and at low space velocities of 5,000 to 10,000 1/h.
- Plasma catalysts can provide significant NO_x conversions at temperatures as low as 100°C.

23.4.2 SCR (Urea) Catalyst

Plasma has been also studied in combination with selective catalytic reduction (SCR) of NO_x using ammonia and/or urea

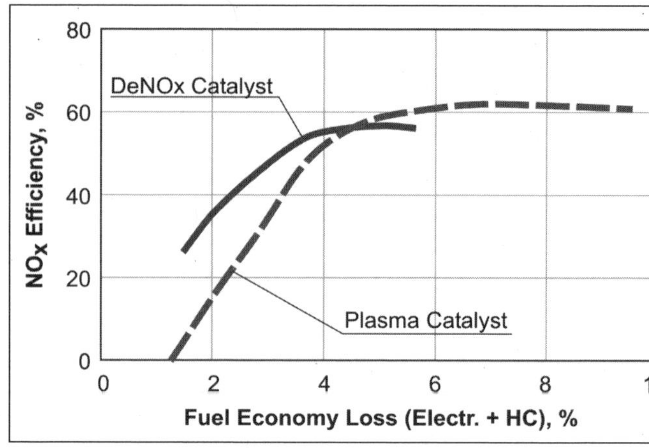

Figure 23.7 *Modeled performance of plasma and DeNOx catalysts (2.5-L TDI engine, 5-L catalysts, projected FTP-75 cycle, plasma: 5.5:1 HC_1:NO_x ratio, DeNOx: varying HC:NO_x ratio).*

[15]. The objective is to improve the cold start NO_x performance of SCR systems for diesel cars. It is known that the performance of SCR catalysts can be enhanced by increasing their inlet NO_2 concentration (see Section 19.3). Therefore, oxidation catalysts have been placed in front of SCR catalysts to oxidize NO. Replacing the oxidation catalyst with plasma is expected to enable higher NO_x reductions at low temperatures.

Preliminary reports indicate that the low-temperature SCR catalyst efficiency can be significantly improved by plasma [16]. The estimated energy costs are lower in comparison to the cost of heating the oxidation catalyst. Potential problems in the application of plasma to SCR are related to the formation of ammonium nitrate aerosols, which would increase PM emission.

23.5 Particulate Matter Control by Plasma

23.5.1 PM Oxidation by Plasma

In comparison to the wealth of publications on NO_x control by plasma, little research has been done on the application of NTP for the control of DPM emissions. For some time it was not obvious whether plasma devices can have an effect on DPM emissions at all. Research studies carried out in the late 1990s indicate that diesel particulates can be oxidized, and PM trapping devices can be regenerated, using NTP. More precisely, the diesel particulates can be effectively oxidized by some reactive species generated in plasma, such as O and OH radicals or NO_2. An important advantage of plasma is the potential to oxidize particulates at low exhaust temperatures.

DPM, when measured on the molecular scale, is characterized by large particle sizes. For this reason, the oxidation of particulates by plasma is possible only if the PM residence time in the reactor is much longer than the residence time normally required for gaseous species. In other words, in a typical plasma reactor, gas-phase oxidation of PM would be incomplete at energies suitable for the control of gas-phase species [11]. This observation is not surprising; in fact, a similar residence time argument is made to explain why carbonaceous PM can be catalytically oxidized in a diesel particulate filter but not in a diesel flow-through catalyst. To increase the residence time without increasing the reactor size beyond reasonable limits, the particulate has to be removed from the gas stream (trapped) and held in the reactor until its oxidation is completed. Two reactor configurations have been proposed to achieve this objective in a plasma device:

- Single-stage reactor, where a packed-bed plasma reactor acts as the PM trapping device
- Two-stage reactor, where a "classic" diesel particulate filter (DPF) is positioned downstream of the plasma generator.

The role of plasma in the two-stage configuration is to generate NO_2 and possibly ozone, which can oxidize particulates in the downstream filter. The plasma chemistry in this case is similar to what was seen in the reactors designed for NO_x reduction.

The chemistry in the single-stage plasma PM-reducing device, however, appears to be quite different. The presence of carbonaceous particulates in the reactor influences plasma reactions. When particulates are present, less hydrocarbon oxidation is observed and less NO_2 is generated. Furthermore, PM reductions can be achieved at $HC_1:NO_x < 1$, a mode where NO_2 production is not very efficient. Although the reaction mechanisms are not understood, it is speculated that particulates can be oxidized by different plasma species than NO_2, presumably by O and OH radicals [7].

23.5.2 Single-Stage Packed-Bed Plasma Reactor

In the packed-bed configuration, the PM is trapped in the NTP reactor itself. NTP packed-bed reactor packings, in the form of beads or pellets typically made of such ferroelectric materials as barium titanate, magnesium niobate, lead titanate, or lead zirconate can be designed to trap particulates, primarily through electrostatic impaction. In addition, a separate pair of electrodes may be provided upstream of the reactor to increase the trapping efficiency by electrically charging soot particles entering the reactor [17].

One of the first reports on the use of a packed-bed plasma device for PM control (Figure 23.8) was published by Fanick *et al.* [18, 19]. The tests were conducted with a 2.2-L Toyota diesel engine on a chassis dynamometer at 60 mph and 13.1-hp road load. A partial exhaust gas flow setup and a small packed-bed plasma reactor utilizing a barium titanate type of ceramic material were used.

The very high PM removal efficiencies that were observed at low space velocities would suggest that both the SOF and the elemental carbon fractions of diesel particulates were affected by plasma. An accompanying NO_x removal was also recorded during the tests. The authors estimated that the power consumption of 1.1 kW would be needed by a system capable of handling the entire exhaust gas flow from the 2.2-L Toyota engine.

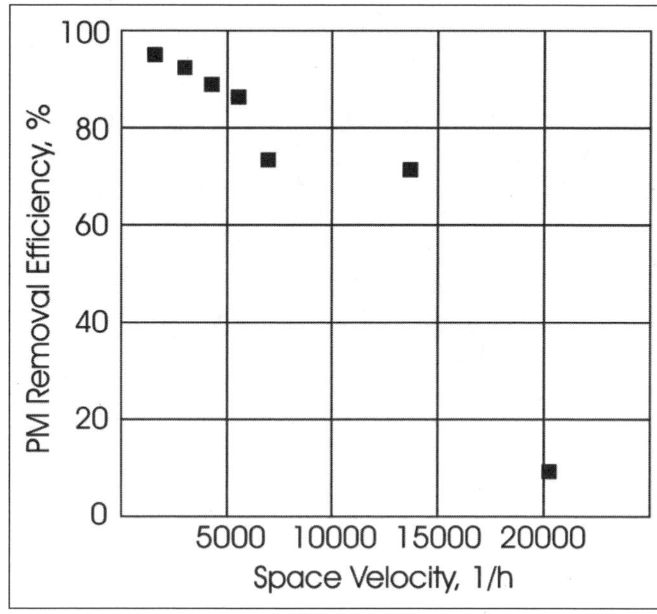

Figure 23.8 *Particulate matter removal efficiency by plasma.*

A full-scale prototype of a plasma PM removal device for a diesel car was developed by AEA Technology [7]. The system, shown in Figure 23.9, utilizes a bed of ceramic pellets placed between two cylindrical electrodes. Perforation and a system of channels in the inner electrode direct the exhaust gas flow through the pellet bed. The reactor has an active volume of 2600 cm^3 and replaces the original muffler on the test vehicle—a 1994 model year LTI Fairway London Taxi equipped with a Nissan 2.7-L IDI diesel engine.

The unit was reported to support low-temperature, complete oxidation of soot during vehicle tests. No blow-through of PM was observed, even after long periods of operation. There was no particle buildup on the packing material, which remained relatively clean during plasma experiments. Vehicle tests were performed with no hydrocarbon enrichment of the exhaust gas.

Emission measurements over the ECE+EUDC test yielded about 50% PM emission reduction at ambient temperature of 25°C and about 70% reduction at 0°C. It was speculated that these results could be increased by optimizing the reactor packing material for higher trapping efficiency. The device provided a 90% conversion efficiency (and a 99.9% steady-state efficiency) for ultrafine particulates 60 nm in diameter. High removals of PAHs were also measured in the plasma device.

The authors determined the upper energy limit required for the NTP low-temperature oxidation of PM at 0.34 kWh/g. This number was calculated from steady-state experiments running 2.34 g/h of PM that was completely oxidized at an input energy deposition of 18 J/L.

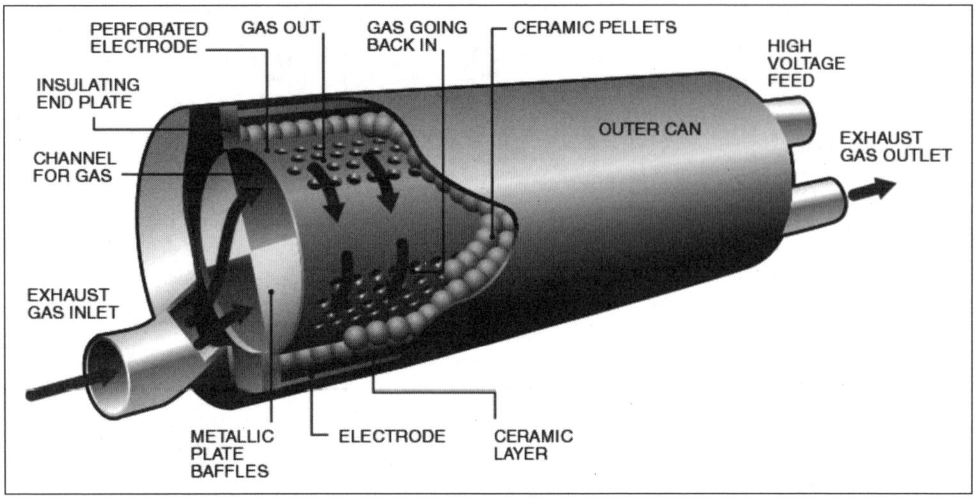

Figure 23.9 *Plasma device for diesel particulate removal. (Courtesy of AEA Technology.)*

23.5.3 Two-Stage Reactor: NTP-DPF Configuration

The two-stage reactor configuration (Figure 23.10) corresponds to the CRT filter layout (see Section 22.3), where NO_2 is generated by plasma, rather than over an oxidation catalyst. The plasma-generated NO_2 facilitates low-temperature regeneration of the particulate filter [20]. In this configuration, plasma-generated ozone may also play a role in the filter regeneration. Ozone was reported to be particularly effective at causing low-temperature combustion of soot accumulated in diesel filters [21].

The system presented in Figure 23.10 includes a NO_x reduction catalyst downstream of the filter. This diagram is an example of wishful thinking about simultaneous reduction of NO_x and PM emissions by plasma. In reality, it is not yet clear if such a system is feasible. As already mentioned, it is believed that the generation of NO_2, which is essential for catalytic NO_x reduction, is impaired by particulates due to the competition between NO and carbon for the oxygen radicals in plasma. As an alternative approach, it was suggested that the simultaneous NO_x and PM control could be realized by designing the plasma catalyst reactor for maximum NO_x reduction and periodically switching into the thermal (hot) plasma operation mode in order to thermally oxidize particulates [11].

23.6 Future Issues

Plasma studies are being conducted by a number of government and private sector organizations in the United States and in Europe. The major areas that need further research or require improvement can be grouped as follows:

- Improvement in the NTP reactor design with possible savings in electrical energy consumption; control of electromagnetic interferences required for commercial applications
- Understanding of the reaction mechanisms in plasma, in plasma catalysts, and reactions in the presence of diesel particulates
- Increasing the NO_x conversion efficiency and decreasing the catalyst size (i.e., increasing the space velocity)
- Refining PM removal reactors and increasing the removal efficiencies
- Studying the possibilities of simultaneous NO_x and PM removal.

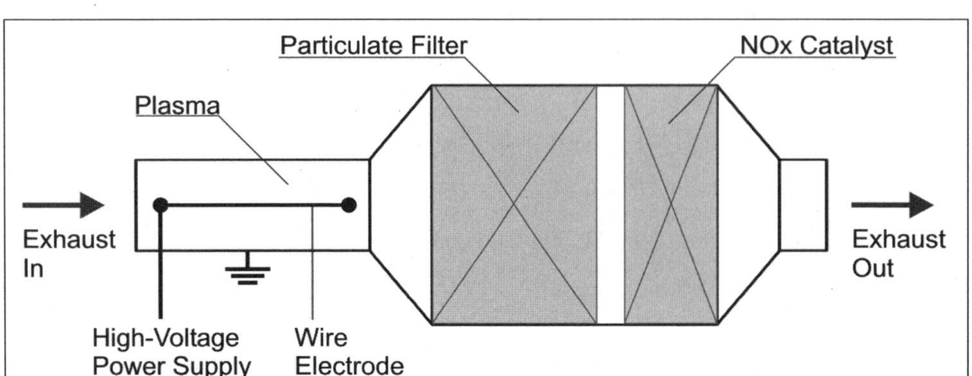

Figure 23.10 *Plasma regenerated particulate filter system concept.*

References

1. Penetrante, B.M., and S.E. Schultheis (Eds.), *Non-Thermal Plasma Techniques for Pollution Control: Part B—Electron Beam and Electrical Discharge Processing,* Springer-Verlag, Berlin, 1993.
2. Hoard, J., "Plasma Assisted Catalysis," Paper presented at SAE Diesel Engine TOPTEC Conference, Indianapolis, IN, September 26–27, 2000.
3. Whealton, J.H., *et al.,* "Non-Thermal Plasma Exhaust Aftertreatment: A Fast Rise-Time Concept," SAE Paper No. 971718, SAE International, Warrendale, PA, 1997.
4. Hemingway M.D., *et al.,* "Evaluation of a Non-Thermal Plasma System for Remediation of NOx in Diesel Exhaust," SAE Paper No. 1999-01-3639, SAE International, Warrendale, PA, 1999.
5. Herling, D., *et al.,* "Evaluation of Corona Reactors of Several Geometries for a Plasma Assisted Nitrogen Oxide Emission Reduction Device," SAE Paper No. 2000-01-2899, SAE International, Warrendale, PA, 2000.
6. Hoard, J., *et al.,* "Comparison of Plasma Catalyst and Lean NOx Catalyst for Diesel NOx Reduction," SAE Paper No. 2000-01-2895, SAE International, Warrendale, PA, 2000.
7. Thomas, S.E., *et al.,* "Non Thermal Plasma Aftertreatment of Particulates—Theoretical Limits and Impact on Reactor Design," SAE Paper No. 2000-01-1926, SAE International, Warrendale, PA, 2000.
8. Penetrante, B.M., R.M. Brusacso, B.T. Merritt, and G.E. Vogtlin, "Sulfur Tolerance of Selective Partial Oxidation of NO to NO_2 in a Plasma," SAE Paper No. 1999-01-3687, SAE International, Warrendale, PA, 1999.
9. Hoard, J., and M.L. Balmer, "Analysis of Plasma-Catalysis for Diesel NOx Remediation," SAE Paper No. 982429, SAE International, Warrendale, PA, 1998.
10. Penetrante, B.M., *et al.,* "Fundamental Limits on NOx Reduction by Plasma," SAE Paper No. 971715, SAE International, Warrendale, PA, 1997.
11. Penetrante, B.M., *et al.,* "Feasibility of Plasma Aftertreatment for Simultaneous Control of NOx and Particulates," SAE Paper No. 1999-01-3637, SAE International, Warrendale, PA, 1999.
12. Lepperhoff, G., *et al.,* "Exhaust Emission Reduction of Combustion Engines by Barrier Discharge—A New Reactor/Generator System," SAE Paper No. 1999-01-3638, SAE International, Warrendale, PA, 1999.
13. Tonkyn, R., *et al.,* "Vehicle Exhaust Treatment Using Electrical Discharge Methods," SAE Paper No. 971716, SAE International, Warrendale, PA, 1997.
14. Penetrante, B.M., *et al.,* "Plasma-Assisted Catalytic Reduction of NOx," SAE Paper No. 982508, SAE International, Warrendale, PA, 1998.
15. Hammer, T., Broer S., "Plasma Enhanced Selective Catalytic Reduction of NOx for Diesel Cars," SAE Paper No. 982428, SAE International, Warrendale, PA, 1998.
16. Hammer, T., Broer S., "Plasma Enhanced Selective Catalytic Reduction of NOx in Diesel Exhaust: Test Bench Measurements," SAE Paper No. 1999-01-3633, SAE International, Warrendale, PA, 1999.
17. Bayliss, K.H., D. Raybone, and S.I. Hall, "Exhaust Gas Purification," European Patent No. EP 0 608 619 A1 (UK Atomic Energy Authority), 1994.
18. Fanick, E.R., *et al.,* "Reduction of Diesel NOx/PM Emissions Using a Non-Thermal Plasma," in *Proceedings of the DOE Diesel Emissions Workshop,* 1995, p. V57.
19. Fanick, E.R., *et al.,* "Simultaneous Reduction of Diesel Particulate and NOx Using a Plasma," SAE Paper No. 942070, SAE International, Warrendale, PA, 1994.
20. Penetrante, B.M., *et al.,* "Plasma Regenerated Particulate Trap and NOx Reduction System," U.S. Patent No. 6,038,854 (Lawrence Livermore National Laboratory), 2000.
21. Twigg, M.V., and I.C. Wishart, "System and Method for Purifying Exhaust Gases," International Patent Application No. WO 00/21646 (Johnson Matthey), 2000.

Index

About the Authors

Addy Majewski holds a Ph.D. in Chemical Engineering from the Wroclaw University of Technology, Poland. He worked at the University in areas of heat transfer and heat exchangers, evaporation, distillation and absorption with chemical reactions.

Since 1991 he has lived and worked in Toronto, Ontario, Canada. His work experience includes development of diesel particulate filters and emission control catalysts for diesel, LPG and natural gas engine applications. Currently he is the president of Ecopoint Inc., a consulting firm that specializes in engine emission control technologies. He is also the founder and editor of DieselNet.com, an internet-based technical information service on clean diesel engine and emission technologies.

Mr. Majewski is a member of the Society of Automotive Engineers (SAE), and a professional engineer in Ontario.

Magdi Khair is a staff engineer in the Department of Emissions Research at Southwest Research Institute. He received his B.S. in Automotive Engineering from Ain Shams Univ.; his M.S. in Thermodynamics from the Univ. of Birmingham; and his M.B.A. from Michigan State Univ.

Mr. Khair is experienced in the areas of engine testing from exhaust emissions control. His prior experience was with AlliedSignal Automotive Catalyst Co. with the development of catalytic aftertreatment for light-duty and heavy-duty diesel engines; Ford New Holland with primary responsibility for the development of the 6.6 and 7.8 liter midrange diesel engines to meet 1991 emissions standards; Bendix Diesel Engine Controls where he led the development of advanced electronically controlled diesel fuel injection systems and also established several cooperative engineering programs with European and North American engine manufacturers; and with Chrysler Corp. where he converted the slant six gasoline engine into an open chamber, pilot injected and electronically controlled diesel engine, supervised a combustion kinetics project, and participated in the design and development of electronic controls for a passenger car turbine engine.

Early in his career, Mr. Khair had professional assignments at Nasr Automotive Mfg. Co., Egypt, and John Brown Engineering Ltd., Scotland. His experience includes extensive engine emissions testing, calibration of fuel injection systems, development of intake manifolds, optimization of intake ports, and development of electronic engine controls. Mr. Khair holds 13 U.S. patents in electronic fuel injection, turbocharging, exhaust gas recirculation, and aftertreatment systems. He has authored and co-authored numerous SAE and ASME papers and is a popular instructor of SAE's Diesel Engine Technology Seminar.

The details of the proof that $\lim_{\alpha} \hat{f}_{\alpha,t}(M)$ is a character are left to the reader. It is clearly independent of the choice of $\{f_\alpha\}$. We designate the character associated in this way with M by χ_M.

We next present an important formula relating f to χ_M.

PROPOSITION 3. For every $f \in L_1(G)$, and every M, we have

$$f(M) = \int_G f(t)\chi_M(t)\, dt.$$

Proof. A set in a locally compact space is said to be σ compact if it is contained in the union of countably many compact sets. It is not very hard to show, and we leave the proof to the reader, that, if $E \subset G$ and $\mu(E) < \infty$, then E is σ compact. It follows that for every neighborhood U of e, E may be covered by a countable set of translations of U. This also holds for the support of f since it is easily seen to be the union of countably many sets of finite measure.

Suppose f is non-negative. Since $\chi_M(t) = \lim_{\alpha} \hat{f}_{\alpha,t}(M)$, for every $\epsilon > 0$ there is an α such that

$$|\hat{f}_{\alpha,t}(M) - \chi_M(t)| < \epsilon.$$

Let U be such that $UU^{-1} \subset U_\alpha$, and let $E_1, E_2, \ldots, E_n, \ldots$ be such that they are disjoint, the support of f is contained in their union, and each is contained in a translation of U. Let

$$f_n(t) = f(t)C_{E_n}(t)$$

where C_{E_n} is the characteristic function of E_n. Then, for $t \in E_n$,

$$\begin{aligned}
&\left|\hat{f}_n(M) - \int_G f_n(t)\chi_M(t)\, dt\right| \\
&\qquad = \left|\frac{\hat{f}_n(M)}{\int f_n(t)\, dt} - \int_G \frac{f_n(t)}{\int f_n(t)\, dt}\chi_M(t)\, dt\right| \int_G f_n(t)\, dt \\
&\qquad = \left|\int_{E_n} [\hat{f}_{\alpha,t}(M) - \chi_M(t)]\frac{f_n(t)}{\int f_n(t)\, dt}\right| \int_G f_n(t)\, dt \\
&\qquad \le \epsilon \int_G f_n(t)\, dt = \epsilon \int_{E_n} f(t)\, dt.
\end{aligned}$$

But, then, by summing over n,

$$\left|\hat{f}(M) - \int_G f(t)\chi_M(t)\, dt\right| \le \epsilon \int_G f(t)\, dt.$$

As a corollary, we obtain the fact that $M_1 \neq M_2$ implies $\chi_{M_1} \neq \chi_{M_2}$.

Let χ be a character in G. For each $f \in L_1(G)$, we consider the complex number $\int_G f(t)\chi(t)\, dt$. It can be shown that this mapping of $L_1(G)$ into the complex numbers is an algebraic homomorphism. The only difficulty is with multiplication. Not all of $L_1(G)$ is mapped into 0. For suppose $f \in L_1(G)$ and $\int_G f(t)\,dt \neq 0$. Then

$$\int_G f(t)\overline{\chi(t)}\chi(t)\,dt \neq 0.$$

It follows that

$$M_\chi = \left[f : \int_G f(t)\chi(t)\,dt = 0\right]$$

is a maximal regular ideal in $L_1(G)$. Moreover, it is not hard to prove that for every character χ, we have $\chi = \chi_{M_\chi}$. It follows from the above that the characters in G are in one-one correspondence with the maximal ideals in $L_1(G)$.

As an example, let G be the circle of real numbers modulo 1. The characters on G are easily seen to be the functions $e^{in\pi t}$, $n = 0, \pm 1, \pm 2, \ldots$. The maximal regular ideals are thus in one-one correspondence with the integers. Thus, for $f \in L_1(G)$ and any maximal regular ideal n, we have

$$\hat{f}(n) = \int_G f(t)e^{int}\,dt.$$

These are simply the Fourier coefficients of f.

The theory we are discussing is accordingly seen to be simply that of Fourier analysis on an arbitrary locally compact abelian group. A surprisingly large portion of the theory holds in this setting, which gives a deeper insight into the nature of things. The interested reader will have no trouble finding books and papers where this subject is developed. In particular, the books by Naimark [1] and Loomis [1] and articles by Gelfand [1], Raikov [1], and Naimark [2], can be recommended.

EXERCISES

1.1 In a Banach algebra, with identity e, show that there is an equivalent norm for which $\|e\| = 1$.

1.2 If X is a Banach space, show that the set of bounded operators of X into itself, with the operator norm, is a Banach algebra.

1.3 Find a sequence $\{p_n\}$ of algebraic polynomials which forms an approximate identity for $L_1(I)$, where I is the interval $[-1, 1]$.

1.4 Find a sequence $\{t_n\}$ of trigonometric polynomials which forms an approximate identity for $L_1(J)$, where J is the interval $[-\pi, \pi]$.

1.5 Give a sequence $\{\varphi_n\}$ of functions of class C^∞ which forms an approximate identity for $L_1(R)$.

1.6 Show that every Banach algebra A is isomorphic to a subalgebra of the algebra of bounded operators of a Banach space into itself.

1.7 If A is a semi-simple Banach algebra and it admits another norm for which it is also a Banach algebra, show that the two Banach algebras are equivalent.

2.1 Verify that the system defined in Sec. 6.2 is indeed a Banach algebra with identity.

3.1 In a topological group, show that the left translations of a base for the open sets containing the identity form a base for all the open sets.

3.2 Define topological group by axioms on a neighborhood system of e.

3.3 In a topological group, show that the component of the identity is an open and closed subgroup.

3.4 Show that if $x_1, \ldots, x_n$ is a finite subset of an open set U in a locally compact group, there are pair-wise disjoint open $U_i \subset U$, $x_i \in U_i$, $i = 1, \ldots, n$.

3.5 Prove the easy-to-prove relation

$$n(E, F)n(F, U) \geq n(E, U).$$

3.6 Show that the set function μ, defined for compact sets in G, may be extended to a left invariant measure on $\mathcal{S}$.

3.7 Give an example of a locally compact topological group whose Haar measure is not totally σ finite.

3.8 If G is a locally compact group with Haar measure μ, then if $\mu(E) < \infty$, show that E may be covered by a countable set of compact sets in G.

3.9 Define a "natural" topology for the multiplicative group $R - \{0\}$ and show that this yields a topological group.

3.10 Determine Haar measure explicitly for this group.

3.11 Do the same as in exercise 3.10 for the multiplicative group of complex numbers.

3.12 Define a topology for the group of rotations of the 2 sphere and show that this is a topological group.

3.13 Define a left invariant measure explicitly for the group of exercise 3.12.

3.14 If G is an abelian locally compact group, give the details of the proof that $L_1(G)$ is an abelian Banach algebra.

3.15 Show that if G is a locally compact abelian group and $f \in L_1(G)$ and U is any neighborhood of e, then the support of f may be covered by a countable set of translations of U.

4.1 Show that if X is a commutative Banach algebra with identity and H is the open set on which the inverse exists, the mapping $\varphi : x \to x^{-1}$ is continuous on H.

4.2 If $f, g \in L_1(R)$ and T is the Fourier transform, show that

$$T(f * g) = Tf \cdot Tg.$$

4.3 Define the Fourier transform for the n dimensional case, and prove the result of exercise 4.2 for this case.

4.4 Given an $A \subset R$, show that the set I of functions $f \in L_1(R)$ for which $T(f)$ vanishes on A is an ideal in $L_1(R)$.

4.5 Are there ideals in $L_1(R)$ which are not of the type described in exercise 4.4?

4.6 Give details of the proofs that if $x, y \in \bar{I}$ and $a, b \in C$, then $ax + by \in \bar{I}$; and that if $x \in \bar{I}$ and $y \in X$, then $xy \in \bar{I}$.

4.7 Give details of the proof that for every continuous linear functional f on the Banach space X, $f((x - ae)^{-1})$ is an entire function.

5.1 If X is the set of complex functions of a real variable of the type

$$x(t) = \sum_{n=-\infty}^{\infty} a_n e^{int},$$

where

$$\sum_{n=-\infty}^{\infty} |a_n| < \infty \quad \text{and} \quad \|x\| = \sum_{n=-\infty}^{\infty} |a_n|,$$

show that X is a Banach algebra with identity.

6.1 If x' is a bounded non-trivial homomorphism of a commutative Banach algebra X, with identity, into the complex numbers, show that its null space is a maximal ideal in X.

7.1 With $\sigma(\pi, p)$ defined as in Sec. 6.7, show that for every $\epsilon > 0$ there is a $\delta > 0$ such that $\|\pi\| < \delta$, $\|\pi'\| < \delta$ implies

$$\|\sigma(\pi, p) - \sigma(\pi', p')\| < \epsilon.$$

7.2 If x is analytic in an open set G, and Γ is a simple closed curve of type C', such that Γ and its interior are in G, show that

$$x(z) = \frac{1}{2\pi i} \int_\Gamma \frac{x(\zeta)}{\zeta - z}\, d\zeta$$

for z in the interior of Γ.

7.4 Show that if $y = \int_\Gamma x(z)\, dz$, Γ of class C', then for every continuous linear functional f on X,

$$f(y) = \int_\Gamma f(x(z))\, dz.$$

9.1 Show that if X is compact, the maximal ideals in $C(X)$ correspond to points.

9.2 Find all the maximal regular ideals in the Banach algebra of continuous functions on R which vanish at infinity.

9.3 Let A be a commutative Banach algebra with maximal regular ideal space $\mathcal{M}$. For every $S \subset \mathcal{M}$, the kernel of S is defined as $k(S) = \bigcap[M : M \in S]$. Then $k(S)$ is a closed ideal. If I is an ideal, its hull is defined as $h(I) = [M : I \subset M]$. Show that for every $S \subset \mathcal{M}$, $\bar{S} = h(k(S))$.

10.1 Give the details of the proof that $\lim_\alpha f_{\alpha,t}(M)$ in proposition 2 is a character.

10.2 In the character group $\hat{G}$ of a group G, a topology is introduced by taking as a base for the topology at the group identity those sets of characters determined by open sets A containing 1 and compact sets $K \subset G$, as those characters which take K into A. Show that $\hat{G}$ is then a topological group.

10.3 Show that $\hat{R}$ is isomorphic with R.

10.4 Show that $\hat{G}$ is discrete if and only if G is compact.

10.5 Show that $\hat{G}$ is compact if and only if G is discrete.

10.6 Show that $\widehat{G \times H} = \hat{G} \times \hat{H}$ both algebraically and topologically.

10.7 For every character χ in G show that the mapping $f \to \int_G f(t)\chi(t)\,dt$ is a homomorphism of $L_1(G)$ into the complex numbers.

10.8 Show that, for every character χ, we have $\chi = \chi_{M_\chi}$.

10.9 Show that the characters for the circle of real numbers modulo 1 are the functions $e^{in\pi t}$, $n = 0, \pm 1, \pm 2, \ldots$.

10.10 Let G be the group of integers modulo 2 and $H = G \times G \times \ldots$, a denumerable product. Discuss the character group of H.

10.11 Identifying sequences of 0's and 1's with real numbers, discuss the characters of the group H in exercise 10.10 as functions on $[0, 1]$. Show that they form a complete orthonormal set for $L_2[0, 1]$.

REFERENCES

Aronszajn, N., and K. T. Smith, *Functional Spaces and Functional Completion,* Annales de l'Institut Fourier, t. VI, 1956, pp. 125-85.

Bary, N. K. *A Treatise on Trigonometric Series,* (translated from the Russian), 2 vols., London: Pergamon, 1965.

Gelfand, I., *Normierte Ringe,* Matematiceskii Sbornik, N.S. 9, 1941, pp. 3-23.

Goffman, C., *Real Functions.* New York: Holt, Rinehart & Winston, Inc., 1953.

Hardy, G. H. *Divergent Series.* Oxford: Clarendon Press, 1949.

Hardy, G. H., and W. W. Rogosinski, *Fourier Series.* New York: Cambridge University Press, 1950.

Hausdorff, F., *Set Theory,* (translated from the German), 2nd ed., New York: Chelsea Publishing Company, 1967.

Kelley, J. L., *General Topology.* Princeton, N. J.: D. Van Nostrand Co., Inc., 1955. *Repr.,* New York: Springer-Verlag, 1975.

Kuratowski, C., *Topology,* 2 vols., (translated from the French), New York: Academic Press, 1966 and 1969.

Loomis, L. H., *An Introduction to Abstract Harmonic Analysis.* Princeton, N. J.: D. Van Nostrand Co., Inc., 1953.

Lorch, E. R., *Spectral Theory.* New York: Oxford University Press, 1962.

Naimark, M. A., (1) *Normed Algebras,* (translated from the Russian), 3rd ed., Gronigen: P. Noordhoff, 1972. (2) *Rings with Involution,* (translated from the Russian), Providence: American Mathematical Society, 1950. [Translation No. 25.]

Natanson, I. P., *Constructive Function Theory,* 3 vols., (translated from the Russian), New York: Frederick Ungar Publishing Company, 1963.

Nehari, Z., *Conformal Mapping,* New York: McGraw-Hill Book Company, 1952.

Raikov, D. A., "Die harmonische Analyse auf kommutativen Gruppen mit Haarschem Mass und die Theorie der Charaktere" (translated from Russian), *Sowietische Arbeiten zur Funktionalanalysis.* Berlin, 1954.

Widder, D. V., *The Laplace Transform.* Princeton; Princeton Press, 1941.

Zeller, K., *Theorie der Limitierungsverfahrung*. Berlin: Springer-Verlag, 1958.

Zygmund, A., *Trigonometrical Series,* Vol. I. New York: Cambridge University Press, 1959. Rev. ed., 2 vols. in one. New York: Cambridge University Press, 1968.

GENERAL REFERENCES

Akhiezer, N. I., and I. M. Glazman, *Theory of Linear Operators in Hilbert Space,* (translated from Russian). New York: Frederick Ungar Publishing Company, 1961.

Banach, S., *Théorie des Opérations Linéaires.* New York: Chelsea Publishing Company, 1973.

Bourbaki, N., *Espaces Vectoriels Topologiques, Éléments de mathématique, livre V.* Paris: Hermann, 1953 and 1955.

Bourbaki, N., *Intégration, Éléments de mathématique, livre VI.* Paris: Hermann, 1952.

Day, M. M., *Normed Linear Spaces.* 3rd ed. Berlin: Springer-Verlag, 1973.

Dieudonné, J., *History of Functional Analysis.* Amsterdam: North-Holland Publishing Company, 1981.

Dieudonné, J., and L. Schwartz, *La dualité dans les espaces (F) et (LF),* Annales de l'Institut Fourier, t. I, 1950, pp. 61-101.

Dunford, N., and J. T. Schwartz, *Linear Operators I.* New York: Interscience, 1958.

Halmos, P. R., *Measure Theory.* Princeton, N. J.: D. Van Nostrand Co., Inc., 1950. *Repr.* New York: Springer-Verlag, 1975.

Lyusternik, L. A., and W. I. Sobolev, *Elements of Functional Analysis,* Frederick Ungar Publishing Company, 1960.

Riesz, F., and B. St.-Nagy, *Functional Analysis,* (translated from the French), New York: Frederick Ungar Publishing Company, 1954.

Saks, S., *Theory of the Integral.* 2nd rev. ed., Warsaw-Lwow: Monografie Matematyczne, 1937.

Taylor, A. E., *Introduction to Functional Analysis.* New York: John Wiley & Sons, Inc., 1958.

Vulikh, B. Z., *Introduction to Functional Analysis* (translated from the Russian). London: Pergamon, 1963.

Zaanen, A. C., *Linear Analysis.* Amsterdam: North-Holland Publishing Co., 1953.

INDEX

CHELSEA

SCIENTIFIC

BOOKS

THEORIE DES OPERATIONS LINEAIRES

By S. BANACH

—1933-64. xii + 250 pp. 5⅜x8. 8284-0110-1.

DIFFERENTIAL EQUATIONS

By H. BATEMAN

CHAPTER HEADINGS: I. Differential Equations and their Solutions. II. Integrating Factors. III. Transformations. IV. Geometrical Applications. V. Diff. Eqs. with Particular Solutions of a Specified Type. VI. Partial Diff. Eqs. VII. Total Diff. Eqs. VIII. Partial Diff. Eqs. of the Second Order. IX. Integration in Series. X. The Solution of Linear Diff. Eqs. by Means of Definite Integrals. XI. The Mechanical Integration of Diff. Eqs.

—1917-67. xi + 306 pp. 5⅜x8. 8284-0190-X.

MEASURE AND INTEGRATION

By S. K. BERBERIAN

A highly flexible graduate level text. Part I is designed for a one-semester introductory course; the book as a whole is suitable for a full-year course. Numerous exercises.

Partial Contents: PART ONE: I. Measures. II. Measurable Functions. III. Sequences of Measurable Functions. IV. Integrable Functions. V. Convergence Theorems. VI. Product Measures. VII. Finite Signed Measures. PART TWO: VIII. Integration over Locally compact Spaces (. . . The Riesz-Markoff Representation Theorem, . . .). IX. Integration over Locally Compact Groups (Topological Groups, . . . , Haar Integral, Convolution, The Group Algebra, . . .). BIBLIOGRAPHY. INDEX.

—1965-70. xx + 312 pp. 6x9. 8284-0241-8.

L'APPROXIMATION

By S. BERNSTEIN and CH. de LA VALLÉE POUSSIN

TWO VOLUMES IN ONE:

Leçons sur les Propriétés Extrémales et la Meilleure Approximation des Fonctions Analytiques d'une Variable Réelle, *by Bernstein.*

Leçons sur l'approximation des Fonctions d'une Variable Réelle, *by Vallée Poussin.*

—1925/19-65. 363 pp. 6x9. 8284-0198-5. 2 v. in 1.

CONFORMAL MAPPING

By L. BIEBERBACH

Translated from the fourth German edition by F. STEINHARDT.

Partial Contents: I. Foundations; Linear Functions (Analytic functions and conformal mapping, Integral linear functions, $w = 1/z$, Linear functions, Groups of linear functions). II. Rational Functions. III. General Considerations (Relation between c. m. of boundary and of interior of region, Schwarz's principle). IV. Further Study of Mappings ($w = z + 1/z$, Exponential function, Trigonometric functions, Elliptic integral of first kind). V. Mappings of Given Regions (Illustrations, Vitali's theorem, Proof of Riemann's theorem, Actual constructions, Potential-theoretic considerations, Distortion theorems, Uniformization, Mapping of multiply-connected plane regions onto canonical regions).

"Presented in very attractive and readable form."—*Math. Gazette.*

—1952-64. vi + 234 pp. 4½x6½. 8284-0090-3. Cloth
8284-0176-4. Paper

BASIC GEOMETRY

By G. D. BIRKHOFF and R. BEATLEY

"is in accord with the present approach to plane geometry. It offers a sound mathematical development . . . and at the same time enables the student to move rapidly into the heart of geometry."—*The Mathematics Teacher.*

"should be required reading for every teacher of Geometry."—*Mathematical Gazette.*

—3rd ed. 1959. 294 pp. 5⅜x8. 8284-0120-9.

TEACHER'S MANUAL. 160 pp. 5⅜x8.
ANSWER BOOK. 38 pp. 5⅜x8.

PLANE TRIGONOMETRY

By L. E. DICKSON

In all his books, advanced and elementary, Professor Dickson is noted for the extraordinary clarity of his writing. This very elementary book is no exception.

MATHEMATICAL PAPERS

By W. K. CLIFFORD

One of the world's major mathematicians, Clifford's papers cover only a 15-year span, for he died at age 34. [Included in this volume is Clifford's English translation of an important paper of Riemann.]

—1882-67. 70 + 658 pp. 5⅜x8. 8284-0210-8.

ESSAI SUR L'APPLICATION DE L'ANALYSE AUX PROBABILITÉS

By M. J. CONDORCET

A photographic reproduction of a very rare and historically important work in the Theory of Probability. An original copy brings many hundreds of dollars in the rare book market.

—1785. Repr. 1971. 191 + 304 pp. 6x9.

A TREATISE ON THE CIRCLE AND THE SPHERE

By J. L. COOLIDGE

—2nd (c.) ed. 1916-71. 602 pp. 5⅜x8. 8284-0236-1.

MODERN PURE SOLID GEOMETRY

By N. A. COURT

—2nd ed. 1964. xiv + 353 pp. 5½x8¼. 8284-0147-0.

SPINNING TOPS AND GYROSCOPIC MOTION

By H. CRABTREE

Partial Contents: Introductory Chapter. CHAP. I. Rotation about a Fixed Axis. II. Representation of Angular Velocity. Precession. III. Discussion of the Phenomena Described in the Introductory Chapter. IV. Oscillations. V. Practical Applications. VI-VII. Motion of Top. VIII. Moving Axes. IX. Stability of Rotation. Periods of Oscillation. APPENDICES: I. Precession. II. Swerving of "sliced" golf ball. III. Drifting of Projectiles. IV. The Rising of a Top. V. The Gyro-compass. ANSWERS TO EXAMPLES.

—2nd ed. 1914-67. 203 pp. 6x9. 8284-0204-3.

THE DOCTRINE OF CHANCES

By A. DE MOIVRE

In the year 1716 Abraham de Moivre published his *Doctrine of Chances,* in which the subject of Mathematical Probability took several long strides forward. A few years later came his *Treatise of Annuities.* When the third (and final) edition of the *Doctrine* was published in 1756 it appeared in one volume together with a revised edition of the work on Annuities. It is this latter two-volumes-in-one that is here presented in an exact photographic reprint, with an afterword by Prof. H. M. Walker.

—3rd ed. 1756-1967. xii + 368 pp. 6x9. 8284-0200-0.

COLLECTED MATHEMATICAL PAPERS

By L. E. DICKSON

—5 vols. Approx. 3,400 pp. 6½x9¼.

HISTORY OF THE THEORY OF NUMBERS

By L. E. DICKSON

"**A monumental work** . . . Dickson always has in mind the needs of the investigator . . . The author has [often] expressed in a nut-shell the main results of a long and involved paper *in a much clearer way than the writer of the article did himself.* The ability to reduce complicated mathematical arguments to simple and elementary terms is highly developed in Dickson."—*Bulletin of A. M. S.*

—Vol. I (Divisibility and Primality) xii + 486 pp. Vol. II (Diophantine Analysis) xxv + 803 pp. Vol. III (Quadratic and Higher Forms) v + 313 pp. 5⅜x8. 1971. 8284-0086-5.
Three vol. set.

STUDIES IN THE THEORY OF NUMBERS

By L. E. DICKSON

A systematic exposition, starting from first principles, of the arithmetic of quadratic forms, chiefly (but not entirely) ternary forms, including numerous original investigations and correct proofs of a number of classical results that have been stated or proved erroneously in the literature.

—1930-62 viii+230 pp. 5⅜x8.

THEORIE GENERALE DES SURFACES

By G. DARBOUX

One of the great works of the mathematical literature.

An unabridged reprint of the latest edition of *Leçons sur la Théorie générale des surfaces et les applications géométriques du Calcul infinitésimal.*

—Vol. I (2nd ed.) xii+630 pp. Vol. II (2nd ed.) xvii+584 pp. Vol. III (1st ed.) xvi+518 pp. Vol. IV. (1st ed.) xvi+537 pp. 8284-0216-7.

GYROSCOPIC THEORY

By G. GREENHILL

This work is intended to serve as a collection in one place of the various methods of the theoretical explanation of the motion of a spinning body, and as a reference for mathematical formulas required in practical work.

—1914-67. vi + 277 pp. + Fold-out Plates. 6½x10¾. 8284-0205-1.

LES INTÉGRALES DE STIELTJES et leurs Applications aux Problèmes de la Physique Mathématique

By N. GUNTHER

The present work is a reprint of Vol. I of the publications of the V. A. Steklov Institute of Mathematics, in Moscow. The text is in French.

—1932-49. iv + 494 pp. 5⅜x8. 8284-0063-6.

TRAITE DES COURBES SPECIALES REMARQUABLES

By F. GOMES TEIXEIRA

A comprehensive treatise, in three volumes, on curves in the plane and in space, and their special properties.

—2nd (corr.) ed. 1908/09/15-71. 1,337 pp. 3 vol. set.

INVARIANTENTHEORIE

By P. A. GORDAN

Two volumes in one. A classical work.

—1885/7-1971. 583 pp. 5⅜x8. 2 vols. in 1. **In prep.**

GYROSCOPIC THEORY

By G. GREENHILL

This work is intended to serve as a collection in one place of the various methods of the theoretical explanation of the motion of a spinning body, and as a reference for mathematical formulas required in practical work.

—1914-67. vi + 277 pp. + Fold-out Plates. 6½x10¾. 8284-0205-1.

LES INTEGRALES DE STIELTJES et leurs Applications aux Problèmes de la Physique Mathématique

By N. GUNTHER

The present work is a reprint of Vol. I of the publications of the V. A. Steklov Institute of Mathematics, in Moscow. The text is in French.

—1932-49. 498 pp. 5⅜x8. 8284-0063-6.

ALGEBRAIC LOGIC

By P. R. HALMOS

"Algebraic Logic is a modern approach to some of the problems of mathematical logic, and the theory of polyadic Boolean algebras, with which this volume is mostly concerned, is intended to be an efficient way of treating algebraic logic in a unified manner.

"[The material] is accessible to a general mathematical audience; no vast knowledge of algebra or logic is required . . . Except for a slight Boolean foundation, the volume is essentially self-contained."—*From the Preface.*

—1962. 271 pp. 6x9. 8284-0154-3.

INTRODUCTION TO HILBERT SPACE AND THE THEORY OF SPECTRAL MULTIPLICITY

By P. R. HALMOS

A clear, readable introductory treatment of Hilbert Space.

—2nd ed. 1957. 120 pp. 6x9. 8284-0082-2.

ELECTRICAL PAPERS

By O. HEAVISIDE

The *Electrical Papers*, together with the *Electromagnetic Theory*, constitute the Collected Papers of the great mathematical physicist.

—2nd (c.) ed. 1892-1970. 1,183 pp. 5⅜x8. 8284-0235-3.
Two vol. set.

ELECTROMAGNETIC THEORY: Including An Account of Heaviside's Unpublished Notes for a Fourth Volume

By O. HEAVISIDE

Third edition, with an introduction by SIR EDMUND WHITTAKER.

Oliver Heaviside continued active scientific work for more than twenty years after the publication of the third volume of his *Electromagnetic Theory*. His unpublished notes, some of which were found only recently, in 1957, contain many new discoveries. An account of some of these discoveries and the ways in which Heaviside's ideas developed during these twenty years is included as two lengthy appendices to the present edition.

—3rd ed. 1971. 1,791 pp. 5¼x8. 8284-0237-X.
Three vol. set.

VORLESUNGEN UEBER DIE THEORIE DER ALGEBRAISCHEN ZAHLEN

By E. HECKE

"An elegant and comprehensive account of the modern theory of algebraic numbers."
—*Bulletin of the A. M. S.*

—2nd ed. 1970. viii + 274 pp. 5⅜x8. 8284-0046-6.

INTEGRALGLEICHUNGEN

By E. HELLINGER and O. TOEPLITZ

—1928-53. 286 pp. 5⅜x8. 8284-0089-X.

GEOMETRY AND THE IMAGINATION

By D. HILBERT and S. COHN-VOSSEN

Translated from the German by P. NEMENYI.

"A fascinating tour of the 20th century mathematical zoo. . . . Anyone who would like to see proof of the fact that a sphere with a hole can always be bent (no matter how small the hole), learn the theorems about Klein's bottle—a bottle with no edges, no inside, and no outside—and meet other strange creatures of modern geometry will be delighted with Hilbert and Cohn-Vossen's book."
—*Scientific American.*

"Should provided stimulus and inspiration to every student and teacher of geometry."—*Nature.*

"A mathematical classic. . . . The purpose is to make the reader *see* and *feel* the proofs. . . . readers can penetrate into higher mathematics with . . . pleasure instead of the usual laborious study."
—*American Scientist.*

"Students, particularly, would benefit very much by reading this book . . . they will experience the sensation of being taken into the friendly confidence of a great mathematician and being shown the real significance of things."—*Science Progress.*

"A person with a minimum of formal training can follow the reasoning. . . . an important [book]."
—*The Mathematics Teacher.*

"A remarkable book. . . . A veritable geometric anthology. . . . Over 330 diagrams and every [one] tells a story."—*The Mathematical Gazette.*
—1952. 358 pp. 6x9.

COLLECTED PAPERS
(Gesammelte Abhandlungen)

By D. HILBERT

The present work contains all of David Hilbert's papers on Mathematics, Physics, and the Foundations of Mathematics. An exception is his work on Integral Equations, which is available in a separate volume [*Integralgleichungen,* in German] and those of his papers on the Foundations of Mathematics that appear as an appendix in his book on the Foundations of Geometry [*Grundlagen der Geometrie,* in German; English translation in prep.].

LEHRBUCH DER THETAFUNKTIONEN

By A. KRAZER

"Dr. Krazer has succeeded in the difficult task of giving a clear deductive account of the complicated formal theory of multiple theta-functions within the compass of a moderate sized text-book . . . Distinguished by clearness of style and general elegance of form."—*Mathematical Gazette.*

—1903-71. xxiv + 509 pp. 5⅜x8. 8284-0244-2.

GROUP THEORY

By A. KUROSH

Translated from the second Russian edition and with added notes by PROFESSOR K. A. HIRSCH.

Partial Contents: PART ONE: The Elements of Group Theory. Chap. I. Definition. II. Subgroups (Systems, Cyclic Groups, Ascending Sequences of Groups). III. Normal Subgroups. IV. Endomorphisms and Automorphisms. Groups with Operators. V. Series of Subgroups. Direct Products. Defining Relations, etc. PART TWO: Abelian Groups. VI. Foundations of the Theory of Abelian Groups (Finite Abelian Groups, Rings of Endomorphisms, Abelian Groups with Operators). VII. Primary and Mixed Abelian Groups. VIII. Torsion-Free Abelian Groups. Editor's Notes. Bibliography.

Vol. II. PART THREE: Group-Theoretical Constructions. IX. Free Products and Free Groups (Free Products with Amalgamated Subgroup, Fully Invariant Subgroups). X. Finitely Generated Groups. XI. Direct Products. Lattices (Modular, Complete Modular, etc.). XII. Extensions of Groups (of Abelian Groups, of Non-commutative Groups, Cohomology Groups). PART FOUR: Solvable and Nilpotent Groups. XIII. Finiteness Conditions, Sylow Subgroups, etc. XIV. Solvable Groups (Solvable and Generalized Solvable Groups, Local Theorems). XV. Nilpotent Groups (Generalized, Complete, Locally Nilpotent Torsion-Free, etc.). Editor's Notes. Bibliography.

—Vol. I. 2nd ed. 1959. 271 pp. 6x9. 8284-0107-1.
—Vol. II. 2nd ed. 1960. 308 pp. 6x9. 8284-0109-8

FOUNDATIONS OF ANALYSIS

By E. LANDAU

"Certainly no clearer treatment of the foundations of the number system can be offered. . . . One can only be thankful to the author for this fundamental piece of exposition, which is alive with his vitality and genius."—*J. F. Ritt, Amer. Math. Monthly.*

—2nd ed. 1960-66. xiv + 136 pp. 8284-0079-2.

ELEMENTARE ZAHLENTHEORIE

By E. LANDAU

"Interest is enlisted at once and sustained by the accuracy, skill, and enthusiasm with which Landau marshals . . . facts and simplifies . . . details."

—*G. D. Birkhoff, Bulletin of the A. M. S.*

—1927-50. vii + 180 + iv pp. 5½x8½. 8284-0026-1.

ELEMENTARY NUMBER THEORY

By E. LANDAU

The present work is a translation of Prof. Landau's famous *Elementare Zahlentheorie*, with added exercises by Prof. Paul T. Bateman.

—2nd ed. 1966. 256 pp. 6x9. 8284-0125-X.

Einführung in die Elementare und Analytische Theorie der ALGEBRAISCHE ZAHLEN

By E. LANDAU

—2nd ed. 1927-49. vii + 147 pp. 5⅜x8. 8284-0062-8.

NEUERE FUNKTIONENTHEORIE, *by E. LANDAU.*

See WEYL

Mémoires sur la Théorie des SYSTÈMES DES ÉQUATIONS DIFFÉRENTIELLES LINÉAIRES, Vols. I, II, III

By J. A. LAPPO-DANILEVSKIĬ

THREE VOLUMES IN ONE.

CELESTIAL MECHANICS

By P. S. LAPLACE

One of the landmarks in the history of human thought. Four volumes translated into English by NATHANIEL BOWDITCH, with an extensive running commentary by the Translator, plus a fifth volume in the original French containing historical material and commentary on the earlier volumes. For the most part, this latter is incorporated into Bowditch's systematic commentary.

"Undoubtedly the greatest systematic treatise ever published."—*Bulletin of the Amer. Math. Society.*

"The four superb volumes [are] much more than a translation; indeed, the extent of Bowditch's own contributions equals, or perhaps exceeds, that of the translation proper . . . Bowditch's commentary restores all the intermediate steps omitted by Laplace . . . The notes also contain full accounts of progress subsequent to the publication of the original volumes. The fifth, supplementary, volume of the French edition . . . is not translated, but most of its content is embodied in the translator's notes."

—*Nature.*

—Vol. I: xxiv+136+746 pp. Vol. II: xviii+990 pp. Vol. III: xxx+910 pp.+117 pp. of tables. Vol. IV: xxvi+1,018 pp. 1829; 1832; 1834; 1839. Reprint, 1967. 6½x9¼. 8284-0194-2. Vols. I-IV.

—Vol. V (in French): ix + 508 pp. 1825-1882. Reprint, 1969. 6½x9¼. 8284-0214-0. Vol. V.

TOPOLOGY

By S. LEFSCHETZ

CONTENTS: I. Elementary Combinatorial Theory of Complexes. II. Topological Invariance of Homology Characters. III. Manifolds and their Duality Theorems. IV. Intersections of Chains on a Manifold. V. Product Complexes. VI. Transformations of Manifolds, their Coincidences, Fixed Points. VII. Infinite Complexes. VIII. Applications to Analytical and Algebraic Varieties.

—2nd ed. 1930-66. 410 pp. 5⅜x8. 8284-0116-0.

SELECTED PAPERS

By S. LEFSCHETZ

DIFFERENTIAL AND INTEGRAL CALCULUS

By E. LANDAU

A masterpiece of rigor and clarity.

"And what a book it is! The marks of Landau's thoroughness and elegance, and of his undoubted authority, impress themselves on the reader at every turn, from the opening of the preface . . . to the closing of the final chapter.

"It is a book that all analysts . . . should possess . . . to see how a master of his craft like Landau presented the calculus when he was at the height of his power and reputation."

—*Mathematical Gazette.*

—3rd ed. 1965. 372 pp. 6x9. 8284-0078-4.

HANDBUCH DER LEHRE VON DER VERTEILUNG DER PRIMZAHLEN

By E. LANDAU

TWO VOLUMES IN ONE.

To Landau's monumental work on prime-number theory there has been added, in this edition, two of Landau's papers and an up-to-date guide to the work: an Appendix by Prof. Paul T. Bateman.

—2nd ed. 1953. 1,028 pp. 5⅜x8. 8284-0096-2.

VORLESUNGEN UEBER ZAHLENTHEORIE

By E. LANDAU

—Vol. I, Pt. 2. * (Additive Number Theory) xii + 180 pp. Vol. II. (Analytical Number Theory and Geometrical Number Theory) viii + 308 pp. Vol. III. (Algebraic Number Theory and Fermat's Last Theorem) viii + 341 pp. 5¼x8¼. * (Vol. I, Pt. 1 is issued as **Elementare Zahlentheorie** (in German) or as **Elementary Number Theory** [in English].) 1927-69. 8284-0032-6.

Three vols. in one.

GRUNDLAGEN DER ANALYSIS

By E. LANDAU

The student who wishes to study mathematical German will find Landau's famous *Grundlagen der Analysis* ideally suited to his needs.

Only a few score of German words will enable him to read the entire book with only an occasional glance at the Vocabulary! [A COMPLETE German-English vocabulary, prepared with the novice especially in mind, has been appended to the book.]

—4th ed. 1965.

VORLESUNGEN UEBER DIFFERENTIAL-GLEICHUNGEN MIT BEKANNTEN INFINITESIMALEN TRANSFORMATIONEN

By S. LIE

Edited by G. Scheffers. A textbook on the integration of ordinary and partial differential equations in which the Lie theory for solving such equations is expounded.

—1891-1967. xiv + 568 pp. 6x9. 8284-0206-X.

THEORIE DER TRANSFORMATIONSGRUPPEN

By S. LIE

The Bible of transformation groups, a work of vast range and great originality. It is written with especial simplicity of language. There are frequent *résumés* throughout the work and an abundance of illustrations.

—2nd (corr.) ed. 1888/93-70. 2,090 pp. 6x9. 8284-0232-9.
Three vol. set

VORLESUNGEN UEBER CONTINUIERLICHE GRUPPEN MIT GEOMETRISCHEN UND ANDEREN ANWENDUNGEN

By S. LIE

An expository work by the originator of the Lie theory of continuous groups.

—1893-71. xii + 810 pp. 6x9. 8284-0199-3.

LES ENSEMBLES ANALYTIQUES

By N. LUSIN

Leçons sur les Ensembles Analytiques et leurs Applications, with a preface by H. Lebesgue and a note by W. Sierpinski.

—1930-71. xv + 328 pp. 5⅜x8. 8284-0250-7.

COMBINATORY ANALYSIS, Vols. I and II

By P. A. MACMAHON

TWO VOLUMES IN ONE.

A broad and extensive treatise on an important branch of mathematics.

—1915/16-64. 660 pp. 5⅜x8. 8284-0137-3. 2 v. in 1.

VORLESUNGEN UEBER DIE ALGEBRA DER LOGIK

By E. SCHRÖDER

One of the classics of logic.

The present edition includes, as an addendum to the third volume, the complete text of the short two-volume work *Abriss der Algebra der Logik.*

—2nd ed. 1966. (1st ed.: 1890-1905; 1909/10.) 2,192 pp. 6x9. 8284-0171-3. (5 vols. in 3.) Three vol. set.

GESAMMELTE MATHEMATISCHE ABHANDLUNGEN

By H. A. SCHWARZ

—2nd (c.) ed. 1890-1972. 726 pp. 6x9. 8284-0260-4. Two vols. in one.

AN INTRODUCTION TO THE OPERATIONS WITH SERIES

By I. J. SCHWATT

Many useful methods for operations on series, methods for expansions of functions, methods for the summation of many types of series, and a wealth of explicit results are contained in this book. The only prerequisite is knowledge of the Calculus.

—1924-62. x + 287 pp. 5⅜x8. 8284-0158-6.

PROJECTIVE METHODS IN PLANE ANALYTICAL GEOMETRY

By C. A. SCOTT

CHAPTER HEADINGS: I. Point and Line Coordinates. II. Infinity. Transformation of Coordinates. III. Figures Determined by Four Elements. IV. The Principle of Duality. V. Descriptive Properties of Curves. VI. Metric Properties of Curves; Line at Infinity. VII. Metric Properties of Curves; Circular Points. VIII. Unicursal (Rational) Curves. Tracing of Curves. IX. Cross-Ratio, Homography, and Involution. X. Projection and Linear Transformation. XI. Theory of Correspondence. XII. The Absolute. XIII. Invariants and Covariants.

VARIATIONSRECHNUNG IM GROSSEN

By H. SEIFERT and W. THRELFALL

The brilliant expository talents of the authors of *Lehrbuch der Topologie* are here applied to an exposition of the theory of Marston Morse.

Topologically, the book is self-contained.

—1938-71. iv + 116 pp. 6x9. 8284-0049-0.

CONGRUENCE OF SETS, and other monographs

By SIERPIŃSKI, KLEIN, RUNGE, and DICKSON

FOUR VOLUMES IN ONE.

ON THE CONGRUENCE OF SETS AND THEIR EQUIVALENCE BY FINITE DECOMPOSITION, by *W. Sierpinski.* 1. Congr. of Sets. 2. Translation of Sets. . Equiv. of Sets by Finite Decomposition. 4. D. into Two Parts. . . . 7. Paradoxical D's. . . . 10. The Hausdorff Paradox. 11. Paradox of Banach and Tarski. 12. Banach Measure. The General Problem of Measure. 13. Absolute Measure. 14. Paradox of J. von Neumann.

THE MATHEMATICAL THEORY OF THE TOP, by *F. Klein.* Well-known lectures on the analytical formulas relating to the motion of the top.

GRAPHICAL METHODS, by *C. Runge.*

INTRODUCTION TO THE THEORY OF ALGEBRAIC EQUATIONS, by *L. E. Dickson.* Dickson's earliest (1903) account of the subject, substantially less abstract than his later exposition. *From Dickson's Preface:* "The subject is here presented in the historical order of its development. The First Part (Chaps. I-IV) is devoted to the Lagrange-Cauchy-Abel theory of general algebraic equations. The Second Part (Chaps. V-XI) is devoted to Galois' theory of algebraic equations . . . The aim has been to make the presentation strictly elementary, with practically no dependence upon any branch of mathematics beyond elementary algebra. There occur numerous illustrative examples, as well as sets of elementary exercises."

—1954/1897/1912/1903-1967. 461 pp. 5¼x8. 8284-0209-4
Four vols. in one

ANALYTIC THEORY OF CONTINUED FRACTIONS

By H. S. WALL

Partial Contents: CHAP. I. The c. f. as a Product of Linear Fractional Transformations. II. Convergence Theorems. IV. Positive-Definite c. f. VI. Stieltjes Type c. f. VIII. Value-Region Problem. IX. J-Fraction Expansions. X. Theory of Equations. XII. Matrix Theory of c. f. XIII. C. f. and Definite Integrals. XIV. Moment Problem. XVI. Hausdorff Summability. XIX. Stieltjes Summability. XX. The Padé Table.

—1948-67. xiv + 433 pp. 5⅜x8. 8284-0207-8.

LOGIC, COMPUTERS, AND SETS

By H. WANG

Partial Contents: GENERAL (EXPOSITORY) SKETCHES (*Chaps. I-VI*) : I. The Axiomatic Method (§ 2. The problem of adequacy, . . . , §6. Gödel's Theorems, . . .). II. Eighty Years of Foundational Studies. III. The Axiomatization of Arithmetic (§2. Grassmann's calculus, . . . , § 6. Dedekind and Frege). V. Computation (§ 1. Concept of computability, . . . , § 6. Control of errors in calculating machines). CALCULATING MACHINES (*Chaps. VI-X*) : VI. A Variant to Turing's Theory. VII. Universal Turing Machines. VIII. The Logic of Automata. IX. Toward Mechanical Mathematics. FORMAL NUMBER THEORY (*Chaps. XI-XV*) : XII. Many-Sorted Predicate Calculi. XIII. Arithmetization of Metamathematics (§ 1. Gödel numbering, . . . , § 4. Arithmetic translations of axiom systems). XIV. Ackermann's Consistency Proof. . . . IMPREDICATIVE SET THEORY (*Chaps. XVI-XX*) : XVI. Different Axiom Systems . . . XVIII. Truth Definitions and Consistency Proofs . . . PREDICATIVE SET THEORY (*Chaps. XXI-XXV*) : XXII. Undecidable Sentences. XXIII. Formalization of Mathematics. . . .

Originally published under the title: *A Survey of Mathematical Logic.*

—1962-71. x + 651 pp. 6x9. 8284-0245-0.

A HISTORY OF THE CALCULUS OF VARIATIONS IN THE 18th CENTURY

By R. WOODHOUSE

"Those interested in the beginnings of the Calculus of Variations should read the book by Lagrange and the—unfortunately, very rare—book by Woodhouse. Euler's masterpiece [*Methodus . . .*] will then offer him very little difficulty. This work of Euler's is distinguished by the abundance of its examples and is, to this day, one of the most interesting books that the mathematical literature has to offer."—C. CARATHEODORY, in *Variationsrechnung und Partielle Differentialgleichungen erster Ordnung.*

—I. TODHUNTER

—1811-1965. ix + x + 154 pp. 5x8⅜. 8284-0177-2.

BEGINNER'S BOOK OF GEOMETRY

By G. C. YOUNG and W. H. YOUNG

An elementary beginner's book by two distinguished mathematicians. Unlike Birkhoff's *Basic Geometry,* which is intended for teenagers, this book is truly for young children.

"The right one from a psychological point of view . . . Its main object is to awaken the pupil's mind to the ideas by which we classify the properties of space; this is done by directions in paper-folding, in dissection of areas, in the construction of solid models, and the like. At the same time, various theorems are stated and proved."—*Nature.*

—2nd ed. 1905-70. xvi + 222 pp. 4⅝x6⅜. 8284-0231-0.

THE THEORY OF SETS OF POINTS

By W. H. YOUNG and G. C. YOUNG

—1906-71. Appr. 402 pp. 5⅜x8.

THE THEORY OF GROUPS

By H. J. ZASSENHAUS